Cambridge Technicals Level 3

Laboratory Skills

Stephen Hoare, Paul Hatherly,
Roya Vahdati-Moghaddam,
Debbie Brunt, Mike Hill

HODDER EDUCATION
AN HACHETTE UK COMPANY

Although every effort has been made to ensure that website addresses are correct at time of going to press, Hodder Education cannot be held responsible for the content of any website mentioned in this book. It is sometimes possible to find a relocated web page by typing in the address of the home page for a website in the URL window of your browser.

Hachette UK's policy is to use papers that are natural, renewable and recyclable products and made from wood grown in sustainable forests. The logging and manufacturing processes are expected to conform to the environmental regulations of the country of origin.

Orders: please contact Bookpoint Ltd, 130 Park Drive, Milton Park, Abingdon, Oxon OX14 4SE. Telephone: (44) 01235 827720. Fax: (44) 01235 400454. Email education@bookpoint.co.uk Lines are open from 9 a.m. to 5 p.m., Monday to Saturday, with a 24-hour message answering service. You can also order through our website: www.hoddereducation.co.uk

ISBN: 978-1-471-87482-6

© Stephen Hoare, Paul Hatherly, Roya Vahdati-Moghaddam, Debbie Brunt, Mike Hill

First published in 2017 by
Hodder Education,
An Hachette UK Company
Carmelite House
50 Victoria Embankment
London EC4Y 0DZ

www.hoddereducation.co.uk

Impression number   10 9 8 7 6 5 4 3 2 1
Year   2021 2020 2019 2018 2017

All rights reserved. Apart from any use permitted under UK copyright law, no part of this publication may be reproduced or transmitted in any form or by any means, electronic or mechanical, including photocopying and recording, or held within any information storage and retrieval system, without permission in writing from the publisher or under licence from the Copyright Licensing Agency Limited. Further details of such licences (for reprographic reproduction) may be obtained from the Copyright Licensing Agency Limited, Saffron House, 6–10 Kirby Street, London EC1N 8TS.

Cover photo © psdesign1 – Fotolia
Illustrations by Barking Dog Art
Typeset in Din-Light 10/12 pts by Aptara Inc.
Printed in Slovenia by DZS

A catalogue record for this title is available from the British Library

Contents

About this book ... iv
Acknowledgements ... vi

Unit 1 Science fundamentals ... 1

Unit 2 Laboratory techniques ... 37

Unit 3 Scientific analysis and reporting ... 65

Unit 4 Human physiology ... 99

Unit 5 Genetics ... 124

Unit 6 Control of hazards in the laboratory ... 146

Unit 7 Human nutrition ... 167

Unit 8 Cell biology ... 188

Unit 11 Drug development ... 207

Unit 13 Environmental surveying ... 228

Unit 14 Environmental management ... 255

Unit 18 Microbiology ... 281

Unit 21 Product testing techniques ... 302

Glossary ... 319
Index ... 330

About this book

This book helps you to master the skills and knowledge you need for the OCR Cambridge Technicals Level 3 Laboratory Skills qualification.

This resource is endorsed by OCR for use with specification Cambridge Technicals Level 3 Laboratory Skills. In order to gain OCR endorsement, this resource has undergone an independent quality check.

Any references to assessment and/or assessment preparation are the publisher's interpretation of the specification requirements and are not endorsed by OCR. OCR recommends that a range of teaching and learning resources are used in preparing learners for assessment. For more information about the endorsement process, please visit the OCR website, www.ocr.org.uk.

Using this book

ABOUT THIS UNIT

In this unit, you will learn about how the organs and systems of the human body interact to carry out the basic functions of life – acquiring materials, distributing them around the body, releasing and using energy. You will understand how the systems are controlled and regulated by homeostasis, and how the body is protected by the immune system.

Know what to expect when you are studying the unit.

LO1 Understand the importance of health and safety and quality systems to industry

1.1 To use aspects of good laboratory practice throughout all practicals

2.1 Techniques to separate and identify substances present in a mixture

Prepare for what you are going to cover in the unit.

How will I be assessed?

You will be assessed through a series of assignments and tasks set and marked by your tutor.

Find out how you can expect to be assessed after studying the unit.

How will I be graded?

Find out the criteria for achieving Pass, Merit and Distinction grades in internally assessed units.

LO1 Understand the chemical structures of elements and compounds

Understand all the requirements of the qualification, with clearly stated learning outcomes and assessment criteria fully mapped to the specification.

GETTING STARTED

What is important and why? (5 minutes)

Think about why health and safety and quality systems are important. Why do they exist? What would happen if they were not followed? Share your thoughts with the group.

Try activities to start you off with a new learning outcome.

INDEPENDENT ACTIVITY

PAIRS ACTIVITY

GROUP ACTIVITY

Carry out tasks that help you to think about a topic in detail and enhance your understanding.

CLASSROOM DISCUSSION

A better model for atomic structure (10 minutes)

The popular image of an atom shows a central nucleus and electrons whizzing round it like planets around the sun. Discuss why this is misleading. How do models such as electron shells and electron orbitals help our understanding?

Take the opportunity to share your ideas with your group.

KEY TERMS

Polymer – A large molecule formed from one or more smaller molecules or monomers.
Monomer – The individual molecule or molecules that form a polymer.
Repeat unit – A way of showing the pattern of monomers in the polymer.

Understand important terms.

KNOW IT

1 Which type of blood vessel carries blood (a) away from the heart (b) through tissues (c) towards the heart?
2 Give three functions of the blood.
3 What does an electrocardiograph measure?

Answer quick questions to test your knowledge about the learning outcome you have just covered.

L03 Assessment activities *P3 M2*

Try the types of questions you may see in your externally assessed exam.

TOP TIP
✔ The digestive system could be represented on the flow chart as a straight tube.

Start preparing for your internally assessed assignments by carrying out activities that are directly linked to Pass, Merit and Distinction criteria. Top Tips give you additional advice.

Read about it

Suggests books and websites for further reading and research.

Acknowledgements

Every effort has been made to trace and acknowledge ownership of copyright. The publishers will be glad to make suitable arrangements with any copyright holders whom it has not been possible to contact. The authors and publishers would like to thank the following for permission to reproduce copyright illustrations:

Fig 1.7 © Gunilla Elam/Science Photo Library; Fig 1.41 © Jacopin/BSIP/Science Photo Library; Fig 2.1 © Rainer Lesniewski/Shutterstock.com; Fig 2.3 Universal Images Group North America LLC/Alamy Stock Photo; Fig 2.4 © Science Photo Library; Fig 5.1 © Science Photo Library; Fig 5.3 © WhiteDragon/Shutterstock; Figs 5.5 and 5.6 © EBI/Creative Commons; Fig 6.3 © Designua/Shutterstock.com; Fig 6.4 © Peter Gardiner/Science Photo Library; Fig 7.5 © studiomode/Alamy Stock Photo; Fig 8.1 © Mariana Ruiz Villarreal; Fig 8.2 Universal Images Group North America LLC/Alamy Stock Photo; Fig 8.4 © Olympus Corporation; Fig 8.7 © Mikael Häggström/Wikimedia Commons; Fig 11.3 © Colin Cuthbert/Science Photo Library; Fig 11.7 © Poznyakov/Shutterstock.com; Fig 13.3 © WRI Aqueduct Water Risk Atlas (aqueduct.wri.org); Fig 13.4 © Pictorial Press Ltd/Alamy Stock Photo; Fig 13.7 © S-F/Shutterstock.com; Fig 13.7b © Ragnar Th Sigurdsson/ARCTIC IMAGES/Alamy Stock Photo; Fig 13.8 © Julia Gavin/Alamy Stock Photo; Figs 13.9 and 13.10 © NASA; Fig 13.11 © J C Williams; Fig 13.12 © Keith Moseley; Fig 14.1 © Designua/Shutterstock.com; Fig 14.3 © Environment Agency; Figure 14.6 © 2016 Global Footprint Network, www.footprintnetwork.org; Fig 14.7 © Merve Guvendik; Fig 14.8 © European Environment Agency (EEA); Figs 18.1 and 18.2 © Microbiology Society; Fig 18.8 © Review on Antimicrobial Resistance; Fig 21.1 © Wyss Institute at Harvard University.

This document contains public sector information [Figure 14.2] licensed under the Open Government Licence v3.0 – http://www.nationalarchives.gov.uk/doc/open-government-licence/version/3/

Unit 01
Science fundamentals

ABOUT THIS UNIT

A thorough understanding of scientific principles and practices is essential for science technicians. Knowledge learnt in this unit will create a solid foundation in the fundamentals of science that you will be able to build on in your further study through your choice of additional optional units. These will provide you with greater depth of knowledge and practice in your chosen specialisms.

LEARNING OUTCOMES

The topics, activities and suggested reading in this unit will help you to:

1. Understand the chemical structures of elements and compounds
2. Understand reactions in chemical and biological systems
3. Understand cell organisation and structures
4. Understand the principles of carbon chemistry
5. Understand the importance of inorganic chemistry in living systems
6. Understand the structures, properties and uses of materials

How will I be assessed?

All Learning Outcomes are assessed through externally set written examination papers, worth a maximum of 90 marks and 2 hours in duration.

As well as the external assessment, it would be helpful if you could undertake a series of time constrained assessments to test your knowledge, understanding and application during the learning process. These should include a range of assessment content such as short questions which test knowledge, understanding and application. They should not demand long theoretical answers but concentrate on your ability to interpret realistic or real information in order to explain or carry out particular scientific tasks.

LO1 Understand the chemical structures of elements and compounds

1.1 The atom is the basic structure, it is made up of subatomic particles

GETTING STARTED

(5 minutes)

What is your mental picture of an atom? Discuss with the group the different types of model or ideas of what an atom is that you have come across.

KEY TERMS

Atom – This is the basic structure of all matter and consists of a nucleus surrounded by electrons.

Nucleus – This contains protons and neutrons surrounded by electrons.

Electron – A negatively charged particle with a mass about 1/1836th that of a proton.

Nucleon – A general term describing both protons and neutrons.

Proton – A nucleon with a single positive charge and a relative mass of approximately 1, or more accurately 1.0073 so the mass is slightly less than that of a neutron.

Neutron – A nucleon with no net electric charge and a relative mass of approximately 1, or more accurately 1.0087 so the mass is slightly more than that of a proton.

Isotopes – These are atoms of the same element (i.e. they have the same number of protons) but with different numbers of neutrons.

Nuclear and atomic diameters

Most of the mass of an **atom** is contained in the **nucleus**, because the **electrons** have such a small mass relative to **nucleons**. However, the nucleus is also more dense, so the diameter of the nucleus is much smaller than that of the atom. Because of this density, it is possible to measure the nuclear diameter fairly accurately. You are more likely to see the term 'atomic radius' than 'atomic diameter' (see section 1.2) and it can be quite difficult to measure the atomic radius exactly. This is because the atomic radius is the distance from the centre of the nucleus to the boundary of the surrounding electrons – exactly where that boundary lies is not always clear and can vary depending on many factors. The atomic diameter is about 20,000 times greater than the nuclear diameter.

Proton number defines the type of atom

If an atom contains 6 **protons** then it is an atom of carbon (C) and if it contains 7 protons it is an atom of nitrogen (N). The proton number (sometimes called atomic number) determines the type of atom (carbon, nitrogen, oxygen, etc.). The sum of the protons and **neutrons** is known as the mass number. Most carbon atoms have 6 protons and 6 neutrons, so they have a mass number of 12. However, about 1% of carbon atoms have 7 neutrons and have a mass number of 13. A smaller percentage have 8 neutrons and have a mass number of 14. These different forms of the same element are known as **isotopes**.

Nuclear notation

We can use nuclear notation to distinguish these isotopes:

$^{12}_{6}C$ $^{13}_{6}C$ $^{14}_{6}C$ $^{14}_{7}N$ $^{15}_{7}N$

In these examples, the upper number is the mass number (sum of protons and neutrons) and the lower number is the proton (or atomic) number. Don't confuse $^{14}_{6}C$ with $^{14}_{7}N$; just because they have the same mass number, it doesn't mean they are the same element.

In fact, it is not really necessary to show the proton number as well as the chemical symbol, and so the above isotopes would usually be written:

^{12}C ^{13}C ^{14}C ^{14}N ^{15}N

You might also see these written as carbon-12, carbon-13, carbon-14, nitrogen-14 and nitrogen-15.

Attractive and repulsive forces within the nucleus

Let's pause for a moment and consider the structure of the atom. A positively charged nucleus is surrounded by a cloud of negatively charged electrons. But like charges repel; that's not a problem for the electrons, because they are spread over a large volume (look at the section on nuclear and atomic diameters). However, the protons are packed closely together in the dense nucleus. Newton's Law of Gravity says that the gravitational force between any two objects is proportional to their mass. This applies whether the pair of objects is an apple dropping from a tree in Newton's garden and the earth towards which it is falling, or two protons in the nucleus. But the mass of the protons is very small, so the gravitational force must also be very small. The electromagnetic force repelling the protons is much stronger. Therefore, there must be an even stronger force that is holding the nucleus together.

We can get some kind of an estimate for the strength of these forces from a simple 'back of the envelope' calculation. Being able to make such estimates is an important skill in all sciences, allowing you to quickly get a feel for the magnitudes of phenomena, and to identify anything missing.

> ### 💬 CLASSROOM DISCUSSION
> #### Scientific estimation (10 minutes)
> Let's do a couple of simple calculations on the attractive and repulsive forces that we know about between protons in an atomic nucleus.
>
> Atomic nuclei are of the order of fm (1×10^{-15} m) in diameter, so we can use this number as a typical separation between protons, which we'll write as r.
>
> To estimate the gravitational attraction, we also need the mass of a proton, m in kg (1.7×10^{-27} kg) and Newton's equation for the gravitational force, F_g which is:
> $$F_g = \frac{Gm_1m_2}{r^2}$$
> where G is the gravitational constant and is approximately 6.7×10^{-11} N·m² kg⁻² and m_1 and m_2 are the masses of the two protons.
>
> A very similar equation expresses Coulomb's law for the electrostatic force between two charged particles, namely:
> $$F_e = \frac{k_e q_1 q_2}{r^2}$$
> where k_e is Coulomb's constant which is about 9×10^9 N·m² C⁻² and q_1 and q_2 are the charges on the protons, which are both about 1.6×10^{-19} C.
>
> Split your group into two, with each carrying out one of the calculations. Finally, calculate the ratio of the electrostatic and gravitational forces, i.e. $F_e \div F_g$.

You should find that the electrostatic repulsion between protons is about 1×10^{36} times the gravitational attraction, confirming that there must be something else holding the protons together in the nucleus.

This force is known as the strong nuclear force, and, at the scale of the atomic nucleus, is an attractive force between nucleons, some 10^{38} (about 100x) the strength of the electrostatic repulsion.

There is one other nuclear force, known as the weak nuclear force, which is manifest in some radioactive processes and plays a crucial role in nuclear fission and fusion.

So, if the gravitational force = 1, then:
- Weak nuclear force = 1×10^{25}
- Electromagnetic force = 1×10^{36}
- Strong nuclear force = 1×10^{38}

A question we might well be asking ourselves is: if the strong nuclear and electromagnetic forces are so much stronger than gravity, why aren't these forces felt in everyday experiences? (For example, why aren't we held onto the Earth by electrostatic or strong nuclear forces?)

👥 GROUP ACTIVITY
(15 minutes)
In groups, think about and discuss reasons for gravity dominating our experience of forces at a human scale. Factors you might like to consider are:
- How do the different forces vary with distance? (i.e. what are the ranges of the forces?)
- What do the forces depend on? (e.g. mass in the case of gravity)

(You may need to do some research on these questions for the strong force.)

1.2 Elements are based on atomic structure and can be classified by the Periodic Table

You should always have a copy of the periodic table at hand because it contains so much useful information. Not just things like **atomic number** or **relative atomic mass**, but also the way in which the elements are arranged into **groups** and **periods**. In fact, the periodic table was

> ### 🔑 KEY TERMS
> **Atomic number** – Also called the proton number, this is the number of protons in the nucleus of an atom. It is also the number of electrons in the atom.
>
> **Relative atomic mass** – The mass of an atom relative to 1/12th the mass of an atom of carbon-12. It is more convenient to use this relative unit rather than trying to determine the mass of individual atoms or particles in grams; it also becomes more useful when calculating amounts of substance (you will cover this in Unit 2). The same principle is used to define relative molecular or formula mass.
>
> **Group** – This refers to the columns in the periodic table. Elements in the same group have similar chemical properties.
>
> **Period** – This refers to the 'rows' in the periodic table. Each period has one more shell of electrons than the previous period.

3

originally produced when it was noticed that the properties of some elements were similar to each other and that these properties repeated in a pattern as the mass of the element increased, i.e. they were periodic (repeating).

Organisation of elements within the table

The modern periodic table is in increasing order of atomic number, but the elements are also arranged according to their electronic structure into the s, p, d and f blocks.

The periodic table is also organised into groups – the columns in the table.

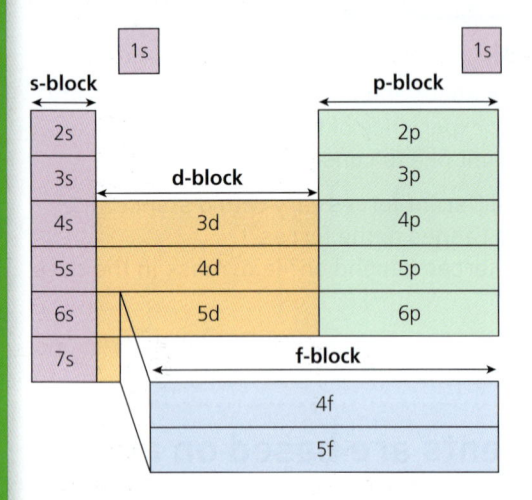

▲ Figure 1.1 The periodic table showing the arrangement of the s, p, d and f blocks

Groups

Elements in the same group have similar properties. This is particularly true with groups 1 (the 'alkali metals'), 2 (the 'alkaline earths'), 7 (the 'halogens') and 8 (the 'noble gases'). Chemical reactions depend on the number of outer shell electrons and each element has the same number of outer shell electrons as its group number. For example, lithium, sodium and potassium all have one outer shell electron and they react in a very similar way with water. However, lithium is the least reactive and potassium is the most reactive. We can understand this trend if we understand the factors that affect it.

Periods

Each period in the table corresponds to a shell of electrons. Evidence for this comes from **ionisation energy**. It is possible to remove electrons one at a time and measure the amount of energy required – these are called successive ionisation energies. In the case of magnesium, there is a big jump between the 2nd and 3rd ionisation energies. This is because the first two electrons are being removed from the 3rd shell. However, the next electron has to be removed from the 2nd shell, so it is much more tightly held.

In Figure 1.2, the ionisation energies are for the following reactions:

1st ionisation energy (IE): $Mg(g) \rightarrow Mg^+(g) + e^-$

2nd IE: $Mg^+(g) \rightarrow Mg^{2+}(g) + e^-$

3rd IE: $Mg^{2+}(g) \rightarrow Mg^{3+}(g) + e^-$

4th IE: $Mg^{3+}(g) \rightarrow Mg^{4+}(g) + e^-$

And so on.

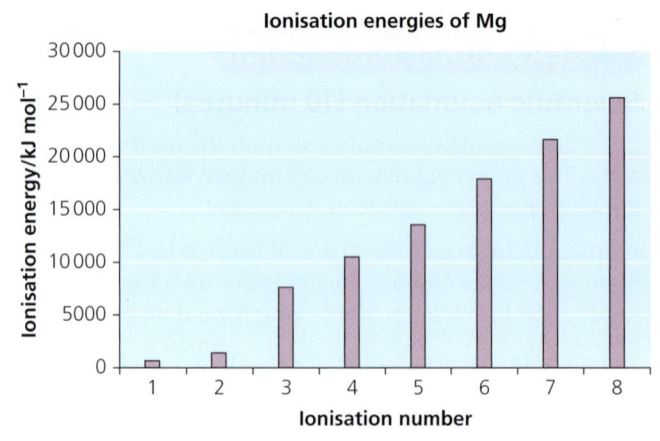

▲ Figure 1.2 The first and successive ionisation energies of magnesium – notice the jump between 2nd and 3rd showing that the 3rd ionisation removes an electron from the next shell in towards the nucleus

GROUP ACTIVITY

The periodic table (15 minutes)

Think about the following and, as a group, work out how and why the periodic table is arranged the way it is and how we can use that information to explain the way in which chemicals behave and react.

- The periodic table is organised by increasing atomic number.
- Atomic mass isn't always the same as mass number – where do isotopes fit in?
- The atomic radius depends on the number of shells and the strength of attraction between the positive nucleus and the negative electrons. So as you go down a group the atomic radius increases but as you go across a period the atomic radius decreases. Think about what effect this will have on trends in the properties and reactivity of elements across a period and down a group.

KEY TERM

Ionisation energy – This is the energy needed to remove an electron from an atom in the gas phase. It is a measure of how tightly the electron is held by the nucleus – the strength of attraction between the positive nucleus and the negative electron.

1.3 Elements react together to form compounds

Reactions and forming **compounds** is pretty much what chemistry is all about. If we understand how compounds are formed, we can understand why some compounds are formed and why others aren't, or work out how to make new compounds or better ways of making existing compounds.

> 🔑 **KEY TERM**
>
> **Compound** - A substance that contains more than one type of atom. Those atoms can be held together in two main ways – by ionic or covalent bonds.

Ionic bonding

The noble gases (Group 8 or Group 0, depending on which periodic table you look at) are all very unreactive and generally exist as single atoms. They also have a 'full' outer shell, meaning two s electrons in He and two s electrons plus six p electrons in the others. We can ignore the d and f orbitals and just concentrate on what are known as the valence electrons. Therefore, we conclude that this structure of a 'full' outer shell is very stable and that atoms will form bonds in ways that allow them to achieve this electron arrangement.

Ionic compounds are formed between a metal and a non-metal. Figure 1.3 illustrates how this happens.

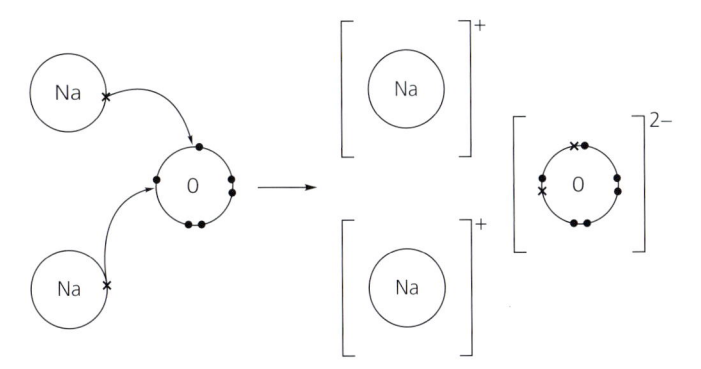

▲ **Figure 1.3** The ionic compound Na_2O is formed when sodium ions lose electrons to form sodium ions and oxygen atoms gain electrons to form oxide ions

> 👥 **GROUP ACTIVITY**
>
> **Ionic bonds (15 minutes)**
>
> Take some examples of ionic compounds (your tutor will be able to help you select some). Work out how each compound is formed from a metal that loses one or more electrons and a non-metal that gains one or more electrons.

What are the forces that hold the particles together in an ionic solid?

Can you work out the rules for combining metals and non-metals based on their group number?

Covalent bonding

Covalent bonds are made between atoms by sharing unpaired valence electrons to form electron pairs. This often allows an atom to effectively fill its outer shell with electrons, which is a very stable configuration. The simplest example is in the hydrogen molecule, H_2. Another example is the chlorine molecule, Cl_2. Figure 1.4 shows how these bonds can be represented in a 'dot and cross' diagram.

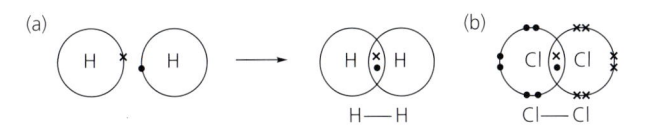

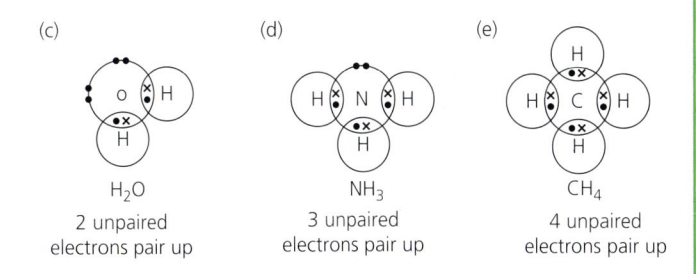

▲ **Figure 1.4** Dot and cross diagrams showing how sharing of electrons forms bonding pairs in H_2, Cl_2, H_2O, NH_3 and CH_4

> 💬 **CLASSROOM DISCUSSION**
>
> **A better model for atomic structure (10 minutes)**
>
> The popular image of an atom shows a central nucleus and electrons whizzing round it like planets around the sun. Discuss why this is misleading. How do models such as electron shells and electron orbitals help our understanding?

> 🔍 **RESEARCH ACTIVITY**
>
> **Shapes of orbitals (30 minutes)**
>
> A Google image search will bring up lots of images of atomic orbitals. Look at the different shapes. If you can understand the shape of 2s and 2p orbitals, how do the corresponding orbitals in the n = 3 and higher shells work? Where do the electrons go? Does this help explain why simple models are sometimes better?
>
> Now look at images of different molecular orbitals. Do these help you visualise molecules?

KNOW IT

1. **a** Use the periodic table to help you complete Table 1.1.

   Table 1.1

Isotope	^{204}Pb	^{206}Pb	^{207}Pb	Unknown isotope
Number of protons				83
Number of neutrons				126
Number of electrons				83

   b Identify the unknown isotope.
2. The periodic table is arranged in rows and columns. State the name given to:
   - **a** rows
   - **b** columns.
3. Use dot and cross diagrams to show how HCl and NaCl are formed and use these to explain the different types of bonding.

LO2 Understand reactions in chemical and biological systems

Chemistry is about reactions and interactions. It is important to understand the difference between a physical change, e.g. ice melting, salt (sodium chloride) dissolving in water to produce sodium chloride solution, and a chemical reaction, e.g. sodium hydrogen carbonate reacting with hydrochloric acid to produce carbon dioxide and sodium chloride solution. In the next sections we will look at the importance of mixtures, e.g. alloys. Then we will look at a range of different types of chemical reactions and the ways in which they are used to create a wide range of different substances.

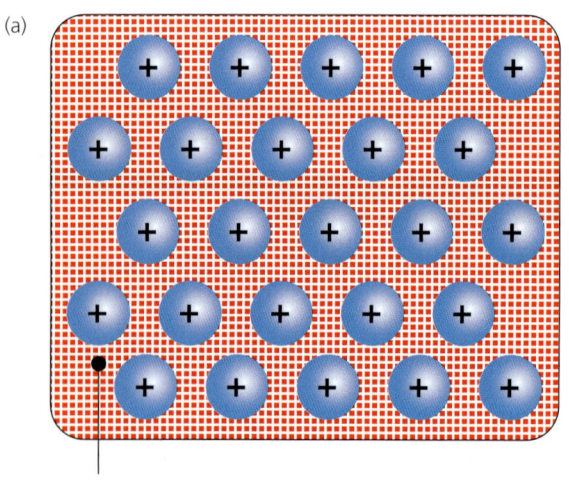

GETTING STARTED

Chemical reactions: why and how? (10 minutes)

Why do chemical substances react? How do they react? How can we tell the difference between mixing and reacting? Finding answers to these questions has formed the basis of most chemical research – and will continue to do so. Take a few minutes to think about what you know about chemical reactions and share your thoughts with the group.

2.1 Chemicals interact and react with each other

Mixtures and alloys

We are often able to mix two or more substances together without a chemical reaction taking place, yet the properties of the mixture can be very different to any of the constituent substances. This idea is of great importance when we combine metals together, creating alloys.

What is a metal?

In most molecules, electrons are tightly bound to the atoms. In metals though, electrons are less tightly bound, so when we bring a collection of metal atoms close enough together in a solid, electrons become free to move between atoms (See figure 1.5). In other words, they become *delocalised* and cannot be said to belong to any one atom.

The physical and electrical properties of a metal depend on the amount of electron delocalisation, bonding between atoms and the three-dimensional arrangements of atoms. In consequence, metals have a vast range of properties, from being liquid at room temperature (mercury) to the element with the highest melting point (tungsten, at 3422°C).

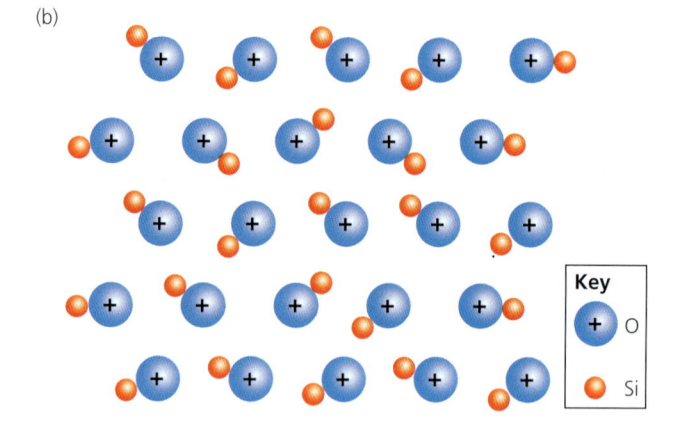

▲ **Figure 1.5** Comparison of (a) metallic and (b) non-metallic solids (e.g. quartz). In the former, electrons are free to move between atoms. In the latter, electrons are bound to individual atoms

Some metals are found in nature in their pure state, indicating low reactivity, or ability to form compounds. For example, whilst rare, gold is found as dust or larger 'nuggets' around the world. This rarity, combined with its lack of reactivity, has led to it being valued as currency, in jewellery and in electronics where other properties such as high conductivity and ductility (the ability to be drawn into fine wires) are important. More often though, metals are found as compounds (ores), such as oxides and sulphides, requiring chemical or electrical processes to 'refine' them.

Many pure metals are useful; for example, copper and aluminium for electrical cables and some domestic purposes (e.g. cooking utensils). However, many pure metals have undesirable properties, such as corroding (e.g. rust in iron), poor elastic or mechanical properties (the hardness and elasticity of copper depends on its history of being 'worked') or inability to be formed. For example, the high melting point of tungsten makes casting impractical and the hardness of titanium makes machining difficult. To overcome these problems, pure metals are often mixed in various quantities, forming, in effect, 'solid solutions' or alloys.

Alloys

A typical characteristic of an alloy is that it comprises a significant percentage of a 'base' metal (for example, copper in the case of brass) with smaller percentages of one or more metals or non-metals (again, in the case of brass, between 10% and 35% zinc). Typical compositions and uses of some of the more common alloys are shown in Table 1.2.

Commonly, the relatively small amount of additional material, such as the small percentage of tin in copper to produce bronze, causes a large change in metal properties. Again, as an example, bronze is significantly harder than either of its constituent metals. The causes of this are complex, but in essence, the presence of different atoms within the metal structure means that:

- planes of atoms cannot slide over one another as easily due to the presence in the planes of more than one type of atom, frustrating the motion. This can make the material much less malleable, hence more suitable for tool use
- the presence of different atoms in the metal structure creates strains in the structure, in effect 'stress reinforcing' the alloy. The presence of such reinforcement means that much greater external forces need to be applied to deform or fracture the metal, leading to enhanced strength.

A special case of alloys are those formed with mercury, commonly known as amalgams. Whilst largely of historical interest, a common use was as dental amalgam for the restoration of decayed or damaged teeth.

A typical dental amalgam consists of about 50% mercury combined with about 25% silver, about 14% tin, about 8% copper, with the remainder other trace elements. A crucial property is that the amalgam can be prepared at room temperature

Table 1.2 Some common alloys, their composition and uses

Alloy	Components	Typical uses
Brass	copper (65–90%), zinc (10–35%)	door locks and bolts, brass musical instruments
Bronze	copper (78–95%), tin (5–22%), plus manganese, phosphorus, aluminium or silicon	statues, musical instruments, springs and marine components (phosphor bronze)
Cast iron	iron (96–98%), carbon (2–4%), trace of silicon	metal structures such as bridges and heavy-duty cookware
Nichrome	nickel (80%), chromium (20%)	heating elements in electrical appliances
Pewter	tin (80–99%), with copper, lead and antimony	ornaments, formerly used for tableware
Solder	older solders: tin (50–70%), lead (30–50%), copper, antimony and other metals newer solders: various, but include tin (99%), copper (1%)	electronic circuit manufacture, plumbing
Steel	iron (80–98%), carbon (0.2–2%), plus other metals such as chromium, manganese and vanadium	metal structures, engineering components
Steel (stainless)	iron (50%+), chromium (10–30%), plus smaller amounts of carbon, nickel, manganese, molybdenum and other metals	medical tools, scientific equipment, tableware

and is initially in a very weak state, allowing it to be inserted into dental cavities and moulded. Over a period of a few hours, the atoms in the amalgam rearrange themselves in a process of ageing, creating a hard, inert material.

Suspensions and colloids

In metals and alloys, materials interact at an atomic level to create substances with new properties. We can though mix substances together so that they (a) don't chemically react and (b) don't combine at an atomic scale. Notice that this is also distinct from a solution, in that the two components of the colloid or suspension exist in distinct phases (for example, particles of soot in air, or oil mixed with water) whereas in a solution, the solute (that which is being dissolved) is homogeneously mixed with the solvent (such as salt or sugar in water).

The term we use for such mixtures depends on both the size of the particles in the mixture and the stability of the mixture. (Is it prone to 'settling' or otherwise spontaneously separating the components?)

For mixtures in which particles are uniformly distributed in a fluid (a gas or a liquid), and separate out slowly under gravity, we use the term *suspension*. Suspensions are usually characterised by relatively large particle sizes, often larger than one micrometre.

Conversely, *colloids* are mixtures of particles in a fluid, where separation does not occur. Often, colloids are characterised by small particle sizes, typically less than one micrometre.

Notice that the material the particles are embedded in doesn't have to be a liquid such as water. As an example, smoke is a suspension of soot particles in air.

GROUP ACTIVITY

Suspensions and colloids (30 minutes)

As a group, research a number of different types of colloid and suspension, and think about their applications. For example, what examples and applications can you discover for colloids and suspensions in the following?

- food and food preparation
- medicine and biology
- waste treatment
- science and technology

2.2 Reactions

Chemical reactions involve breaking bonds and making new bonds and so we need to think about what is happening to the electrons. For this reason, it sometimes helps to write an equation as an ionic equation. So the reaction between an acid and a base to produce a salt and water: $HCl + NaOH \rightarrow NaCl + H_2O$ can be written:

$H^+ + Cl^- + Na^+ + OH^- \rightarrow Na^+ + Cl^- + H_2O$

Just like in algebra, we can cancel out anything that appears on both sides, so the reaction simplifies to:

$H^+ + OH^- \rightarrow H_2O$

In fact, this simplified ionic reaction is the same for the reaction between any acid and any base taking place in water!

Oxidation and reduction

A simple definition of oxidation is addition of oxygen, or removal of hydrogen. The latter is important in biology, because most biological oxidations actually involve removal of hydrogen rather than addition of oxygen. It follows that reduction is the addition of hydrogen or removal of oxygen. We generally talk about redox (*red*uction / *ox*idation) reactions, because you can't have one without the other. We can understand this more easily if we look at half equations. For example, the reduction of iron oxide by carbon in a blast furnace has the overall equation:

$2Fe_2O_3 + 3C \rightarrow 4Fe + 3CO_2$

Here, the iron is being reduced:

$4Fe^{3+} + 12e^- \rightarrow 4Fe$

While the carbon is being oxidised:

$6O^{2-} + 3C \rightarrow 3CO_2 + 12e^-$

By separating the two halves of the reaction, we have added electrons on the left-hand side of the first half equation and on the right of the second half equation to make the charges balance. This illustrates another definition of redox: oxidation is loss of electrons and reduction is gain of electrons. Look again at the three equations and see how the two definitions (addition or removal of oxygen and loss or gain of electrons) both apply.

This definition of redox is useful in situations where there is no oxygen or hydrogen involved and is particularly useful in some biological reactions. Just remember OILRIG – Oxidation Is Loss, Reduction Is Gain (of electrons).

Addition
An addition reaction is one where one molecule combines with another molecule forming a larger molecule – there are no other products. These reactions are common in organic chemistry, particularly in molecules with carbon-carbon double bonds (see section 4).

Substitution
A substitution reaction involves replacement of one functional group by another, so there will often be two reactants and two products. Again, they are common in organic chemistry – see section 4 for more information.

Polymerisation
Polymerisation involves the formation of a large molecule from one or more types of small molecule. The small molecules are called monomers and the polymer usually consists of a repeating pattern of monomers joined by covalent bonds. Polymerisation is important in the production of plastics, but it is also found widely in biology. Most biological molecules are polymers: carbohydrates, such as cellulose and starch, are polymers of simple sugars such as glucose; proteins are polymers of amino acids; nucleic acids (DNA and RNA) are polymers of nucleotides (see section 4.2).

Radical reactions
A radical is an atom or a molecule that contains an unpaired electron, making it highly reactive. You may sometimes see a radical written with a dot (e.g. Cl•), indicating the unpaired electron, but this isn't compulsory. If two atoms are joined by a covalent bond that splits equally, with each atom keeping one of the pair of electrons forming the bond, the products will be radicals. This is true whether it is a Cl – Cl bond in chlorine (Cl_2) or a C – H bond in methane (CH_4). Radical reactions are usually substitution reactions, such as the chlorination of methane. Other important examples are the ozone cycle and the destruction of ozone catalysed by chlorine radicals.

PAIRS ACTIVITY

The chlorination of methane, a radical chain reaction (10 minutes)

The chlorination of methane can be summarised as:

$Cl_2 + CH_4 \rightarrow HCl + CH_3Cl$

This starts with an initiation step, forming two radicals. UV light provides the energy to break the Cl – Cl bond. The next two steps are the propagation steps. Try and work out what the steps are.

KEY TERM
Anion – A negatively charged ion, e.g. Cl^- or CO_3^{2-}. Anions are usually formed by non-metals, or compounds of non-metals.

Displacement
Single displacement reactions involve replacement of one element in a compound by a more reactive element and are a type of redox reaction:

$Cu + 2AgNO_3 \rightarrow 2Ag + Cu(NO_3)_2$

Copper is more reactive than silver, so it replaces silver in the silver nitrate solution.

Another example, this time involving **anions**, is:

$Cl_2 + 2NaBr \rightarrow 2NaCl + Br_2$

The more reactive chlorine replaces bromine. If you split both of these reactions into half equations, you will see which things gain electrons (reduced) and which lose electrons (oxidised) i.e. they are redox reactions.

A double displacement reaction involves exchange of bonds between two reactants to form a new compound. If you mix two solutions, e.g. NaCl and KNO_3, then you will just produce a solution containing a mixture of all four ions. But, if one combination of ions is insoluble, then a precipitate will form:

$AgNO_3 (aq) + NaCl (aq) \rightarrow AgCl (s) + NaNO_3 (aq)$

A simplified ionic equation shows this is actually $Ag^+ (aq) + Cl^- (aq) \rightarrow AgCl (s)$. This reaction forms the basis of the test for halides (chloride, bromide and iodide) that you will encounter in section 5.1 of Unit 2. These double displacement reactions can be useful ways of preparing insoluble salts by mixing two soluble salt solutions. They are not redox reactions.

2.3 Rate of reaction can be affected by various factors

It doesn't matter whether you are designing a new chemical production process, or preparing a sample of an experimental new drug in the lab, it is not enough to know how to make something – you also need to know how quickly it can be made. Therefore, we need to understand the factors that affect the rates of chemical reactions.

Table 1.3 Factors affecting the rate of chemical reactions

Physical state	Usually, a reaction will occur in the liquid or gas phase where the reactants are free to move and collide.
Temperature	Increasing the temperature increases the kinetic energy of the reactants. This means they are more likely to collide and, having collided, are more likely to have enough energy to react.
Pressure	Increasing the pressure of a gas or concentration of a solution means the particles are closer together so they are more likely to collide. In each case, the rate of reaction will increase.
Solvents	Increasing the solubility of reactants will increase the rate of reaction, so polar solvents such as water or ethanol are good for reactions involving polar substances, such as inorganic salts or polar organic molecules such as sugars or some amino acids. Non-polar solvents such as hexane work better with non-polar compounds such as long chain hydrocarbons.
Catalysts and enzymes	Catalysts (and enzymes are simply biological catalysts) all increase the rate of reaction by providing an alternative reaction pathway that reduces the energy of the transition state and so reduces the activation energy. Catalysts work in several different ways. Some bind the reactants to a surface, bringing them closer together and also weakening some bonds. This is how a catalytic converter in a car works. Others actually participate in the reaction but are regenerated at the end, such as the chlorine radical in the chlorination of methane described in section 2.2.
Surface area	Reactions between solids and gases or solids and liquids occur at the surface of the solid. By increasing the surface area, e.g. by grinding a solid into a fine powder, the rate of reaction will increase. Another approach is used in car catalytic converters where the catalyst is deposited on a honeycomb support to increase the surface area.
Electromagnetic radiation	Electromagnetic radiation can increase the kinetic energy of the reactants. Certain frequencies are absorbed by molecules and cause bending or stretching of bonds, weakening them, e.g. UV light provides the energy to break the Cl–Cl bond in the chlorination of methane.
Light intensity	In reactions where electromagnetic radiation in the visible region (i.e. visible light) is a source of energy, then increasing the light intensity will increase the rate of reaction. One example is photosynthesis, where light is absorbed by chlorophyll molecules, leading to excitation of electrons. These higher energy electrons provide the energy to convert carbon dioxide and water into glucose. Therefore, increasing the light intensity will increase the rate of photosynthesis.

Chemical reactions (including ones in biological systems) all require the reactants to collide before they can react. So, any factor that increases the chance of collision will increase the rate of reaction. However, collision on its own isn't enough. The reactants need to have sufficient energy so that, when they do collide, reaction can take place. We call this minimum energy the activation energy.

Table 1.3 outlines some of the factors that affect the rate of chemical reactions.

> **KNOW IT**
>
> 1 a Give two examples of an alloy.
>   b For each alloy you mention, give one example of its use.
> 2 Describe the difference between a suspension and a colloid.
> 3 Give two examples of a redox reaction.
> 4 Give two examples of a polymerisation reaction.
> 5 What is needed for a displacement reaction to occur?
> 6 State two factors that will increase the rate of a chemical reaction.

LO3 Understand cell organisation and structures

GETTING STARTED

Levels of organisation (5 minutes)

What we now know as 'biology' started as 'natural history' – studying how plants and animals work. Then the sciences of anatomy and physiology developed, leading to a more detailed understanding of the processes that controlled living organisms. Now we can investigate details of cells invisible to the eye, or even the light microscope. Think about how studying these different levels of organisation gives us different insights and helps us get a better understanding of how living things work.

We can study and understand biology at different levels of organisation. Starting with the whole organism, we can move upwards to study the ways in which organisms interact in populations and ecosystems. Alternatively, we can look at the way in which organisms work, in increasing levels of detail. The cell is the fundamental unit of all organisms. We need to understand the structure and organisation of the cell to get a proper understanding of how cells work together and also understand the environment in which the chemical reactions of the cell take place.

3.1 Types of cells

There are two types of cell: prokaryotic cells and eukaryotic cells. Eukaryotic cells are complex and include all animal and plant cells, whereas prokaryotic cells are simpler and smaller, and include the bacteria. Both types of cell have cell **membranes**, cytoplasm and DNA. However, eukaryotic cells have membrane-bound organelles, such as mitochondria or chloroplasts, and the DNA is contained within a nucleus in a complex with histones and other proteins known as chromatin. In prokaryotic cells, the DNA just floats freely in the cytoplasm and is not associated with proteins.

Table 1.4 Comparison of eukaryotic and prokaryotic cells

Eukaryotic cells	Prokaryotic cells
Nucleus	No nucleus
True chromosomes (DNA associated with histone proteins)	DNA not associated with histones, plasmid is present instead
Can be unicellular or multicellular	Unicellular
Animal and plant cells	Bacterial cells
10–100 µm (larger)	1–10 µm (smaller)
DNA is linear	DNA is circular
Membrane bound organelles	No membrane bound organelles
Transcription occurs in the nucleus	Transcription occurs in the cytoplasm
Ribosomes are large (80S)	Ribosomes are small (70S)
Motility flexible	Motility rigid
Reproduction sexual or asexual	Reproduction asexual

3.2 Components of the cell and their role in the cell

Many components (organelles) are enclosed by a membrane and some, such as mitochondria, chloroplasts and the nucleus, are enclosed by a double membrane called an envelope. Other organelles, such as ribosomes, are not enclosed by a membrane, although they can be associated with or attached to a membrane.

PAIRS ACTIVITY

Cell components (15 minutes)

Your tutor will provide you with unlabelled diagrams of a prokaryotic cell, a plant cell and an animal cell. Use the table on page 12 to help you label the diagrams, using the following abbreviations (these are also used in the table):

Pro = found in prokaryotes, **Eu** = found in eukaryotes, **Pl** = found in plants, **An** = found in animals

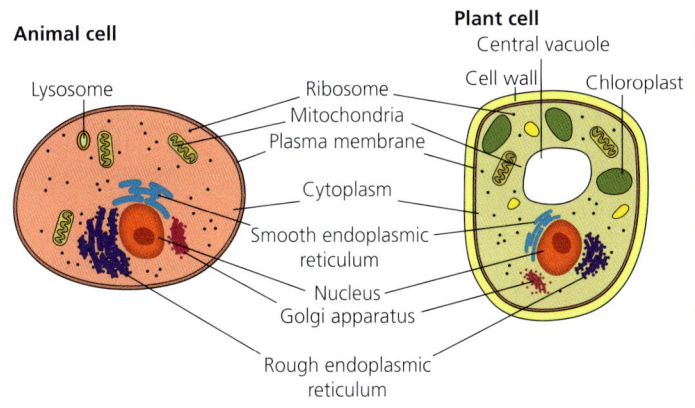

▲ Figure 1.6 Components of plant and animal cells

Animal cell labels: Lysosome, Ribosome, Mitochondria, Plasma membrane, Cytoplasm, Smooth endoplasmic reticulum, Nucleus, Golgi apparatus, Rough endoplasmic reticulum

Plant cell labels: Central vacuole, Cell wall, Chloroplast

You need to be able to distinguish between prokaryotic and eukaryotic cells based on drawings or electron micrographs.

KEY TERMS

Membrane – All membranes consist of a lipid bilayer together with proteins and other components. They are selectively permeable and can control movement of substances across the membrane as well as being the sites of many important processes in the cell.

Phospholipid – A molecule containing a glycerol molecule covalently bound to two fatty acid molecules and a phosphate. It has a hydrophilic head group (because of the phosphate) and a hydrophobic tail (because of the fatty acids).

Lipid bilayer – A double layer of phospholipids with the hydrophobic tails arranged towards the middle and the hydrophilic head groups on the outside; it forms the basis of all biological membranes.

Cell components (Pro = found in prokaryotes, Eu = found in eukaryotes, Pl = found in plants, An = found in animals)

Component	Description
Cell wall Pro, Eu, Pl	Cell walls are based on carbohydrates: plant cell walls are made from cellulose, fungal cell walls are made from chitin and bacterial cell walls have a different carbohydrate that also incorporates peptide cross-links. Cell walls provide strength and rigidity, for protection and support. If animal cells take up too much water they burst, whereas the cell wall prevents this.
Plasma membrane Pro, Eu, Pl, An	Contains the contents of the cell and controls entry and exit of substances into and out of the cell. • The **phospholipid bilayer** is permeable to small, non-polar molecules such as oxygen, carbon dioxide and most lipids. • It is impermeable to large polar molecules such as sugars, amino acids and proteins as well as charged ions. • These substances can only cross the plasma membrane if there are specific carrier proteins or ion channels to facilitate (help) this. • Specific receptors on the plasma membranes of some cells bind hormones and other molecules that act as signals in the body, allowing communication between cells.
Cytoplasm Pro, Eu, Pl, An	The cytoplasm refers to everything inside the plasma membrane but outside the nucleus (in eukaryotes).
Mitochondria Eu, Pl, An	Mitochondria (the singular is mitochondrion) are sometimes referred to as the 'powerhouses' of the cell because their main role is production of ATP in aerobic respiration. Respiration is the process where the energy from 'fuel' molecules like glucose is released and used to make ATP, the 'energy currency' of the cell. Almost all processes that require energy obtain that energy from ATP. Mitochondria are enclosed by a double membrane (envelope). The inner membrane is folded into structures called cristae and is responsible for the final stages of aerobic respiration where most of the ATP is made. Inside this is the matrix, a fluid containing many of the enzymes involved in aerobic respiration.
Chloroplasts Eu, Pl	Like mitochondria, chloroplasts are enclosed by an envelope and contain membranes called thylakoids arranged in stacks called grana (singular is granum). The chloroplast is the site of photosynthesis, the process whereby plants and fungi use light energy to make complex organic molecules from carbon dioxide and water. The thylakoid membranes contain chlorophyll and other pigments that absorb light energy as well as proteins involved in production of ATP. The rest of the chloroplast consists of a fluid called the stroma, which is where the other reactions of photosynthesis take place.
Golgi apparatus Eu, Pl, An	This consists of a series of flattened sacs bound by single membranes and filled with fluid. This is where proteins and glycoproteins (containing sugar molecules attached to the protein) that will be secreted (exported) from the cell are processed and packaged. Vesicles (small membrane-bound sacs) bring proteins from the endoplasmic reticulum to the Golgi, where they are processed and then packaged into secretory vesicles that bud off the Golgi and fuse with the plasma membrane, releasing their contents to the outside of the cell.
Lysosome Eu, Pl, An	This is the cell's recycling facility. When proteins and other cell components get worn out they are moved into lysosomes. Digestive enzymes break these down into their constituents, e.g. amino acids that can be re-used to make new proteins. Lysosomes are also involved in digestion of invading pathogens (bacteria and viruses) that are taken into the cell by the process of phagocytosis. It is important that these enzymes are kept separate from the rest of the cytoplasm because of the damage they could do, which is why trauma (damage or wounding) can be so harmful.
Endoplasmic reticulum (ER) Eu, Pl, An	This is a system of membrane-bound flattened sacs that fills a large part of the cytoplasm. The rough ER has ribosomes attached. Proteins that will be released from the cell or incorporated into the plasma membrane are made on these attached ribosomes and they are then folded and transported in the ER to the Golgi. The smooth ER does not have attached ribosomes and is responsible for making the lipids that the cell needs.
Ribosomes Pro, Eu, Pl, An	These are the smallest of the organelles and are the site of protein synthesis. Some float free in the cytoplasm and make the proteins needed within the cell. Ribosomes attached to the ER form the rough endoplasmic reticulum (RER) and make proteins destined for export from the cell. Ribosomes use the information coded in an mRNA molecule to assemble the correct order of amino acids in the protein.
Nucleus Eu, Pl, An	The nucleus contains the genetic information, in the form of DNA. When a gene is switched on (expressed) the information in DNA is transcribed into messenger RNA (mRNA) which is used as the template by the ribosomes to make the protein coded for by that gene. The nuclear membrane surrounds the nucleus and controls movement of substances in and out of the nucleus. Chromatin is a DNA-protein complex that is coiled and super-coiled to form the chromosomes. The proteins in chromatin are also involved in regulating the expression of genes – which ones are switched on or off as needed.

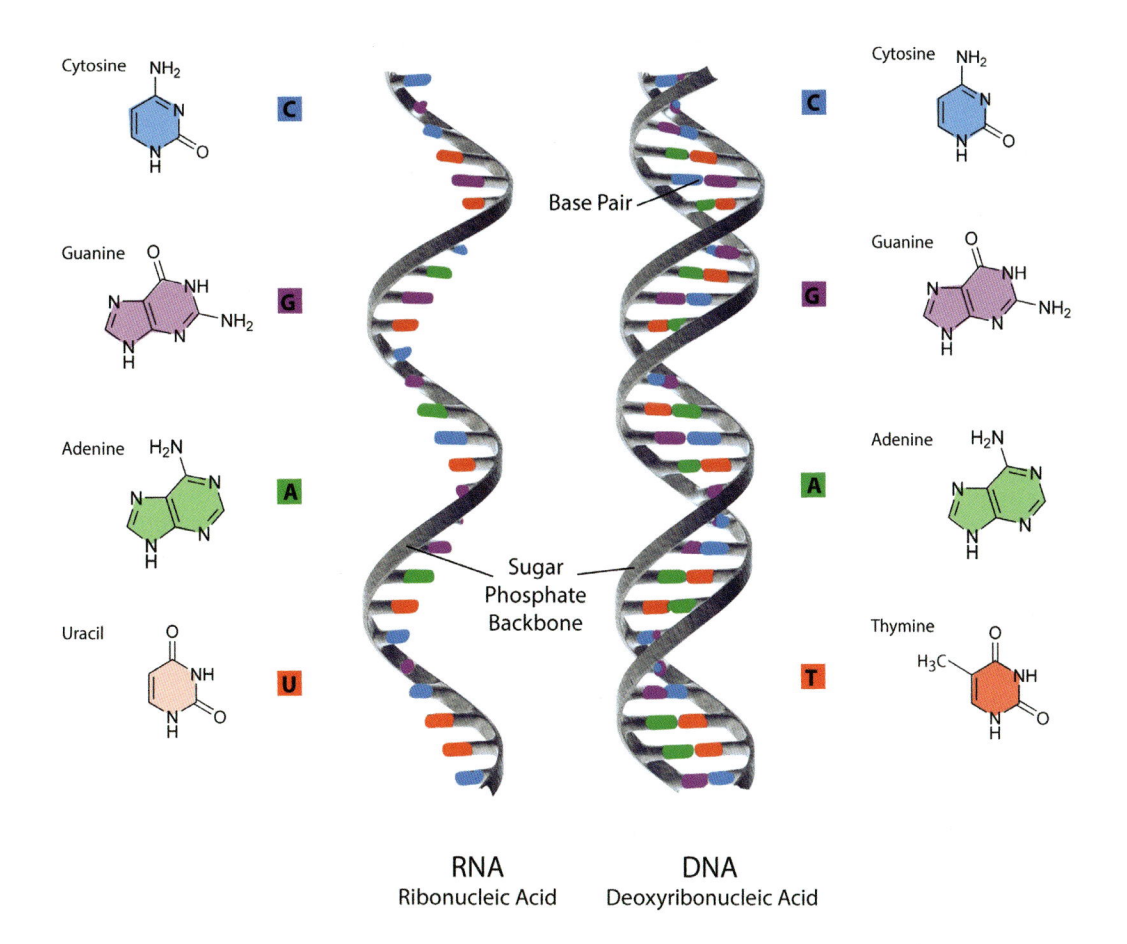

▲ Figure 1.7 The structure of DNA and RNA

DNA, RNA

Nucleic acids are polymers in which there is a backbone of alternating sugar and phosphate. In deoxyribonucleic acid (DNA) the sugar is deoxyribose, whereas in ribonucleic acid it is ribose. Nitrogen-containing bases are attached to each sugar molecule in the polymer. DNA consists of a double helix, where two strands of sugar phosphate backbone are held together by hydrogen bonds between the bases – see Figure 1.7.

The bases in DNA are thymine (T), adenine (A), guanine (G) and cytosine (C). Their structures mean that T will only hydrogen bond with A and G with C. This is known as base pairing and holds the strands together. Base pairing is also important when DNA is copied (replicated) when cells divide and when the information stored in DNA is transferred to mRNA for directing synthesis of a protein.

RNA has a similar structure to DNA, except that the sugar is ribose and T is replaced with a similar base, uracil (U).

The roles of DNA and RNA in protein synthesis are covered in section 4.4.

3.3 Understand how tissue types are related to their function

A tissue is a group of similar cells that perform a particular function. Don't confuse a tissue with an organ, because organs usually contain several different types of tissues working together to perform a function or set of functions. So, for example, the lung (an organ) consists of epithelial tissue, connective tissue and muscle tissue.

In each of the following tissues, there is a clear link between the structure of the tissue, e.g. the shape of the cells, and the function of the tissue.

Epithelial

Epithelial tissues usually cover the outside of a structure, such as the skin, cornea of the eye as well as the lining of the lungs, gut and bladder.

Epithelial tissues are normally involved with absorption or exchange – of gases, ions, molecules, etc. This is why they are found in lungs (gas exchange) and the gut (uptake of digested food molecules). The layer of cells is usually quite thin and the cells themselves can be cuboid or flattened.

Connective

Connective tissues support or bind other tissues together. They contain relatively few cells, embedded in a matrix that contains strong collagen fibres (the same protein found in tendons and ligaments) as well as elastic fibres.

Muscle

Skeletal muscle is responsible for movement of the limbs. The cells are long and arranged in bundles. The cells can appear striped under the microscope, which is why skeletal muscle is sometimes called striated muscle ('striated' means striped).

However, there is another type of muscle called smooth muscle or involuntary muscle – involuntary because we don't have conscious control over it. Smooth muscle is found in arteries and bronchioles (airways in the lungs) where contraction and relaxation can control flow or blood or air. Other smooth muscle is found in the gut, where it is responsible for the rhythmic contractions that push food through. As the name suggests, smooth muscle is not striped like skeletal muscle. The cells are not as long and nor are they organised into large bundles like skeletal muscle.

Bone

Bones are actually organs, because they contain bone tissue as well as blood vessels, marrow, nerves and epithelial tissue. Bone tissue itself is a type of connective tissue that has become mineralised. Cells, known as osteoblasts, produce a matrix of collagen fibres and release calcium, magnesium and phosphate ions that produce a crystalline mineral. This combination of collagen matrix and bone mineral produces a structure that is strong and relatively light. As well as supporting the body, bones act as attachment points for muscles.

Nerve

Nervous tissue consists of specialised cells called neurones that carry nerve signals throughout the body. Sensory neurones transmit signals from receptors in the periphery to the central nervous system (CNS, consisting of the brain and spinal cord) while motor neurones transmit signals from the CNS to muscles throughout the body. Neurones within the CNS are responsible for co-ordinating responses to external stimuli (touch, taste, vision, etc.).

Ovary and testis

Strictly speaking, the ovaries in females and testes in males are the reproductive organs, consisting of more than one type of tissue working together. They are both gonads, i.e. they produce the gametes (egg and sperm in mammals) as well as being endocrine glands that produce the sex hormones: testosterone in males and oestrogen, progesterone and testosterone (to a lesser extent) in females. As well as producing the gametes (egg cells), the ovaries are involved in the regulation of the menstrual cycle in human females.

GROUP ACTIVITY

Observing tissue types (1 hour)

You should be able to identify the different tissue types from microscope slides or photomicrographs. Your tutor will provide you with a number of different slides or photomicrographs for you to identify. Make annotated drawings of the different types showing the characteristic features that help identification.

KNOW IT

1 Give two similarities and two differences between:
   a plant cells and animal cells
   b eukaryotic cells and prokaryotic cells.
2 a Give the full names and single letter abbreviations of the four bases found in DNA.
   b Which bases pair together in DNA?
3 Explain the difference between a tissue and an organ.
4 Give the names of two types of muscle.

LO4 Understand the principles of carbon chemistry

GETTING STARTED

The properties of carbon (5 minutes)

If it were not for the unique properties of carbon, life on Earth (if it existed at all) would be very different. Carbon forms many different compounds with itself or other elements. Carbon chemistry is often known as 'organic chemistry' because all compounds important for life contain carbon. Think of all the biological molecules you know about – what do they have in common?

4.1 Carbon forms a vast number of different types of compounds with other elements due to the nature of the carbon atom

The nature of the carbon atom

The first thing to note about carbon is that it has four bonding electrons in the valence shell. This means that it can form four bonding pairs, i.e. four covalent bonds. However, these can be single, double or even triple bonds – meaning one, two or three bonding pairs shared between two carbon atoms.

Alkanes, alkenes and alkynes

These are all hydrocarbons, meaning they contain only carbon and hydrogen, so they are non-polar compounds. Each forms a **homologous series**. However, alkanes contain only carbon – carbon single bonds. Alkenes contain carbon – carbon double bonds and alkynes contain carbon – carbon triple bonds.

The way in which these bonds are formed is dealt with in section 4.2.

Table 1.5 below shows the names, formulae and general formulae of a number of alkanes, alkenes and alkynes.

Table 1.5 Information about alkanes, alkenes and alkynes

Alkanes (general formula $C_nH_{(2n+2)}$)		Alkenes (general formula C_nH_{2n})		Alkynes (general formula $C_nH_{(2n-2)}$)	
Name	Formula	Name	Formula	Name	Formula
Methane	CH_4	--	--	--	--
Ethane	C_2H_6	Ethene	C_2H_4	Ethyne	C_2H_2
Propane	C_3H_8	Propene	C_3H_6	Propyne	C_3H_4
Butane	C_4H_{10}	Butene	C_4H_8	Butyne	C_4H_6
Pentane	C_5H_{12}	Pentene	C_5H_{10}	Pentyne	C_5H_8

Actually, the table is misleading because alkenes and alkynes with four carbons or more can differ in the position of the double or triple bond. There are naming conventions that take account of this, but for now we are trying to keep things simple!

Table 1.6 Functional groups

Aldehydes and ketones	These all contain the carbonyl group, where carbon is bonded to oxygen via a double bond. In aldehydes (e.g. propanal), the carbonyl carbon is attached to at least one hydrogen; in ketones (e.g. propanone) the carbonyl carbon is attached to two other carbons. This fact accounts for the similarity in the reactions of aldehydes and ketones, as well as the differences.
$H_3C-CH_2-C(=O)H$ $H_3C-C(=O)-CH_3$ Propanal Propanone	
Alcohols $CH_3-CH_2-CH_2OH$ $CH_3-CHOH-CH_3$ Top: propan-1-ol, Bottom: propan-2-ol	Alcohols all contain the -OH functional group, known as the alcohol or hydroxyl group. Because of the electronegativity of the oxygen, alcohols are relatively polar. However, the polarity of the homologous series of alcohols decreases as the length of the carbon chain increases. This makes them very useful as solvents in organic chemistry. Alcohols are also very important synthetic intermediates, because they can be oxidised to ketones or aldehydes; aldehydes can then be further oxidised to carboxylic acids. The -OH group can be replaced by a halogen and alcohols also form esters with carboxylic acids. Some alcohols have more than one -OH group. For example, glycerol (systematic name propane-1,2,3-triol) is found in all lipids, where it forms esters bonds with fatty acids. See section 4.2.
Carboxylic acids $H_3C-C(=O)OH$ Ethanoic acid	Carboxylic acids all contain the –COOH (carboxyl) functional group. As well as being weak acids, they can be reduced to aldehydes or alcohols and form esters with alcohols. The carboxyl group is found in all amino acids. Ethanoic acid (acetic acid) is found in vinegar.
Esters $H_3C-C(=O)-O-CH_3$ Methyl ethanoate	Esters are formed by a condensation reaction between an acid and an alcohol. As well as being important in biology, the ester bond is found in polyesters, an important group of polymers. Esters are commonly found in fragrances and flavourings.

> **KEY TERM**
>
> **Homologous series** – A series of compounds with the same general formula, but differing in one respect – usually each one has an extra CH_2 than the one before it in the series, i.e. the carbon chain is longer. All of the groups of compounds we will study in this section form their own homologous series.

4.2 Carbon compounds can be represented using empirical and structural formulae

In section 4.1 we saw how most compounds contain carbon – carbon single bonds (e.g. the alkanes) where one pair of electrons is shared. In some compounds, neighbouring carbon atoms can form double bonds (e.g. the alkenes) where two pairs of electrons are shared, or triple bonds (e.g. the alkynes) where three pairs of electrons are shared.

3-Methylbutanoic acid

Commonly known as isovaleric acid, 3-methylbutanoic acid is a major component of foot odour. Table 1.7 shows its various formulae.

Table 1.7 Formulae of 3-methylbutanoic acid

Empirical formula	$C_5H_{10}O_2$
Molecular formula	$C_5H_{10}O_2$ (often the empirical and molecular formulae are the same)
Structural formula	$(CH_3)_2CHCH_2CO_2H$ or $(CH_3)_2CHCH_2COOH$
Displayed formula	(displayed structure of 3-methylbutanoic acid)
Skeletal formula	(skeletal structure of 3-methylbutanoic acid)

Carbon compounds can be large, e.g. long chains, and/or complex with ring structures that may also incorporate other atoms, particularly oxygen or nitrogen.

Polymers

Many **polymers** are formed by addition reactions (e.g. poly(ethene)) starting from a **monomer** with a double bond. Others are made by condensation reactions (e.g. polylactate); see Table 1.8. The square brackets in the polymer structure show that the polymer is just a repeat of the part shown – 'n' simply means a large number of repeats. The horizontal lines at either end of the **repeat unit** represent the bonds made to the next unit on either side. It is generally easier to understand how different polymers are formed if the monomer is drawn based on ethene, as shown in Table 1.8. The different functional groups (methyl-, phenyl- and chloro- in these examples) modify the properties of the polymer. Table 1.8 shows the IUPAC names for the polymers, as well as their common names.

> **KEY TERMS**
>
> **Empirical formula** – The simplest whole number ratio of the different atoms in the compound. The empirical formula of any compound can be obtained from the percentage by mass of each element in the compound. Different compounds can have the same empirical formula. For example, methanal (the simplest aldehyde), ethanoic acid and glucose all have the empirical formula CH_2O.
>
> **Molecular formula** – Lists the number of each type of atom in a compound. If you know the empirical formula of a compound and its mass (obtained by mass spectrometry) you can calculate the molecular formula. So, the molecular formula for methanal is CH_2O (the same as the empirical formula), for ethanoic acid it is C_2H_4O and for glucose it is $C_6H_{12}O_6$.
>
> **Structural formula** – Shows the structure of the molecule. There are a number of conventions that you need to follow. The structural formula of methanal is HCHO, ethanoic acid is CH_3COOH and glucose is complicated! One form has the structural formula $CH_2OHCHOHCHOHCHOHCHOHCHO$, which is why displayed formulae are often more useful!
>
> **Displayed formula** – Shows the arrangement of atoms in a molecule. Each atom is shown by its chemical symbol and covalent bonds are shown by single, double or triple straight lines. You may also see simplified displayed formulae where not every single bond is shown, e.g. using –CH_3 to represent a methyl group.
>
> **Skeletal formulae** – These show just the bonds as lines and leave out the carbons.
>
> **Polymer** – A large molecule formed from one or more smaller molecules or monomers.
>
> **Monomer** – The individual molecule or molecules that form a polymer.
>
> **Repeat unit** – A way of showing the pattern of monomers in the polymer.

Table 1.8 Some polymers, their monomers and their repeat units

Polymer name	Monomer structure	Repeat unit
Poly(ethene) (common name = polythene)	H₂C=CH₂	−[CH₂−CH₂]−ₙ
Poly(propene) (common name = polypropylene)	H₂C=CH(CH₃)	−[CH₂−CH(CH₃)]−ₙ
Poly(phenylethene) (common name = polystyrene)	H₂C=CH(C₆H₅)	−[CH₂−CH(C₆H₅)]−ₙ
Poly(chloroethene) (Common name = polyvinyl chloride, PVC)	H₂C=CHCl	−[CH₂−CHCl]−ₙ
Polylactate (a condensation polymer)	HO−CH(CH₃)−C(=O)−OH	−[O−CH(CH₃)−C(=O)]−ₙ

4.3 Carbon compounds form different types of isomer

Isomers are compounds that have the same molecular formulae but different arrangement of the atoms in the molecule. This means that they have different chemical and/or physical properties – sometimes very different.

Structural isomers

These are isomers that have the same molecular formula (i.e. the same number and type of atoms) but a different arrangement of those atoms in space.

There are various types of structural isomers, as shown in Table 1.9.

Geometric isomers

Geometric isomers are a type of stereoisomer and rely on the fact that rotation about a carbon – carbon double bond is not possible. If there are substituents (i.e. something other than hydrogen) on each carbon, the substituents can be on the same side of the double bond (*cis-*), or on opposite sides (*trans-*), as shown in Table 1.10.

Table 1.9 Types of structural isomers

Type of isomer	Nature of isomerism	Examples Molecular formula		
Chain	Carbon chains can be branched as well as straight.	C_4H_{10}	$CH_3-CH_2-CH_2-CH_3$ Butane	$CH_3-CH(CH_3)-CH_3$ 2-Methylpentane
Positional	The carbon chain is the same, but the position of functional groups varies.	C_3H_7Br	$H_3C-CH_2-CH_2Br$ 1-Bromopropane	$H_3C-CHBr-CH_3$ 2-Bromopropane
		C_3H_8O	$CH_3-CH_2-CH_2OH$ Propan-1-ol	$CH_3-CHOH-CH_3$ Propan-2-ol
Functional group	The isomers have different functional groups, i.e. they belong to different families (homologous series).	C_3H_6O	H_3C-CH_2-CHO Propanal (aldehyde)	$(H_3C)_2C=O$ Propanone (ketone)

Table 1.10 Geometric isomers

Geometric isomers	
cis-but-2-ene	trans-but-2-ene

Table 1.11 The enantiomers of the amino acid alanine

This works in simple situations like this, but can't be applied to more complicated situations where there are more than two substituents. The *E-/Z-* system addresses this. You don't need to know how the rules work, just that a *cis*-isomer is a *Z*-isomer and a *trans*-isomer is an *E*-isomer – *sometimes!*

Optical isomers

Optical isomers are another type of stereoisomer. They are called 'optical' isomers because of their effect on plane polarised light. These isomers arise when a carbon atom has **four** different substituents – these are known as chiral. Such compounds have two isomers that are mirror images of each other. These are called enantiomers. Look at the two enantiomers of the amino acid alanine in Table 1.11. It helps to imagine the line as a mirror. There is no way that you can rotate the left-hand structure to be superimposable on the right-hand structure.

Optical isomers are very important in biology. All 20 naturally occurring amino acids found in proteins, except for glycine, have a chiral α-carbon and all have the same arrangement of groups about that chiral carbon; they are all L-isomers (so you will often see L-amino acids, L-alanine, and so on). Many synthetic chemicals are a mixture of the two isomers, known as a racemate or racemic mixture. This is often of great importance in pharmaceuticals (see Unit 11), because the targets for drugs (medicines) are often enzymes or other proteins and usually only one of the two isomers will bind or be effective.

4.4 Carbon compounds can form large complex molecules

Biochemicals

It used to be thought that compounds from living organisms had some special 'vital force' that distinguished 'organic' chemicals from 'inorganic' chemicals. We have come a long way since then, but the name 'organic chemistry' is still used for the chemistry of carbon-containing compounds. Although there is nothing 'magic' about the carbon-containing chemicals found in living organisms, their enormous breadth and variety earns them a special name: biochemicals.

Amino acids, peptides and proteins

Proteins are polymers of **amino acids** formed, again, by condensation reactions. The bond formed is known as a **peptide** bond. Proteins have many functions:

- enzymes
- structural proteins, e.g. collagen found in connective tissue (see LO 3.3) or keratin found in skin and hair
- carrier proteins, such as haemoglobin
- antibodies
- and many, many more!

Peptides are important as hormones or regulator molecules, e.g. insulin and glucagon involved in control of blood glucose.

Lipids

Lipids are the only group of biological **macromolecules** that are not polymers. Most lipids are esters of fatty acids and glycerol. If glycerol makes ester bonds with three fatty acids, it is known as a triglyceride. Oils (from plants) and fats (from animals) are triglycerides and are mostly used as energy stores and, in mammals, for insulating the body. They are also important energy sources in the diet.

Some lipids only have two fatty acids making ester bonds. The third –OH group on the glycerol makes a phosphate ester and this is known as a phospholipid (see section 3.1).

> ### 🔑 KEY TERMS
>
> **Protein** – A polypeptide with a recognisable three dimensional structure.
>
> **Amino acid** – A molecule with both an amino group and a carboxyl group. There are 20 naturally occurring amino acids found in proteins and all have the amino and carboxyl groups attached to the same carbon, the α-carbon (hence α-amino acids). The same carbon also has a hydrogen and another substituent – the side-chain, or R-group – which is different in each different amino acid.
>
> **Peptide** – A compound containing two or more amino acids joined together by peptide bonds.
>
> **Macromolecule** – A very large molecule, containing thousands or more atoms. They are often, but not always, polymers and the term is used particularly to describe large biological molecules.

Carbohydrates

Carbohydrates (Table 1.12) are important energy sources, for example starch, sucrose ('sugar', refined from sugar cane), lactose (found in milk) or maltose (used as an energy source for yeast in the brewing industry). However, carbohydrates also have structural roles, the most common of which is cellulose, found in plant cell walls.

Principles of protein synthesis

Protein synthesis was mentioned in section 3.2 and involves nucleic acids. The structure of nucleic acids was also covered in section 3.2, so it would be worth reviewing that before you continue.

DNA

The sequence of bases in the DNA molecule codes for the sequence of the protein.

RNA – messenger, ribosomal and transfer

mRNA carries the information contained in the DNA in the nucleus to the ribosomes, located in the cytoplasm.

A transfer RNA (tRNA) molecule contains, within its sequence of bases, an anti-codon. There are, therefore, 64 different tRNAs. Each tRNA will be loaded with the amino acid that corresponds to its anti-codon. tRNA binds to mRNA by a specific base-pairing interaction between the **codon** and **anti-codon**.

Transcription

The instructions to make proteins are contained within our DNA. Transcription is a process where DNA in the nucleus is copied over to messenger RNA (known as mRNA) with the help of an enzyme RNA polymerase. The new mRNA strand contains the codes for a specific protein. The mRNA is then carried out of the nucleus into the cytoplasm, ready for the next stage.

Translation

In the cytoplasm a ribosome brings together the mRNA and a number of different transfer RNA (tRNA) molecules. Different tRNA molecules can only 'fit' with particular sequences in the mRNA, known as codons. Each different tRNA molecule brings with it a different amino acid, so that a chain of amino acids are constructed that correspond to the codons in the mRNA. The amino acids are linked together by the ribosome to form a polypeptide chain, which goes to on to become a new protein.

KEY TERMS

Codon – A sequence of three bases in DNA or mRNA that codes for a single amino acid. There are four bases, so there are 4 x 4 x 4 = 64 possible combinations, meaning 64 different codons – more than enough to code for all 20 naturally occurring amino acids.

Anti-codon – The complementary sequence to the codon, following the base-pairing rules.

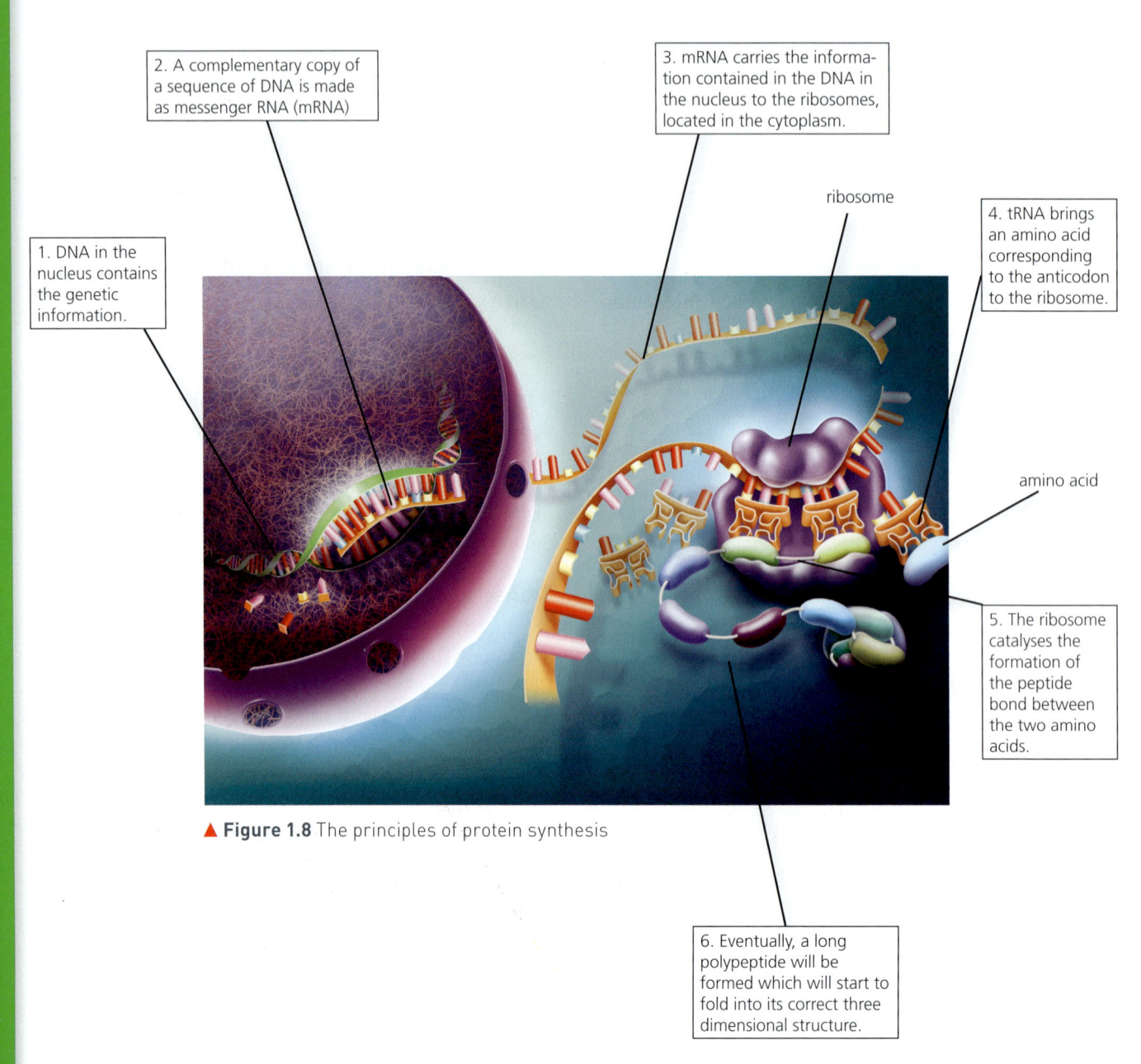

1. DNA in the nucleus contains the genetic information.

2. A complementary copy of a sequence of DNA is made as messenger RNA (mRNA)

3. mRNA carries the information contained in the DNA in the nucleus to the ribosomes, located in the cytoplasm.

ribosome

4. tRNA brings an amino acid corresponding to the anticodon to the ribosome.

amino acid

5. The ribosome catalyses the formation of the peptide bond between the two amino acids.

6. Eventually, a long polypeptide will be formed which will start to fold into its correct three dimensional structure.

▲ Figure 1.8 The principles of protein synthesis

Table 1.12 Types of carbohydrates

Type of carbohydrate	Examples	Notes
Monosaccharides	Glucose	Monosaccharides contain just one sugar molecule and include glucose, fructose and galactose.
Disaccharides	Maltose	Disaccharides are formed by condensation reactions between two monosaccharides – the bond formed is known as a glycosidic bond. Common disaccharides include maltose (containing two glucose molecules), sucrose (containing glucose and fructose) and lactose (containing glucose and galactose).
Polysaccharides	Amylose, Amylopectin (Starch), Glycogen, Cellulose (fibre)	Polysaccharides are polymers of monosaccharide monomers. The most common are starch (straight chain amylose or branched chain amylopectin), glycogen and cellulose, all of which are polymers of glucose.

GROUP ACTIVITY

Protein synthesis animation (15 minutes)

It is much easier to understand the principles of protein synthesis if you watch a good animation. Your tutor will help the group find something online. Make annotated sketches of some of the key stages to help you understand and remember the process.

KNOW IT

1 State the general formula of the following:
   a alkanes
   b alkenes
   c alkynes
2 Draw as many isomers as you can of compounds with the molecular formula $C_6H_{10}O_2$. For each isomer, draw the structural, displayed and skeletal formula.
3 Give the names of three disaccharides.
4 Describe the role of the following in protein synthesis:
   a mRNA
   b tRNA

4.4 Carbon compounds can form large complex molecules

LO5 Understand the importance of inorganic chemistry in living systems

Organic chemistry is such a huge subject and can be quite overwhelming – but it is central to biochemistry, the chemistry of life. However, we also need to understand the importance of inorganic chemistry in living systems.

5.1 Inorganic chemistry is the study of elements and compounds which do not include carbon-hydrogen bonds

In section 1.2 we saw how the periodic table is arranged into groups, with elements in the same group having similar properties. We also considered how the periodicity in the table explains trends in the properties of elements in the same group. We looked at ionic bonding in section 1.3. All of that will be of help here, so take a moment to review those pages.

There is one more concept we need to consider – oxidation state. In section 2.2 we looked at redox reactions in terms of loss and gain of electrons. Oxidation state is related to this and, in many ways, makes it easier to see what is happening in redox reactions and to construct the half equations that explain things more clearly.

Oxidation state

This is a measure of how oxidised or reduced an element is. Consider iron – as well as the metal Fe, iron can also exist as Fe^{2+} ions and Fe^{3+} ions. Converting Fe^{2+} to Fe^{3+} involves removal of an electron, which (remember OILRIG) is an oxidation.

There are some simple rules that allow us to work out the oxidation state of any element or atom in a compound.

- All uncombined elements (e.g. O_2, Fe, Cl_2, Na) have an oxidation state of 0 (zero), meaning they are neither oxidised nor reduced.
- The sum of the oxidation states of atoms or ions in a neutral compound is zero.
- Simple ions have an oxidation state equal to their charge, so Na^+ is +1, Cl^- is -1, Fe^{2+} is +2 and Fe^{3+} is +3.
- In compound ions, such as CO_3^{2-} or NH_4^+, the sum of the oxidation states equals the charge on the ion.

- Some elements (almost) always have the same oxidation state in their compounds:
  - Group 1 metals are always +1.
  - Group 2 metals are always +2.
  - Fluorine always has an oxidation state of -1.
  - Chlorine has an oxidation state of -1 except in compounds with oxygen (chlorates) or fluorine.
  - Oxygen has an oxidation state of -2, except in peroxides where it is -1 or F_2O where it is +2.
  - Hydrogen has an oxidation state of +1, except in hydrides where it is -1.

Using oxidation states

One use for oxidation states is in the naming of compounds. You may sometimes see names like iron(II) sulfate and iron(III) chloride. The Roman numerals refer to the oxidation state of the iron, so we can immediately see that these are actually $FeSO_4$ and $FeCl_3$.

Another use is to work out whether a reaction is a redox reaction or not. Go back to section 2.2 and use oxidation states to work out what is happening in those reactions.

Metals and metal ions

We saw, from section 1.3, that metals generally form ionic compounds by losing electrons to generate positive ions. You will learn about techniques used to identify various metal ions in Unit 2, section 5.1.

You learned about displacement reactions in section 2.2 and how not all mixtures of ionic solutions produce a reaction – sometimes the result is just a mixture of soluble ions. That means that we can often study ions (**cations** or anions) in isolation, without worrying about the counterion.

> 🔑 **KEY TERM**
>
> **Cation** – A positively charged ion, e.g. Na^+ or NH_4^+ (the ammonium ion). Cations are usually formed by metals, but not always, as you can see from the ammonium ion.

Inorganic compounds

In section 1.3 (ionic bonding) we saw that relatively few non-metals formed simple ions – it was mostly just the halides and oxygen. The majority of anions are actually compound ions, such as CO_3^{2-} or SO_4^{2-}. Table 1.13 describes a range of inorganic compounds.

Table 1.13 Inorganic compounds

Oxides, e.g. CO_2, N_2O, MgO	Most oxides are ionic compounds. But one of the most important oxides, carbon dioxide, is a covalent compound. Animals produce CO_2 in respiration and plants use it to produce sugars in photosynthesis. Oxides of nitrogen, such as N_2O, are also covalent compounds and can be harmful pollutants.
Peroxide H_2O_2	This is one of the few compounds where oxygen has an oxidation state of -1 rather than -2. It is also a powerful oxidising agent, because it reacts to reduce the oxidation state of oxygen from -1 to the more usual -2. In biology, hydrogen peroxide is used by cells in the immune system to kill other cells in the body that are infected with a virus.
Hydroxides, e.g. NaOH, KOH, $Ba(OH)_2$, $Fe(OH)_2$	Inorganic hydroxides are inorganic bases, sometimes called alkalis. They are often used in titration (see Unit 2, section 3.1) to determine the concentration of an unknown acid. Sodium hydroxide (NaOH) is most widely used, but barium hydroxide is useful in titrating weak organic acids.
Hydracids, e.g. HF, HCl, H_2S	These are all sources of H^+ ions. HF is highly corrosive and highly toxic (due to the fluoride ions produced). Hydrogen sulfide, H_2S, is a highly toxic gas (it inhibits respiration in mitochondria). It is a product of breakdown of organic matter in the absence of oxygen – which is why rotting vegetation can smell of bad eggs. Some types of bacteria can even use hydrogen sulfide as a food source!
Nitrates	Nitrates contain the NO_3^- ion and are widely used because most nitrates are soluble. This is the basis of the silver nitrate test for halides (see section 5.1 in Unit 2). In biology, nitrates are the main source of nitrogen in plants, which use nitrogen to produce amino acids and nucleic acids.
Phosphates	The phosphate ion is crucial in biology. ATP, the 'energy currency' of the cell, is adenosine triphosphate. Hydrolysis of ATP to produce ADP (adenosine diphosphate) releases the energy needed for almost all cellular processes. Nucleic acids have a sugar-phosphate backbone in the polymer. Phospholipids are phosphate esters of diglycerides and form the phospholipid bilayer that is the basic structure of all biological membranes. Activation of many enzymes or regulatory proteins involves adding or removing a phosphate group. We could go on – the list is a long one!
Sulfates	More useful in the chemistry lab than in biology, sulfates are found in sulfuric acid and in copper (II) sulfate solutions used in tests such as those using Fehling's reagent (for a reducing sugar) or Tollen's reagent (test for an aldehyde). Barium sulfate is widely used as a component of pigment in white paint (usually in combination with titanium dioxide) and as an optical brightener in printing papers. It is also used as a radiocontrast agent in X-ray imaging – as a heavy metal (high atomic number), barium is relatively opaque to X-rays. The 'barium meal' is a suspension of barium sulfate. Although, like most heavy metals, barium is toxic, the low solubility of barium sulfate means that very little is absorbed by the body.

Bioinorganic chemistry – biological functions of metal ions

We all know that a balanced diet needs a good source of vitamins and minerals and we should know that 'minerals' means something different in nutrition than it does in geology. But what exactly are 'minerals' and why are they important? This section addresses those questions.

Minerals in this context refers to inorganic salts or ions. This includes some anions mentioned in the previous section on inorganic compounds. However, most of the inorganic ions that are important in biology are metal cations and many of them are enzyme **co-factors**. Table 1.14 shows the biological functions of metal ions

> **KEY TERMS**
>
> **Co-factor** – A non-protein 'helper' for an enzyme that is involved in the enzyme mechanism in some way. Co-factors can be inorganic (usually metal ions) or organic molecules (or metallo-organic if they contain metal atoms or ions).

> **KNOW IT**
>
> 1 What is the oxidation state of manganese in the following compounds?
>   a $MnSO_4$
>   b $KMnO_4$
> 2 Nitrogen forms many compounds with oxygen. Give an example of the following:
>   a a compound nitrogen and oxygen only that is a pollutant
>   b an ion containing nitrogen and oxygen only that is essential for plant growth
> 3 Give an example of a metal ion used in medicine.
> 4 Describe the importance of Ca^{2+} ions in the body.

5.1 Inorganic chemistry is the study of elements and compounds which do not include carbon-hydrogen bonds

Table 1.14 Biological functions of metal ions

Ni^{2+} Hydrogenase, hydrolase	Hydrogenases occur widely in prokaryotes and are used to catalyse redox reactions, particularly in bacteria using inorganic chemicals as an energy source, such as the sulfate-reducing bacteria. These types of bacteria are interesting, not just because they don't 'feed' on organic matter, but because they have potential for use in cleaning up toxic contamination. These enzymes contain nickel ions, often in complexes with sulfur and iron.
Fe^{2+}, Fe^{3+}, Cu^{2+} Oxygen transport and storage, electron transfer	This is a huge field! Haemoglobin contains haem, a **prosthetic** group, that has an Fe^{2+} ion at its centre to which oxygen binds. A similar protein in muscle, myoglobin, can act as a store of oxygen. A property of **transition metals** is their variable oxidation state. This is important in the electron transport chains in mitochondria (aerobic respiration) and chloroplasts (photosynthesis). Electrons are passed along the chain in a series of redox reactions. Cytochromes also contain a haem group where the iron atom switches between Fe^{2+} and Fe^{3+} as electrons are transferred. Other electron transport proteins involve copper switching between Cu^+ and Cu^{2+}.
Na^+, K^+ Osmotic balance, charge carrier	The Na^+/K^+-ATPase enzyme is found in the plasma membrane of all animal cells that use energy from ATP to pump sodium ions out of cells while pumping potassium ions into cells. This has many different functions. It can act to maintain the osmotic balance of the cell. In nerve cells the Na^+/K^+-ATPase maintains the resting potential of the cell, so that the inside is about 60mV more negative than the outside. The reversal of this membrane potential is the basis of transmission of the nerve signal. In other cells the Na^+ concentration gradient that is established is used to provide the driving force for uptake of other substances, such as glucose or amino acids, by co-transport with sodium ions.
Ca^{2+} Structural, charge carrier	Calcium compounds provide rigidity in bone (see section 3.3) as well as shells (snails, shellfish) and teeth. Ca^{2+} ions have widespread functions in intracellular signalling, such as control of muscle contraction or movement of microtubules and the cytoskeleton or formation of spindle fibres during mitosis and meiosis. Ca^{2+} ions play a part in the fusion of vesicle membranes with the plasma membrane (see section 3.2).
Mn^{2+} Oxidase, structural, photosynthesis	Many metalloproteins contain manganese ions, including the enzyme manganese superoxide dismutase (Mn-SOD). This enzyme is found in almost all organisms living in aerobic conditions. Mn-SOD is responsible for breaking down superoxide (O_2^-), a toxic species that can be produced in mitochondria. Manganese helps the body form connective tissue and bones. Manganese is essential for photosynthesis. The first step in photosynthesis involves photolysis of water, carried out by the manganese-containing Photosystem II. Light energy is used to split water into oxygen molecules, H^+ ions and electrons. Without this, we wouldn't have enough oxygen to breathe!
Li^+ Treatment of hypertension, bipolar disorder	Lithium salts, such as lithium carbonate or lithium citrate, are used as mood-altering drugs in treatment of bipolar disorder. The exact mechanism of action is not known, although it probably involves modification of various neurotransmitters in the brain.
Pt^{2+} Treatment in chemotherapy	There are several platinum (Pt^{2+}) containing drugs used to treat cancer, of which the first was cisplatin. These all act by binding to DNA and interfering with mitosis, leading to apoptosis (programmed cell death). Like many anti-cancer drugs, they rely on the fact that cancer cells are rapidly dividing and so targeting them will kill the cancer before killing the patient.

> 🔑 **KEY TERMS**
>
> **Prosthetic group** – A co-factor that is tightly bound to the enzyme, sometimes even by covalent bonds.
>
> **Transition metal** – A metal in the d block. Transition metals can have multiple oxidation states. They can also accept dative covalent bonds co-ordinate. These two features make transition metals useful as catalysts in chemistry and as cofactors in biochemistry.

LO6 Understand the structures, properties and uses of materials

6.1 The properties of a material determine its uses, and can be explained by its chemistry

Mechanical properties of materials

We often talk about mechanical properties of materials in fairly qualitative terms. For example, we might refer to a material as 'strong' or 'flexible'. Scientifically, it is important that we have a good definition of such terms, allowing us to design experiments and tests, and to choose materials for particular tasks

Some definitions of properties of materials

It's important to be clear on some definitions of mechanical properties, especially where some of these may have different or imprecise definitions in everyday language.

Strength (compression and tension) is the ability of a material or object to withstand a load without mechanical failure (e.g. breaking or permanently changing shape).

Elasticity is a measure of an object's ability to return to its original shape after being deformed. The term **Elastic modulus** is used to describe the specific elastic properties of a material. If a material is deformed too far, it no longer behaves elastically and undergoes **plastic deformation** (i.e. permanently changes shape) or breaks.

Stiffness is a measure of the rigidity of a structure or object to resist deformation. Its counterpart is **flexibility**. This is similar to the ideas of **elasticity** and **elastic moduli**, which are more often specific properties of materials.

Malleability and **ductility** refer to a material's ability to **plastically deform** under stress, such as being stretched or hammered. Ductile materials can be stretched to form fine wire; malleable materials can be hammered or pressed into thin sheets or formed into objects. As examples, gold is both malleable and ductile while lead is malleable, but not ductile.

Brittleness can be regarded as a counterpart to malleability and ductility. While a malleable or ductile material readily undergoes plastic deformation, a brittle material does not, and fails (i.e. breaks) with little or no permanent deformation.

Hardness is closely related to all the above properties, and is usually defined as a material's ability to resist, for example, scratch, fracture or indentation damage.

Density is simply a measure of the amount of material in a particular volume and is usually defined as the mass of material per cubic metre. For example, water has a density of about 1000 kg m^{-3}. A common term is **relative density**, which measures the density as a multiple of that of water. For example, gold has a relative density of 19.3, meaning it is 19.3 times as dense as water.

Atomic models of mechanical properties

Materials vary greatly in their mechanical properties from hard, inelastic, but brittle materials such as glass and ceramics, to soft, elastic materials such as rubber. How can we understand these properties in terms of the atomic and molecular models of matter we've developed in earlier sections?

We saw in section 2 that polymers are long-chain molecules, made up of repeating units. In an elastic polymer, such as rubber, the chains are arranged randomly in the unstretched state, as shown in Figure 1.9 (i). In this state, the shape of the individual chains is determined by the individual bonds and by interactions (e.g. electrostatic attractions) between chains, known as cross-links. These interactions are indicated by dotted lines in Figure 1.9. If the rubber is now stretched, the chains are drawn into alignment, partially overcoming the individual inter- and intra-molecular interactions (Figure 1.9 (ii)).

Whilst the material behaves elastically, the deformation in Figure 1.9 isn't permanent. Once the stretching force is removed, the interactions between chains reassert themselves and the structure relaxes back to its original form. If, however, the stretching forces become too large, the interactions between chains become insufficient to hold the network of chains together, and individual chains begin to move independently. This leads to plastic (i.e. permanent) deformation and ultimately failure.

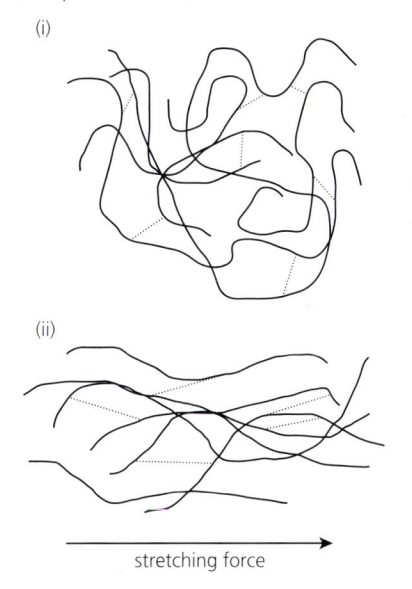

▲ **Figure 1.9** (i) Unstretched rubber polymer, showing random chains of molecules. (ii) Stretched rubber polymer. The chains are now drawn into alignment. Weak inter-chain interactions are indicated by dotted lines

The rubber polymer illustrated above is an amorphous material, in that the molecules are not ordered. By contrast, atoms in metals are arranged in an ordered, crystalline fashion, as shown in Figure 1.10 (i). The interatomic, electrostatic forces between atoms maintain the ordered array, in the absence of any external force.

If an external force is now applied, the crystal structure is distorted, as Figure 1.10 (ii) shows. The external forces cause the layers of atoms to begin sliding over one another, as shown in the highlighted, red column of atoms. So far, the distortion between layers is less than the separation of the atoms, so if the external force is removed, the interatomic forces return the array to its original configuration. The material as a whole has therefore behaved elastically.

If, however, the external force is large enough, the layers of atoms can slide over one another into a new, stable configuration. In Figure 1.10 (iii), the layers have slipped by one atom each. If you look carefully, you can see that this is essentially identical to the original, and hence is stable. The whole material has now undergone a plastic deformation.

Quantifying mechanical properties

In the previous section, we discussed the ideas and mechanisms of mechanical properties qualitatively. We will now look at some of these more quantitatively and see how they relate to the qualitative descriptions.

Hooke's Law

Suppose we have a simple spring hung vertically, and we then hang different weights to it, causing it to extend (Figure 1.11). Qualitatively, we readily see that the more weight we attach, the greater the extension of the spring. Can we quantify this?

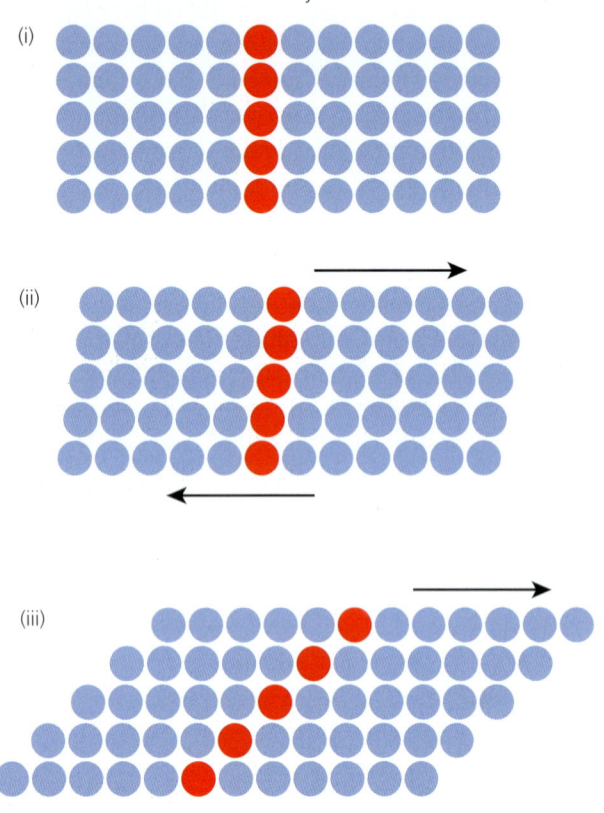

▲ Figure 1.10 (i) An unstressed metal atom array. To illustrate distortions, one of the columns of atoms has been highlighted. (ii) A small force is now applied to the array, distorting the lattice elastically. If the force is removed, the array returns to its configuration in (i). (iii) If the external force is increased further, the rows of atoms now 'slip' into a new, stable configuration. The array (and hence the material) has now undergone an irreversible plastic deformation

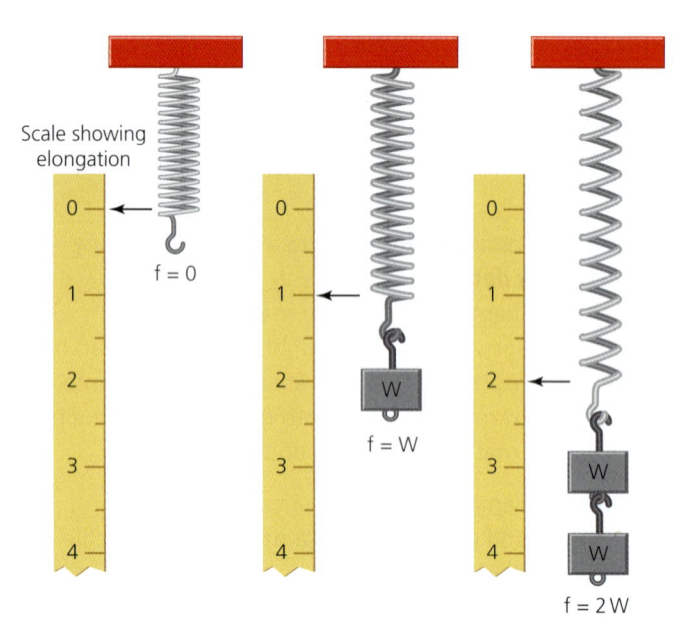

▲ Figure 1.11 An illustration of a simple Hooke's Law experiment

CLASSROOM DISCUSSION

Hooke's Law 1 (10 minutes)

The following data is taken from a Hooke's Law experiment.

Table 1.15 Data from a Hooke's Law experiment

Mass (g)	Force (N)	Extension (cm)
0	0.000	0.0
10	0.098	1.2
20		2.5
30		3.3
40		4.6
50		6.0
60		7.1
70		7.6
80		9.4
90		10.7
100		11.9

Complete the table (remember that the force of the weight is the acceleration due to gravity (9.8 ms^{-1}) multiplied by the mass in kg) and plot a graph of the force on the x-axis versus the extension on the y-axis.

What do you observe?

You should be able to draw a straight trendline through the points on your graph. (You need to estimate a straight line, as there is some uncertainty on the points due to inaccuracies in measurement.) This is telling us that there is a linear relationship between the force applied and the extension, of the form:

$$F = kx$$

Where F is the force applied in newtons and x is the extension. The quantity k is a constant for a particular spring and has units of N kg^{-1} (newtons per kilogram). For a particular spring, we can find the value of k from the graph we've just drawn.

PAIRS ACTIVITY

Hooke's Law 2 (10 minutes)

Figure 1.12 below is the result of an investigation using another spring. The gradient of the trendline on this graph (i.e. its slope) is the constant k.

Extension of a spring versus force applied

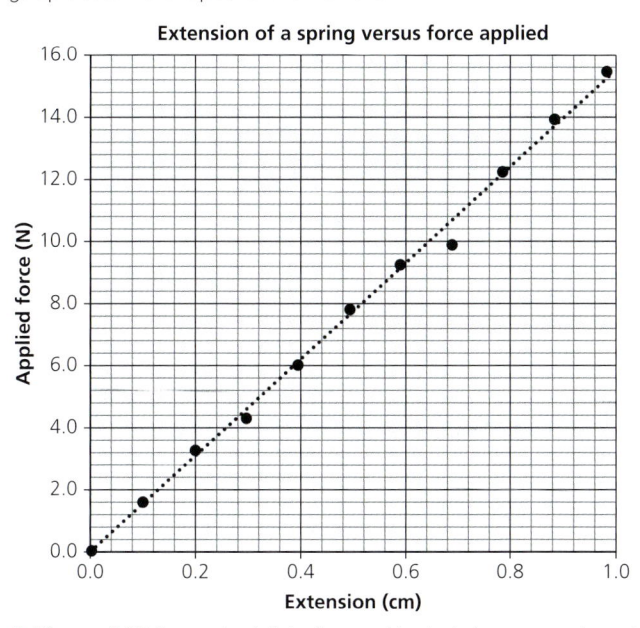

▲ Figure 1.12 A graph of data from a Hooke's Law experiment

Find the gradient of the trendline.

Is this spring stiffer than the first spring? Would you expect the constant k of the first spring to be greater or smaller? (Optional – check your deduction by measuring the slope of the trendline on your first graph.)

Young's modulus

The idea of Hooke's Law and the spring constant k is very useful in understanding the elastic properties of individual springs. However, the value of k is different for every spring or object, even if they're made of the same material. Is there a way we can define an elastic property which is specific to a particular material?

Two useful concepts to help us define a specific quantity are **stress** and **strain**.

KEY TERMS

In normal speech, the terms **stress** and **strain** are often used interchangeably. However, in science and engineering, they have well-defined meanings.

Stress – The force applied per unit cross-sectional area of material. For example, if we have a wire with a cross-sectional area of 1mm^2 (1 × 10^{-6} m^2) and the wire is stretched by a force of 1N, the stress is:

$$\frac{1N}{1 \times 10^{-6} m^2} = 1 \times 10^6 \, Nm^{-2}$$

The units Nm^{-2} are those of pressure, pascals (Pa). Hence, in this case, the stress is 1 × 10^6 Pa, or 1 MPa (megapascal).

Stress is often given the symbol σ.

Strain – The ratio of the extension of an object divided by its total length. In the example above, if the wire were 2 m in length and extended by 2 mm (2 × 10^{-3} m) when 1 N is applied, the strain would be:

$\frac{2 \times 10^{-3} \text{ m}}{2 \text{ m}}$ (note: because strain is a simple ratio, it has no units)

Strain is often given the symbol ε.

Using stress and strain, we can now define an elastic quantity which is specific to a material and is independent of the shape of an object. This is Young's modulus, E, defined as:

$E = \frac{\sigma}{\varepsilon}$ Pa.

The values of the Young's modulus for a number of materials are given in Table 1.16.

As you can see, the values in Table 1.16 are for when the materials are behaving elastically. When we looked at our atomic models, we saw that if we apply too much force, a material will behave inelastically, permanently deform and ultimately break. We can use some of the ideas of Young's modulus, stress and strain to quantify some of these inelastic properties.

Table 1.16 The Young's modulus for some common materials, expressed in GPa (1 GPa = 1×10^9 Pa)

Material	Young's modulus (GPa)
Diamond	1200
Mild steel	210
Copper	120
Titanium	116
Granite	55
Bone	18
Lead	18
Concrete	16.5
Oak	12
Nylon	2
Rubber	0.02

Figure 1.13 shows a typical plot of stress versus strain for a metal, where we have gone beyond the elastic limit.

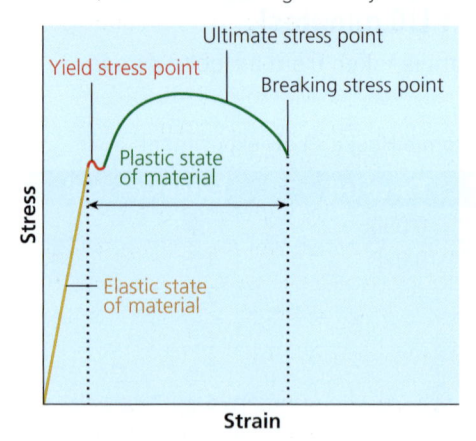

▲ **Figure 1.13** Schematic stress-strain graph for mild steel

In the elastic region (shown in orange), the material behaves according to Hooke's Law. At the yield point, shown in red, the atomic planes can start slipping over one another, and we enter the plastic region, where the material deforms permanently. As more strain is applied, the stress increases to a maximum value, which is the ultimate tensile strength of the material. Beyond this, increasing the strain no longer increases the stress – rather, continued application of strain leads to a 'run-away' effect, ultimately causing the material to fail at the breaking point.

Physico-chemical properties of materials

In this section, we will look at some properties of materials which are relevant to both physics and chemistry. In particular, we will be interested in those properties which involve a change of physical state; namely **melting**, **boiling** and **sublimation**. To help us, we will also need the idea of **vapour pressure**.

🔑 KEY TERMS

Melting – The process by which a solid converts to a liquid. Normally, this means that the inter-atomic or inter-molecular forces either reduce or are no longer sufficient, such that individual atoms or molecules are able to move freely. The forces are still sufficient though to prevent the liquid boiling into the vapour phase.

Boiling – The process by which a liquid converts into its gas state when the vapour pressure of the liquid is equal to the ambient pressure. For example, at the pressure of Earth's atmosphere at sea level, water boils at 100°C, meaning that at this temperature, the vapour pressure of water is 1 atmosphere.

Sublimation – A process not often seen in everyday circumstances. Sublimation occurs when a material goes straight from the solid phase into the vapour, without melting into a liquid. Examples of where this occurs are frozen carbon dioxide ('dry ice'), naphthalene and water under reduced pressure (e.g. water ice on the planet Mars).

Vapour pressure – All liquids are evaporating all the time – that is, atoms or molecules in the liquid are escaping into the gas phase. The vapour pressure of a liquid is the pressure exerted by these atoms or molecules when in equilibrium with the liquid. In other words, the rate at which atoms or molecules are escaping from the liquid is equal to the rate at which they are returning.

A chemical substance can exist in one of three phases – a solid, liquid or gas. For instance H_2O can exist as ice (solid), water (liquid) or steam (gas). In a solid the molecules of the substance are close to each other and tightly bound together in a regular pattern. The molecules can vibrate but they do not move past each other and this is why a solid retains its shape.

In a liquid the molecules of the substance are less tightly bound together; they can vibrate but they can also move past each other. This is why a liquid does not retain its own shape but will sit in a container. There is no overall pattern to the arrangement of molecules.

In a gas the molecules of the substance are not bound together at all. They are all completely separate and they are free to vibrate and move around in any direction. This is why a gas will escape a container if it is not 'air tight'.

Temperature is a measure of the internal energy of a substance. As you heat a solid you give it more energy and its molecules vibrate more. At **melting point** the molecules have enough energy to become less tightly bound and begin moving past each other, so that the solid changes into a liquid. As you continue heating the liquid you give it more energy and its molecules vibrate even more than before. Once you reach **boiling point** the molecules have so much energy that they can break away from each other completely, and so the liquid becomes a gas. If you continue heating the gas then the molecules continue to vibrate and also move around faster and faster.

Different substances have different melting and boiling points, which are due to their chemical properties. The degree to which an increase in energy leads to a rise in temperature is known as the heat capacity of the material. The heat capacity per unit mass (1 kg) is called the specific heat capacity. Different substances have different specific heat capacities. Whilst you may not be examined on specific heat capacity, you may find the following helps you to further understand the processes of melting and boiling.

Boiling a kettle 1 – specific heat capacity of water

In boiling water in a kettle, we are raising water from room temperature to its boiling point at 100°C. Assuming that room temperature is about 20°C, we therefore heat the water by 80°C.
Suppose now our kettle has a heating power of 2 kW, and we find that it heats 0.5 l (equivalent to 0.5 kg) of water from room temperature to boiling in 1.5 minutes (90 seconds). The amount of energy we've put into the water is:

2 kW × 90 s = 180 kJ (kilojoules, remembering that energy = power × time)

This is the energy required to heat a particular amount of water by a given number of degrees. A more general quantity for water is the *specific heat capacity*, the energy required to raise the temperature of 1 kg of water by 1°C.

In our example, we heated 0.5 kg by 80°C using 180 kJ of energy. Therefore, to heat 1 kg by 80°C, we will need twice as much energy, i.e. 360 kJ. Assuming the specific heat capacity doesn't change with temperature, we can now find the amount of energy required to heat the water by 1°C:

$$\text{Energy to raise 1 kg of water by 1°C} = \frac{360\,kJ}{80\,°C}$$
$$= 4.5 \text{ kJ °C}^{-1}$$

Hence, the specific heat capacity we've calculated for water, c, is:

$c = 4.5$ kJ kg^{-1} °C^{-1}

In general, the energy, E, required to raise the temperature of a mass, m, of a material from a temperature T_1 to T_2 is:

$E = cm(T_2 - T_1)$

GROUP ACTIVITY

Discussion (10 minutes)

The value we've just found for the heat capacity of water, 4.5 kJ kg^{-1} °C^{-1}, is different from the value of about 4.2 kJ kg^{-1} °C^{-1}.

Think about why this might be. In particular, think about what assumptions you've made, and what modifications you might make to the experiment to make the value you find more reliable.

Boiling a kettle 2 – latent heat

GROUP ACTIVITY

Discussion (10 minutes)

Discuss what happens when the water in the kettle reaches boiling point. In particular, think about the following:

1 What happens to the temperature when the water boils?
2 Is energy still being supplied to the water from the kettle heating element?
3 If so, what's happening to the energy?

In the above activity, you should come to some perhaps unexpected conclusions. In particular, water boils at 100°C on Earth at sea level (1 atmosphere pressure) and putting more energy into boiling water doesn't raise its temperature. Instead, the energy is going into converting the water from the liquid state. This energy is referred to as the *specific latent heat of vaporisation*, *L*, and is expressed in units of energy per kilogram. Notice that there's no reference to temperature in the units, as there's no change in temperature.

As an example, we find that the kettle we've used above containing 0.5 kg of water boils dry 10 minutes after it's started to boil. What is the specific latent heat of vaporisation of water?

To start, we can find the amount of energy supplied to the boiling water as:

Energy = power × time (in seconds)

$= 2 \text{ kW} \times 600 \text{ s}$

$= 1200 \text{ kJ}$

This is the energy to convert 0.5 kg of water from liquid to vapour. Hence the specific latent heat of vaporisation is:

$L = \dfrac{1200 \text{ kJ}}{0.5 \text{ kg}}$

$= 2400 \text{ kJ kg}^{-1}$

As with our value for the specific heat capacity, this differs from the data tables quantity of about 2300 kg kg^{-1} for the same reasons you've discussed above.

Latent heats occur in all phase transitions – in water, the energy required to melt ice at 0°C is known as the latent heat of fusion, and likewise, substances which sublime (e.g. ice at low atmospheric pressure, such as on Mars) have a specific latent heat of sublimation.

Electrical properties of materials

In many areas of science and engineering, the electrical properties of materials are of vital importance. For example, what materials should we choose to carry electrical power over long distances?

In this section, we will look at the underlying physical principles behind electrical properties and how these properties can be used in practice.

An atomic scale model of electrical conduction

We are familiar with the idea of an electric current as some kind of flow of energy, driving electrical devices such as electric kettles or mobile phones. At this level, the electric current seems a fairly abstract concept, so we need to look at its origins at an atomic level.

The first point to recall is that individual atoms are composed of central, positively charged nuclei surrounded by negatively charged electrons. It is these electrons which are key to many examples of electrical conduction.

In section 6.1, we saw that metals are composed of ordered arrays of atoms. What the discussion didn't include was the idea that some of the electrons cease to be bound to individual atoms, and can wander freely throughout the array. Figure 1.14 (i) shows this, with the atoms in a fixed array and electrons free to move between them.

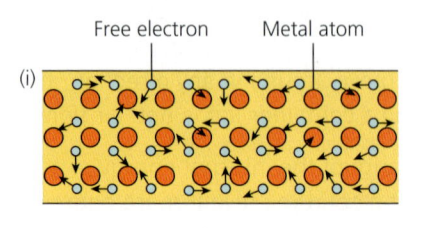

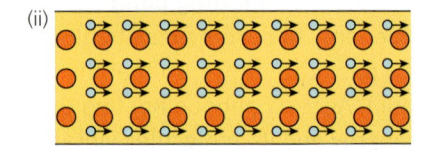

▲ **Figure 1.14** (i) An ordered array of metal atoms, with free electrons moving randomly between them. (ii) The same array, with an electric field applied, showing the flow of electrons in response to the field

Under normal circumstances, the electrons move randomly, and there is no net flow. However, if an electric field is applied, the electrons, since they are negatively charged, are attracted to the positive end of the material, resulting in a flow of charge and hence an electric charge.

Conventional current

In Figure 1.14, we see that the electrons are flowing from the negative to the positive end of the material. However, in electrical circuits and electrical engineering, current is shown flowing from positive to negative! The reason for this is historical – electricity and electric current were discovered and investigated long before the discovery of the electron, and investigators, realising there was something they couldn't see flowing, chose a direction for that flow. They happened to choose 'wrong', but ultimately it doesn't matter. Physically, the flow of negative particles in one direction is equivalent to the flow of positive particles in the other.

With an understanding at the atomic level of why metals conduct electricity, we can now see why many materials, such as other crystalline materials like quartz, and most polymers, do not conduct. In the most simple cases, these materials do not have any free electrons to carry the current.

Electrical resistance

The above discussion on conduction might, at first sight, suggest that electrons flow without hindrance in metals. This excludes though the atoms making up the metal lattice.

With no field present, electrons randomly moving in the lattice will also randomly interact with and scatter off the atoms. Now, if an electric field is applied, electrons will tend to move in the direction shown in Figure 1.15, but will still scatter, effectively slowing their drift towards the positive end.

▲ **Figure 1.15** The path of an electron in a metal with an electric field applied, and with the electron scattering from a metal atom at each change of direction. Individual atoms have been omitted for clarity

The effect of this process is to reduce the flow of current for a given electric field, in other words, increasing the material's resistance.

Insulators and semiconductors

A large number of materials will not allow electrons to flow with an electric field applied. Such materials are called insulators.

Another class of materials are called semiconductors. These have a higher electrical resistance than metals but lower electrical resistance than insulators. By increasing the temperature, the resistance of a semiconductor decreases so that it begins to behave more like a conductor. Semiconductors are incredibly important materials because they are the building blocks of all electronics - which is to say all computers, mobile phones, games consoles and any other electronic devices.

Voltage, current, resistance and Ohm's Law

You are no doubt familiar with the terms voltage and current as applied to electricity in everyday situations. We have also seen how we can understand electricity at an atomic level. We will now look at what we mean by **charge**, **voltage** and **current** and how these properties can be used in electrical applications.

> **KEY TERMS**
>
> **Charge** – The electric charge is a fundamental property of many subatomic particles, and is a measure of how strongly a particle feels the electromagnetic force. By convention, electrons are regarded as having a negative charge. The unit of charge is the coulomb (C); the charge of a single electron is 1.6×10^{-19} C. Charge is often given the symbol Q.
>
> **Voltage** – Fundamentally, the voltage is a measure of the change of energy of a charged particle when it moves through an electric field, and, scientifically, is often referred to as potential difference. The unit of voltage is the volt (V), which can be expressed as J C^{-1} (joules per coulomb). This means that if a material with a charge of 1 C moves through a potential difference of 1 V, it will gain 1 J of energy. Voltage is often given the symbol V.
>
> **Current** – A measure of the rate at which electric charge is flowing past a particular point in a circuit. The unit of current is the ampere (A), often shortened to 'amp', and can be expressed as C s^{-1} (coulombs per second). A current of 1 A therefore means that in 1 s, 1 C of charge flows past a given point in a conductor. Current is often given the symbol I. So the current, I, expressed as the flow of charge is written as the change in current, ΔQ in a unit of time Δt: $I = \Delta Q/\Delta t$

Let's think now about what happens when we apply a voltage across a conducting material (Figure 1.16). The voltage is, in effect, a force driving the electrons. We now measure the current as a particular rate of flow of charge.

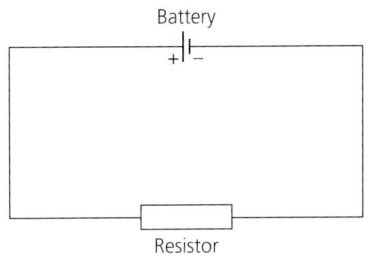

▲ **Figure 1.16** A simple circuit showing a battery applying a voltage across a conducting material, a resistor

If we now investigate what happens as we increase the voltage, we find that (for most normal materials) the current increases in direct proportion. That is:

$V \propto I$ (voltage is proportional to current) or,

$V = R \times I$ where R is the electrical resistance; an electronic device with a well-defined resistance is referred to as a resistor.

This relationship is known as Ohm's Law, named after the 19th-century physicist Georg Ohm who first quantitatively investigated this effect.

The units of resistance, R, is V I^{-1} (volts per ampere), often referred to as ohms, and often written as 'Ω'. For example, if 1 V drives a current of 1 A through a material, the material has a resistance of 1 ohm, or 1 Ω.

Resistors

Resistors are very common electrical and electronic components, used for a multitude of purposes in circuits, including:

- restricting or controlling currents, protecting sensitive components
- creating well-defined voltages (potential dividers)
- timing (in combination with other types of component).

A number of different technologies are used in manufacturing resistors, depending on their application. Here are some examples.

- Carbon film for low power (e.g. 0.25 W) applications. Typically these are physically small, so suitable for high density circuits.
- Wire wound for high power applications (10s to 100s of W) – used in power control applications where high currents need to be handled and in high power amplifier applications.
- Metal film for low power, high tolerance and high stability applications, for example, timing circuits and precision electronic instruments.

Kirchhoff's Current Law and combining resistors

So far, we've looked at currents flowing in simple circuits, like that of Figure 1.16. What happens though, if we have a network of resistors, such as Figure 1.17?

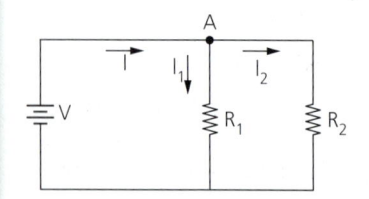

▲ **Figure 1.17** A simple resistor network, showing resistors in parallel

In Figure 1.17, a voltage V is applied across the resistors R_1 and R_2, and consequently a current I flows from the voltage source. At point A, the current splits between the two resistors, with I_1 flowing in R_1 and I_2 flowing in R_2.

Notice that the voltage across both resistors is identical and is equal to V.

An important rule in electronic circuits is Kirchhoff's Current Law, which, simply expressed, states that:

At a node in a circuit (point A in our example), the total current flowing into the node is the same as the total current flowing out.

In our example, we can see that Kirchhoff's Current Law tells us that:

$$I = I_1 + I_2$$

Notice that this is the same as saying that the charge carriers (electrons) are neither created or destroyed; that is, charge is conserved.

A useful outcome of Kirchhoff's Current Law is that we can, in principle, replace the two resistors in Figure 1.17 with a single resistor, R (Figure 1.18).

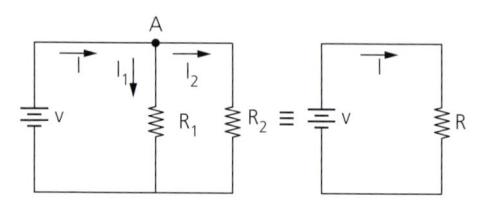

▲ **Figure 1.18** The equivalence of the circuit in Figure 1.17 to a circuit with a single resistor

To find the value of the single resistor R, we can apply both Ohm's Law and Kirchhoff's Current Law.

Thinking first about the original circuit in Figure 1.17:

From Ohm's Law (rearranged to find the current, given the voltage and resistance):

The currents are:

$$I_1 = \frac{V}{R_1}$$

and

$$I_2 = \frac{V}{R_2}$$

From Kirchhoff's Current Law:

$$I = I_1 + I_2$$

Substituting in the expressions we've just found from Ohm's Law:

$$I = \frac{V}{R_1} + \frac{V}{R_2}$$

Now let's look at the equivalent circuit with one, equivalent, resistor in Figure 1.18.

Ohm's Law is sufficient here, to tell us that:

$$I = \frac{V}{R}$$

Comparing with the equation above, we have the result that:

$$\frac{V}{R} = \frac{V}{R_1} + \frac{V}{R_2}$$

Which becomes, remembering that we can cancel the Vs on both sides of the equation:

$$\frac{1}{R} = \frac{1}{R_1} + \frac{1}{R_2}$$

This relationship generalises, so that if there were three resistors in parallel we would have:

$$\frac{1}{R} = \frac{1}{R_1} + \frac{1}{R_2} + \frac{1}{R_3}$$

PAIRS ACTIVITY

Resistors in series (30 minutes)

We've found how we can calculate the equivalent resistance of two resistors in parallel. What about if we put the resistors in series (i.e. one after another) as shown in Figure 1.19?

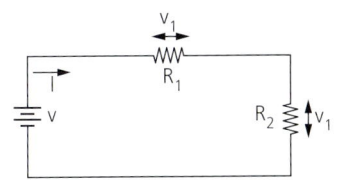

▲ **Figure 1.19** Resistors in series

In this activity, we will derive an expression for a single equivalent resistor for resistors in series, and therefore show that:

$$R = R_1 + R_2$$

There are two important points to help you:

1 Kirchhoff's Current Law tells us that the currents are the same in all parts of the circuit – that is, the current in R_1 is equal to that in R_2, and equal to I.
2 A corresponding law, Kirchhoff's Voltage Law, tells us that the voltages across individual components add up to the total voltage applied – that is,

$$V = V_1 + V_2$$

The steps to take are:

1 Apply Ohm's Law to the original circuit.
2 Apply Ohm's Law to the equivalent circuit.
3 Equate the two situations, as we did for the parallel resistors case.
4 Cancel out any common terms.

Note: This relationship also generalises - for instance if there were three resistors in series then the equation would become:

$$R = R_1 + R_2 + R_3$$

Energy and power in electrical circuits

When we defined voltage and current, we saw that the units of voltage, V, were J C^{-1} (joules per coulomb), or the amount of energy transported or dissipated per unit charge. Similarly, current, I, is defined as the amount of charge passing a point per unit time, or C s^{-1} (coulombs per second).

If we now multiply the current and voltage, we see that the units become:

J C^{-1} × C s^{-1} J s^{-1} since C cancels with C^{-1}.

This is the unit of power, the watt (W), with 1 W defined as the dissipation of 1 joule per second. Hence, we can identify the product of the voltage applied across a material and the current flowing in the material with the power, P, dissipated in that material (usually as heat) as:

$$P = VI$$

And therefore, if the power is applied for a time, t, the total energy dissipated, E, is:

$$E = Pt = VIt$$

In an electronic component such as a resistor, the power is dissipated as heat. If the value of the resistance is known, we can apply Ohm's Law to relate the power to either the voltage or current, and the resistance. For example, suppose we know the voltage applied and the resistance. Then, we can find the current using Ohm's Law:

$$I = \frac{V}{R}$$

Substituting this into the equation for power above:

$$P = V \times \frac{V}{R} = \frac{V^2}{R}$$

PAIRS ACTIVITY

Power in circuits 1 (10 minutes)

We've just seen how we can use Ohm's Law to relate the power dissipated in a circuit to resistance and the voltage across the resistor. Suppose though we know the current, I, flowing in the circuit rather than the voltage.

Use Ohm's Law and the power equation ($P = VI$) to derive an equation relating power to current and resistance.

GROUP ACTIVITY

Power in circuits 2 (20 minutes)

As an example of power and energy in a circuit, let's look back at the kettle we used in section 6.2, which has a power of 2 kW.

Assuming it's connected to UK mains, which has a voltage of 240 V, we can calculate some additional properties of the kettle.

1. What is the resistance of the heating element in the kettle?
2. What current is drawn by the kettle?
3. Suppose the kettle were to be run for one hour continuously. How much energy would it use?

Think of another typical electrical appliance and repeat the above.

Including the Pairs Activity above, we now have three expressions for electrical power:

$P = VI$, $P = I^2R$, and $P = V^2/R$.

We have already stated that energy is power multiplied by time, i.e. $E = Pt$, and therefore we also have three expressions for electrical energy, by simply substituting in for P:

$E = VIt$, $E = I^2Rt$ and $E = V^2t/R$

KNOW IT

1. Name the four fundamental forces of nature, and place them in order of strength.
2. What distinguishes metals from non-metals?
3. Why are the properties of alloys often different from those of the pure metals?
4. Define three parameters which affect the strength of materials.
5. Explain why we define Young's modulus rather than using the simple Hooke's Law.
6. What do you understand by *specific heat capacity and specific latent heat*?
7. State Ohm's Law.
8. State the equations for calculating resistors in series and resistors in parallel.
9. State Kirchhoff's Current Law.

Assessment practice questions

You will be tested on each of the LOs in the exam. Below are practice questions for you to try, followed by some top tips on how to achieve the best results.

1. This question is about atomic structure.
   a. How many protons are there in the nucleus of the ^{35}Cl isotope? (1 mark)
   b. How many neutrons are there in the nucleus of the ^{37}Cl isotope? (1 mark)
   c. How many electrons surround the nucleus in a Cl$^-$ ion? (1 mark)
   d. How many electrons are there in a Cl$_2$ molecule? (1 mark)

2. This question is about the structure of chemicals.
   a. In the periodic table, what is meant by a group? (1 mark)
   b. Explain why elements in the same group have similar chemical properties and why there are trends in those properties, using examples to support your explanation. (4 marks)

3. This question is about polymers.
   a. What is meant by the term 'monomer'? (1 mark)
   b. For each of the following polymers, draw the structure of the monomer and one repeat unit of the polymer.
      i. poly(ethene) (polythene)
      ii. poly(chloroethene) (polyvinylchloride, PVC)
      iii. polylactate (6 marks)

4. This question is about isomers.
   a. What is meant by the term 'isomers'? (1 mark)
   b. The diagram shows the structural formula of an alkene.

   H—C(H)(H)—C(H)(H)—C(H)(H)—C(H)=C(H)—C(H)(H)—C(H)(H)—H

   i. What type of isomer will this structure form? (1 mark)
   ii. Draw the isomer. (2 marks)

5. This question is about cell structure.
   a. For each of the following, state whether you would find them in only animal cells, only plant cells or both animal and plant cells:
      i. ribosomes
      ii. cell wall
      iii. endoplasmic reticulum (3 marks)
   b. Describe the relationship between the structure and the function of the following:
      i. chloroplasts
      ii. Golgi apparatus (6 marks)

6. The human body consists of different types of tissues.
   a. Describe the key features of the following types of tissue:
      i. epithelial
      ii. muscle (4 marks)
   b. State two sites where epithelial tissue occurs in the body. (2 marks)

7. This question is about alloys.
   a. Explain what you understand by an *alloy*. (1 mark)
   b. Give three reasons why we may wish to create an alloy. (3 marks)

8. This question is about the strength of materials.
   a. Explain what you understand by the following (1 mark each):
      i. strength
      ii. plastic deformation
      iii. ductility
      iv. hardness

9. This question is about Young's modulus. A sample of a material is 100 cm long and has a cross-sectional area of 0.5 mm^2. In a mechanical test, it is found that when a force of 30 N is applied, the sample extends by 1 mm. Calculate the Young's modulus of the material. (4 marks)
   Bonus mark – Can you identify the material from tables of Young's moduli?

10. This question is about electrical circuits and resistors.
    a. A voltage of 10 V is applied across an unknown resistor, and the current flowing is measured to be 3 mA. Calculate the value of the resistor, expressing your answer in Ω and kΩ. (2 marks)
    b. State the equation allowing you to find the combined value of two resistors in parallel. (1 mark)
    c. We want to increase the current in the circuit in part (a) to 10 mA. A colleague suggests you do this by putting another resistor in parallel with the first. Calculate the value of this new resistor. (3 marks)

6.1 The properties of a material determine its uses, and can be explained by its chemistry

35

> **TOP TIPS** ✓
> - There is a lot to learn and a lot to remember, but make sure you understand all the basics – if you understand those, you will remember facts more easily, or be able to work things out even if you can't quite remember.
> - Try to relate the content of this unit to the practical work you undertake and then apply it to the other units that you study.
> - You may need to keep referring back to this unit as you work through your course.
> - Remember that it is easier to learn and retain information if you can see a need for that information or application of that learning in your work and daily life.

Read about it

There are many online resources that will help you explore areas of interest.

Royal Society of Chemistry

www.rsc.org/Learn-Chemistry

Various resources covering chemistry and its applications.

Association of the British Pharmaceutical Industry (ABPI)

http://abpischools.org.uk

The ABPI Resources for Schools website has a wide range of resources, many of them interactive, covering a wide range of topics including:

- interactive periodic table game (test your knowledge and understanding)
- cells and organelles
- chemistry of life.

Wellcome Trust

The Wellcome Images website has a large collection of images, including microscope slides:

https://wellcomeimages.org

Nuffield Foundation

The Nuffield Foundation supports science education in many ways, including through provision of teaching and learning resources. The Nuffield approach is based around context-based, hands-on practical learning. Two places to start are:

www.nuffieldfoundation.org/science-education

www.nuffieldfoundation.org/teachers

Virtual Microscope

If you are interested in exploring cells and tissues under the microscope, but you don't have access to suitable facilities, the virtual microscope (a project of the Open University) is worth investigating:

www.open.ac.uk/blogs/dkm/

Materials and electronics

Most professional bodies have a wide range of resources to help you. Good starting points are:

The Institution of Mechanical Engineers, for information and resources on materials and mechanical properties:

www.imeche.org/get-involved/special-interest-groups/structural-technology-and-materials-group/information-resources/weblinks

Resources and information on copper alloys:

www.copper.org/resources/properties/microstructure/

For using resistors in practice, some good informative video clips are available. For example:

www.youtube.com/watch?v=CZgqGTxL9cA

Unit 02
Laboratory techniques

ABOUT THIS UNIT
The aim of this unit is to provide learners with a good grounding in working in a laboratory. This is a general skills unit and covers the skills required by technicians working in any kind of scientific laboratory, including working for an industrial company, the NHS, contract analysis of environmental samples and working in the education sector. You will learn about the roles and duties of a scientific technician and the systems used to ensure the effective operation of a laboratory. You will understand the importance of health and safety in the laboratory and know how to carry out and record the outcomes of standard laboratory procedures.

LEARNING OUTCOMES
The topics, activities and suggested reading in this unit will help you to:
1. Understand the importance of health and safety and quality systems to industry
2. Be able to separate, identify and quantify the amount of substances present in a mixture
3. Be able to determine the concentration of an acid or base using titration
4. Be able to examine and record features of biological samples
5. Be able to identify cations and anions in samples
6. Be able to use aseptic technique

How will I be assessed?
All Learning Outcomes are assessed through externally set written examination papers, worth a maximum of 90 marks and 2 hours in duration.

Learners should study the basic experimental design requirements, influences and user needs within the taught content in the context of a range of real laboratory experiments. Exam papers for this unit will use examples of real experiments as the focus for some questions; however, it is not a requirement of this unit for learners to have any detailed prior knowledge or understanding of particular products used. Questions will provide sufficient product information to be used, applied and interpreted in relation to the taught content. During the external assessment, learners will be expected to demonstrate their understanding through questions that require the skills of analysis and evaluation in particular contexts.

LO1 Understand the importance of health and safety and quality systems to industry

GETTING STARTED

What is important and why? (5 minutes)

Think about why health and safety and quality systems are important. Why do they exist? What would happen if they were not followed? Share your thoughts with the group.

1.1 To use aspects of good laboratory practice throughout all practicals

Good laboratory practice (GLP) is important in all types of laboratory, but particularly in those laboratories that are subject to additional regulation. GLP helps to ensure that you, and others, can have confidence in the results of your laboratory work. GLP is not a standard for laboratory safety. We will look at both quality systems and health and safety in the course of this learning outcome and you will get the chance to research further aspects that are relevant to your own experience.

KEY TERMS

Good laboratory practice (GLP) – Embodies a set of principles providing a framework for planning, performing, monitoring, recording, reporting and archiving. Don't confuse GLP with standards for laboratory safety.

Precise – Data points are close together, although there could be a random error.

Accurate – The result is close to the true or reference value for the measurement. The accuracy of any measurement will depend on the quality of the apparatus and the skill of the person using it.

Reference value – A value accepted as being very close to the true value, for example a 'standard weight' could have been measured on a balance with very little error and so is very close to the true value.

Uncertainty – Is the amount of error that is inherent in any measurement. It will depend on the sensitivity of the equipment you are using.

Some useful concepts

You are likely to come across the following terms at various times in your work, so it will be good to make sure you understand them from the outset.

- Repeatability – An experiment is repeatable when the same person repeats it using the same equipment and gets results that are close together.
- Reproducibility – An experiment is reproducible when it is carried out by different people using different equipment and the results are close together.
- Systematic error – This happens when there is a problem with the apparatus (e.g. incorrect calibration), so the results are **precise** (close together) but not **accurate**.
- Reliability – This is assessed by comparing an individual result with a **reference value** or with a mean of a large number of results. The reliability of any experiment is increased by repeating and reproducing the results.

The choice of measuring equipment and the importance of calibration

PAIRS ACTIVITY

How accurate? (10 minutes)

Work in pairs. Your tutor will provide the class with a selection of glassware and other measuring equipment. Most equipment will be accurate to ± half the smallest division, so a balance that weighs to the nearest 0.1 g will have an uncertainty or error of ±0.05 g. Volumetric glassware will be marked with the uncertainty, e.g. ±0.1cm^3. What can you tell about the accuracy of the different types of equipment that you will use? It might be helpful for the class to make a wallchart of the different types that can be referred to throughout future practical work.

It is important to choose the measuring equipment appropriate to the task and measurement. For example, in a titration you can calculate the number of moles in a given volume of a solution of unknown concentration – you will learn more about this in LO 3. To do this you need to measure mass and volume accurately. Therefore, you would use a 100 cm^3 or 250 cm^3 volumetric flask to make up standard solutions rather than a 1 dm^3 measuring cylinder because the volumetric flasks measure volume more accurately than a large measuring cylinder. If you need to weigh out 15 mg of solid, you need a balance that weighs to the nearest 0.5 mg.

KEY TERMS

Percentage error – The percentage error or percentage uncertainty is:

$$\frac{\text{uncertainty (or error)}}{\text{reading or measurement}} \times 100$$

The overall percentage error for any procedure can be estimated by adding up the percentage errors of each individual measurement, although there are more complex methods of calculating the overall percentage error.

True value – An ideal or perfect value of a measurement (mass, volume, temperature, etc.) that can never be known exactly.

PAIRS ACTIVITY

Percentage error (10 minutes)

The error or uncertainty of any experiment will depend on a number of factors. Calculating the **percentage error** can be a useful way of checking whether the measurements you are making or the methods you are using are appropriate.

If you combine measurements, you must combine the uncertainties. For example, in a titration using a burette with divisions at 0.1 cm^3 intervals, the volume added (titre) is calculated as the final reading minus the initial reading. Each one has an uncertainty of ±0.05 cm^3, so the total uncertainty is ±0.1 cm^3.

Working in pairs, calculate the total percentage error in the titration experiments shown in Table 2.1. Three different sodium hydroxide (NaOH) solutions were made up and 25 cm^3 transferred to a conical flask with a pipette. The NaOH was added to the flask using a burette until the end point was reached (see section 3 for an explanation of what this means) – the volume added is the titre. The titration was repeated three times to obtain a mean titre. Hint: Calculate the percentage error at each step and then work out the overall percentage error.

What does this tell you about percentage error in experiments? Share your thoughts with the group.

Calibration is important for some types of measuring apparatus. Volumetric glassware is calibrated during manufacture and marked with the maximum error (e.g. ±0.05 cm^3). However, a thermometer might give precise readings, i.e. very close together, but may be several degrees away from the **true value** because it has not been properly calibrated (see the Pairs activity 'A simple calibration method').

Calibration methods will depend on the equipment being used, but they usually involve use of a standard. Calibration is essentially the process of comparing measurements. One measurement is of a known size or correctness, e.g. a reference standard. The other is on the device or instrument being calibrated.

Calibration is required for:

- any instrument that has moving parts, such as an analogue ammeter or voltmeter, because movement can upset the balance of the instrument
- anything that can be affected by temperature change
- electronic equipment, as the performance of various electronic components can change over time; for example, the glass membrane in a pH electrode can be affected by deposits (dirt, oils and grease, protein, inorganic materials) and should be calibrated before each use.

PAIRS ACTIVITY

A simple calibration method (10 minutes)

Calibration of a thermometer can be done by using two set points: freezing point and boiling point of water. Mix ice and water and leave it for a few minutes; the thermometer should read 0°C. Then place the thermometer in boiling water; the thermometer should read 100°C. Any deviation from these values should be allowed for when taking temperature readings with the thermometer.

Table 2.1 Titration experiments

Equipment used	Uncertainty	Measurement	Expt 1	Expt 2	Expt 3
Balance	0.1 g intervals	Mass NaOH/g	0.4	40	4
Volumetric flask	±0.2 cm^3	Vol flask size/cm^3	100	250	250
Pipette	±0.06 cm^3	Pipette size/cm^3	25	25	25
Burette	±0.05 cm^3	Mean titre/cm^3	21.5	19.3	5.0

Calibration is only as good as the reference standard used, so it is important to take care of standards, for example by preventing cross-contamination of standard solutions, corrosion of standard masses, etc.

GROUP ACTIVITY

Use and calibration of equipment (10 minutes)

In this exercise you are going to use a piece of equipment to make a measurement before and after calibration. Choose one or more of the following:

- a piece of glassware
- a thermometer or thermocouple
- a pH meter
- a balance
- a timer
- another piece of equipment provided by your tutor.

Make several measurements using the equipment and record your results. Then calibrate the equipment, making notes of the method you use. Finally, repeat the measurements originally made. What conclusions can you draw? Discuss with the group.

The assessment and management of risk (risk assessments; safety precautions/minimising risk)

Every activity you undertake in the lab must be accompanied by a risk assessment: risks must always be managed while working in the lab. A risk assessment must be carried out before starting a task, but you must also be aware of risks throughout the course of any practical work. Risks may change or new ones occur and you must recognise them and know how to deal with them.

You will find more information about this subject in sections 1 and 2 of Unit 6.

EXTENSION ACTIVITY

Health & safety legislation

You will be expected to explain why key features of the Health and Safety at Work Act 1974 (HSWA 1974) and Control of Substances Hazardous to Health (COSHH) Regulations are relevant to examples of scientific and laboratory workplaces.

For more information on legal responsibilities for health and safety, as well as information about COSHH in laboratories, visit:

www.rsc.org/learn-chemistry/collections/health-and-safety

Recognise laboratory hazards

You need to be familiar with the types of laboratory hazard that you might encounter. These will depend on the type of lab you are working in, but could include the following:

- pathogenic bacteria and other infectious agents
- harmful enzymes
- chemicals
- flammable substances
- sharp materials (broken glassware, knives, needles, pins)
- radioactive material.

There is more information about this in Unit 6.

The labelling of hazardous substances is controlled under a European Union regulation that covers classification, labelling and packaging of substances and mixtures, known as the CLP Regulation. This adopts the United Nations' Globally Harmonized System (GHS) for classification and labelling. The GHS Hazard Pictograms (see Figure 2.1) are similar to those that have been in use for several years, so if you become familiar with these, you will most likely be able to understand the hazards associated even with dusty old bottles found at the back of a fume cupboard!

◀ Figure 2.1 The GHS Hazard Pictograms. The exclamation mark pictogram has replaced the old 'harmful / irritant' symbol that you might still see on labels

The Health and Safety Executive (HSE) is responsible for enforcement of HSWA 1974 and associated regulations. It also has a website that is an excellent source of information about all matters related to health and safety in the workplace: www.hse.gov.uk.

Other sources of information to help identify laboratory hazards include:

- safety data sheets supplied with chemicals
- hazard and precautionary statements found on supplier websites and in good catalogues.

GROUP ACTIVITY

Hazards in your laboratory (30 minutes)

As a group, make an assessment of all the hazards in the laboratory where you are taught. Identify any substances where full hazard data is not available and make a list that could be used for further research.

Risk assessment

Risk assessment is a part of the risk management process that was discussed earlier in the section on following health and safety regulations. Risk assessment involves identifying a hazard (the potential to cause harm, discussed in section 2 of Unit 6) and then deciding on the likelihood of exposure to that hazard.

You need to determine the likelihood of a hazard actually causing harm and how serious the consequences could be.

For example, if a highly dangerous substance is contained so effectively that there is almost no possibility of coming into contact with it, the risk will be low. On the other hand, the probability of exposure to a less hazardous substance might be much higher and will pose a greater risk, if it is less well contained.

PAIRS ACTIVITY

Risk assessment (15 minutes)

Working in pairs, make a risk assessment for a piece of practical work that you are about to undertake. Document this and then explain the features of your risk assessment to the group.

Use the information in Unit 6 to help you in this activity.

There is an online tutorial on the Royal Society of Chemistry website that covers the whole process of risk assessment: the content is appropriate to anyone working as a laboratory technician.

www.rsc.org/learn-chemistry/resource/res00001314/risk-assessment

Control measures to minimise risks

After assessing the risk, control measures should be put in place systematically. There is more information about this in Unit 6.

The use of appropriate sampling techniques

At some point, all scientific investigations require the collection and recording of data. Often, the data will be collected at a particular place and time in field studies, or use a particular selection from a set of specimens in a laboratory setting. The process by which these factors are chosen and accounted for in both investigations and in analysis and reporting is **sampling**.

KEY TERMS

Sample – A sample is simply a subset of a population selected for a particular investigation. How this sample is chosen can impact the outcomes of an investigation. For example, a **complete sample** is a subset of all the items or events in a population which have a particular characteristic (e.g. 'all black cats' selected from the population of 'cats'). Conversely, a **unbiased sample** draws items or events from a population, irrespective of their particular properties.

Population – In the context of scientific sampling, a population is a set of objects or events which are of interest to the investigation. Population sizes can range from small (e.g. the number of rare plants or animals in a habitat; the rate of occurrence of supernovae in our galaxy) to very large or effectively infinite (e.g. the total number of stars in the galaxy or the number of possible ways a pack of cards can be shuffled).

Sampling can be a complex issue, involving some thought to avoid unintentional biases in outcomes, and to ensure a sufficiently large sample is examined in order to be able to draw statistically significant conclusions.

Whole sample

Analysing a whole **population** in a scientific investigation is relatively rare, as most often science is concerned with drawing generalisations. It is also rarely practical to analyse an entire population, as

- populations can be very large
- some parts of a population may be inaccessible.

PAIRS ACTIVITY

Whole populations (10 minutes)

Think about circumstances where sampling a whole population may be possible or applicable. You might like to consider the following:

- wildlife or nature conservation
- medical research

Random sample

As discussed above, using a whole population is relatively rare in science. In many cases, selecting a random sample from a population is a good solution, ensuring a manageable number of specimens or records.

As an example, suppose we wish to investigate the effect of a pesticide on a field of crops. Clearly, we cannot study each plant in the field, since (1) the number of plants is very large and (2) in doing so we may well destroy the crop – i.e. the sampling may well be destructive.

We now need to decide how large a sample we need. Factors which affect this include:

- the need to ensure statistical significance (i.e. the sample is large enough)
- the need to ensure the investigation can be carried out in a cost-effective and timely manner. (What is your budget, how long do you have and do you have the personnel and laboratory capacity?).

There are many methods which can now be used to collect the sample. In our example, individual specimens could be selected at random co-ordinates within the whole field (Figure 2.2 (a)). Other methods include assigning a number to each member of a whole population and then drawing the sample from a randomly generated list.

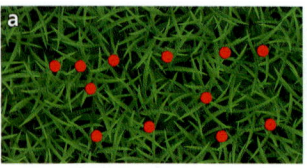

▲ **Figure 2.2** Examples of sampling of specimens from a field of crops. (a) random sampling, with specimens taken at random co-ordinates in the field and (b) representative sampling, with specimens taken from a small, defined area (a quadrat)

The sample selected should be free of human bias (you have not made the choice of specimens), but, especially if the number of specimens is small, there may be accidental biases.

GROUP ACTIVITY

Random sampling (10 minutes)

In the example of random sampling in Figure 2.2 (a), discuss:

1. whether there may be some accidental bias in the sample
2. why any such bias might be occurring
3. how the bias could be reduced or eliminated.

Representative sample

Alongside random sampling is the idea of *representative sampling*. In this case, we do not attempt to draw from the entire population, but instead select a subset of the population and draw the sample from that. An assumption in this is that the subset is *representative* of the whole population – in other words, all the conditions and properties of the subset are the same as the whole population.

An example of representative sampling is shown in Figure 2.2 (b). In this case, rather than taking specimens at random locations in the whole field, a small area (a quadrat – see Unit 13) is defined, and the specimens taken from that.

Investigations where representative sampling may be appropriate include:

- those where the whole population is either unknown or inaccessible. For example:
  - We cannot observe all the stars in the galaxy, therefore any astronomical investigations inevitably use a representative sample.
  - It may not be desirable to sample randomly in a delicate natural environment, as this may cause wider damage compared to taking a representative sample from a small area.
- those where random sampling might introduce unintended bias. For example:
  - If other variables are present, and need to be accounted for. In the crops example, this might be variations in soil composition across the field. We might therefore choose our representative sample to only include those areas where the soil has given characteristics.

GROUP ACTIVITY

Sampling methods (30 minutes)

Table 2.2 below outlines some characteristics of the sampling methods we've discussed.

Table 2.2 Characteristics of sampling methods

Type	Population size	Advantages	Disadvantages	Application
Whole	Small	Looks at an entire population	Whole population may be too large or inaccessible	Conservation of rare species. Medical research into rare conditions
Random	Large			Biological studies
Representative	Large			Astronomy

As a group, complete the table with examples from your own experiences and investigations. If you disagree with any of the applications listed, which sampling method would you choose instead, and why?

Sample integrity

In many investigations, especially those in environmental and food sciences, samples will be collected 'in the field' (i.e. away from the laboratory) and some travel and time will inevitably elapse before they can be analysed. Additionally, different people may be involved at each stage. For the outcomes of an investigation to be valid, samples must be:

- recorded
- labelled
- appropriately stored and transported (in the case where samples are collected; this does not apply where merely measurements or observations are made).

The recording of procedures and information used for the collection of high quality data

The recording of all procedures and information in sample collection is essential to any investigation. The nature of these will clearly depend on the precise investigation, so you will always need to think and plan carefully ahead of your study.

You will also need to think about how you will record the information. The most common method is using a written log book but other methods, such as an audio or video recording, may be more appropriate or relevant. Whatever method is used, you should always ensure that the record:

- is as complete as possible
- is accurate
- is permanent (write in pen, ensure you back up electronically recorded information regularly)
- cannot be tampered with (e.g. mark blank sections of log books so information cannot be inserted later; ensure your logs are dated; ensure 'master copies' of electronic logs are securely stored and 'read only').

This applies also to recording data. In this case, you will also need to ensure you record not just the data you want, but other data as well which may also have an influence. In our crops example, you might be recording plant height; an additional variable might be plant health, which might affect your results in the final analysis. Further details about defining procedures and thinking about variables are in Unit 13.

Labelling, storing and transporting samples

If samples are physically collected (e.g. individual plant specimens in our crops investigation, or water samples in a pollution investigation), it is essential that you have:

- procedures defined to label the specimen appropriately, cross-referenced to log book data where appropriate
- facilities to store the specimens securely
- the capability to transport the specimens securely to the laboratory.

As with procedures, labelling, storage and transport will vary with the type of investigation, but a few general points should be noted:

- Labelling should include at least what the specimen is, where and when it was collected and what investigation it relates to.
- Labelling may also contain a 'serial number' which can be cross-referenced to more extensive logged information.
- Storage should at least aim to:
  - prevent the specimen from being contaminated
  - prevent the specimen from 'cross contaminating' other specimens
  - preserve and protect the specimen in a state suitable for the laboratory investigation.

- Storage involves both the packaging of specimens in the field for transport and storage of specimens in the laboratory. It is therefore essential that suitable facilities are identified in the laboratory before an investigation commences.
- Transport of specimens should be carried out as rapidly as possible, especially in cases where the quality of the specimens may be time-sensitive (e.g. biological materials).

GROUP ACTIVITY

Sample integrity – what would you do? (30 minutes)

In our example of the crops investigation, we want to study how the application of a pesticide has affected growth. As a group, discuss and decide upon the following:

- What would you need to record and collect in order to carry out the investigation?
- If you are collecting specimens:
  - Do you need entire plants?
  - What information would you label the specimens with?
  - What kind of packaging would be suitable for in-the-field storage?
  - How rapidly would you need to transport specimens?
  - How would you store the specimens in the laboratory?

Reporting findings in detail and in an appropriate format

In science, it isn't enough to collect and analyse data; reporting and presenting your outcomes is vital for the whole scientific process.

GROUP ACTIVITY

Why do we need to report science? (10 minutes)

In small groups, think of three reasons for reporting on a scientific investigation and report back to your whole group.

Any investigation will have both descriptive or narrative elements as well as quantitative results and outcomes. How you'll be preparing and presenting your results for this unit will vary, but two common methods are:

- a scientific report or paper
- an oral or conference presentation.

Which method you choose, and how you present the information, will depend on the target audience. Factors to consider will include:

- the use of language (formal or informal)
- the use of jargon or technical terms.

The scientific report

A report of this type is usually prepared as a written, formal paper and would normally be targeted at the scientific community. It's also important to remember that scientific reports are often 'peer-reviewed', meaning they are scrutinised by colleagues in the scientific area of study (usually anonymously) to ensure scientific rigour is maintained.

A typical structure of a scientific report is shown in Table 2.3 on the next page, together with some guidelines for writing the content.

The scientific presentation

Scientific presentations are often given at conferences or public events, and are considerably less formal than a written report. They are, however, no less important, providing you with a way to communicate your ideas to an audience, and give the audience a chance to think about your work.

There are no hard and fast rules for giving presentations, though there are plenty of guidelines available for what makes a good presentation (and, indeed, what not to do!). Here are a few tips.

- Prepare your material carefully and logically. Give a clear narrative which should have four parts: (a) Introduction, (b) Method, (c) Results, (d) Conclusion/Summary.
- Don't include too much material. Have one or two key points you want to get over, and concentrate on those. As a rough guide, a single slide in a presentation should take between 1–2 minutes to talk about.
- Likewise, have only a few conclusion points.
- Talk to the audience, not the screen. This can be daunting, especially in a large conference theatre. A good way of addressing this is to find a friend in the audience and imagine you're just talking to them.
- Design your graphics to be clear. Remember, some of your audience might be at the back of the room, so clarity of slides is of the essence. In particular:
  - Don't use small or 'fancy' fonts for text.
  - Keep graphics simple and bold.
  - Use colour to highlight particular points.
  - Don't be afraid to include humour, if appropriate.

Table 2.3 Typical structure of a scientific report

Section	Description
Title	Provide a brief, but informative, title. For example, 'A field and laboratory investigation into iron pollution in the River Cherwell'.
Abstract	A very brief indication of: • the aim of the report • what you did • what you found • what you concluded. Abstracts are often compiled together, or provided online, so it's important for the abstract to give sufficient information to a potential reader.
Introduction	This section provides the context for the report. It should: • give a scientific context for the investigation, e.g. briefly say what other scientists have found • state why the topic is important or useful • state the primary aims and objectives of the study • explain any abbreviations or special terms.
Method	Here, you set out what you did in sequence. You should: • indicate what materials, techniques or equipment you used • explain the procedures and protocols used • provide sufficient information for the reader to potentially replicate the study.
Results	This is a key section, allowing you to present your findings. You should: • provide a clear narrative of your results and their analysis • include clearly titled and labelled graphs, tables and figures as appropriate.
Discussion	Here, you explain what the results mean. You should: • indicate whether the results were consistent or inconsistent with your expectations • discuss any statistical significance • indicate how the study could be improved or extended.
Conclusion	This need not be a long section. Its purpose is to: • briefly restate the main results • briefly explain the significance of the findings.
References	It's important that you put your investigation into the context of the scientific field. For that reason, you should provide references to other papers and other work carried out previously. You should give this as a list at the end of the report, but you should also 'cite' the work in the text of your report, e.g. 'Smith *et al* (2010) found that …'

- Avoid too much mathematics.
- Don't treat your slides as 'scripts' for you to read out. Think of the slides as a 'framework' which you build the talk around. Your audience will hear what you say – they don't need to read it too.
- Practise. This goes without saying – one or two 'dry runs' will help you get the timings right and help identify any points you need to be clearer on. You might find it helpful to do this in pairs.
- Finally, remember to be courteous to your audience. Remember to thank the audience for their attention at the end and answer any questions (even difficult ones) calmly and courteously.

GROUP ACTIVITY

Other methods of communication (20 minutes)

We looked at two methods of communicating your scientific outcomes. As a group, are there other methods which would be suitable for your investigation?

You might like to think about the following:

- Are there target audiences which are not addressed by the two methods we've looked at?
- What technological solutions are there to help you reach wider audiences?

Safe storage of materials

Safe storage, particularly of chemicals, starts as soon as the material is received. All chemicals should be supplied with a Safety Data Sheet that will give information about any hazards associated with the material and information about required storage conditions. For more information, see Unit 6.

GROUP ACTIVITY

Ask the technician (30 minutes)

Your tutor will organise a tour of your laboratory and/or prep room together with the technician from the department. Before doing so, as a group you should prepare a list of questions to ask about the rules that are followed concerning:

- safe storage of materials
- record keeping
- safe handling and disposal of hazardous materials.

KNOW IT

1. Explain the difference between repeatability and reproducibility.
2. You are preparing 250 cm^3 of a solution containing 25 g dm^{-3} of sodium chloride. You use a balance that is accurate to ±0.1 g and a 250 cm^3 volumetric flask that is marked ±0.2 cm^3. Calculate the total percentage error in preparing the solution.
3. What are the key stages in carrying out a risk assessment for laboratory work?
4. Why would you use a small sample from a population rather than the population itself? (Give two reasons.)
5. Give examples of where you would use random and representative sampling in science.
6. Explain why well designed protocols are necessary in science for labelling and storage of samples.
7. Give three reasons for communicating science.

LO2 Be able to separate, identify and quantify the amount of substances present in a mixture

Separation is central to most forms of analysis and many separation methods used for analysis can be scaled up as a way of purifying substances. As well as separating the components of a mixture, we sometimes need to quantify them.

GETTING STARTED

Quality and quantity (5 minutes)

List as many types of substance as you can that you might need to analyse. For each one, state whether it would be enough to know what is present (qualitative analysis) or whether you would need to know how much of each substance is present (quantitative analysis). Discuss your thoughts with the group.

2.1 Techniques to separate and identify substances present in a mixture

The various types of **chromatography** are all based on substances in a mixture having different affinities for two phases: one mobile, the other stationary. It is important, in this context, that we use the term **adsorption** for the process where a substance, either a gas or in solution, binds to a material in the stationary phase; this is not the same as absorption, which is a term often heard. In absorption, one substance is taken in or absorbed by another, rather like a sponge absorbing water.

KEY TERMS

Chromatography – Separation of the components of a mixture dissolved in a liquid or gas (the mobile phase) carrying it through a structure holding the stationary phase.

Adsorption – When a substance (e.g. a gas, liquid or solute) binds to or attaches to another, usually solid.

KEY TERMS

Polar – A polar (hydrophilic) substance will dissolve in or mix with water.

Adsorbent – Often used to describe the stationary phase in chromatography because substances become adsorbed to it during separation.

Elution – To wash out. In column chromatography this means 'washing out' a substance that has become adsorbed to the column (stationary phase).

Eluent – The solvent (mobile phase) used to wash substances out of a column.

Eluate – The mobile phase, containing dissolved substances, as it emerges from a column.

Paper chromatography to separate mixtures of coloured and colourless components

In paper chromatography, the stationary phase is the paper, which consists largely of cellulose, a **polar** substance. The stages are as follows:

1. A pencil line is drawn about 1 cm from one short edge of the paper (the origin).
2. The sample or samples are applied in solution at points on the origin and allowed to dry.
3. The sheet is placed vertically in a container with a shallow layer of solvent, so that the origin is above the level of the solvent.
4. The container is covered or sealed to prevent evaporation and the solvent will be drawn up the paper.
5. When the solvent reaches the origin, substances in the sample will dissolve and begin to move.
6. Substances with greater affinity for the paper will move more slowly than substances with a greater affinity for the solvent and so the mixture will become separated.
7. After the solvent has moved far enough up the paper (usually almost to the top), the sheet is removed from the container.
8. The position that the solvent reached (the solvent front) is marked.

Efficient separation requires choice of solvent (mobile phase) so that the different components of the mixture will have different affinities for the mobile phase (i.e. different solubility in the mobile phase). If all components are equally soluble, they will all move the same distance and if insoluble, they will remain at the origin. Therefore, different solvents or mixtures of solvents with a range of polarities are used depending upon the substances being separated.

Paper chromatography can be used to separate the components of pen dye or plant pigments; fortunately, all the components are different colours. The result of the separation is a series of different colour spots between the origin and the solvent front, so they are easy to see. For substances where the spots are invisible, UV light can be used to make them visible. Even if the components don't fluoresce, they might show up as dark patches on the light background of the paper. Other methods used to locate the separated components involve spraying the paper with a dye that will bind to the substances being analysed but not to the paper – ninhydrin is often used to stain the spots when a mixture of amino acids is separated.

Thin layer chromatography to separate mixtures of coloured and colourless components, e.g. plant pigments, pharmaceuticals

Thin layer chromatography (TLC) is very similar to paper chromatography as the method and the principles are the same. The difference lies in the stationary phase. Instead of paper, TLC uses a thin layer of an **adsorbent** such as silica gel, alumina or powdered cellulose attached to a flat, inert support such as glass or plastic.

TLC has several advantages over paper chromatography: the runs are faster, the separation is better, there is a choice of different adsorbents and the TLC sheet or plate is self-supporting, unlike paper.

Stationary and mobile phases in chromatography

There are other types of chromatography that you will encounter, particularly column chromatography and gas chromatography, but they all use a stationary phase (solid support) and a mobile phase (solvent). Column chromatography uses similar stationary phases to TLC, but a much wider range is available, particularly those used in the separation of biologicals. In all cases, the sample is applied to the top of the column and **eluted** with a suitable mobile phase, the **eluent**. The advantage of column chromatography is that the **eluate** can be collected in small amounts (fractions) and different substances will be in the different fractions because they elute from the column at different times. This allows column chromatography to be used for purification (preparative scale) as well as for analysis.

Calculate R_f values and make comparisons (standards, literature values)

As well as separating the components of a mixture, paper chromatography and TLC can also be used to identify the components. To do this we calculate the retention factor or R_f value, which measures how far a particular component moves relative to the solvent. To calculate the R_f value, we measure the distance between the origin and the solvent front and the distance between the origin and the leading edge of an individual spot and use the following formula:

$$R_f = \frac{\text{distance moved by substance}}{\text{distance moved by solvent}}$$

If we perform the analysis under standard conditions, we can use published literature values for R_f of various substances to identify them in a mixture. Alternatively, we can run pure substances as standards alongside the mixture and use R_f values to match unknowns in the mixtures with the standards.

> **GROUP ACTIVITY**
>
> ### Laboratory class (60 mins)
>
> Carry out paper chromatography and thin layer chromatography (TLC) for at least four different samples for each method. You should then interpret the results from a given chromatogram. You must be able to describe how to carry out the techniques to get accurate and repeatable results. You should have the opportunity to:
>
> 1. separate the components of mixtures
> 2. calculate R_f values, to use a locating agent
> 3. identify an unknown substance.
>
> You should be able to calculate R_f values for spots on given TLC or paper chromatograms and to identify substances from their TLC or paper chromatograms.

2.2 Alternative qualitative and quantitative techniques

Paper chromatography and TLC both use difference in solubility or affinity for the mobile and stationary phases to separate the components of a mixture. **Gas chromatography (GC)** and **high performance liquid chromatography (HPLC)** provide an enhanced form of this approach. However, other types of separation make use of different properties. For example, electrophoresis separates substances based on differences in charge, or charge and size. **Gel electrophoresis** is widely used in separating samples of DNA.

> ### 🔑 KEY TERMS
>
> **Gas chromatography (GC)** – A separation technique using an inert carrier gas as the mobile phase and a thin layer of liquid or polymer on an inert solid support as the stationary phase.
>
> **High performance liquid chromatography (HPLC)** – A type of column chromatography that uses very small particles and high pressures to achieve better separation. It can be used for analysis or, on a larger scale, for purification.
>
> **Gel electrophoresis** – Use of an electric field for separation of compounds based on differences in their charge and size.

The use of electrophoresis for the separation of the components of a mixture that are charged in DNA analysis

DNA molecules have a negative charge due to all the phosphate groups in the sugar-phosphate backbone. If they are placed in an electrical field they will be attracted to the positive terminal (anode). This is usually carried out in a gel made of agarose (a type of polysaccharide) or polyacrylamide. The gel forms a matrix that allows the DNA molecules to pass through, but shorter molecules can move more easily through the matrix and so move faster. See LO 3.1 in Unit 5 for more information about the use of gel electrophoresis in molecular biology.

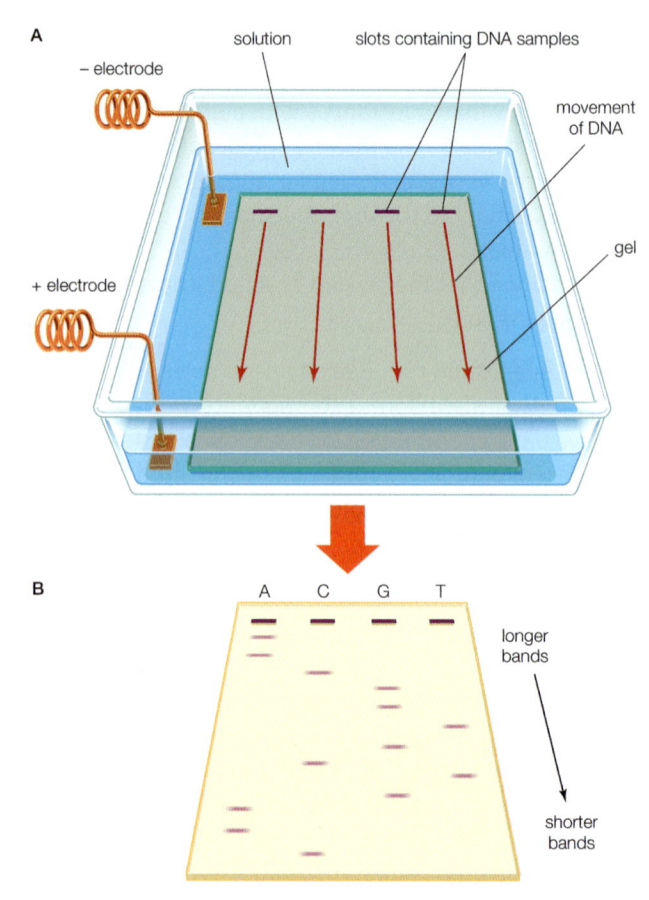

▲ **Figure 2.3** Gel electrophoresis uses an electric field to separate DNA fragments. Shorter DNA fragments move more quickly through the gel. After separation, the gel contains a number of bands formed by the different length DNA molecules

UNIT 2 LABORATORY TECHNIQUES

48

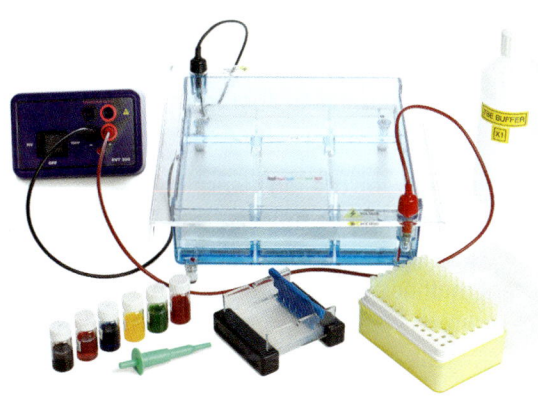

▲ **Figure 2.4** Equipment for gel electrophoresis. Elements of a kit to demonstrate the principles of electrophoresis using coloured dyes (to represent DNA fragments). The largest component is a gel bath with a buffer solution. The dyes are put into small wells in the gel and a current is applied across the cell. The different components of each dye move at different rates through the gel, based on their size and electric charge. This is essentially the same process as the one used in analysing fragments of DNA

Retention times when using gas chromatography (GC) and high performance liquid chromatography (HPLC)

Principles of HPLC and GC

HPLC is a variant of column chromatography that uses a stationary phase with a much finer particle size. This generates a high back pressure, which requires use of powerful pumps and pressure resistant columns. The sample being analysed or purified can't simply be applied to the top of the column – the system is sealed and under pressure and the sample is introduced by injection through a valve or port. The advantages are much greater speed and higher resolution, and HPLC has become one of the standard methods of separation in analytical laboratories. It can also be scaled up to operate as a purification method on a much larger scale, handling grams or even kilograms of substance.

A more recent development of HPLC, ultra-high performance liquid chromatography (or UPLC), is used to separate mixtures prior to analysis by mass spectrometry.

The other main type of chromatography using stationary and mobile phases is gas chromatography (GC). In GC, the mobile phase is an inert or unreactive carrier gas that is pumped through the column – a narrow glass or stainless steel tube, usually coiled inside an oven to maintain a relatively high temperature. This means that the substances being separated must be volatile, i.e. can be vaporised without decomposing. The stationary phase is usually a microscopic layer of high boiling point liquid or polymer on an inert solid support that is packed into the tube.

In capillary GC the column is a narrow bore capillary tube where the stationary phase coats the inside of the capillary tube. The sample is injected into the gas stream and the components of the mixture being separated will interact with the stationary phase to different degrees and so will emerge from the column at different times. The substances emerging from the column are detected, usually by a flame ionisation detector (FID) or a thermal conductivity detector (TCD). These work with a very wide range of substances, although other detection methods are available for specific applications.

The time taken between injection and detection of a particular component, whether in HPLC or GC, is known as the retention time. We saw how R_f values can be used in TLC to identify unknowns based on standard published literature values. The same is true of retention time in HPLC and, particularly, in GC. However, conditions used for the analysis must be identical to those used when determining the standard values, including use of the same column (not just the same stationary phase), mobile phase, temperature, etc.

Another way to confirm the identity of a substance in a mixture is to add a purified sample of that substance (a standard) to the sample when it is injected onto the column (GC or HPLC). This technique is known as spiking. If the substance in the mixture co-elutes with the standard (i.e. has the same retention time), it is strong evidence of identity.

Producing calibration standards (using serial dilution) to enable quantitative analysis

Detection methods used in GC (e.g. FID or TCD) and HPLC (usually absorption of UV light) will give a response that is proportional to the amount of substance. The output from the detector will usually be a trace showing a series of peaks (the **chromatogram**). The first step in quantitative analysis is to determine the area of each peak – the area under the curve, usually by computerised integration. To convert this into a quantity in grams, milligrams or micrograms needs a calibration curve. Firstly, we have to prepare a series of standards containing known amounts of the substance being studied (the analyte) and secondly, inject different amounts of the standard and measure the response (area under the curve) to calibrate the system.

However, the quantities involved can be very small – concentrations of parts per billion even – and we can't make an appropriate standard solution by weighing a few micrograms or less. For this reason, we generally use serial dilution to prepare standard solutions of appropriate concentration. This involves accurately weighing an amount of substance – how much depends on the nature of the substance and the type of balance available. Remember that a two decimal place balance will have an uncertainty of ±0.005 g. Assume we can weigh 0.100 g of substance accurately. Dissolve this in 100 cm^3 of solvent (in this example, water) and we have a solution with a concentration of 0.001 g cm^{-3}. Take 1 cm^3 of this solution and make it up to a volume of 10 cm^3; this results in a solution that contains 0.1 g cm^{-3} of substance. Repeat this procedure a few times and we soon have a very dilute solution indeed, as you can see from Table 2.4.

Serial dilution allows us to prepare a range of concentrations that we can use to produce a calibration curve for the detector. We plot a graph of response against concentration and draw a line of best fit. In this way the amount of that particular substance in a sample chromatogram can be quantified. We must use the linear portion of the calibration curve – very often the detector will become saturated and give the same response even when the concentration of **analyte** increases.

A new calibration curve must be prepared for each different analyte or detection method.

Positive identification of the components of a mixture when a chromatograph is linked to a mass spectrometer (GC-MS and LC-MS)

We have seen how GC can separate many different types of compound. However, the detection methods, such as FID, do not give any information about the identity of the compound emerging from the GC column – we have to rely on retention time to do that.

Mass spectrometry (MS) can identify the amount and type of compound by ionising the sample and then measuring the ratio of mass to charge (m/z) of the ions produced. This can be done in various ways:

- By measuring how far the ions were deflected by a magnetic field; more massive ions are deflected less.
- More recent methods such as time of flight (TOF) measure how long it takes for ions to reach the detector; more massive ions are slower and take longer.

This provides the molar mass (M_r) of the compound, but additional information about the structure of the compound can be obtained from the way in which the ion breaks up (fragments) in the mass spectrometer.

By feeding the output of a GC column into a mass spectrometer, it is possible to identify the compounds being separated much more accurately than simply by retention time. This technique, known as GC-MS, is used widely because of the compact nature of the equipment, its speed and relatively low cost. Applications include airport screening for drugs and explosives, fire forensics and space exploration. Probes containing miniaturised GC-MS have been sent to Mars, Venus and Titan.

EXTENSION ACTIVITY

LC-MS

A similar approach to GC-MS can be taken with liquid chromatography. In LC-MS, a very small capillary column is used for separation of complex mixtures such as proteins and peptides. Research how this method works and its applications. Report your findings to the group.

KEY TERMS

Chromatogram – The trace produced by the detector attached to the bottom of the chromatography column.

Mass spectrometry (MS) – A technique that creates positive ions from a sample and then separates them according to their mass-to-charge ratio.

Analyte – The solution of unknown concentration in a titration.

Table 2.4 Using serial dilution to prepare standard solutions of appropriate concentration

Step	Amount taken	Taken from	Make up to	Final concentration
1	1 cm^3	1.00 g cm^{-3} stock solution	10 cm^3	0.1 g cm^{-3}
2	1 cm^3	Step 1	10 cm^3	0.01 g cm^{-3}
3	1 cm^3	Step 2	10 cm^3	0.001 g cm^{-3}
4	1 cm^3	Step 3	10 cm^3	0.0001 g cm^{-3}
5	1 cm^3	Step 4	10 cm^3	0.00001 g cm^{-3}

> **KNOW IT**
>
> 1 What separation technique would you use for each of the following?
>   a analysis of a mixture of amino acids in a food additive
>   b analysis of the components in a perfume or flavouring
>   c a DNA sample from a crime scene
> 2 In an analysis of a mixture of dyes using TLC, the solvent front moved 10.2 cm from the origin. Calculate the R_f values of spots that moved 3.6 cm, 5.9 cm and 8.1 cm.
> 3 Explain what is meant by 'retention time'.
> 4 You are analysing samples of effluent (waste) from a chemical plant for the presence of a coloured pollutant. Describe how you would use a calibration curve to obtain quantitative data.
> 5 Explain how MS can be used to identify compounds that are separated by GC.

LO3 Be able to determine the concentration of an acid or base using titration

GETTING STARTED

What's so special about concentration? (5 minutes)

Think about why you might want to measure the concentration of an acid or base. Of course, you need to know whether an acid is concentrated or dilute if you are going to use it in an experiment such as a chemical reaction – some reactions need concentrated acids; others would react explosively with concentrated acid. But why do you need to measure accurate concentrations? Are you interested primarily in the concentration or the amount of acid or base in a solution? Share your ideas with the group.

Calculating concentrations

Before starting the practical aspects of titration, including how to make up solutions of known concentration, briefly consider how to calculate concentration. This is important in all aspects of chemistry, but particularly important here.

The units of concentration are g dm^{-3} or mol dm^{-3}. This tells us all we need to know about calculating concentrations, particularly if you remember the relationship between moles and grams. You will find yourself using the following equations on a regular basis.

1 Relationship between number of moles (n), mass in grams (m) and relative formula mass (M_r).

$$n = \frac{m}{M_r}$$

2 Relationship between concentration in g dm^{-3}, mass in grams (m) and volume (V in dm^3). This should be obvious, really – the units are the clue!

$$conc = \frac{m}{V}$$

3 Relationship between concentration in mol dm^{-3}, number of moles (n) and volume (V in dm^3) – again, the units are the clue.

$$conc = \frac{n}{V}$$

If we rearrange this, we get an equation for number of moles (n) if we know the concentration and volume V in mol dm^{-3}.

$$n = conc \times V$$

For each equation, if you know two of the terms, you can calculate the third. You will put this to use in the following sections – and possibly every day of your working life in the laboratory! Some people like to use formula triangles to help with this.

KEY TERMS

Standard solution – The solution of known concentration in a titration.

pH – A logarithmic scale that measures acidity (concentration of hydrogen ions) of a solution. Pure water has a pH of 7.0 at room temperature. A strong acid has a pH of 1–2 whereas a strong base will have a pH of 13–14. Reducing the pH by one whole unit means the hydrogen ion concentration increases ten times.

Equivalence point – The point of neutralisation where the number of moles of acid and base are equal. This should ideally be the point at which the indicator changes colour.

Indicator – A substance that changes from one colour to another or from coloured to colourless depending on whether it is in acidic or basic solution.

2.2 Alternative qualitative and quantitative techniques

PAIRS ACTIVITY

Calculating concentrations (15 minutes)

Work in pairs and try the following calculations – they are the sort you will need to use in most titrations.

A bottle of ethanoic acid was found in the laboratory but the label was damaged and the concentration couldn't be read. A 10 cm^3 sample was taken from the bottle and diluted with water to a total volume of 250 cm^3. This was then titrated against a **standard solution** of NaOH.

1. What mass of NaOH (Mr = 40.1) is required to make 250 cm^3 of a 0.040 mol dm^{-3} standard solution?
2. 25 cm^3 of the diluted ethanoic acid required 21.3 cm^3 of the 0.040 mol dm^{-3} for neutralisation. Calculate the number of moles of ethanoic acid in the 25cm^3 of solution.
3. Based on your answer to 2, calculate the concentration of the 25 cm^3 sample, and hence the concentration of the original bottle of ethanoic acid.

Calculation advice

- Use the same number of decimal places as there are in the information given – the trailing zero in 0.040 ml dm^{-3} is there for a reason!
- Don't round any intermediate values as this can introduce rounding errors; leave the rounding to the very end.

When an equal quantity of acid and base react together to form a salt and water, this is known as neutralisation. If a strong acid, such as hydrochloric acid, reacts with a strong base, such as sodium hydroxide, in exactly equal proportions, the mixture will be neither acid nor base but neutral, **pH** 7.0; this is known in titration as the **equivalence point**. At this point an **indicator** will change colour. We use this in the process of titration to calculate the concentration of a solution of acid or base.

In titration we have one unknown – the number of moles in the analyte. We titrate this against the standard solution. From the formulae above, you can see that because we know the concentration of the standard solution, we must know the number of moles in a given volume. It is easier to see how this works by using examples, so let's concentrate for now on the practical methods we need to use.

3.1 Techniques to determine the concentration of an acid or base using titration

This would be a good time to refresh your memory about the meaning of terms such as *precision*, *accuracy*, *uncertainty* and *percentage error* that were covered earlier in section 1.1.

Choice of appropriate measuring equipment

Table 2.5 Measuring equipment

Equipment	Uncertainty	Notes
Burette	May be marked on the barrel, otherwise: Class B has 0.1 cm^3 divisions ±0.05 cm^3 per reading Total = ±0.1 cm^3 (see section 1.1 for explanation)	Delivers different amounts of liquid and allows you to add very precise amounts one drop at a time when necessary. Usually used to deliver the standard solution. Take care when reading the burette to always read from the bottom of the meniscus (see Figure 2.5 on the next page).
One-mark (volumetric) pipette	Will be marked, usually on the stem.	Used to measure a single volume, e.g. 10 cm^3 or 25 cm^3. Used to transfer single volume, e.g. 10 cm^3 or 25 cm^3 of analyte to the flask. Follow safety procedures (no mouth pipetting).
Balance	Two decimal place balance = ±0.005 g	Required for weighing out solids to make up solutions for analysis. Reduce weighing errors by: • using weighing paper or a weighing boat to transfer all of the solid to the volumetric flask • weighing into a beaker and dissolving the solid in a small volume of water. Transfer this to the volumetric flask and rinse the beaker with water into the flask. Try not to spill any!

> **KEY TERM**
>
> **End point** – The point in a titration where the indicator changes colour.

Choice of appropriate indicators

An acid-base indicator will change colour (or change from coloured to colourless) at some point between about pH 3 and pH 9. The choice of which indicator to use will depend on the **end point** of the titration. At the end point, there will be a large change in pH with only a small addition from the burette. However, the pH range over which this occurs will depend on the type of titration, as you will see from Figure 2.6.

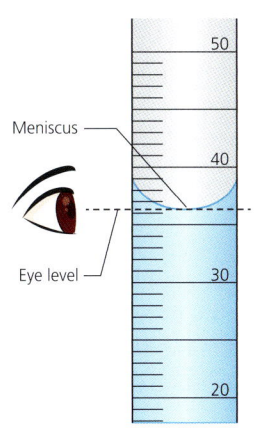

Calculation of mass required to make a solution of a given concentration

From the exercise at the start of this learning outcome, you should now be able to calculate how to work out the mass of substance you need to make a solution of a given concentration. This will normally be in mol dm^{-3}. To calculate the number of grams required, you need the M_r of the substance you are going to weigh out. Normally you will find this on the label of the bottle it came in, but you can also work it out by using the A_r data from the periodic table.

Having calculated the amount of substance to weigh, you then have to weigh it accurately and prepare a solution. Solutions should always be prepared using purified water – either distilled or deionised, depending on what is available.

Standard solutions are prepared using a volumetric flask. These are calibrated with a single mark in the neck of the flask to hold exactly 100 cm^3 or 250 cm^3 (other sizes are available, but not as widely used). The method is as follows:

- Add solid to the flask.
- Add enough water to come to the bottom of the neck, but well below the mark.

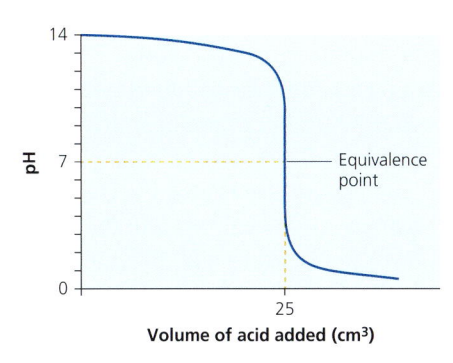

◀ **Figure 2.5** Always read volumetric and graduated glassware from the bottom of the meniscus and at your eye level. This reading is 36.5 cm^3

◀ **Figure 2.6** Change in pH during acid/base titration illustrating the choice of indicator for different types of acid/base titrations. The indicator used should change colour at a pH value corresponding to the vertical section of the curve. There are no suitable indicators for use with weak acid / weak base titration

Strong acid / strong base Bromothymol blue (blue in alkali, yellow in acid)

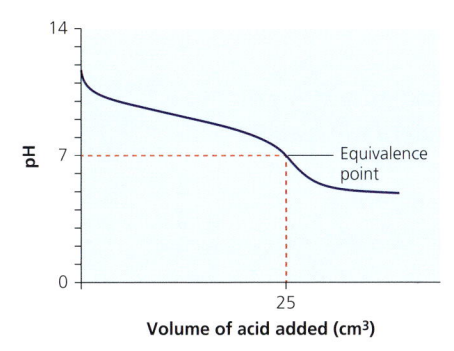

Weak acid / weak base

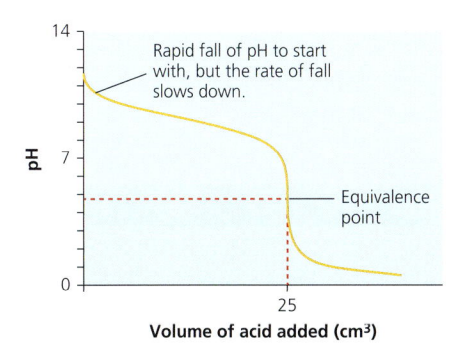

Strong acid / weak base Methyl orange (yellow in alkali, red in acid)

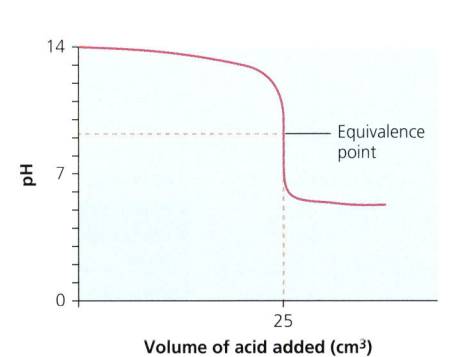

Weak acid / strong base Phenolphthalein (pink in alkali, colourless in acid)

53

- Stopper the flask and agitate gently to dissolve the solid.
- When the solid is dissolved, add water up to the mark ensuring the bottom of the meniscus is level with the mark (see Figure 2.5).

It is advisable to add the last few cm³ dropwise from a dropping or Pasteur pipette. If you have weighed the solid into a beaker, then dissolve the solid in about half the volume the flask holds and transfer the solution into the flask. Rinse out any remaining solution from the beaker into the flask before making up to the mark as described.

> **KEY TERMS**
>
> **Secondary standard** – A solution used as a standard in titration that has been calibrated against a primary standard.
>
> **Primary standard** – A reference material of high purity, such as sodium carbonate or potassium hydrogen phthalate, is used to prepare a standard solution of known concentration.
>
> **Titre** – The volume of **standard solution** needed to neutralise the analyte (i.e. to reach the end point of the titration).

Primary and secondary standards

So far we have talked about titrating an analyte of unknown concentration against a standard solution. We would usually use something like sodium hydroxide or hydrochloric acid as the standard solution. In fact, these should be described as **secondary standards**. They are useful for performing titrations but it might be difficult to prepare solutions of known concentration with them. For example, sodium hydroxide pellets absorb water during storage and so any mass we use will contain an unknown mass of water. These secondary standards must be calibrated by titration against a **primary standard**. This is a substance such as sodium carbonate or potassium hydrogen phthalate that is available as an anhydrous solid in high purity so that we can accurately prepare a primary standard solution of known concentration. For the greatest accuracy, Certified Reference Standards can be purchased from organisations such as LGC Standards, part of the UK's designated National Measurement Institute for chemical and bioanalytical measurement.

Performing the titration

Having selected the right equipment for the job and prepared the standard solution that has been calibrated against a primary standard, we then have to perform the titration. For example, use a standard solution of 0.050 mol dm⁻³ hydrochloric acid to determine the concentration of 250 cm³ of sodium hydroxide solution of unknown concentration. The analyte is usually placed in the flask and the standard solution in the burette.

Stage 1 Approximation:

1 Put 25 cm³ of the analyte (sodium hydroxide solution) into a 250 cm³ conical flask using a one-mark pipette.
2 Add the appropriate indicator (bromothymol blue); the solution should be blue.
3 Put the flask on a white tile or sheet of paper so that the colour change of the indicator is easier to see.
4 Fill the burette with the hydrochloric acid and put it on its stand above the flask.
5 Take an initial reading from the burette and note it in your laboratory notebook.
6 Run in liquid from the burette into the flask, swirling the flask, quite rapidly, to mix the liquids.
7 Stop the flow as soon as the indicator changes colour (in this case to yellow).
8 Take another reading from the burette and note this. Calculate the volume delivered by the burette (final reading minus initial reading); this is the **titre**. This is used as a rough estimate of how much standard is needed.

Stage 2 Accurate titration:

1 Empty the flask and rinse it. It doesn't matter if the flask is still wet when you add the next lot of analyte – titration measures the number of moles in the flask not the concentration.
2 Refill the burette if necessary and repeat the titration.
3 At 2 cm³ before the approximate volume worked out in stage 1 for the end point, start to add the standard dropwise until the indicator just turns colour.
4 Repeat the procedure until you get concordant titres, i.e. within 0.1 cm³ of each other.
5 Calculate the mean titre – you can include any titres that are within 0.3 cm³ of each other when calculating the mean but not those outside this range.
6 You should now have a table of results in your laboratory notebook that looks like Table 2.6.

Table 2.6

Titration	Rough	1	2	3	4	5
Final burette reading/cm³	23.00	44.90	22.05	44.15		
Initial burette reading/cm³	0.00	23.00	0.00	22.05		
Titre/cm³	23.00	21.90	22.05	22.10		

There was a difference between titrations 1 and 2 of 0.15 cm^3 (i.e. they were not concordant) so titration 3 was carried out. This was concordant with titration 2, so we can stop. As all three are within 0.3 cm^3 we can use them all to calculate the average titre – in this case it is 22.02 cm^3.

Calculation of concentration in mol dm^{-3} given the concentration of one solution

1 With an average titre (22.02 cm^3) calculate the number of moles (n) of hydrochloric acid used to neutralise the sodium hydroxide (don't forget the volume must be converted to dm^3 by dividing by 1000).

n(HCl) = 0.050 × (22.02/1000) = 0.0011 mol

2 Use the chemical equation to work out how many moles of acid and base react:

NaOH + HCl → NaCl + H$_2$O

In this case 1 mol NaOH reacts with 1 mol HCl

Therefore, n(NaOH) = 0.0011 mol

3 This was contained in 25 cm^3 (0.025 dm^3) of solution, so the concentration of NaOH is given by:

conc = 0.0011/0.025 = 0.044 mol dm^{-3}

? THINK ABOUT IT
Uncertainty (10 minutes)
Think about all the steps in this process, from making up standard solutions to performing the titration. What errors do you think could be made? Even if you don't make any errors, there will always be uncertainty in the measurements. What are they? Work out the percentage uncertainty at each stage.

? THINK ABOUT IT
Sources of error (5 minutes)
We often use titration to find the concentration of an unknown solution (the analyte). But we are actually finding the number of moles in the flask. We can use that information to work out the concentration of the original solution because we have found the number of moles in the analyte and we know the original volume.

Think about what would happen if you diluted the analyte, e.g. by rinsing the pipette into the flask with distilled water, or by washing down the sides of the flask with distilled water before you start the titration. Would either of these affect the result? What other things might affect the result? Think about all the possible sources of error.

3.2 Alternative techniques offering enhanced accuracy and sensitivity

It can, sometimes, be quite difficult to judge the end point of a titration based on a colour change. If you look back at Figure 2.6 you will see that it is impossible to use an indicator to judge the end point of a titration between a weak acid and weak base, as the pH of the solution changes gradually. Study Figure 2.6 carefully - you will be expected to be able to describe the general shape of the titration curves for strong acid/strong base, strong acid/weak base and weak acid/strong base.

There are more accurate methods that can be used and these are discussed below.

pH meter
A pH meter uses an electrode that generates a voltage that is proportional to the hydrogen ion concentration connected to a meter that converts this voltage to a pH value: for accuracy the meter must be calibrated at least daily.

Calibration of a pH meter
This is done using buffer solutions, usually one at pH 4.00 and one at pH 10.00. These should be prepared regularly using sachets of buffer powder dissolved in deionised or distilled water according to the instructions on the sachet. It is important to rinse the electrode between measurements to prevent cross-contamination.

Autotitration
An autotitrator will automate the whole process of titration from adding the standard (titrant), monitoring the reaction (e.g. pH change), recognising the end point and storing the data. Then it will perform the necessary calculations and display, print or store the results.

Autotitrators use electrodes (e.g. the pH electrode for acid base titration) to identify the end point and are programmed to make small additions of titrant in the region of the end point so that a rapid change in pH for a small addition of titrant allows the end point to be pinpointed. This process is not only more accurate than manual titration, but it can also be adapted for high throughput automated systems. Besides acid / base reactions, other types of autotitrator are available that can be used with redox and other reactions.

> **KNOW IT**
>
> 1 State the most suitable piece of equipment to perform the following:
>   a Prepare 250 cm³ of a solution of known concentration.
>   b Add exactly 25 cm³ of solution to a flask.
>   c Accurately add sufficient acid of known concentration to a flask to reach the end point.
> 2 Name an indicator that would be suitable for the following titrations:
>   a a strong acid and weak base
>   b a weak acid and strong base
>   c a strong acid and strong base
> 3 Describe how you would carry out a titration to determine the concentration of ammonia (a base) in a household cleaning solution.
> 4 Explain why it is difficult to titrate a weak acid and a weak base.
> 5 25 cm³ of NaOH solution of unknown concentration required 15.5 cm³ of H_2SO_4 to neutralise it. Calculate the concentration of the NaOH solution. Hint: You need to remember that every mole of H_2SO_4 produces 2 moles of H^+ ions.

L04 Be able to examine and record features of biological samples

GETTING STARTED

How much detail is necessary? (5 minutes)

Think about the types of biological sample you might encounter and how you would examine them – the naked eye might not be sufficient. Discuss this with the group – what level of detail do you need to see and, therefore, what level of magnification do you need?

4.1 Techniques to examine and record features of biological samples

In this section we will consider visual observations made at different magnifications (see Table 2.7).

Accurate recording of observations; calculating magnification and scale; use of a graticule

Besides developing drawing and observational skills, you need to have a good understanding of magnification and how to calculate it, so that you can relate what you see in the microscope to what you learn about in class.

One important equation you need to know is:

$$\text{Magnification} = \frac{\text{size of image}}{\text{size of object}}$$

Bear in mind that the sizes have to be in the same units (e.g. mm, µm, etc.) and that magnification should normally be quite a large number – if it isn't, then you've made a mistake somewhere!

You can rearrange this equation to enable you to calculate the actual size of the object if you know the magnification of the microscope and the size of the image.

Table 2.7 Examples of visual observations at different magnifications

Method	Example
Hand lens/magnifying glass (Magnification from ~2x to 30x)	• parts of plants (leaf, stem, seeds, flowers) • small animals (e.g. insects) • record features • measure distances, lengths, etc. • use reference standards, textbook drawings, identification key to interpret the image
Stereo microscope (Magnification up to ~100x)	• similar applications to hand lens, but greater detail • uses reflected light • surface detail of samples • living specimens • use in dissection
Light microscope Magnification up to ~2 000x Resolution 200 nm (200 x 10^{-9} m)	• uses transmitted light • examination of tissue sections • blood smears • tissue squashes (e.g. root tip to show nuclear division)

Measuring the size of the image can be done in different ways:

- Use a scale or ruler on the stage of the microscope – this only works at low magnification.
- Use an accurate scale engraved on the slide or cover slip. These are expensive, so handle carefully!
- Use a graticule – a scale in the eyepiece of a microscope. These need to be calibrated using an object of known size.

GROUP ACTIVITY

Class practical: microscopy (2 hours)

You will carry out a range of observations using different types of equipment available in your laboratory. Record your observations, including drawings and measurements where appropriate, in your laboratory notebook.

1. Examine one or more samples provided by your tutor using a hand lens and/or stereo microscope. Make drawings of the object(s) and use a measurement scale. Identify the object(s) by comparison with standard examples. Write up your notes so that someone else could follow the procedure.
2. Identify the features of the light microscope you are using and record these in your laboratory notebook. Your tutor will provide you with a selection of stained tissue slides. Make drawings of the features you can see. For each drawing calculate the magnification that you are using and record this in your notebook.
3. Prepare a slide of onion skin and draw the cells identifying the key features. Record the process in your notebook.
4. Use a graticule and calibration slide to measure, for example, the width of a human hair. Record the process in your notebook.

Discuss your observations with the class. Think about the limitations of the light microscope and record your thoughts in your notebook.

4.2 Alternative techniques offering enhanced visual examination

Alternative techniques allow the study of microscopic features too small to see with a light microscope as well as features hidden from view or difficult to access (see Table 2.8).

Table 2.8 Different techniques that offer enhanced visual examination

Method	Application
Electron microscopy Magnification up to ~10 000 000× Resolution 50 pm (50 × 10^{-12} m)	- Uses electron beam rather than visible light. - Examination of ultrastructure of cells. - Transmission electron microscope (TEM) uses very thin sections as the electron beam passes through the sample. - Scanning electron microscope (SEM) shows the surface structure and can show a representation of the three-dimensional shape of the sample. - Biological samples must be chemically fixed, dehydrated, embedded in a stabilising polymer and ultra-thin sections prepared (for TEM). - May also require treatment with heavy atom labels to enhance contrast. - These treatments can cause artefacts (something that is not natural and is a result of the preparation method).
X-ray analysis	- Uses X-rays (high energy electromagnetic radiation). - Reveals internal detail, e.g. the skeleton. - May require the use of contrast agents, e.g. barium meal, to reveal detail of soft tissues. - X-rays can be harmful to the subject and/or operator so exposure must be limited.
Ultrasound	- Uses high frequency sound. - Reveals internal details or hard-to-access structures. - Can be used to show soft tissues, e.g. in observing and monitoring the foetus during pregnancy.

PAIRS ACTIVITY

Method evaluation (30 mintues)

Working in pairs, prepare an evaluation of the advantages and disadvantages of light microscopy. As well as your own experience of using light microscopy, you should do some additional online research. The following website would be a good place to start:

www.jic.ac.uk/microscopy/about.html

You can read about other microscopy methods such as confocal microscopy and high content imaging in section 2.1 of Unit 8.

EXTENSION ACTIVITY

Revealing what is hidden

Ultrasound and X-ray analysis are widely used in medicine. There are many other examples of their application in fields as diverse as airport security and engineering.

Choose an area that interests you and research the current application and future potential for ultrasound and/or X-ray analysis. Prepare a short report on your findings and report back to the group.

KNOW IT

1. State the most suitable method to examine the following types of biological specimen:
   a. a section through a cell to show the organelles such as mitochondria and ribosomes
   b. the parts of a flower, such as anther, stigma and petals
   c. an anaesthetised insect such as a small beetle so you can draw details of legs and antennae
   d. a section of a plant stem stained to show the vascular tissues (xylem and pholoem)
2. Calculate the following:
   a. the magnification of an electron micrograph where a bacterial cell (actual length 1 µm) appears 23 mm long
   b. the size of the image of a red blood cell (actual diameter 8 µm) viewed in a light microscope at 400× magnification
3. Give two features of each of the following techniques that explain their usefulness in biology:
   a. ultrasound
   b. X-ray analysis
   c. electron microscopy

LO5 Be able to identify cations and anions in samples

You need to be able to identify the anions and cations listed by carrying out a range of practical tests. You should be able to identify the ions present in an unknown sample from the results of such tests.

GETTING STARTED

Why test for ions? (10 minutes)

Think about the inorganic ions and compounds you learned about in Unit 1. Why might you want to test for their presence in samples? When would it be good enough to know if an ion was present or absent? When might you need to know exactly how much of an ion was present?

5.1 Techniques to identify cations and anions in samples

PAIRS ACTIVITY

Class practical: identification of cations and anions (1 hour)

Your tutor will provide you with a range of test solutions. Some of these will be labelled so that you can check your methods are working; others will be unknown and will give you practice at identification of unknowns. Work in pairs to identify the solutions. Think about:

- the tests you need to carry out
- the order in which you should perform the tests.

Flame tests for cations

To carry out a flame test you need a solution of the ion, ideally in concentrated HCl (the chloride salts are most volatile, so give the strongest colour). Take a nichrome wire and clean it by dipping into concentrated HCl and then heating in the hottest part of a Bunsen flame. Then dip the cleaned wire in the solution to be tested and heat in a hot Bunsen flame.

Table 2.9 Flame test colours

Metal	Symbol	Colour of flame
Barium	Ba^{2+}	apple green
Calcium	Ca^{2+}	red
Copper	Cu^{2+}	blue
Lithium	Li^+	carmine red
Potassium	K^+	lilac
Sodium	Na^+	yellow

Chemical tests for cations

Some metal ions give coloured hydroxide precipitates and so addition of dilute sodium hydroxide can help in identification. However, it can help identification if you observe the behaviour with excess sodium hydroxide or aqueous ammonia solution (see Table 2.10 on the next page).

Table 2.10 Behaviour of metal ions with NaOH, excess NaOH or aqueous ammonia

Metal	Symbol	With NaOH	Excess NaOH	Aqueous ammonia
Aluminium	Al^{3+}	White precipitate	Precipitate redissolves.	White precipitate; does not dissolve in excess ammonia.
Copper(II)	Cu^{2+}	Blue/turquoise precipitate	Precipitate does not redissolve.	Blue precipitate that dissolves in excess ammonia to give deep blue solution.
Iron(II)	Fe^{2+}	Dark green precipitate that darkens in air (oxidation to Fe^{3+})	Precipitate does not redissolve.	Dark green precipitate; does not dissolve in excess ammonia.
Iron(III)	Fe^{3+}	Brown precipitate	Precipitate does not redissolve.	Brown precipitate; does not dissolve in excess ammonia.
Lead	Pb^{2+}	White precipitate	Precipitate redissolves.	White precipitate; does not dissolve in excess ammonia.

You will have noticed that aluminium and lead cannot be distinguished using NaOH. However, addition of potassium iodide solution gives a yellow precipitate of lead(II) iodide if Pb^{2+} ions are present.

Chemical tests for anions

Carbonate (CO_3^{2-})

Heating a solid carbonate strongly will cause it to decompose, releasing CO_2 which can be identified by bubbling it through limewater (a solution of calcium hydroxide, $Ca(OH)_2$). If CO_2 is produced, the limewater will turn cloudy.

An alternative is to add acid to a solid or solution. The solid will dissolve and effervesce (fizz) as CO_2 is produced; the solution will just effervesce. The production of CO_2 can be confirmed by bubbling the gas produced through limewater.

Halide ions

The tests for chloride, bromide and iodide all follow the same procedure. Add dilute nitric acid to the sample followed by silver nitrate solution – if a halide is present, there will be a precipitate. The colours (white, cream and yellow) can be difficult to distinguish, so the next step is to add dilute ammonia and, if the precipitate does not redissolve, add concentrated ammonia (see Table 2.11).

Sulfate (SO_4^{2-})

Add dilute hydrochloric acid followed by a few drops of barium chloride or barium nitrate solution. Sulfate ions will produce a dense white precipitate of barium sulfate.

> **? THINK ABOUT IT**
>
> **In what order do you do the tests? (10 minutes)**
>
> When presented with an unknown sample, it is important that you carry out the tests in a particular order. It is also important to add dilute acid to the sample before carrying out several of the tests – this is mentioned in the descriptions. Use the following information to work out why the order is important:
>
> - Silver carbonate, silver sulfate and barium carbonate are all insoluble.
> - Carbonates react with acid to form carbon dioxide and water.

Table 2.11 Silver nitrate test for halide ions

Ion	Symbol	Appearance with $AgNO_3$	Add dilute NH_3	Add concentrated NH_3
Chloride	Cl^-	White precipitate	Precipitate redissolves.	--
Bromide	Br^-	Cream precipitate	Precipitate does not redissolve.	Precipitate redissolves.
Iodide	I^-	Yellow precipitate	Precipitate does not redissolve.	Precipitate does not redissolve.

5.2 Alternative techniques offering improved separation, sensitivity and quantification

The methods covered in section 5.1 can be used to demonstrate the presence or absence of an ion. This may be sufficient in some cases. However, it is often important to know how much of each ion is present – it requires a quantitative method. Sensitivity is also an issue, particularly when identifying and quantifying trace elements in water, soil, foodstuffs, etc. There are two methods you need to know about that can be used for this: ion chromatography and atomic emission spectroscopy (AES).

Ion chromatography

Ion chromatography or ion exchange chromatography uses a similar principle to water softeners. A column is packed with beads of resin with either positive or negative charges on the surface of the beads, so that ions of the opposite charge will bind. Different ions will bind more or less strongly and can be eluted from the column by changing the pH or concentration of salt solution. Ions being eluted from the column are detected because they cause changes in the conductivity of the eluate.

Atomic emission spectroscopy (AES) and inductively coupled plasma-atomic emission spectroscopy (ICP-AES)

AES is based on the same principle as the flame test for ions covered in LO 5.1, but is much more sensitive and quantitative. The sample is heated in a flame and the wavelength of the light emitted identifies the element while the intensity of the light is proportional to the number of atoms in the sample. ICP-AES is an even more sensitive technique.

> **EXTENSION ACTIVITY**
>
> **Quantitative techniques (1 hour)**
>
> Research ion chromatography, AES and ICP-AES. Create a presentation that explains the application and value of each technique.

> **KNOW IT**
>
> 1. Name the cations that give the following colours in a flame test:
>   a blue   b lilac   c yellow
> 2. Answer the following:
>   a Name two cations that give a white precipitate with NaOH that redissolves in excess NaOH.
>   b Describe a test you would use to distinguish between these two cations.
> 3. Two solutions both gave precipitates when tested with silver nitrate, suggesting that they could contain chloride, bromide or iodide ions.
>   a Describe how you would ensure that carbonate ions did not interfere with the test.
>   b Describe the additional step you would take to identify exactly which halide ion was present.
> 4. State the detection method used in ion chromatography.
> 5. Give two advantages of AES over a simple flame test.

LO6 Be able to use aseptic technique

Aseptic technique underpins all work in microbiology. It is possible that you might have to handle **pathogens** at some time in your working life. However, most of your work is likely to be with non-pathogenic cultures. Nevertheless, it is sensible to treat all cultures as potentially pathogenic – they may have become contaminated or mutated into a **pathogenic** form.

> **KEY TERMS**
>
> **Aseptic** – Free from infection or source of infection; also means methods to prevent infection.
>
> **Pathogen pathogenic** – A pathogen is a disease-causing organism, usually a microorganism.

> **GETTING STARTED**
>
> **Who is more dangerous, me or the microorganism? (5 minutes)**
>
> Well, obviously it is more important that you protect yourself from any risk of infection by a potentially harmful culture. But what about protecting the culture from infection? What sources of infection might there be? What problems could these cause?

6.1 The purpose of working in an aseptic or clean room whilst maintaining sterility and cleanliness

You should be able to recognise the types of practical work where aseptic technique is essential, e.g. cell and tissue culture, preparation of medical test kits, pharmaceutical production, microbiology, medical and surgical procedures.

You will also need to explain the reason for using aspects of aseptic technique.

Sterilisation methods

Sterilisation is usually done in an autoclave (essentially a type of pressure cooker) that uses pressurised steam at 121°C. If held for 15 minutes under these conditions, all microorganisms including spores will be killed. Sterilisation can also be achieved using radiation, although this is less common in a laboratory environment.

Sterilisation of liquids can also be done by filtration through a 0.45 µm or 0.2 µm membrane filter.

Sterilisation of wire loops is by heating in a Bunsen flame until red hot (disposable sterile loops are also available) whereas glass spreaders and metal forceps are sterilised by flaming in 70% alcohol.

Decontamination of surfaces and equipment

Decontamination is usually by use of a chemical disinfectant. Many disinfectants used in the past are hazardous substances and no longer used. Common disinfectants now available are:

- VirKon, used for work surfaces, discard pots and skin **disinfection**
- hypochlorite (bleach), used for discard pots
- alcohol, used for skin disinfection.

Avoiding contamination of material by the environment and by people

This is an important aspect of aseptic technique. Contamination of a culture by air-borne or waterborne microorganisms can ruin experiments, possibly destroying weeks, months or years of work. If cultures become contaminated by pathogens, then they represent a much greater risk to those working with them.

Preventing people coming into direct contact with pathogens

Working with pathogens involves a higher level of risk and so requires a higher level of risk prevention. However, all cultures are potentially pathogenic and so even the most basic level of risk prevention must reduce the risk of contact with the culture.

Controlled airflow cabinets

Controlled airflow cabinets serve to protect the culture from contamination as well as protecting the user from the culture. They incorporate filters and directed airflow that helps to contain any hazard within the cabinet. It is essential to follow recommended procedures so as not to disrupt the airflow – if you do this, then there is a greatly increased risk of contamination.

GROUP ACTIVITY

Aseptic procedures (1 hour)

The class should prepare a laboratory manual covering aseptic technique and procedures to follow. Divide into groups; each group should cover one of the five aspects outlined in this section. Research the subject and prepare your group's contribution to the manual.

You will find the following online resources useful starting points for your research:

Nuffield Foundation in partnership with the Society of Biology have produced notes on aseptic techniques:

www.nuffieldfoundation.org/practical-biology/aseptic-techniques

The Microbiology Society (formerly known as the Society for General Microbiology) has produced a manual for schools: 'Basic Practical Microbiology'. You should use this as a guide, but remember that it is aimed at secondary schools. In your working life you may be exposed to greater risks and need to employ more advanced techniques. You can download a PDF copy using the link above.

You will need to work with other groups in the class to produce a single resource that can be used in the practical work you will encounter in the next section.

6.2 To follow standard aseptic procedure to streak a plate

Estimation of the purity of a culture

Preparing a streak plate of yeast cells or bacteria is a way of checking the purity of a culture. The procedure progressively dilutes the **inoculum** (bacteria or yeast) so that when we **inoculate** the plate, then the colonies grow well separated from each other after incubation. When this happens, each colony is formed from one single cell in the inoculum. If more than one shape or colour of colony is present on the plate, then it shows the culture contains more than one type or species of microorganism.

The technique can also be used to start a new culture by picking a single colony of yeast or bacteria and growing them in nutrient broth.

GROUP ACTIVITY

Class practical, preparing a streak plate (1 hour plus 2–3 days incubation time)

You need to:

- practise streaking a plate
- be able to explain how to carry out the technique correctly
- be able to interpret the appearance of streaked plates in terms of the purity of the culture used to make the streak plate.

Your tutor will provide you with the materials needed for this activity. Be sure to follow the aseptic procedures you have prepared in the activity in section 6.1.

Make drawings of the plates after incubation and annotate them to show any contamination present.

You will find details of a method to use on the Nuffield Foundation website:

www.nuffieldfoundation.org/practical-biology/making-streak-plate

6.3 To follow standard aseptic procedure in tissue culture

So far in LO6 we have looked at culturing microorganisms such as yeasts and bacteria. However, culture of cells (plant and animal) is also an important area that you need to be familiar with. Animal cell culture is widely used in medical and pharmaceutical research. Plant cell culture is used in research but also as a way of propagating plants. Traditionally gardeners or horticulturists have propagated plants by taking cuttings and similar methods. However, methods such as these can produce tens of new plants from one original plant. Micropropagation is a way of generating whole plants from small pieces of plant material and can produce hundreds, even thousands, of new plants.

Aseptic procedures are important in tissue culture, so you need to be able to apply what you have learned about aseptic technique to a tissue culture method. You should be able to explain how aspects of aseptic technique are important in the process.

GROUP ACTIVITY

Class practical, cloning cauliflower (1 hour plus approx. 3 weeks incubation time)

Your tutor will provide you with the materials and method needed to clone cauliflower.

You can get more information about this technique on the Science & Plants for Schools (SAPS) website:

www.saps.org.uk/secondary/teaching-resources/706-cauliflower-cloning-tissue-culture-and-micropropagation

KNOW IT

1. What are the two main reasons for carrying out proper aseptic procedures when handling microorganisms?
2. Describe two methods of sterilisation used when handling microorganisms.
3. Describe the steps that you would take to prevent contamination of cultures when working with microorganisms or cells.
4. Explain why micropropagation is a useful technique in horticulture.

KEY TERMS

Sterilisation – A process that eliminates or kills all forms of life; including viruses.

Disinfection – The destruction, inhibition or removal of microbes that may cause disease or other problems.

Inoculum – The substance, usually a microorganism, used to inoculate a culture.

Inoculate – Introduction of a microorganism into a culture medium (e.g. in a flask) or onto an agar plate.

Assessment practice questions

You will be tested on each of the LOs in the exam. Below are practice questions for you to try, followed by some top tips on how to achieve the best results.

1 This question is about an environmental research project. An environmental research project aims to investigate the biodiversity of invertebrates in a small river.
   a What kind of sampling would you choose to collect your samples? Justify your answer. (2 marks)
   b State four items of information you would record on sample labels. (4 marks)
   c A colleague has been asked to present your findings to a local nature club. State four pieces of advice you would give your colleague in regard to giving a presentation to non-scientists. (4 marks)

2 Illyria Enviro is a small start-up company providing environmental monitoring services, including water testing and microbial analysis. You are responsible for health and safety training for new recruits.
   a Give two reasons that you would give to management as to why health and safety training is important for new recruits. (2 marks)
   b Outline the steps you would undertake to prepare a risk assessment for Illyria Enviro. (4 marks)

3 You have been asked to analyse a mixture of amino acids by TLC.
   a Describe the method you would use. (4 marks)
   b After performing the separation the solvent has moved 8.8 cm from the origin. Table 2.12 shows the R_f values of some amino acids.

Table 2.12

Amino acid	R_f value
alanine	0.38
arginine	0.20
glutamic acid	0.30
glycine	0.26
histidine	0.11
leucine	0.73
lysine	0.14
methionine	0.55
tryptophan	0.66
tyrosine	0.45
valine	0.61

   c Calculate the R_f values for the following amino acid spots and use the data in Table 2.12 to identify each amino acid:
      i Amino acid A moved 1.6 cm.
      ii Amino acid B moved 6.3 cm.
      iii Amino acid C moved 5.3 cm. (3 marks)

4 Potassium hydrogen phthalate (KHP, molar mass = 204.23 g mol^{-1}) is often used as a primary standard to standardise sodium hydroxide solutions for use in titration.
   a Calculate the mass of KHP needed to make 250 cm^3 of a 0.05 mol dm^{-3} solution of KHP. (5 marks)
   b Prepare a standard operating procedure (SOP) for a technician to follow when preparing a standard solution of KHP. Include details of all the necessary glassware. (6 marks)

5 Illyria Enviro have been asked to analyse a water sample for various cations and anions.
   a Describe tests you could perform to show the presence of the following. Make sure you mention any precautions you would have to take to ensure your analysis is reliable.
      i sulfate (SO$_4^{2-}$)
      ii copper(II) (Cu^{2+}) (4 marks)
   b Aluminium and lead both give white precipitates with NaOH that do not redissolve in excess NaOH. However, addition of potassium iodide solution gives a yellow precipitate of lead (II) iodide if Pb^{2+} ions are present. Write the ionic equation for this reaction between Pb^{2+} ions and potassium iodide. Include state symbols. (2 marks)
   c Outline the experimental procedure you would use to determine the concentration of Pb^{2+} ions. (3 marks)

6 Illyria Enviro has asked you to prepare a training module for new technicians who will be working in the microbiology lab.
   a Outline the risks that workers might be exposed to. (4 marks)
   b Outline the precautions that should be taken to reduce risk to workers and reduce contamination of samples. (4 marks)
   c Describe how you would use a streak plate to:
      i assess the purity of a bacterial culture (4 marks)
      ii prepare a culture of just one type of bacterium. (2 marks)

TOP TIPS

- ✔ Remember to think carefully when choosing your sampling method.
- ✔ Make sure your sample size is sufficient.
- ✔ When labelling and recording specimens, be sure you note all relevant information clearly.
- ✔ Have a written protocol in place for labelling and handling specimens and data.
- ✔ Remember to think about the nature of your audience when preparing scientific communications.
- ✔ When choosing a separation technique, think about the nature of the substances being separated, the quantities involved and detection methods you can use.
- ✔ When performing a titration, remember you are calculating the number of moles in the flask – the exact concentration is not important. Once you know the number of moles, you can use the original volume to calculate the concentration.
- ✔ When testing for cations and anions, the order in which you carry out the tests can be important in order to rule out ambiguous results.
- ✔ Good aseptic technique is about protecting you from potential pathogens and also about protecting the culture from contamination.

Read about it

The Microbiology Society education website:

www.microbiologysociety.org/education/index.cfm

Nuffield Foundation in partnership with the Society of Biology have produced a series of instructions for standard techniques useful in microbiology:

www.nuffieldfoundation.org/practical-biology/standard-techniques

Further information on sampling techniques

Some general information and guidance:

https://en.wikipedia.org/wiki/Sampling_(statistics)

Guidance on sampling from the Field Studies Council:

www.field-studies-council.org/centres/dalefort/outdoorclassroom/resources-for-teachers-students/biology/sampling-techniques-resources.aspx

Guidance on sample labelling and integrity

Some common mistakes in labelling:

www.labmanager.com/laboratory-technology/2010/11/five-common-mistakes-in-lab-labeling

Many labelling protocols are devised for the health sector. See, for example:

www.gosh.nhs.uk/health-professionals/clinical-guidelines/blood-tests-requesting-labelling-and-sampling-requirements

Articles and information about sample integrity

General information and guidance

www.aweimagazine.com/article.php?article_id=681

Water and wastewater

www.epa.vic.gov.au/~/media/Publications/IWRG701.pdf

UK Biobank sample handling and storage protocol

http://ije.oxfordjournals.org/content/37/2/234.full

Unit 03

Scientific analysis and reporting

ABOUT THIS UNIT

Scientists are often required to collect and analyse information and data, and to present it in a form that can be easily understood. Scientific experiments require the accurate and reliable gathering of data and ultimately its interpretation.

The techniques presented in this unit underpin the work of scientists in the collection, analysis and presentation of data and information. The unit will develop your knowledge and understanding of a range of useful analytical techniques that can be applied in experimental and investigative settings.

LEARNING OUTCOMES

The topics, activities and suggested reading in this unit will help you to:

1. Be able to use mathematical techniques to analyse data
2. Be able to use graphical techniques to analyse data
3. Be able to use keys for analysis
4. Be able to analyse and evaluate the quality of data
5. Be able to draw justified conclusions from data
6. Be able to use modified, extended or combined laboratory techniques in analytical procedures
7. Be able to record, report on and review scientific analyses

How will I be assessed?

All Learning Outcomes are assessed through externally set written examination papers, worth a maximum of 60 marks and 1 hour 30 minutes in duration.

This unit is externally examined. Teachers are advised to obtain sample examination papers available from the OCR website. Examination papers typically contain six questions covering the Learning Outcomes presented in the unit specification. Problems are presented to learners using a range of styles, including short answer, calculation, fill the blanks, matching, true/false, longer essay type questions, etc. Problems are presented in a scientific context.

LO1 Be able to use mathematical techniques to analyse data

Scientific investigations invariably require data and information to be expressed quantitatively, that is, in numerical form. Whilst you will be familiar with many arithmetic and mathematical techniques and functions, it's important that we take a look at these in the context of science, especially in situations where you will need to choose the best way of presenting results.

GETTING STARTED

Numbers everywhere (5 minutes)

In science, we most often need to handle and record situations involving numerical data. Sometimes, the numerical aspect isn't obvious. Think of one or two situations in your everyday life where numbers and their interpretation are important. Examples might include travelling, nutrition and housing.

KEY TERMS

Mean – In statistics, the mean is simply the average of all the quantities in a data set. This is calculated by adding together all the quantities and then dividing by the number of quantities. For example, if a data set contains five numbers; 2, 6, 2, 7, 3, then the mean is $\frac{(2 + 6 + 2 + 7 + 3)}{5} = 4$.

Mode – The mode is simply the most common quantity in a set of data. In our example it is 2.

It's important to note that the mean, median and mode may be the same, though this is not universally the case.

Median – The median is similar to the mean, except it splits the data set in half, so that there are the same number of quantities above the median as below. In our example above, if we arrange our numbers in order, we have 2, 2, 3, 6, 7. The number 3 now sits in the middle of the set, splitting it in half, so 3 is the median.

1.1 Application of basic arithmetic techniques

In this first section, we'll look at how the simplest arithmetic techniques can help us in analysing and interpreting scientific results, and indeed, some of the pitfalls which may be encountered.

Finding a mean, median, mode

A common requirement in scientific analysis is reducing a collection of numbers to some representative value. Often, this may simply be an 'average' or, more scientifically, the **mean**.

Formally, suppose we have a collection of n quantities, $a_1, a_2, ..., a_n$. The mean (technically, the arithmetic mean) a_{mean} is defined as:

$$a_{mean} = \frac{1}{n}\sum a_i = \frac{1}{n}(a_1 + a_2 + ... + a_n)$$

In words, the arithmetic mean is equal to the sum of all the quantities divided by the number of quantities.

As an example, Table 3.1 on the next page shows the ages at death of some people from the 12th to 14th centuries CE, for whom reliable records exist (primarily the nobility and higher levels of society). The ages have been sorted from youngest to oldest.

The sum of all the ages is 920, and there are 20 ages listed. The mean age is therefore:

$$\text{mean age} = \frac{920}{20} = 46 \text{ years}$$

The mean isn't necessarily the best representation of a set of data. For example, although the average is 46 years, if you examine the table, you will see that the most common age at death is 48. This quantity, the most common value in a data set, is termed the **mode**. In this case, the mode is quite close to the mean, though you should be aware this isn't necessarily the case.

Finally, it's sometimes useful to identify a number which splits the data set in two equal parts – that is, there are the same number of values above as below. This quantity is referred to as the **median**.

For example, looking at our age data, we know that there are 20 ages. To split this evenly in two, we want 10 ages below the median and 10 above. The 10th age is 45 years, hence the median is higher than this. The 11th is 47 years, so the median lies between these two values – simply defined as the average:

$$\text{median} = \frac{(45 + 47)}{2} = 46 \text{ years}$$

We therefore have 10 people dying younger than 46, and 10 older.

You will also notice that there is a spread of numbers either side of the mean. Later in this section, we will look at the implications of such a spread, and how we can quantify and interpret such parameters.

Table 3.1 Ages of death of people from the 12th to 14th centuries ce, from youngest to oldest

Name	Age (years)
Eleanor de Braose: 1226–1251	25
Isabella de Braose: 1222–1248 (wife of Dafydd)	26
Elinor de Montfort: 1252–1282	30
William de Braose: 1198–1230	32
Tangwystl: 1168–1206 (mistress of Llywelyn Fawr)	38
Humphrey de Bohun: 1225–1265	40
Dafydd ap Llywelyn: 1208–1246 (Prince of Wales)	42
Maud de Fiennes: 1254–1296	42
Eve Marshall: 1203–1246	43
Gwladys: 1206–1251 (Princess of Wales)	45
Joanna: 1190–1237 (daughter of King John of England; wife of Llywelyn Fawr)	47
Ralph Mortimer 1198–1246 (husband of Gladwys)	48
Gruffydd: 1196–1244 (Prince of Wales)	48
Humphrey de Bohun: 1249–1298	49
Roger Mortimer: 1231–1282	51
Edmund Mortimer: 1251–1304	53
Llywelyn ap Gruffydd: 1228–1282	54
Margaret de Fiennes: 1269–1333	64
Llywelyn Fawr: 1173–1240 (Prince of Wales)	67
Maud de Braose: 1224–1300	76

GROUP ACTIVITY

Mean, median and mode (15 minutes)

In our example, the mean, median and mode are quite close. Table 3.2 shows the hypothetical scores in a class test.

Table 3.2

Student Scores in a class test	Score /10
Alice	4
Bob	3
Chris	7
Dee	7
Eve	3
Fred	1
George	7
Hannah	9
Ivy	2
Jack	2

Find the mean, mode and median of this data set. Think about why they differ and which quantity could best represent the outcome of the test.

KEY TERMS

Rounding – This is the process of simplifying a number while keeping its value close to the original value. For example, 63 rounded to the nearest ten is 60.

Significant figures – We can be more formal in rounding by specifying a number of significant figures, or the number of figures which carry real meaning. For example, you might calculate the distance of a journey to be 1343.348 metres, but it can only be measured to the nearest 10 metres. This means that it's only sensible to quote the calculation to no better than this, i.e. only the first three figures in the calculation are significant. Our distance is therefore 1340 metres to three significant figures.

Rounding of values and use of significant figures

When carrying out calculations, you may frequently be presented with results with a large number of digits. For example, suppose we take several measurements of the length of an object using a standard ruler, with millimetre markings. We decide we can make our readings to the nearest mm, and our measurements are:

100.0 mm, 101.0 mm, 99.0 mm, 100.0 mm, 101.0 mm, 100.0 mm, 101.0 mm

Using a typical calculator capable of displaying eight decimal places, we find the average to be 100.28571429 mm.

GROUP ACTIVITY

Is this a sensible value? (10 minutes)

As a group, discuss the result above, thinking in particular about if it is sensible to quote our answer with eight decimal places.

You may find it useful to note that 0.00000001 mm is about one tenth of the diameter of an atom.

For our calculations and results to be sensible, and to reflect our inability to measure quantities such as length exactly (i.e. to an arbitrarily high precision), we apply the procedure of **rounding** and **significant figures** to our calculation.

A significant figure in a number is one which carries sensible meaning. For example, the population of the UK according to the 2011 census was 63,182, 178 people. For most practical purposes, measuring to the nearest person isn't very meaningful; we might only be interested to the nearest million. To do this, we apply two simple rules:

- If the number you are rounding to is followed by 5, 6, 7, 8 or 9, round the number up. For example, if we want to round 636 to the nearest 10, we note that the 'tens' digit (3) is followed by a 6. We therefore round the 'tens' digit up to 4 and set the 'units' to 0.
- If the number you are rounding to is followed by 0, 1, 2, 3 or 4, round the number down. For example, if we want to round 632 to the nearest 10, we note that the 'tens' digit (3) is followed by a 2. We therefore keep the 'tens' digit as 3, and again set the 'units' to 0.

Rounding our population to the nearest million, we note that the digit following the 'millions' is 1. We can therefore set all digits following the millions to 0 (i.e. we are rounding down), so we can say that the population of the UK in 2011 was 63 million.

We can extend the idea of rounding to decimal places, allowing us to express numbers to a certain number of significant figures. In our measurements of the length of an object, we can justify expressing the average to one decimal place. This is because our original measurements are given to four significant figures, e.g. 100.0 mm.

Applying the rules above, we want to round to the first decimal place. We therefore need to look at the next place, which is 8. We therefore round the first decimal place, a 1, up to 2, so we would quote our result as:

100.2 mm to four significant figures.

Significant figures and decimal places

Sometimes, you might need to quote a quantity to a given number of decimal places. For example, to write 10.23478 to two decimal places, we truncate the number to two figures after the decimal point – in this case, it would be 10.23. The same rules for rounding apply, so rounding our number to three decimal places would give 10.235.

Be careful not to confuse significant figures and decimal places:

- For significant figures, you're counting digits both *before* and *after* the decimal place. 10.23478 to three significant figures is 10.2.
- For decimal places, you're only counting those *after* the decimal place. 10.23478 to three decimal places is 10.235.

PAIRS ACTIVITY

Rounding and significant figures (10 minutes)

In pairs, express the following numbers to the specific number of significant figures. You may well need to think carefully about some of them.

100.28571429 mm to three significant figures

100.28571429 mm to five significant figures

4.549 to three significant figures

4.549 to two significant figures

59 999.3 to five significant figures

59 999.3 to three significant figures

59 999.8 to five significant figures

The SI system of units and standard form

SI units

In science and engineering, it is important that we understand what the quantities we measure or calculate mean. This allows us to not only interpret them correctly, but also avoid combining quantities in meaningless ways – for example, it is meaningless to add a distance to a mass.

KEY TERMS

SI – Commonly known as the metric system, SI (Système International d'Unités) is an international set of standards of measurement and units, adopted by most countries, at least for science and engineering purposes.

Standard Form or Scientific Notation – This is a means of writing quantities enabling (1) a clear statement of significant figures and (2) both large and small quantities to be easily represented and understood. A number in scientific notation is written as:

$$m \times 10^n$$

where m is a number greater than zero but less than 10 and n is the power of ten by which m must be multiplied. In our example for significant figures, the distance of 1343.348 metres can be written as 1.34×10^3 metres in scientific notation.

To achieve a consistent meaning to quantities, we use the **SI** system universally in science and engineering, and in many other walks of life.

The SI system

The SI (Système International d'Unités) scheme of units is a coherent system of units defined using seven base units. The base units, together with their current definitions, are outlined in Table 3.3.

Table 3.3 The base SI units

Unit name	Unit symbol	Quantity name	Definition (date in brackets, most recent re-definition)
metre	m	length	(1983): The distance travelled by light in vacuum in 1/299792458 second.
kilogram	kg	mass	(1889): The mass of the international prototype kilogram.
second	s	time	(1967): The duration of 9192631770 periods of the radiation corresponding to the transition between the two hyperfine levels of the ground state of the caesium-133 atom.
ampere	A	electric current	(1946): The constant current which, if maintained in two straight parallel conductors of infinite length, of negligible circular cross-section, and placed 1 m apart in vacuum, would produce between these conductors a force equal to 2×10−7 newtons per metre of length.
kelvin	K	thermodynamic temperature	(1967): 1/273.16 of the thermodynamic temperature of the triple point of water.
mole	mol	amount of substance	(1967): The amount of substance of a system which contains as many elementary entities as there are atoms in 0.012 kilogram of carbon-12.
candela	cd	luminous intensity	(1979): The luminous intensity, in a given direction, of a source that emits monochromatic radiation of frequency 5.4×1014 hertz and that has a radiant intensity in that direction of 1/683 watt per steradian.

Additionally, the following units are common enough to be given names and are derived from the base units.

Table 3.4 Some derived SI units

Name	Symbol	Quantity	Expressed in terms of other SI units	Expressed in terms of SI base units
radian	rad	angle		m • m^{-1}
steradian	sr	solid angle		m^2 • m^{-2}
hertz	Hz	frequency		s^{-1}
newton	N	force, weight		kg • m • s^{-2}
pascal	Pa	pressure, stress	N/m^2	kg • m^{-1} • s^{-2}
joule	J	energy, work, heat	N·m	kg • m^2 • s^{-2}
watt	W	power, radiant flux	J/s	kg • m^2 • s^{-3}
coulomb	C	electric charge or quantity of electricity		s • A
volt	V	voltage (electrical potential difference), electromotive force	W/A	kg • m^2 • s^{-3} • A^{-1}
ohm	Ω	electric resistance, impedance, reactance	V/A	kg • m^2 • s^{-3} • A^{-2}
degree Celsius	°C	temperature relative to 273.15 K		K
lumen	lm	luminous flux	cd·sr	Cd
becquerel	Bq	radioactivity (decays per unit time)		s^{-1}

Most of the units are defined in terms of fundamental constants, such as the speed of light or properties of particular atoms. However, the kilogram is still defined in terms of a physical object, although a more satisfactory definition is being developed.

Unit prefixes and standard form

In many cases, quantities written using the base or derived units are either very large or very small numbers. For example, a typical size of a bacterium is about 0.000001 m. Writing and handling such numbers can become very cumbersome and prone to error. To help overcome this, we can write the number in a more informative form, known as **Standard Form or Scientific Notation**.

Standard Form or Scientific Notation

In standard form, we express the number of digits before or after the decimal place as a power of ten. This is best seen in some examples.

Table 3.5 Examples of standard form

Decimal form	Standard form
1	1×10^0
0.000001	1×10^{-6}
2000	2×10^3
12345	1.2345×10^4

All the numbers take the form:

$$m \times 10^n$$

m is commonly referred to as the *mantissa* and n is the *exponent*.

To write a number in standard form (e.g. 123.45), write the mantissa by moving the decimal point so that the mantissa is between 1 and 10. In the example, move the decimal place two spaces to the left to give the mantissa 1.2345.

The exponent is the power of ten you have to multiply the mantissa by to get the original number. You can find this by counting the number of places you moved the decimal place by in step 1, with steps to the left being positive and to the right negative. In the example, the exponent is 2.

Write the number in standard form, in this case 1.2345×10^2.

As another worked example, if our number were 0.0012345, we move the decimal three places to the right to give a mantissa between 1 and 10, namely 1.2345 again. Our decimal is moved to the right, so the exponent is –3 and our number is:

$$1.2345 \times 10^{-3}.$$

The SI system includes a set of prefixes to put ahead of the main units to express both large and small multiples.

Table 3.6 SI unit prefixes

Factor	Name	Symbol	Factor	Name	Symbol
10^{12}	tera	T	10^{-1}	deci	d
10^9	giga	G	10^{-2}	centi	c
10^6	mega	M	10^{-3}	milli	m
10^3	kilo	k	10^{-6}	micro	μ
			10^{-9}	nano	n

In our example of the size of a bacterium, we originally wrote the size as 0.000001 m, or 1×10^{-6} m in standard notation. Using a prefix, we can also write this as 1 mm.

As we see in our example, the prefixes are written ahead of the main unit. Hence, 0.001 W would be written as 1 mW (read as 1 milliwatt). A few other rules also need to be noted:

1. Prefixes mustn't be combined. For example, 1 W should not be written as 1 mkW (milli-kilo watt).
2. The base unit 'kilogram' already contains a prefix and the above rule still applies. For example, a mass of 1×10^{-6} kg would be written as 1 mg, rather than 1 mkg.
3. The interval between most prefixes is factors of 10^3. Some other factors, such as 'c' (centi-) as in cm and 'h' (hecta) as in hectares are retained, but their use is not recommended in science and engineering.

GROUP ACTIVITY

Unit conversions (15 minutes)

Whilst useful to represent large and small quantities, when carrying out calculations it's advisable to convert quantities to the base or derived SI units in order to ensure the calculations are correct. For each of the quantities below, convert them to their base or derived units, writing the result in standard form.

Wavelength of yellow light, 600 nm

Mass of a bacterium, 1 pg

Power output of a typical power station, 1.5 GW

Further conversions

We've converted numbers from decimal form to standard form in the section above. If necessary, numbers can also be converted back, simply by applying the method outlined in reverse. For example, suppose we have the standard form number 1.2345×10^3. To convert to a decimal form, note that the 3 in the exponent will mean moving the decimal point in the mantissa three places to the right. Our number therefore becomes 1234.5. Similarly, if our number were 1.2345×10^{-3}, converting to a decimal will involve moving the decimal point three places to the left (inserting leading zeros as necessary), giving 0.0012345.

EXTENSION ACTIVITY

Converting from standard form to decimal notation (10 minutes)

Convert the following numbers from standard form to decimal:

2.3×10^6

4.5×10^{-3}

3.6×10^1

1.3×10^0

5.2×10^{-1}

Another useful conversion is from decimal to fractional form, which can help in finding common factors or simplifying mathematical expressions. This is most easily done with numbers containing relatively few decimal places, as follows:

1 Express the number as an integer divided by a power of 10. For example:

$$0.2525 = \frac{2525}{10\,000}$$

2 Look for a common factor between numerator and denominator – in this case, we can divide both by 5, to give:

$$0.2525 = \frac{505}{2000}$$

3 Repeat step 2 until there is no common factor. In this case, we can divide by 5 again to give:

$$0.2525 = \frac{101}{400}$$

EXTENSION ACTIVITY

Converting from decimal notation to fractional form (10 minutes)

Convert the following numbers from decimal to fractional form, expressing them in the simplest fraction.

0.42

0.625

0.784

0.892

1.2 Use of simple mathematical techniques

Whilst not a course on mathematics, there are a number of mathematical techniques we will need to use, and which we need to put into a scientific context. These include:

- handling percentages in experimental uncertainties and yields
- handling equations, including substitution of variables and rearranging equations
- calculation of surface areas and volumes
- calculating rates of change
- statistical operations such as variances, standard deviations and errors.

Before looking at specific topics, it's worth remembering that in mathematics we have to ensure we use the correct order of operations. As an example, consider the arithmetic expression

$$2^4 \times (2 + 3) + 6$$

1 Calculate all expression in brackets, e.g. in our example, we calculate (2 + 3) = 5 first.
2 Calculate orders or powers – we can now evaluate $2^4 = 16$, so our expression becomes $16 \times 5 + 6$.
3 Perform multiplications and/or divisions, in order from left to right. In our case, we have $16 \times 5 + 6 = 80 + 6$.
4 Finally, perform additions and/or subtractions in order from left to right. Our final result is now 80 + 6 = 86.

A word of caution

You may often see the above set of rules written as 'BODMAS' (Brackets, Order, Division, Multiplication, Addition, Subtraction). Be careful – this seems to imply that division takes precedence over multiplication, and addition over subtraction. This isn't the case – division and multiplication are equal, as are subtraction and addition.

For example, consider: 10 – 3 + 2. If we applied BODMAS literally, we'd do the addition first, giving 10 – 5 = 5 which is incorrect. However, noting that addition and subtraction are equal in precedence, we should perform the operations in the order they appear, reading left to right, doing 10 – 3 = 7 first before adding 2 to give 9, the correct answer.

PAIRS ACTIVITY

What has gone wrong? (20 minutes)

A student carried out the following calculation and made several errors (the correct answer is 160).

$7 + 6 \times 5^2 + 3 = 328$

Discuss in pairs the errors the student may have made, and advice you'd give to ensure the errors aren't repeated.

Percentage yields and errors

In chemistry, it's often required to synthesise compounds from known amounts of reagents. Ideally, we wish to carry out the synthesis and extraction with the highest efficiency to obtain the maximum yield. In practice, we will lose material due to, for example, incomplete reactions and losses during extraction.

As an example, consider the nitration of benzene to nitrobenzene:

$C_6H_6 + HNO_3 \rightarrow C_6H_5NO_2 + H_2O$

Normally, this would be carried out with the nitric acid in excess, so the yield of nitrobenzene would be governed by the starting amount of benzene.

Suppose we start with 0.1 mol of benzene, which, recalling that the molecular mass of benzene is 78, corresponds to 7.8 g. Ideally, since the reaction equation tells us one molecule of benzene yields one molecule of nitrobenzene, we'd produce 0.1 mol of nitrobenzene, or 12.3 g, noting that the molecular mass of nitrobenzene is 123.

In this case then, the **theoretical yield** is 12.3 g.

An experiment actually yields 9.5 g (the **actual yield**).

🔑 KEY TERMS

Theoretical yield – In a chemical reaction, the maximum yield of a reaction, based on the equation and amount of materials used.

Actual yield – The amount of product actually extracted following a reaction.

Percentage yield – The ratio of actual yield to percentage yield, expressed as a percentage.

Percentage error – The ratio of the difference between the theoretical and actual yields, and the theoretical yield, expressed as a percentage.

Our **percentage yield** is therefore calculated as:

$$\text{percentage yield} = \frac{\text{actual yield (g)}}{\text{theoretical yield (g)}} \times 100\%$$

Now we need to **substitute** the values for the actual yield and theoretical yield into the equation:

$$= \frac{9.5\ g}{12.3\ g} \times 100\% = 77.2\%$$

Related to the percentage yield is the **percentage error**. In our example above, we'd calculate this as:

$$\text{percentage error} = \frac{\text{theoretical yield (g)} - \text{actual yield (g)}}{\text{theoretical yield (g)}} \times 100\%$$

$$= \frac{12.3 - 9.5\ g}{12.3\ g} \times 100\% = 22.8\%$$

Notice that the percentage error is also (100 – percentage yield)%.

Calculation of surface area and volume

It's often necessary to calculate a surface area or volume of an object or sample. For simple shapes and objects this is relatively easy. For example:

- A square area of land in a field study is 100 m on each side. The area is therefore the length × the width of the field, or 10 000 m².
- A cube of metal, 0.1 m on each side, is used in a materials experiment. Its volume is the length × the height × the depth of the cube = 0.0001 m³

For other common shapes, such as cylinders and spheres, formulae exist for the areas and volumes, some of which are given in Table 3.7.

Table 3.7 Common shapes, such as cylinders and spheres; formulae exist for the areas and volumes

Object	Picture	Surface Area	Volume
Sphere		$4\pi r^2$	$\frac{4}{3}\pi r^3$
Cube		$6l^2$	l^3
Rectangular prism (box)		$2(lw + lh + wh)$	lwh
Right circular cylinder		$2\pi r^2 + 2\pi rh$	$\pi r^2 h$

An important point to watch when calculating areas and volumes is to be consistent with units. For example, the dimensions of a cylinder might be quoted as 0.5 m long and 5 cm radius. Before calculating the area or volume, ensure both parameters are given in the same units – generally, it's recommended to convert to SI, in this case, m.

1.3 Complex mathematical techniques

Calculating rate

It is very common in science to find that a change in one quantity depends on the change in another quantity. The 'rate' of change of one quantity relative to the other can be calculated.

As an example, imagine a car journey that begins at 1pm, travelling 25 miles from York to Leeds, arriving in Leeds at 1.30; and then carries on from Leeds to Sheffield, arriving in Sheffield at 2:30pm. Sheffield is 60 miles from York. For the Leeds-Sheffield part of the journey we can calculate the rate of change of distance with time, as follows:

	Distance from York (miles):	Time (hours)
York	0	1pm
Leeds	25	1:30pm
Sheffield	60	2:30pm

$$\text{Average rate of change} = \frac{\text{change in first value}}{\text{change in second value}}$$

$$= \frac{(60 - 25) \text{ miles}}{(2:30 - 1:30) \text{ hours}}$$

$$= \frac{35 \text{ miles}}{\text{hour}}.$$

You will notice that the rate of change of distance with time is what we normally call 'speed'.

If we plot a graph of the points 'Distance from York (miles)' on the x-axis and 'Time (hours)' on the y-axis we will find that the 'average rate of change' is equal to the gradient of a straight line between drawn between these points.

Changing the subject of an equation

To change the subject of an equation let's look at Ohm's Law, which is $V = I \times R$. In this form 'V' is the subject of the equation. The key thing to remember is that **whatever you do to one side of an equation, you must do to the other side too.** So, to make 'I' the subject, we do the following:

Divide both sides by R: $\frac{V}{R} = I \times \frac{R}{R}$

Remember that anything divided by itself is equal to 1, so that $R/R = 1$,

and so: $\frac{V}{R} = I \times 1$

Hence: $I = \frac{V}{R}$

Using these principles you should be able to rearrange any of the equations you come across in this course.

Geometric progression (serial dilutions), uncertainty and error

A common procedure in science, especially in chemistry and biology, is the preparation of highly dilute solutions with high precision. Take, for example, the need to dilute a particular reagent from a concentrated stock solution by a factor of 1000 (e.g. dilute 1 mol dm^{-3} HCl to 0.001 mol dm^{-3} HCl). We could just take 1 cm^3 of the stock solution and dilute it in 0.999 dm^3 of deionised water to make a total of 1 dm^3. However, think for a moment about the impact of uncertainty on the measurements of volume.

Litres or dm^3?

You will have noticed that we're writing volumes as dm^3 (cubic decimetres) rather than litres, though you might think they're identical. The reason for this is that, strictly speaking, the litre is not an SI unit (though its use is permitted as a derived unit). Further confusion could also occur since the litre was originally defined as the volume of water with a mass of 1 kg. This cannot be reconciled with a consistent set of units. (See https://en.wikipedia.org/wiki/Litre for further discussion.)

Therefore, to ensure your units are consistent with SI, quote volumes using base SI units – for most purposes, dm^3.

GROUP ACTIVITY

What is the uncertainty?

In the above example, we're diluting 1 mol dm^{-3} HCl to 0.001 mol dm^{-3} HCl in a single step, by diluting 1 cm^3 of the stock solution in 1 dm^3 of deionised water.

Suppose we can measure the volume of HCl taken from the stock solution to ±0.2 cm^3, in other words, our volume will actually be between 0.8 and 1.2 cm^3.

Calculate the range of final concentrations, and therefore the percentage uncertainty in this concentration.

You should conclude that the uncertainty is unacceptably large. We can address this by the method of serial dilutions, taking advantage of the properties of geometric progressions.

Geometric progressions and serial dilutions

In general, a geometric progression is a sequence of numbers where each term is a multiple of the previous term.

For example:

- The sequence 2, 4, 8, 16… is geometric with the ratio between terms being 2.
- The sequence 1000, 100, 10… is geometric with the ratio between terms being 0.1.
- However, the sequence 2, 4, 6, 8, 10… is NOT geometric. The ratio is not constant.

In general, we can represent a geometric sequence as a power series:

a, ar, ar^2, ar^3… where a is the first term and r is the common ratio.

PAIRS ACTIVITY

1 Identify the initial term, a and common ratio, r in the following sequences:
   - 3, 15, 75, 375…
   - 2000, 1000, 500, 250…
2 Construct the geometric progression (first five terms) when $a = 6$ and $r = 3$.

Coming back to our dilution example, instead of a single dilution, suppose we performed three dilutions, of a factor of 10 each. Referring to the general geometric progression, a is 1 (the starting concentration) and r is 0.1. Our sequence is then:

$1, 1 \times 0.1 = 0.1, 1 \times 0.1^2 = 0.01, 1 \times 0.1^3 = 0.001$

In practice, this means we do the following:

1 Dilute the 1 mol dm^{-3} stock solution by a factor of 10, by taking 100 cm^3 of stock and diluting in 0.9 dm^3 of deionised water to make a total of 1 dm^3 of solution. We now have a mol dm^{-3} solution.
2 Repeat step 1, using our new solution, obtaining 0.01 mol dm^{-3}.
3 Repeat once more, obtaining 0.001 mol dm^{-3}.

We can see the advantage of this method when we look at the uncertainties, remembering we can measure our volumes to ±0.2 cm^3 (0.0002 dm^3). In each dilution, we are taking 0.1 dm^3, so our percentage uncertainty in each step is 0.2%, compared to the 20% you calculated in the one-step scenario. Even going through three steps, and recalling how uncertainties add, we find our total uncertainty is:

% uncertainty = $100 \times \sqrt{3 \times 0.002^2} = 0.35\%$

Significantly less than the one-step method.

Calculating variance and standard deviation

KEY TERMS

Variance – In statistics, the variance in a set of quantities is the square of the expected value of how much a measurement in a set of data differs from the mean.

Standard deviation – In statistics, this is a measure of the spread of a set of data about the mean value, and is simply the square root of the variance.

Population – In statistics, a population is the collection of all objects or measurements. For example, all the people in the UK represent a population.

Sample – A sample is simply a subset of a population selected for a particular investigation. How this sample is chosen can impact the outcomes of an investigation. For example, you might only be able to study the people in your college or institution for an investigation. You then have to make assumptions and corrections to allow you to extend your conclusions to the whole population.

Earlier in this section, we met the idea of the mean, median and mode of a set of data, as measures of representative values for that set. Other statistical quantities, the **variance** and **standard deviation**, give measures of how much the quantities in a data set vary from the mean.

Variance of a data set

Suppose we have a set of data comprising n observations, $(a_1, a_2, a_3, … , a_n)$

The variance, σ^2 is defined as the mean value of the squares of the differences between the values in the data set and the mean of the data set, μ. In symbols:

$$\sigma^2 = \frac{1}{n} \sum_{i=1}^{N} (a_i - \mu)^2$$

As an example, look at Table 3.8, the table of ages from the medieval period we saw earlier.

We have 20 numbers in our data set and we found the mean to be 46 years. Applying the formula above, the variance is:

$$\sigma^2 = \frac{1}{20} \sum_{i=1}^{20} (a_i - 46)^2 = 162.8 \text{ years}^2$$

Table 3.8 Ages of death of people from the 12th to 14th centuries CE, from youngest to oldest

Name	Age (years)
Eleanor de Braose: 1226–1251	25
Isabella de Braose: 1222–1248 (wife of Dafydd)	26
Elinor de Montfort: 1252–1282	30
William de Braose: 1198–1230	32
Tangwystl: 1168–1206 (mistress of Llywelyn Fawr)	38
Humphrey de Bohun: 1225–1265	40
Dafydd ap Llywelyn: 1208–1246 (Prince of Wales)	42
Maud de Fiennes: 1254–1296	42
Eve Marshall: 1203–1246	43
Gwladys: 1206–1251 (Princess of Wales)	45
Joanna: 1190–1237 (daughter of King John of England; wife of Llywelyn Fawr)	47
Ralph Mortimer 1198–1246 (husband of Gladwys)	48
Gruffydd: 1196–1244 (Prince of Wales)	48
Humphrey de Bohun: 1249–1298	49
Roger Mortimer: 1231–1282	51
Edmund Mortimer: 1251–1304	53
Llywelyn ap Gruffydd: 1228–1282	54
Margaret de Fiennes: 1269–1333	64
Llywelyn Fawr: 1173–1240 (Prince of Wales)	67
Maud de Braose: 1224–1300	76

Population or sample variance?

Our calculation above is technically known as the *population variance*, since we've assumed the data set comprises the entire **population** we're interested in (similarly, µ is technically the *population mean*).

In reality, we often do not have access to the complete set (it might be either very large, or numbers cannot be recorded) and we therefore normally have a **sample** of a population.

GROUP ACTIVITY

What population is the sample drawn from? (5 minutes)

Consider the age data in Table 3.8. Clearly, this isn't data for everybody who lived in the medieval period and can therefore be thought of as a sample.

- What population is this sample drawn from?
- Is the sample representative of the population?
- Are there other populations the sample could be considered as coming from?

(Hint: Think about the demographic and the time period.)

KEY TERMS

Bessel's Correction – This is a correction to the calculation for standard deviation, introduced by the mathematician Freidrich Bessel, to account for potential bias when taking a sample from a population.

Normal or Gaussian distribution – This is a common distribution of quantities about a mean, and has the useful property that it is symmetric about the mean, so that the mean, median and mode are identical. It is also commonly known as a bell curve.

The calculation of sample variance is similar to that of the population variance. We now though define n as the size of the sample rather than of the population, μ_s as the sample mean and s^2 as the sample variance:

$$S^2 = \frac{1}{n-1} \sum_{i=1}^{N} (a_i - \mu_s)^2$$

The most obvious difference is the division by $n-1$ rather than n, as we might have expected. This difference, known as **Bessel's Correction**, arises from a more detailed statistical analysis of populations and samples, and allows for possible bias in the sample.

In the situation above, the sample variance is:

$$S^2 = \frac{1}{20-1} \sum_{i=1}^{20} (a_i - 46)^2 = 171.37 \; years^2$$

Care does need to be taken in choosing the correct formula. Specifically, the sample variance should be used only where the population mean is unknown, and is estimated from the sample available.

Standard deviation of a data set

Closely related to the variance is the standard deviation, given the symbol σ or s, and is simply the square root of the variance. Therefore:

- For entire populations, the population standard deviation is

$$\sigma = \sqrt{\frac{1}{N} \sum_{i=1}^{N} (a_i - \mu)^2}$$

- For samples, the sample standard deviation is

$$S = \sqrt{\frac{1}{n-1} \sum_{i=1}^{n} (a_i - \mu_s)^2}$$

In our example of ages at death in the medieval period:
- The standard deviation assuming our data set is the entire population is

$$\sigma = \sqrt{\frac{1}{20} \sum_{i=1}^{20} (a_i - 46)^2} = 12.76 \text{ years}$$

- The standard deviation assuming our data is merely a sample of a larger population is

$$S = \sqrt{\frac{1}{20-1} \sum_{i=1}^{20} (a_i - 46_s)^2} = 13.1 \text{ years}$$

Notice that the sample variance and standard deviation are larger than those for the population.

Significance of the standard deviation

In many cases, experimental or observed data follow closely a so-called **normal or Gaussian distribution**, as shown in Figure 3.1.

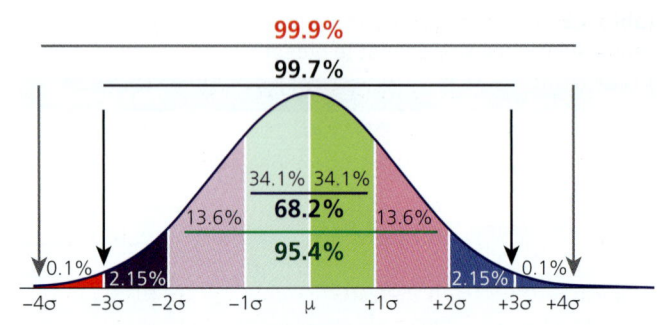

▲ Figure 3.2 A generic normal distribution, showing the relationship between the mean, µ, the standard deviation, σ, and the percentages of areas under the curve between different intervals

Looking at the two distributions for the heights of men and women in Figure 3.1 (b), we note that:

- the mean height of men is greater than that of women
- the distribution of heights for women is smaller than that for men.

The second of these points relates to the standard deviation, whose significance to the normal distribution is illustrated in Figure 3.2.

In Figure 3.2, we note that the standard deviation is a measure of the width of the curve, and therefore of the range of values of the population. In our example in Figure 3.1, we can therefore interpret the narrower distribution curve for the heights of women as reflecting a smaller standard deviation.

Another important point of significance is that the standard deviation gives us a measure of probability. Figure 3.2 shows that the area under the curve between (µ-σ) and (µ+σ) is 68.2%. Noting that the area under the whole curve represents 100% of the population; we can interpret this as meaning that if we were to pick a random item, person or event from the population, we would have a 68.2% chance that it would have a value between (µ-σ) and (µ+σ). As we go to a wider range ((µ-2σ) and (µ+2σ)) we find the probability increases to 95.4%, and so on.

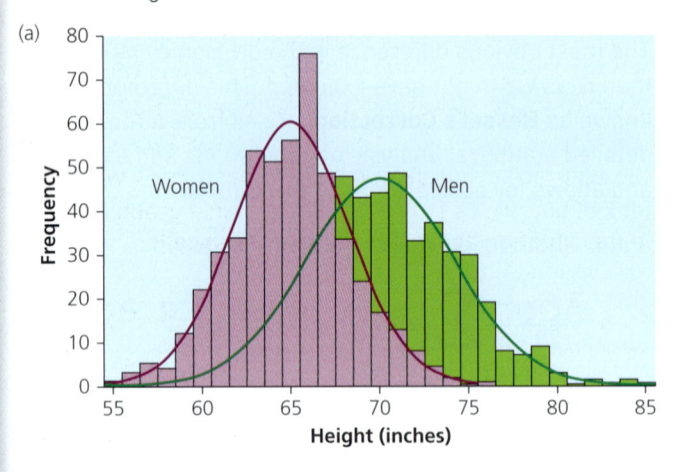

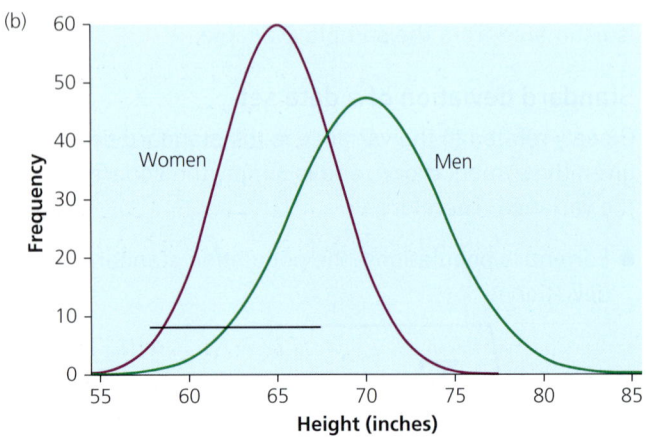

▲ Figure 3.1 (a) Histograms showing the observed distributions of the heights of men and women. (b) The same data with histograms removed, showing the underlying normal distributions fitted to the observed data

GROUP ACTIVITY

Medieval age and probability (20 minutes)

Our data for ages of death in the medieval period gave a mean of 46 years and a standard deviation of 13.1 years.

Assuming the population follows a normal distribution:

1. What is the probability that you would have lived to be between 32.9 and 59.1 years? (Hint: This is ± 1 standard deviation from the mean.)
2. What is the probability that your age would have exceeded 59.1 years? (Hint: What area of the curve is greater than 1 standard deviation?)
3. Roughly, what is the probability that your age would not exceed 88 years? (Hint: Roughly, how many standard deviations is this above the mean?)
4. Do you think assuming a normal distribution is a sensible choice for this data set? (Hint: Think about the factors affecting death rates in the medieval period.)

KNOW IT

1. For a given data set $(x_1, x_2, ... , x_n)$ write mathematical expressions for the mean and standard deviation.
2. State the base SI units. Which of them is currently not defined in terms of fundamental physical constants?
3. A cube has a volume of $2\,m^2$. What would be the radius of a sphere with the same volume?

LO2 Be able to use graphical techniques to analyse data

A well-drawn graph is a powerful way to present data. It communicates results in a more meaningful way than a simple table. It also allows us to observe trends or make inferences from our data and to interpret the data to obtain more information or insight.

GETTING STARTED

Good graph/bad graph (5 minutes)

Your tutor will provide you with a selection of graphs of different types from various sources. Think about which ones are easy to understand and which ones are not so clear. Discuss with the group what you think makes a good graph and what you need to avoid when making a graph.

KEY TERMS

Continuous data – A set of data is said to be continuous if the values of that set can take on any value within the range, e.g. height, weight, light intensity, temperature.

Independent variable – A variable (often denoted by x) whose value does not depend on that of another variable. In an experiment, the independent variable is usually what we change.

Dependent variable – A variable (often denoted by y) whose value depends on that of another variable. In an experiment, we usually measure the independent variable.

Discontinuous data – A set of data is said to be discontinuous (or discrete) if the values are unconnected, i.e. distinct and separate, e.g. shoe size or number of organisms (you can't have fractions of an organism).

In section 2.1 we will consider what type of graph is most appropriate. Section 2.2 deals with how we actually go about drawing graphs. In section 2.3 we will see how we can use a graph to both assess and improve the accuracy of our data. Finally, in section 2.4 we will look at how we can interpret the data in a graph and obtain additional information from the graph.

2.1 Appropriate choice of graph, chart or diagram related to data

The type of graph we use will depend on the nature of our data as well as what we hope to get from the graph.

Scatter graph

Scatter graphs are used when investigating the relationship between two variables that can be measured in pairs, for example the age and height of children in a school. The graph can then be used to establish if there is a relationship between the variables. This could be a positive correlation, negative correlation or no correlation at all.

Line graph

Line graphs are used to show **continuous data**. The **independent variable** should be on the x-axis. You should join the points with straight lines. You can draw a smooth curve or line of best fit if you think that intermediate values will fall on the curve / line.

Bar chart

Bar charts and histograms are used when the **dependent variable** (usually shown on the y-axis) is discrete, i.e. only whole numbers (fractions are impossible) and your data involves frequencies, such as number of students – you can't have half a student!

If you have an independent variable that is non-numerical (e.g. blood group) or you have **discontinuous** (discrete) numerical **data**, then you should use a bar chart. These can be made up of lines, or blocks of equal width, that do not touch. The lines or blocks can be arranged in any order, although it can help make comparisons if they are arranged in order of increasing or decreasing size.

Histogram

These are sometimes called frequency diagrams. The independent variable is usually on the x-axis and is grouped into classes. For example, the height of students in a class could be measured and grouped into 5 cm classes. Height is a continuous variable – students could be any height within a range – and so blocks are drawn touching. The axis is labelled with the class boundaries, e.g. 160 cm, 165 cm, 170 cm, 175 cm, 180 cm, etc. and the y-axis would show the number within each class represented by the height of the bar.

Pie chart

Pie charts can be used when you need to show proportions or percentages. If you are drawing a pie chart by hand, you need to calculate the angle of each sector – divide the percentage by 100 and multiply by 360°, or just multiply the proportion by 360°. However, it is usually much easier to put your data into Excel and let that draw the pie chart!

Kite diagram

A kite diagram is used to show the density and distribution of species in a habitat spread over a distance. You would use a kite diagram to display the results of a survey along a transect.

2.2 Draw linear and non-linear graphs from data

There are some general rules when drawing graphs, besides choosing the appropriate type of graph. You should always give your graph an informative title. Your graph should be a size that makes good use of the paper. As a general rule, the plotted points should cover 50% of the area of the page. If the numerical scale starts at zero you should use a broken axis. The axes should be labelled with the quantity and units in the same way as for tables.

Now that we have looked at the various types of graph, you should be able to decide what is most appropriate for any particular data set and produce a neat and informative graph.

Continuous data

Continuous data is usually measured with some sort of device (thermometer, ruler, pH meter, etc.). If we heat a water bath and measure the temperature every minute, we can plot these points on a graph. As the temperature will increase continuously we can draw a line of best fit.

Discontinuous data

Discontinuous data is normally counted and it is not possible to have intermediate values, for example number of students in a class, number of trees growing in a wood, number of woodlice used in a respiration experiment. The data is plotted as integer values (you can't have a fraction of a student, tree or woodlouse) as a scatter graph or bar chart.

PAIRS ACTIVITY

Graph it (10 minutes)

You tutor will provide you with several sets of data. Working in pairs, choose one set that you think is continuous data and one set that you think is discontinuous data. Now, present each data set in an appropriate graph – one member of each pair should make one graph (continuous data), the other member the other graph (discontinuous data).

Once the graphs are done, share them with the group (this can be done anonymously) and make constructive criticisms of the graphs. Was the type of graph appropriate? Was it clear and of appropriate size? Were the axes correctly labelled? Was the data accurately plotted?

2.3 Apply accuracy and precision to a graph

As a general rule, you should record all raw data to a number of decimal places and significant figures appropriate to the least accurate piece of equipment used in the measurement. Use the same number of decimal places and significant figures for all raw data. You should usually record processed data (e.g. mean, percentage) to the same number of decimal places as the raw data, or to one decimal place more.

Uncertainty and error are factors in any experiment – we covered this in section 1.1 of Unit 2.

Use of error bars (range bars)

We normally repeat experiments several times to obtain a mean value. For example, if we are studying the relationship between the rate of an enzyme reaction and temperature, we will measure the rate of reaction several times at each temperature and take a mean. We can then plot the mean rate at each temperature against temperature. However, it can be informative to show not just the mean, but also the uncertainty in the mean. To do this, we calculate the standard deviation (see section 1.2) for each mean. When plotting each point on the graph, we can plot the mean and add error bars showing the uncertainty of that mean. This can be done in different ways, but plus / minus one standard deviation (± SD) or plus / minus one standard error (± SE) are usual. See Figure 3.3 a) on the next page for an example.

The power of this technique comes when comparing data sets. Take the example in Figure 3.3 b) showing the results of an experiment comparing a range of inhibitors on the rate of an enzyme reaction. This is presented as a bar chart with each bar showing the mean and error bars showing ± SD of the mean.

Where error bars overlap we conclude that any differences in the mean are due to chance; where there is no overlap between error bars we conclude that the differences in the means are significant. From this example, we can conclude that compounds C and D are inhibitors of the enzyme but compounds A and B are not.

This example uses a bar chart, but the approach is equally useful when using histograms or line graphs with multiple data sets where we want to assess whether any differences are statistically significant.

Note that you will sometimes see error bars referred to as range bars, but error bars is the generally accepted term. Strictly speaking, range bars show the range (maximum and minimum) of the data values. Error bars will be symmetrical about the mean but range bars may not be – think about why this is true.

Identification of outliers

Plotting all the data points, like in a scatter graph, can show outliers for particular values of the independent variable. However, it is more likely that we will already have identified anomalous data points and excluded them when calculating a mean. Outliers could represent an error in the experimental procedure or they could simply be due to random variation in the data. By plotting a graph, however, it is easier to spot outliers and investigate their cause, particularly if the outlier is caused by an anomalous result (see section 4.2 for more about this).

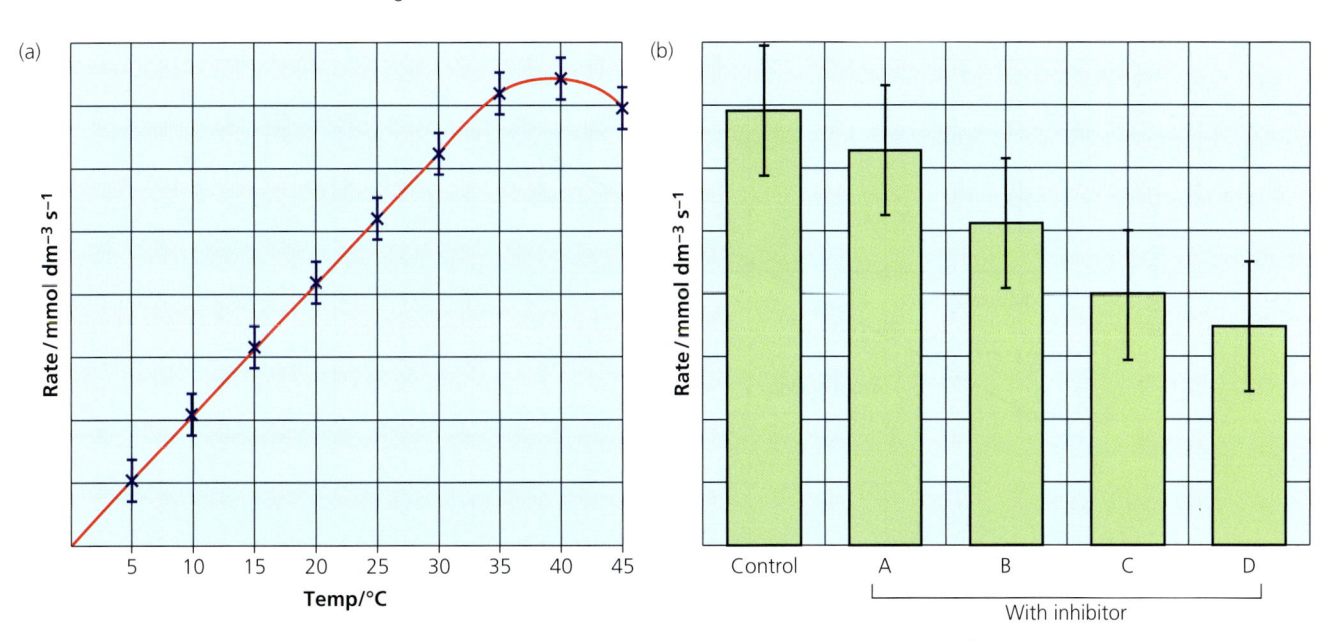

▲ **Figure 3.3** a) Graph of rate of enzyme reaction against temperature showing error bars; b) Effect of various compounds on enzyme activity; error bars in A and B overlap with the control (no inhibitor) whereas error bars in C and D do not, so we conclude there is a significant difference in enzyme activity with C and D

2.4 Interpreting data through graphs

Having plotted a graph and decided that there are no anomalies or outliers, we can then use the graph to get additional information or insight.

PAIRS ACTIVITY

Data interpretation (15 minutes)

Interpreting tables or graphs of data can be challenging, but it is an essential skill. Working in pairs, use the selection of graphs provided by your tutor and practise drawing conclusions from them.

- Describe what you see.
- Describe the general shape of the graph.
- Describe trends (increase, decrease, levelling off).
- Think about what the graph tells you about the underlying process.

Some useful examples of graphs are:

- a scatter plot to show a positive or negative correlation between two variables
- a line graph to show that the rate of a chemical reaction doubles for each 10°C rise in temperature
- a graph of the rate of photosynthesis against light intensity
- a graph of rate of enzyme reaction against substrate concentration
- a graph of rate of an enzyme reaction against temperature.

Find values by interpolation and extrapolation

A graph of the rate of a chemical reaction at 10°C intervals between 20°C and 110°C would most likely be a straight line. This would allow us to predict the rate of reaction at intermediate temperatures, such as 65°C; this is interpolation.

If we wanted to predict the rate of reaction at 125°C or -15°C we could do that by extending the line above and below our existing data points – this is known as extrapolation (see Figure 3.4).

? THINK ABOUT IT

Is extrapolation always valid? (5 minutes)

Figure 3.4 is an example of a graph where extrapolation is likely to be valid. Can you think of other graphs where extrapolation would not be valid? What assumptions do we make when we extrapolate? Is interpolation more reliable?

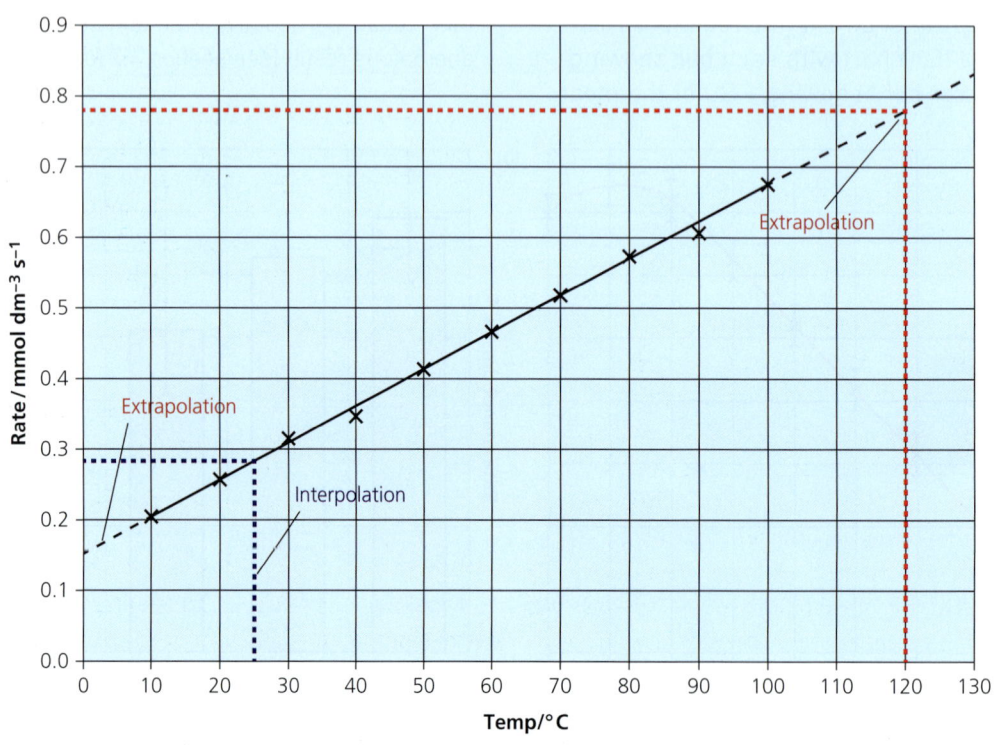

▲ **Figure 3.4** Extrapolation involves extending a line or curve beyond the plotted points and using this to predict values outside the plotted range e.g. at 15°C and 125°C. Interpolation involves estimating a value between two plotted points e.g. at 65°C

Determine intercepts for graphs

If we have a straight line graph, we can extend the line (extrapolate) until it intercepts the y-axis. A straight line graph will always follow the equation:

$y = mx + c$

where y is the dependent variable and x is the independent variable. The slope of the line will be m (see the next paragraph) and c will be the intercept on the y-axis. Depending on what our data relates to, we can make deductions from the intercept.

Calculating the gradient of a line

We can calculate the gradient for any straight line graph by taking two points on the line. If we divide the change in y by the change in x, then we get the slope of the line – see Figure 3.5.

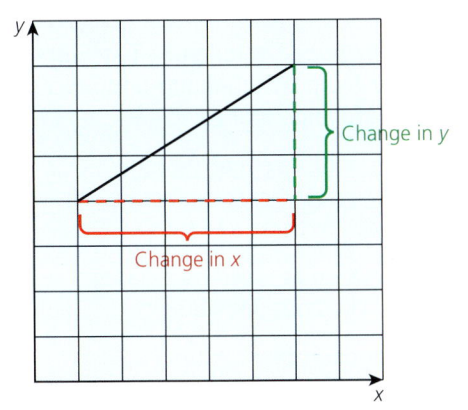

▲ **Figure 3.5** Calculation of the gradient of a line: divide the change in y by the change in x

If our graph is one of distance travelled in metres plotted against time in seconds, then the slope of the line will give us the velocity in $m\ s^{-1}$.

Some situations are more complicated than this. Think about an enzyme reaction where we measure the disappearance of a substrate. The rate of an enzyme reaction is proportional to the concentration of the substrate. This means that, as the substrate concentration falls, the rate of reaction falls. We need to calculate the initial rate, i.e. the rate right at the start of the reaction. We do this by drawing a tangent to the curve of substrate concentration plotted against time and then calculating the gradient of the tangent. This gives us the rate of the reaction in, say, $mg\ s^{-1}$. This means we can repeat the experiment at different temperatures, measure the rate of reaction at each temperature and plot a graph of rate against temperature.

KNOW IT

1. Define these terms:
   a. independent variable
   b. dependent variable
   c. continuous data
   d. discontinuous data
2. What type of graph would you use for the following data?
   a. percentage distribution of five different species in a habitat
   b. size of aphid populations on three types of deciduous tree in a wood
   c. blood glucose concentrations measured using blood samples and urine samples
   d. volume of carbon dioxide produced against time in respiring seeds with and without an inhibitor of respiration

LO3 Be able to use keys for analysis

GETTING STARTED

The problem of identification (5 minutes)

Think about someone starting out in bird-watching. They might be able to tell the difference between a sparrow and a blackbird, or know vaguely what a finch looks like. But the 'Book of British Birds' they were bought as a present is organised by groups of birds. So they might be able to turn to the section on finches and work out if it is a greenfinch or a goldfinch they are looking at. But where would they start in trying to identify the unfamiliar bird that doesn't really look like anything they've seen before? Discuss the times when an identification chart might be useful in identifying a collected specimen and when you might need something more systematic to help.

KEY TERM

Dichotomous – A key where there are two branches (alternatives) at every step.

3.1 To use and construct a key to identify collected specimens

Keys have been around for hundreds of years as ways of helping to identify specimens, but they can often require a great deal of knowledge about the specimens concerned. However, you can understand the principles involved using simpler keys and then apply those principles to more complex keys if needed.

81

Using a **dichotomous** key involves following a series of questions. At each step you are given a choice of two alternatives and you have to decide which best fits the sample you are trying to identify. Having done that, you will either have made an identification or you will be directed to another question.

PAIRS ACTIVITY
Using a key (10 minutes)
Your tutor will provide you with a key and one or more specimens. Use the key to identify the specimens.

Constructing a key is like playing the game of 'Twenty questions' or 'Animal, vegetable or mineral'. Each question has to have a 'yes / no' answer or just two alternatives. The answer to each question will either lead to an unambiguous identification or another question.

PAIRS ACTIVITY
Constructing a key (10 minutes)
Your tutor will provide you with a set of specimens and you have to construct a key that will allow another person to identify the specimens. Suitable specimens include different types of leaf, pictures of different insects or even a range of snack foods or liquorice allsorts.

Interactive computer-based keys can be very powerful, particularly as they can include various multimedia resources to help in identification.

EXTENSION ACTIVITY
Identification keys on iSpot
The iSpot website has links to a number of online resources to help you practise using keys:

www.ispotnature.org/keys

3.2 Use a key to compare the quality of primary data to secondary data

You should now appreciate how useful a good key can be in helping you identify specimens. There are many situations in which this would be helpful. For example, a forensic entomologist will be able to estimate the time since death of a corpse based on the types of insect larvae present. An ecologist will need to identify the various species present when carrying out a survey.

One application that is worth looking at in more detail is the use of keys in identifying what are known as indicator species. An indicator species can be used to monitor pollution, for example water pollution or air quality. Some invertebrates can live in very polluted water while others can only survive in unpolluted water. Some species of lichen can only grow in unpolluted air while others can tolerate higher levels of pollution. Identifying the species present, using a suitable key, can help you judge the level of pollution.

You can then compare the degree of pollution based on the presence or absence of different indicator species with other indicators, such as concentration of pollutant chemicals in water or air samples.

GROUP ACTIVITY
Using invertebrate indicator species to monitor water pollution (1 hour)
The Nuffield Foundation, together with the Royal Society of Biology, has produced details of a fieldwork exercise sampling water from polluted and unpolluted sources and comparing the indicator species:

www.nuffieldfoundation.org/practical-biology/monitoring-water-pollution-invertebrate-indicator-species

This online resource can also be used as a paper-based exercise if fieldwork is not possible.

You could research other examples of the use of indicator species and find or create appropriate identification keys to use with them.

Think about the ways in which the use of indicator species might give a different picture of the degree of pollution when compared with other sources of data, such as chemical analysis of water or air samples.

3.3 Classification system – rationale for classification of living things

The need for classification of living things is rather like the need to classify a stamp collection – pretty incomprehensible to outsiders! But, when you think about it, it makes sense. We assign things (stamps or living things) to classes or categories as a way of understanding them and making sense of their similarities and differences.

However, classification of living things provides much more insight and understanding than classifying a stamp collection. First, we need to understand the idea of a species. That on its own is not a simple task. There are various differences between superficially similar organisms and they all help us to allocate organisms to different species.

One widely accepted definition of a species is that if two organisms can produce fertile offspring, then they are members of the same species. This is true most, but not all, of the time. We can learn a lot from rules, and sometimes learn more from rules that are broken. We need to understand the classification of living organisms and how they are named because this is the basis of understanding the evolutionary relationships between organisms.

3.4 Binomial nomenclature

The science of taxonomy is all about naming and classifying groups of organisms. There are two main classification schemes. One of these systems was developed by 18th-century Swedish botanist, physician and zoologist (the term 'scientist' hadn't been invented then) Carl Linnaeus. Taxonomists have modified his system and still argue about the details. What has not changed, though, is the binomial nomenclature that Linnaeus developed ('binomial' means 'two names'). Every organism is assigned to a species that has a two-part name. The first part describes the genus, with the second narrowing it down to the species. Table 3.9 has some examples.

Table 3.9 Examples of binomial nomenclature

Common name	Genus	Species
Human	*Homo*	*sapiens*
Neanderthal man	*Homo*	*neanderthalis*
Chimpanzee	*Pan*	*troglodytes*
Wolf	*Canis*	*lupus*
Domestic dog	*Canis*	*lupus familiaris*

Some points to note:

- Every species is give a name in two parts: the genus and then the species.
- The genus name always has an initial capital, the species name does not.
- Sometimes the species name is not enough – for example, wolves and pet dogs are both members of the species *Canis lupus*, although we distinguish domesticated dogs by adding another name on the end of the binomial species name.

- You will often see the genus name abbreviated, so the bacterium *Escherichia coli* is usually written as *E. coli*.
- The names are always in Latin, or at least latinised. It saves arguing over whose language to use!
- Species names are usually written in italics – like these have been.

KNOW IT

1. Explain what is meant by a dichotomous key.
2. Construct a dichotomous key to help identify the following types of animal:
   - an earthworm
   - a fish
   - a spider
   - a house fly
   - a dog
   - a human
3. The domestic cat is a member of the genus *Felis*. Give the full species name of the following members of the genus:
   - domestic cat (*catus*)
   - German wild cat (*silvestris*)
   - Indian jungle cat (*chaus*)

LO4 Be able to analyse and evaluate the quality of data

GETTING STARTED

Evaluating data (5 minutes)

What do we need to do to be able to analyse and evaluate data from an experiment? How do we know that we are able to draw valid conclusions? What should we consider when analysing data? Have you encountered any problems when analysing the results of any experiments you have done? Share your thoughts with the group.

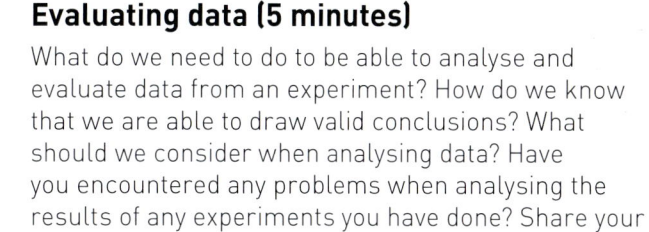

As well as being able to interpret a graph – understand what it tells us about the experimental results – we have to assess the quality of the underlying data. In part this requires an evaluation of the sources of error and uncertainly in the experiment itself. Analysis of the data, either through simple inspection of raw data or graphed data, can also provide insight. If necessary, statistical analysis can be carried out to further improve our understanding of the data and what it is telling us.

4.1 Define and apply terms commonly used in experimental analysis and evaluation

In Unit 2 we covered various concepts and terms used in experiments that form the basis of analysis and evaluation of data and the conclusions we can draw from that data. There are various terminology you must be able to use including: accuracy, precision, repeatability, reproducibility, uncertainty and validity. Accuracy, precision, repeatability and reproducibility are defined in Unit 2, LO1. Validity concerns the overall suitability of an experiment for the investigation in question. Uncertainty is the interval within which the true value of a measurement will lie. The uncertainty of a measurement is often expressed in the form 103cm ±1cm, where 103cm is the measurement and 1cm is the uncertainty. This means that the actual value of the measurement is definitely somewhere between 102cm and 104cm.

4.2 Discuss the quality of data

> **KEY TERMS**
>
> **Anomaly / anomalous result** – A value that is out of line with other results caused by some error rather than uncertainty in the measurement. Anomalous results are normally excluded when calculating mean values.
>
> **Primary data** – The results that we obtain from an experiment or measurement that we carry out ourselves.
>
> **Secondary data** – The results obtained by others in independent but comparable experiments or measurements, for example published in textbooks or research papers.

Identify relationships between variables

We saw in section 2.1 that we can use a scatter graph to demonstrate correlation between two variables. From this we can tell two things:

- if the correlation is positive (increasing one variable increases the other) or negative (increasing one variable decreases the other)
- the strength of the relationship.

It is possible to estimate the strength of the relationship by looking at how close the plotted points are to the line of best fit. We can also use a statistical method known as regression analysis to calculate a value for the strength of the relationship.

However, the visual method only works well when there is a linear relationship between the variables. If the volume of a substance doubles, then the mass will also double. In some chemical reactions, doubling the concentration of one reactant will double the rate of reaction; we say the rate of reaction is proportional to the concentration of reactant. These are both linear relationships. In other cases the relationship is non-linear. For example, the area of a square quadruples when the length of the side doubles. The rate of some chemical reactions is proportional to the square of the concentration of reaction. In such cases, a more complex form of regression analysis will identify the relationship between the variables.

Level of uncertainty of data, including anomalous results

We covered aspects of this in Unit 2. In section 1.1 of that unit we considered the importance of percentage error and uncertainty when choosing measuring equipment. Then in section 3.1, when looking at titration methods, we considered the sources of uncertainty and how these could affect the results of our experiments.

Besides the uncertainty that is inherent in any measuring technique, we should also consider anomalous results. In a titration, if we obtain titres of 21.5 cm^3, 21.4 cm^3, 21.9 cm^3 and 21.4 cm^3, then it should be obvious that the titre of 21.9 cm^3 is an **anomalous result** and should not be included when calculating a mean titre.

Plotting a graph can also help to identify anomalous results by identifying outliers – see section 2.3.

Sources of error

We covered various sources of error and uncertainty in Unit 2 and it is important that these topics are considered when designing experiments so as to make them more accurate and more valid. However, we should also consider error and uncertainty when interpreting results. How will potential sources of error influence our results? What are the sources of error?

Some of these were covered in Unit 2, but we will cover some additional points here – particularly ones that might only become apparent when we come to analyse and interpret our data.

Accuracy and precision of measurements

It is worth looking again at the two concepts of accuracy and precision and how we can use these to analyse and evaluate our data. A useful analogy is one of target shooting. The intention is to hit the bullseye. If we have six shots and all six hit the bullseye, then the shots were accurate (close to the bullseye) and precise (close to each other). On the other hand, if all six hit the target close to each other, but far away from the bullseye, then they are inaccurate but precise. Six shots that are relatively close to the bullseye but spread around will be accurate but imprecise. Finally, six shots that are spread all over the target well away from the bullseye will be inaccurate and imprecise. This is illustrated in Figure 3.6.

High accuracy Low accuracy High accuracy Low accuracy
High precision High precision Low precision Low precision

▲ Figure 3.6 The target analogy of accuracy and precision

Another thing we should consider when assessing the accuracy of our results – the **primary data** – is how closely they match the results obtained by others – the **secondary data**. This could be by comparison with standards (such as calibration standards) or with published data.

Measurement error

This refers to errors or uncertainty when taking measurements and can have a variety of causes. Some errors are random, others are systematic and it is possible for an experiment to be affected by both types of error. Human error can also play a part in measurement error. For example, an instrument might require a reading to be taken with the eye correctly aligned to the scale; if this is not the case, then the reading will be inaccurate (see Figure 2.5 in Unit 2).

Random error

Random errors will usually affect the precision of measurements (spread of values) causing the measurements to fluctuate around the true value. However, they will not change the mean. This shows why it is important to take multiple measurements / readings and calculate a mean.

Systematic error

Systematic errors can sometimes be difficult to trace. The results might appear to be precise (closely spaced), but might still be inaccurate (far from the true value). Think about the burette used in titration (section 3.1 in Unit 2). An accurate volume is obtained when reading from the bottom of the meniscus (see Figure 2.5 in Unit 2). If we read instead from the top, then the volumes will always be inaccurate by a certain amount, even though they appear precise.

Instrument error

When we talk about instrument error we generally mean uncertainty in the measurement and this will depend on the particular instrument. However, errors can also be introduced by instrument faults or calibration errors – see section 2.2 in Unit 2 for an example of how to prepare a calibration curve.

We can improve the quality of our data and reduce the uncertainty associated with our measurements by using an instrument with greater sensitivity, resolution or magnification.

Range and interval

Instruments are often designed only to measure values within a certain range. Using them outside this range can lead to inaccuracy. We should also consider the range of measurements we take in any experiment. Are we covering a representative range of values? Should we take measurements above and below the range we originally considered? Would it improve the quality of the data if we took more intermediate readings?

Repeatability and reproducibility

If we repeat an experiment several times and obtain the same results, then we have obtained repeatable results. If the experiment is repeated by someone else, in a different laboratory or using different methods or equipment, then the experiment is reproducible. We should always ensure that our experiments are repeatable as a matter of course. Acceptance by the wider scientific community will only come if our experiment is reproducible.

The scientific method describes the way in which scientists go about understanding and explaining the world:

- Observation or experiment leads to questions, e.g. why does it happen like that?
- We formulate a theory or hypothesis to suggest an explanation.
- We test the hypothesis, usually by doing an experiment to prove or disprove the hypothesis.
- Then we analyse the results and draw conclusions from our data.

- If the conclusions support the hypothesis, we might want to refine or improve the hypothesis.
- This will lead to further experiments that give further insight and lead to further refinement of the hypothesis.
- If at any stage an experiment disproves the hypothesis, we have to re-evaluate the hypothesis and come up with a new one.
- We report our findings in a way that allows others to carry out their own experiments to independently test our **hypothesis**.
- Our experiments must be repeatable and reproducible for our conclusions to be credible.

In section 5 we will look at how we go about drawing valid conclusions from data.

> **CLASSROOM DISCUSSION**
>
> **The scientific method (20 minutes)**
>
> Think about experimental work that you have done. How does it fit into the scientific method? Can you think of ways it could be improved?

> **KNOW IT**
>
> 1 Think about the different types and sources of error we have covered. What type of error would the following represent and how could they affect the validity of our conclusions?
>   a use of the wrong buffer solution when calibrating a pH meter
>   b missing the end point of a titration
>   c use of a 250 cm^3 measuring cylinder to make up 50 cm^3 of a standard solution
>   d measuring the rate of an enzyme reaction when the enzyme is dissolved in pure water
> 2 Explain the differences between the following pairs of terms:
>   a dependent and independent variable
>   b accuracy and precision
>   c repeatable and reproducible

> **KEY TERMS**
>
> **Hypothesis –** A proposed explanation for a phenomenon or observation. In science a hypothesis is no use unless we can test it, i.e. design an experiment.

LO5 Be able to draw justified conclusions from data

> **GETTING STARTED**
>
> **Cause and effect (5 minutes)**
>
> Think about the following and share your thoughts with the group:
> - There is a strong positive correlation between rate of a chemical reaction and the concentration of a reactant. Can we conclude that the increase in reactant concentration causes the increase in rate of reaction?
> - There is a strong positive correlation between ice cream sales and death from drowning (yes, really!). Does this prove that ice cream consumption causes drowning? What would be a more likely explanation?

We can draw justified conclusions if we follow some basic rules:

- Design experiments (including surveys or sampling methods) to minimise uncertainty and error.
- Use statistical methods to help interpret data and draw conclusions.
- Compare our conclusions with other sources (scientific publications, expert reports, etc.) and look for any supporting or conflicting evidence from these secondary sources.

5.1 Conclusion given and justified

We saw in section 2.1 how we can use a scatter graph to see if there was a correlation between two variables. However, correlation does not imply causation. This means, just because there is a correlation between two variables, it does not show that one variable causes the other.

Comparison between primary and secondary sources of information

Understanding how our own experiments and data (primary source) fit in with published sources and the body of knowledge in the area where we work (secondary sources) is essential in helping us obtain valid conclusions from our data. Most progress in science comes from experiments built on the foundations of the work that has gone before. This provides the basis of our knowledge and understanding and also provides inspiration for our own experiments. Then, it forms a framework for us to analyse and interpret our results and draw valid conclusions from them.

Identification of conflicting evidence

If our data and conclusions are in agreement with secondary sources, we can feel more confident in those conclusions.

However, it is also important to look for ways in which our results conflict with the accepted view. Does that mean our results are wrong? Have we drawn the wrong conclusions? Or have we discovered something new?

Having made the comparison, we need to understand the reasons for any difference between our results and secondary sources.

- Were the experimental methods comparable?
- Have we overlooked sources of error and uncertainty?
- Is our experimental design robust – does it ask the right questions?

Further evidence required to make the conclusion more secure

If we are satisfied that our experimental methods are valid, we can analyse the results and ask some important questions:

- Do we need to perform additional work to confirm the validity of our conclusions?
- Should we repeat the experiment to provide more data points? This will improve the reliability of our data.
- Are the trends in our data significant or do they represent random variation?
- Have any anomalous results been discounted?

Finally, we need to ensure that the conclusions we draw are not only based firmly on good quality data, but that they are also supported by science. We must avoid concluding that increased ice cream consumption is a cause of drowning!

GROUP ACTIVITY

Review of experimental work (1 hour)

Review the experimental work undertaken by the group during this course. Try to identify the following:

1 sources of error
2 steps taken to minimise sources of error
3 improvements that could be made

Once you have done this, assess the validity of your conclusions in each case. Prepare a short report summarising your findings.

EXTENSION ACTIVITY

Bad science (2 hours)

The scientific method has been developed and refined over many decades and is considered to be the only way that scientific knowledge and understanding can be developed and built on. However, things sometimes go wrong.

Choose an example of where the scientific method broke down and led to invalid conclusions. Write a brief report highlighting the shortcomings of the work, identifying points that should have been challenged and how the work could have been improved.

The MMR vaccine work by Andrew Wakefield (see BMJ 2011;342:c7452, http://www.bmj.com/content/342/bmj.c7452) is one of the more high-profile examples, but your tutor may be able to provide others.

KNOW IT

1 Explain the difference between primary and secondary sources of data.
2 Explain the importance of comparing our results with those of others, e.g. in the published literature.
3 In an experiment to investigate factors affecting photosynthesis there was a positive correlation between the rate of photosynthesis and light intensity. Another experiment showed a positive correlation between light intensity and rate of respiration in seeds. Evaluate whether these experiments show that the rate of photosynthesis and rate of respiration are both directly proportional to the light intensity. Suggest a flaw in the experimental design that could explain the results of the seeds experiment. Hint: How would you change the light intensity? Could this affect the results?

LO6 Be able to use modified, extended or combined laboratory techniques in analytical procedures

GETTING STARTED

From school to the real world (5 minutes)

Think about your earlier years at school and the type of experiments you carried out, then think about the techniques you have used in your course. Some of you may have had work experience in industrial laboratories of different sorts, or visited well-equipped analytical laboratories. How did the procedures you saw differ from the ones you used in school or the ones you use now? Some techniques will be more sophisticated, automated or computerised but others might be recognisable. Share your experiences and thoughts with the group.

6.1 Modify microscopic analytical techniques according to need

Section 4 of Unit 2 covers basic techniques in light microscopy as well as techniques such as electron microscopy that offer greatly enhanced resolution of detail. In Unit 8 you will get the opportunity to use various staining techniques widely used in cytology (the study of cell structure). Here we will look at staining procedures for various applications and how to prepare permanent slides.

Use of alternative staining procedures in microscopy

Staining techniques usually increase the contrast between different parts of a tissue section, or stain specific features, structures or substances. Some of these depend on the affinity that some dyes have for, say, lipids or proteins. Other immunology-based techniques use highly specific antibody-antigen interactions to label particular proteins or glycoproteins in cells and tissues.

Preparation of permanent slides

It is possible to make a thin section of a plant stem or leaf to observe the structures. However, it will very soon decompose and be useless. It is often convenient to be able to prepare a slide that can be observed right away, but can also be re-examined days, weeks, months, even years later. This could be so that the section can be used for educational purposes, or for future reference in a piece of work that extends over a long period. Think of some other applications where it would be important to be able to prepare and work with archive material – forensics or pathology, for example.

GROUP ACTIVITY

Preparation of stained permanent slides – research and practical activity (2 hours)

The techniques that you can use will depend on the facilities available to you, so you will need guidance from your tutor.

Working in groups, research one or more methods of staining and preparing permanent slides of animal or plant tissue. You will find some examples in section 2.1 of Unit 8.

Preparation of permanent slides usually involves:
- fixing the specimen to prevent decay
- dehydrating the specimen
- replacing the water with paraffin (a type of wax)
- cutting into thin sections using a microtome
- staining the sections
- clearing the section to make it transparent
- adding mounting medium to stabilise the section and a coverslip to provide physical protection.

You should keep detailed notes of the procedures you use in your laboratory notebook.

6.2 Adaptation of chromatographic techniques

Section 2 of Unit 2 covers simple chromatographic methods such as thin layer chromatography (TLC), as well as more advanced methods such as gas chromatography (GC) and high performance liquid chromatography (HPLC). It would be useful to review that section before moving on here.

Use of column chromatography and thin layer chromatography (TLC) as a preparative and quantitative technique

Column chromatography and TLC both rely on the same principle – those components in a mixture that have low affinity for the solid support (stationary phase) will spend longer in the mobile phase and so will move further up the TLC plate or be eluted more quickly. In theory, TLC could be used as a preparative technique. However, only very small quantities of sample can be applied to a TLC plate. In theory you could scrape a spot off the plate and obtain material by elution; the quantity is unlikely to be enough to be useful.

Column chromatography is a different matter, because it is possible to apply a relatively large amount of sample to even quite a small column. In industry, the method can be used with very large columns indeed.

You could research preparative methods based on different types of column chromatography. Besides simple adsorption methods, you are likely to come across a number of others such as:

- gel filtration (size exclusion)
- ion exchange
- affinity
- immunoaffinity.

Column chromatography can be useful in quantitative analysis. Some form of detection is needed to quantify the substances being eluted from the column, such as:

- UV or visible light absorption
- electrical conductivity.

To make the method truly quantitative you need to prepare a calibration curve. This was dealt with in section 2.2 of Unit 2.

Use of TLC as a quantitative technique by elution or densitometry

We said that TLC is not a very practical preparative technique. However, it is possible to use it quantitatively. There are two options:

- Scrape all the solid phase (usually silica gel) where each spot is, elute the component from the solid phase and measure its concentration.
- Treat the plate with a suitable dye or stain and then use densitometry to quantify the spots.

Densitometry is a way of measuring the optical density (light absorption or darkness) of spots on a plate. A densitometer will carry out the measurement and can be linked to a computer.

Think about how you might measure the concentration of a component eluted from a spot or convert a densitometer reading into an amount of substance.

6.3 Use of alternative titration techniques

Titration was covered in Section 3 of Unit 2.

Acid base titrations work because an indicator shows the end point or equivalence point – where the acid and base neutralise each other. They do this because there are an equal number of moles of acid and base. Other types of titration have an equivalence point, but the way in which that is determined depends on the type of titration.

Precipitation titrations

Instead of the neutralisation reaction in an acid base titration, this method depends on a precipitation reaction such as the reaction between Ag^+ ions and Cl^- ions to produce insoluble AgCl. You might think you could just add silver nitrate solution to a sample containing Cl^- ions, isolate and weigh the precipitate and then use this to calculate the number of moles of Cl^-. You could probably do a simple calculation to work out why this wouldn't be very accurate in most cases. Instead, you need an indicator to show the equivalence point.

Determination of chloride

There are several methods, but the most common is probably the Mohr method where potassium chromate acts as indicator. Ag^+ ions react preferentially with Cl^- ions to form AgCl. Once all the Cl^- ions in the sample are used up, then Ag^+ ions will react with chromate ions (CrO_4^{2-}) to form blood-red silver chromate (Ag_2CrO_4) which plays the same role as an acid-base indicator.

A few points to bear in mind:

- The solution needs to be near neutral as silver hydroxide forms at high pH and the chromate ions react with H^+ ions at low pH – what effect would each of those have on the result?
- Carbonates and phosphates also react with Ag^+ ions, so they need to be absent.

There is a class practical for determination of chloride ions in seawater using microscale apparatus described on the Royal Society of Chemistry website: www.rsc.org/learn-chemistry/resource/res00000538/finding-out-how-much-salt-there-is-in-seawater

Redox titrations

Redox titrations all involve a redox reaction – these were covered in section 2.2 of Unit 1. What distinguishes each type is the method of determining the end point of the titration.

Potassium dichromate

Potassium dichromate(VI) solution turns from orange to green when dichromate acts as an oxidising agent. For example, dichromate will oxidise Fe^{2+} to Fe^{3+}, so this method can be used to determine the amount of Fe^{2+} in a sample. However, the colour change isn't clear enough to allow us to determine the end point accurately, so most dichromate titrations need a redox indicator – sodium diphenylamine is most commonly used as it gives an intense purple colour when the end point is reached.

Iodine solution

Iodine can be reduced to iodide; conversely, iodide can be oxidised to iodine. This is the basis of what is known as iodometric titration. The indicator for this type of titration is starch, which forms a blue-black colour with iodine (this is used in biology as a test for starch in foodstuffs). The end point occurs when either the blue-black colour disappears (iodine converted to iodide) or appears (iodide converted to iodine).

Another application of iodine is to use it to calculate the iodine number of fats and oils. Iodine will react with the C=C double bonds present in unsaturated fats or fatty acids. If excess iodine is added to a fat or fatty acid, it will react with any C=C double bonds. The amount of unreacted iodine can be determined by titration with sodium thiosulfate. From this, the amount of iodine that reacted with the fat can be worked out, giving the iodine number. This is the mass of iodine in grams that reacts with 100 g of the oil and is a measure of the degree of unsaturation of the fat.

In practice, the test uses iodine monochloride (ICl) or monobromide (IBr). After the ICl / IBr has reacted with the oil, potassium iodide solution is added. This converts any unreacted ICl / IBr to I_2, which is then titrated with sodium thiosulfate solution.

Sodium thiosulfate

This is commonly used together with iodine as a way of determining the concentration of oxidising agents, for example in water samples. An excess of iodide is added to the sample being tested. Any oxidising agents present oxidise the iodide to iodine. The amount of iodine produced can then be measured by titration with a standard sodium thiosulfate solution.

Stoichiometry in redox reactions

Acid base titrations are usually quite straightforward. 1 mole of HCl requires 1 mole of NaOH for neutralisation, whereas 1 mole of H_2SO_4 requires 2 moles of NaOH.

Working out the stoichiometry (reacting proportions) of redox reactions can be more challenging. It requires an understanding of oxidation states and redox half equations. However, you don't need to be able to work out the stoichiometry for yourself as you can usually look it up.

Once you know the number of moles reacting, the titration calculation is just like any other.

Compleximetric techniques

These methods rely on the fact that many metal ions will form coloured complexes. This is particularly true with transition metals, but other metal ions can be quantified using complex formation.

EDTA

EDTA forms stable complexes, known as chelates, with metal ions such as Ca^{2+}, Mg^{2+}, Sr^{2+}, Ba^{2+}, Mn^{2+}, Cu^{2+}, Zn^{2+}. Although these complexes may be coloured, the formation of the coloured complex is not usually useful to indicate the end point of the titration. Instead, an indicator such as erichrome black T is added. This forms a wine red complex with metal ions, but EDTA has higher affinity for the ions and so displaces the metal ions. Once all the metal ions have been displaced from the indicator by EDTA (the end point), the wine red colour disappears.

This method can be used to determine the hardness of water samples. By changing the conditions of the titration, different metal ions can be measured.

GROUP ACTIVITY

Practical techniques (1–2 hours)

You should be familiar with a range of practical techniques that are widely used in analytical laboratories. Table 3.10 shows some analyses that you should try.

Table 3.10 Widely used analyses

Analyte	Source	Method
Hydrogen peroxide	Hairdressing or mouthwash	Titration with potassium permanganate (similar method to dichromate, but permanganate solutions need to be standardised before use).
Hypochlorite	Bleach	Reduction of hypochlorite with excess iodide followed by titration of the iodine produced using thiosulfate.
Iodine number	Fats and oils	See above.
Peroxide number	Edible oils	Peroxides are intermediates in auto-oxidation of oils (making them rancid or spoiled). The peroxide number is determined by iodometric titration.
Sulfur dioxide	Wine	Sulfur dioxide oxidises iodide to iodine, so can be measured by iodometric titration.
Vitamin C	Fruits, juices	Ascorbic acid (vitamin C) will reduce iodide to iodine, so it can also be measured by iodometric titration.

6.4 Select and use chemical, i.e. analytical techniques with improved specificity

Section 5.1 in Unit 2 covered basic methods for identification of various cations. However, they are not always sufficiently specific and cannot be used quantitatively.

Tests for cations and ions

There are a number of reagents that give specific colours with certain ions that are more specific.

Thiocyanate for iron(III)

The thiocyanate ion (SCN) can be used as a very sensitive test for iron(III) (Fe^{3+}) ions. Addition of sodium, potassium or ammonium thiocyanate solution to a solution containing Fe^{3+} ions gives an intense blood red solution containing a thiocyano complex of the Fe^{3+}.

Adaptation as quantitative techniques

If a qualitative test produces a coloured product, then it can be turned into a quantitative test by measuring the intensity of the colour produced.

Iron(III) by colorimetry (thiocyanate)

The intense blood red thiocyano complex produced in the reaction between SCN^- ions and Fe^{3+} ions can be used for the quantitative analysis of low concentrations of Fe^{3+} ions in solution. There is a method for this on the Royal Society of Chemistry website: www.rsc.org/learn-chemistry/resource/res00000906/challenging-plants-analytical-methods

This also gives you practice at preparing a standard (calibration) curve with a colorimeter and using this to estimate the concentration of Fe^{3+} ions in an unknown solution.

Iron(II) by spectrophotometry (1, 10-phenanthroline)

When Fe^{2+} reacts with 1,10-phenanthroline it forms a deep red complex that can also be used with a colorimeter to estimate the concentration of Fe^{2+} ions in solution.

You can download a method from the Royal Society of Chemistry website: www.rsc.org/learn-chemistry/resource/res00000906/challenging-plants-analytical-methods#!cmpid=CMP00001170

6.5 Use of a combination of techniques for bacterial identification

Section 1.3 in Unit 18 covers this topic and you will need to review aseptic techniques covered in section 6 of Unit 2 before undertaking any practical work.

Colony morphology

When bacteria are grown on an agar plate, different bacteria will produce different-looking colonies. The colony morphology – appearance of the colonies – can help you identify the bacteria. There is a specific terminology used to describe colony morphology:

- Form – The basic shape of the colony.
- Size – The diameter of the colony.
- Elevation – This describes the side view of a colony (turn the Petri dish on end).
- Margin/border – The edge of a colony. What is the magnified shape of the edge of the colony?
- Surface – The appearance of the surface of the colony.
- Opacity – Transparent (clear), opaque, translucent (like looking through frosted glass), etc.
- Colour.

You can look at the different types of colony morphologies on the Microbiology Society website: www.microbiologyonline.org.uk/teachers/observing-microbes

Colony morphology is only the first step in identification.

Staining techniques

Stains are normally used after transferring samples of bacterial colonies from the plate to a microscope slide.

There are various stains used in identification of bacterial cultures. Most stains are positively charged and so bind to the negatively charged molecules (polysaccharides, proteins and nucleic acids) found in bacteria. They can be used alone (a simple stain) or in combination (differential stain). The Gram stain is an example of a differential stain and is one of the fundamental stains – bacteria are classified as either Gram positive or Gram negative.

Negative stains don't actually stain bacteria; instead the bacteria are made visible against a dark background. This can help observation of the size, shape and features of the bacteria under a microscope.

Other stains you might come across include:

- Acid fast stains. These can be used in identifying the mycobacteria that cause tuberculosis (TB) and leprosy.
- Malachite green. This is used to identify endospores, resistant structures produced by bacteria when environmental conditions are unfavourable.
- Flagella stains. These demonstrate the presence and arrangement of flagellae, needed to identify species of motile bacteria.

Gram stain

The Gram stain differentiates bacteria by detecting the peptidoglycan present in the cell wall of Gram positive bacteria. Crystal violet is used to stain the peptidoglycan and then iodine is added to fix the stain permanently to the peptidoglycan molecules. Stained bacteria appear dark blue or violet. Gram negative bacteria only have a very thin peptidoglycan layer – they have an outer membrane of lipopolysaccharides – and so the stain can be washed out of the cells. A counterstain of fuchsin or safranin is then used to stain all bacteria red/pink. As Gram positive bacteria have already been stained dark blue/violet, only the Gram negative bacteria appear red/pink.

Growth and behaviour on differential, selective and enriched media

Differential media use specific nutrients or indicators to help identify the cultures that grow. Some examples include:

- Blood agar. This contains bovine heart blood and becomes transparent in the presence of haemolytic strains of *Streptococcus*.
- MacConkey agar. Colonies able to ferment lactose will produce acid; a pH indicator in the agar stains the colonies pink. Those unable to ferment lactose use peptone (proteins) instead and produce ammonia, which increases the pH so forming white or colourless colonies.

Selective media allow growth of only specific bacteria. Media containing antibiotics such as ampicillin or tetracycline only permit growth of antibiotic-resistant bacteria. This technique is used widely in cloning of bacteria containing recombinant plasmids.

Enriched media can be used to culture organisms that are unable to grow on medium containing just an energy source (e.g. glucose) and mineral salts. They can be useful for cultivation of mutants unable to produce key metabolic intermediates.

KNOW IT

1. Describe how you would prepare a permanent slide of a tissue section that could be archived for future reference.
2. Which technique would be more suitable for large-scale purification of a mixture of compounds, TLC or column chromatography?
3. Describe how you would determine the concentration of chloride ions in a sample of seawater.
4. Describe how you would determine the concentration of Fe^{3+} ions in a solution.
5. State what stain you would use to identify the presence of TB-causing mycobacteria in a culture.

LO7 Be able to record, report on and review scientific analyses

GETTING STARTED

Just for the record (5 minutes)

Think about why we record the work we do. It could be so that we can repeat an experiment if necessary, possibly changing something or improving things. We need to report our work so that others can learn from it, repeat it and maybe also improve on it. We also need to be able to understand and evaluate the work of others so that we can learn from it, repeat it and possibly improve on it.

Think about the implications of all that – for how we record our work and report it, and how we learn from the work of others. What does that say about our own records and how we should read and understand the work of others?

7.1 Methods of recording data

There are a few rules we must always follow about keeping records of our work:

- We must be able to understand the record.
- It should contain enough detail for someone else to understand it and be able to repeat the work.
- It should be a faithful, honest and accurate record of what we did.

Notebooks, logbooks

Always write up your work as you go along; don't make notes on scraps of paper and write it all up later – even if you don't lose the scraps of paper, you probably won't remember all the details.

It is always good practice to date all your notes. In some laboratories there may even be a legal requirement to do so. In any sort of laboratory where the work may eventually lead to a patent application, it will be necessary to have a fully documented and dated record of the work done in order to support the patent application.

Tables

In section 2 we looked at how to present data in graphical format, so you should review that section before proceeding.

Before looking at the use of graphs it is worth considering how we should present data in a table.

- Put the independent variable in the first column; the dependent variable should be in columns to the right.
- Put any processed data such as means, rates, standard deviations in columns to the far right.
- Don't include calculations in the table, only calculated values (the results of your calculations).
- Head each column with the physical quantity and correct units, and separate the units with brackets or a slash (/).
- Don't include units in the body of the table, only in the column headings.
- Use consistent numbers of decimal places or significant figures throughout, even if it means writing 20.0, for example.
- Calculated values, e.g. mean or other processed data, should be given to the same number of decimal places as the raw data, or one greater.

This is illustrated in Table 3.11, recording the results of a series of experiments to investigate the effect of changing the concentration of nitrogen monoxide (nitric oxide, NO) on the rate of reaction with hydrogen.

Table 3.11 Effect of NO concentration on rate of reaction with hydrogen

[NO(g)] / 10^{-4} mol dm^{-3}	Initial rate / 10^{-4} mol dm^{-3} s^{-1}			
	Expt 1	Expt 2	Expt 3	Mean
1.0	0.25	0.24	0.21	0.233
2.0	0.64	0.65	0.63	0.640
3.0	1.51	1.49	1.52	1.507
4.0	2.68	2.66	2.64	2.660
5.0	4.20	4.19	4.21	4.200
6.0	6.00	6.17	6.14	6.103
7.0	8.22	8.21	8.24	8.223

Graphs

It is also good practice to make a graph of your data at the time; sometimes you only notice flaws in the data in this way and know that you might need to repeat some or all of your experiments or observations.

See section 2 for details of the different types of graph and when to use them.

Photographs and sketches, video and audio

They say a picture is worth a thousand words. A photograph or sketch can greatly enhance the record of your work making it easier to follow. With the widespread availability of smartphone cameras, there is no reason why you should not be able to quickly take a photo of an experiment – unless, of course, you are working in a laboratory governed by statute or regulation.

Sometimes you can use the original data source in your notebook – a TLC plate for example, or a chart recorder trace. However, it is often better to keep a photograph of a chromatogram because colours or spots in the original can fade with time.

3D representations, e.g. of crime scenes

A crime scene is a good example of the need to record data and information using some or all of the methods discussed here. Not only does the record have to be faithful and accurate, but it must be possible to present the information in a clear and unambiguous way. Recording 3D information of a crime scene has been a part of forensic practice for many years. More recently, computers have been used to greatly enhance the information that can be stored and replayed. See www.forensicmag.com/article/2007/01/crime-scene-3d-viewpoints-illustrating-what-was-seen-scene for an example of this.

Modelling, e.g. GIS geographical information system for terrain, fossil organisms

A geographical information system (GIS) is designed to capture, store, manipulate and display data related to geographical location. Although GIS is now computerised, an early example of the technique was in 1854. John Snow identified the source of a cholera outbreak in London by plotting where cholera victims lived on a map and linking those locations to the nearby Broad Street pump that was the source of the bacteria.

As well as applications in epidemiology, GIS can be invaluable in any sort of biological fieldwork. You can probably think of many other situations where presenting data on a map can provide unique insights.

The Ordnance Survey has information about GIS on its website at www.ordnancesurvey.co.uk/support/understanding-gis/

7.2 Reporting data, findings and other scientific information

It is important to record your work in a way that is faithful, honest and accurate. You need to report it in the same way. However, when you report your work you must also consider the target audience and present your work in an appropriate way. Also, you need to analyse, evaluate, summarise and discuss your conclusions, supported by appropriate data. You need to put your work into context; show how it links to other work that has gone before and also make suggestions for how it could be developed in the future.

Reporting to a chosen audience (peers, public, scientific community)

Think about the different types of audience you might need to report your work to: colleagues working in the same laboratory or on the same project; other scientists working in the same field, or related fields; members of the public. Consider how you would structure your report and the level at which you pitch it. How much background information do you need to give? How much prior knowledge can you assume?

Also consider issues such as confidentiality or conflict of interest. In some laboratories it might be necessary to keep some aspects of your work secret or confidential – you might be constrained by legal issues or commercial confidentiality.

While we like to think of scientific research being a noble endeavour driven by a thirst for knowledge and seeking after the truth, back in the real world there can be competition and rivalry between individuals, groups or organisations. Also, while it can be important to report negative results – so that others can learn from your mistakes or avoid blind alleys, for example – too much of this could be career suicide!

Reporting by the scientific media (public information scientists; science journalists)

The tabloid press have rightly earned condemnation for distortion and misrepresentation. We would hope that science journalists operate to higher standards. They are certainly more likely to understand at least some aspects of your work – although if you are a theoretical physicist working on quantum entanglement, your work might not be immediately understood by a science journalist with a background in freshwater biology!

GROUP ACTIVITY
Science journalism (1 hour)

Work individually, in pairs or small groups. Within the group as a whole you should each try to cover a different publication or medium – newspaper and magazine articles (online editions are probably easier to work with), scientific TV programmes (catch-up TV services or websites will help) or one of the various online formats. Look at each critically and assess how well it fulfils its purpose. Also think about ways in which the particular format differs in style, content and presentation from a report that you would write on your own work – and also look for points of similarity. Prepare a short presentation for the group, summarising your findings.

Public information scientists have an important role to play in bringing scientific work to the wider public and helping in the public understanding of science. Many research-based organisations, universities, medical research and other science-based charities have their own information scientists whose job it is to present the organisations' work to the press and public in a positive way that can be easily understood.

GROUP ACTIVITY
Public information scientists (20 minutes)

Working as a group, look at several different websites of universities, government agencies, research organisations and charities, learned societies as well as science-based companies (e.g. pharmaceuticals). Think about the content that will have been created with the help of a public information scientist. What are they trying to achieve: educate, influence, explain, justify, etc?

Here are some suggestions for websites to look at. Your tutor can probably provide others.

The National Institute for Health and Care Excellence
www.nice.org.uk

Royal Society of Chemistry
www.rsc.org

Cancer Research UK
www.cancerresearchuk.org

Glaxo SmithKline (GSK)
www.gsk.com, particularly the 'Behind the Science' section: www.gsk.com/en-gb/behind-the-science/

7.3 Evaluating the reporting of data, findings and other scientific information

Just as you may have to present your own work, you will certainly have to consult the work of others. You might need to find out how a particular technique is carried out or repeat an experiment or series of experiments. You may simply need to learn about an unfamiliar area of science for a new job or new project. To get the full benefit and avoid mistakes, you need to evaluate the source of information that you consult and develop a critical eye for the reporting of scientific information.

Here we will deal briefly with some of the factors you should consider.

Status and affiliation of author(s)

A review article written by a well-known professor is likely to be authoritative. An article in a technical magazine written by the sales manager of a company selling scientific instruments may be little more than a thinly disguised sales pitch. But don't be too hasty – an article by a scientist employed by a company may not be completely impartial, but it could still contain a lot of valuable information.

Think about the questions you should ask yourself whenever you read any scientific publication.

Publication or information source in which data reported

Scientific articles and research reports in academic journals such as Nature, Science, the Journal of Biological Chemistry (JBC) or Proceedings of the National Academy of Sciences of USA (PNAS) go through a process known as peer review. The manuscript from the scientists who did the work is sent to other experts in the field for review and comment. This often leads to faults being found in the experimental work or in the arguments. The scientists publishing the work may need to repeat some experiments or carry out others. They might need to revise their manuscript, including their conclusions. The result of this process is that, by the time you read the article or paper, it will have been checked several times. You should be able to trust it.

Other publications will have been through a less rigorous process. This doesn't mean they can't be trusted, just that you need to think a little more about how independent, unbiased or reputable the publication might be.

You should also consider whether you are reading a report based on primary data (the work of the publishing scientists) or secondary data (work done by others). Review articles generally draw on secondary data. Publications of original research will feature primary data, but also some secondary data as background or supporting information.

Finally, what about Wikipedia?

Nature of data and scientific findings reported

That is, validity of study and data, accuracy, quality of science explanations.

In section 4 we covered the approaches that you should take when analysing and evaluating the quality of your own data and that of others. Section 5 covered the ability to draw justified conclusions from data. You need to apply the same principles to any scientific report.

Quality of reporting

When judging a piece of scientific writing, you should consider:

- Clarity: clearness to enable the reader to understand the information accurately
- Conciseness: delivering information in a clear yet brief format, to not confuse the reader
- How appropriate it is to the intended audience: selecting the appropriate information, and ensuring it accommodates the reader.

KNOW IT

1. List four rules that you should follow when presenting data in a table.
2. Describe the advantages of using GIS to analyse and interpret data in either crime statistics, ecology or epidemiology (the study of the patterns, causes and effects of disease).
3. Describe how you would report findings of an experiment differently to a journalist, in an academic report or journal, or to an audience of secondary school students.

Assessment practice questions

You will be tested on each of the LOs in the exam. Below are practice questions for you to try, followed by some top tips on how to achieve the best results.

1. This question is about units, prefixes and conversions.
   a. A typical wavelength of light is about 500 nm. Express this value in the following:
      i. m (1 mark)
      ii. µm (1 mark)
      iii. mm (1 mark)
   b. The momentum of an object, p, is calculated using the equation $p = mv$ where m is the mass of the object and v is its velocity.
      i. Using this equation, state the units of momentum in SI base units. (2 marks)
      ii. Show that the units of momentum are equivalent to Ns (Newton seconds). (2 marks)

2. This question is about areas and volumes. Two designs for a hot water tank are being considered. The first is a cylinder 0.66 m high and 1 m in diameter, and the second is a sphere with a diameter of 1 m.
   a. Calculate the volumes and surface areas of the designs, expressing your answer in m^3 and m^2 and to two significant figures. (4 marks)
   b. Assuming heat loss from such a tank varies linearly with surface area, which design is the most efficient? Explain your reasoning. (2 marks)

3. This question is about statistics and standard deviations. The following data was obtained in a measurement of grass leaf lengths under controlled conditions.

   Table 3.12

Leaf	Length (mm)
1	9.00
2	4.80
3	19.00
4	5.70
5	3.80
6	6.50
7	14.00
8	10.00
9	6.10
10	9.10

   Calculate:
   a. the mean and
   b. the standard deviation.
      (2 marks each)
   c. State any assumptions you've made in your calculations. (2 marks)

4. A learner carried out an experiment to investigate uptake of water by potato slices placed in different salt solutions. The increase or decrease in mass of the potato slices was measured after twenty minutes. This is the learner's table.

   Table 3.13

Increase in weight of potato	Concentration of sodium chloride solution
1.5 g	Water
1 g	5%
500 mg	10%
-0.1 g	15%
-0.5 g	20%

   Identify four conventions that the learner did not follow when they prepared this table. (3 marks)

5. Unsaturated fats are thought to be more healthy than saturated fats. Polyunsaturated fats (with more than one C=C double bond) are considered most beneficial. Jay works in a food testing laboratory and has been asked to determine the iodine number of a 0.2 g sample of vegetable oil. He uses the following solutions:
   - 0.2 mol dm^3 Hanus reagent (iodine monochloride, ICl)
   - 0.1 mol dm^3 sodium thiosulfate

   Outline the scientific principles involved in determining the iodine number. (4 marks)

6. A technician is analysing the iron content of vitamin tablets using 1,10-phenanthroline solution. One tablet is crushed and dissolved in concentrated sulfuric acid.
   a. Explain why 1,10-phenanthroline can be used quantitatively to determine the concentration of iron(II) in the tablets. (3 marks)
   b. The technician prepares a series of standards of iron(II) ammonium sulfate, $Fe(NH_4)_2(SO_4)_2.6H_2O$.
      i. Calculate the mass of iron(II) ammonium sulfate required to produce 1 dm^3 of a stock solution containing 20 mg dm^{-3} of iron.

(Molar mass of ammonium iron(II) sulfate = 392.14 g mol^{-1}; iron = 55.85 g mol^{-1}.)

ii The technician measured the absorbance at 565 nm of a series of iron(II) standard solutions after adding 1,10-phenanthroline. The readings are shown in the table. Plot a calibration graph. (4 marks)

Table 3.14

Concentration of iron (mg dm^{-3})	Absorbance
0.0	–
2.0	0.010
4.0	0.020
6.0	0.031
8.0	0.041
10.0	0.051
12.0	0.061
14.0	0.071

iii What is the concentration in mg dm^{-3} of iron in a sample that gave an absorbance of 0.045? (1 mark)

7 IMM-101 is a new anti-cancer treatment for advanced pancreatic cancer. The results of a clinical trial were published in the British Journal of Cancer and news reports appeared in The Guardian newspaper (see below). Discuss the quality of the two reports, based on:
   a level of **scientific detail** (4 marks)
   b **style** of the reports. (4 marks)

Randomised, open-label, phase II study of gemcitabine with and without IMM-101 for advanced pancreatic cancer

Source: British Journal of Cancer

Background: Immune Modulation and Gemcitabine Evaluation-1, a randomised, open-label, phase II, first-line, proof of concept study (NCT01303172), explored safety and tolerability of IMM-101 (heat-killed Mycobacterium obuense; NCTC 13365) with gemcitabine (GEM) in advanced pancreatic ductal adenocarcinoma.

Methods: Patients were randomised (2:1) to IMM-101 (10 mg ml^{-1} intradermally) + GEM (1000 mg m^{-2} intravenously; n = 75), or GEM alone (n = 35). Safety was assessed on frequency and incidence of adverse events (AEs). Overall survival (OS), progression-free survival (PFS) and overall response rate (ORR) were collected.

Results: IMM-101 was well tolerated with a similar rate of AE and serious adverse event reporting in both groups after allowance for exposure. Median OS in the intent-to-treat population was 6.7 months for IMM-101+GEM v 5.6 months for GEM; while not significant, the hazard ratio (HR) numerically favoured IMM-101 + GEM (HR, 0.68 (95% CI, 0.44 – 1.04, P = 0.074). In a pre-defined metastatic subgroup (84%), OS was significantly improved from 4.4 to 7.0 months in favour of IMM-101+GEM (HR, 0.54, 95% CI 0.33–0.87, P = 0.01).

Conclusions: IMM-101 with GEM was as safe and well tolerated as GEM alone, and there was a suggestion of a beneficial effect on survival in patients with metastatic disease. This warrants further evaluation in an adequately powered confirmatory study.

New drug 'wakes up' immune system to fight one of deadliest cancers

Source: The Guardian.

A new drug that "wakes up" the immune system to attack cancer has extended the lives of people with metastatic pancreatic cancer and has no side-effects, raising hopes for a new and powerful tool against the most intractable form of the disease.

The drug, IMM-101, is considered groundbreaking because pancreatic cancer that has spread to other parts of the body usually kills within a few months.

Unlike other immunotherapy treatments, IMM-101 is not thought to have any side-effects, which have been hugely debilitating for many patients and even led to long-term disability in some cases.

Angus Dalgleish, professor of oncology at St George's, University of London, who led the research is excited by the potential of the immunotherapy drug, although the trial is relatively small, involving 110 people. Only 18% of patients with advanced pancreatic cancer are alive after one year and 4% after five years, so new treatments for the disease are badly needed.

Adding IMM-101 to the standard chemotherapy drug patients receive, gemcitabine, did not benefit those patients whose pancreatic cancer had spread locally, but it improved the lifespan of those whose cancer was metastatic – it had progressed into other parts of the body – and would be expected to die quickly.

The results are published in the British Journal of Cancer. Most of the patients – 85% – had metastatic disease. Those who were just given the standard chemotherapy drug survived for a median of 4.4 months, but those who had the IMM-101 immunotherapy drug as well survived for seven months. But some lived for more than a year and one died after nearly three years.

> **TOP TIPS**
> - Take care when handling and converting units. It's often safest to convert all units into base SI forms.
> - When doing calculations, do one step at a time to generate intermediate results – don't be tempted to do all stages of a calculation in one go.
> - Take care when thinking about and handling statistics – be aware of any assumptions you're making about the data.
> - When presenting results in a table or graph, ask yourself 'Does this help me understand the data?' If you don't understand it, nobody else will!
> - The difference between continuous and discontinuous data is important, particularly when drawing a graph. If you get it wrong, you distort your data and risk drawing invalid conclusions.
> - As well as being able to draw good graphs, you need to be able to interpret other people's graphs. A graph is a powerful way of displaying and interpreting data.
> - Whether you are interpreting your own experiments or trying to learn from others, you need to develop an understanding of the importance of good experimental design.
> - As well as the basic anion / cation tests you need to understand how more advanced methods can be used to quantify as well as identify.
> - Just because you are working in science doesn't mean you don't need English language skills. You may be able to do great work in the lab, but unless you can produce accurate reports suited to your target audience, your work will go unnoticed.

Read about it

The website Types of Graphs covers graphs used in science, but also includes information about other sorts of graphs that might be useful in your work, such as pictographs, organisational charts, flowcharts and others:

www.typesofgraphs.com/

The iSpot website has useful resources for learning about the use of keys:

www.ispotnature.org/keys

The Science Made Simple website has several useful pages. Don't be put off by the fact that it is aimed at school-age children – there is a lot of information that university researchers would do well to pay attention to!

www.sciencemadesimple.com/

The University of Hull has produced a very detailed guide to scientific report writing:

www2.hull.ac.uk/lli/pdf/Scientific%20Reports.pdf

For an insight into science journalism:

www.theguardian.com/science/series/secrets-science-writing

Unit 04
Human physiology

ABOUT THIS UNIT

In this unit, you will learn about how the organs and systems of the human body interact to carry out the basic functions of life – acquiring materials, distributing them around the body, releasing and using energy. You will understand how the systems are controlled and regulated by homeostasis, and how the body is protected by the immune system.

You will also learn about the important health problems that can unfortunately occur in each system, how these problems can be assessed and treated, and their impacts on individuals' lives.

At the end of the unit, you will have knowledge and understanding of the structures of the body systems, the functions they carry out and how they work together to maintain life. You will also gain practical skills in measuring using cardiovascular and respiratory monitoring equipment on human volunteers.

LEARNING OUTCOMES

The topics, activities and suggested reading in this unit will help you to:

1. Understand the structure and functions of the digestive system
2. Understand the role and function of the musculoskeletal systems
3. Be able to assess how the cardiovascular system functions in the body
4. Be able to assess how the respiratory system functions in the body
5. Understand how homeostasis maintains balance within the body
6. Understand the role and function of the immune system

How will I be assessed?

You will be assessed through a series of assignments and tasks set and marked by your tutor.

How will I be graded?

You will be graded using the following criteria:

Learning Outcome	Pass	Merit	Distinction
	The assessment criteria are the Pass requirements for this unit.	To achieve a Merit the evidence must show that, in addition to the Pass criteria, the candidate is able to:	To achieve a Distinction the evidence must show that, in addition to the Pass and Merit criteria, the candidate is able to:
1 Understand the structure and functions of the digestive system	**P1** Describe how food is processed by the digestive system	**M1** Explain the digestive related symptoms of someone with a common digestive disorder	
2 Understand the role and function of the musculoskeletal systems	**P2** Describe the importance of the musculoskeletal system in maintaining structure and movement of the body		**D1** Explain the importance of bone marrow to the skeletal and immune system
3 Be able to assess how the cardiovascular system functions in the body	**P3** Take a range of measurements related to the cardiovascular system, relating the results to its functions	**M2** Explain how common cardiovascular disorders can affect the functions in the body	
4 Be able to assess how the respiratory system functions in the body	**P4** Take a range of measurements related to the respiratory system, relating the results to its functions	**M3** Investigate effects on the respiratory system in different populations	
5 Understand how homeostasis maintains balance within the body	**P5** Outline the importance of regulating body fluids in the body		
6 Understand the role and function of the immune system	**P6** Describe the immune system, outlining its function		**D2** Explain how the immune system functions when a vaccine is administered thereby preventing infection by certain diseases

LO1 Understand the structure and functions of the digestive system *P1 M1*

GETTING STARTED

(5 minutes)

With a partner, agree a list of five reasons why we need to eat.

1.1 The components and functions of the digestive system

Our food contains some very large molecules, which are too large to be absorbed into the bloodstream. In this form, food would be absolutely no use to our body. The digestive system has two functions: **digestion** and **absorption**.

> 🔑 **KEY TERMS**
>
> **Digestion** – Large, insoluble molecules are broken down into much smaller, soluble molecules.
>
> **Absorption** – Soluble molecules are moved into the blood from the lumen of the intestine into the bloodstream.

The digestive system is basically a long tube called the gastrointestinal tract, which runs from the mouth to the anus. The walls of the tract have muscles for pushing the food along and glands which make **enzymes** to digest the food and mucus to lubricate the passage of food. Enzymes are added at several places to digest the food (see section 1.2). The tract is split into a series of organs, which carry out digestion in stages.

Upper gastrointestinal tract

The buccal cavity (mouth) chews the food, adding saliva which makes it easier to swallow.

The oesophagus is a thin tube which pushes the swallowed food down through the chest and into the stomach.

The stomach is a bag-shaped organ which begins digesting proteins.

The stomach contents are released into the duodenum, where bile is added from the liver and more enzymes from the pancreas (see section 1.2).

Lower gastrointestinal tract

The small intestine (ileum) is a thin tube, about seven metres long. The food molecules are gradually digested as they move through it, and absorbed through its walls into the blood vessels on the other side. The bloodstream carries the food molecules away.

Although wider than the small intestine, the large intestine (colon) is only 1.5 metres long. The matter emerging from the small intestine has had all of the digestible food removed by now, and is a mixture of molecules we cannot digest (mostly plant material, called fibre), water and mucus. The only part we would still find useful is the water, which the large intestine absorbs into the bloodstream through its walls, making the waste matter more and more solid. The waste, called faeces, is lost through the anus at the end of the gastrointestinal tract.

> ### 🔑 KEY TERM
>
> **Enzyme** – A protein which acts as a catalyst. Each one makes a particular chemical reaction happen at the temperatures found in the body, and at a much faster rate. Enzymes are not used up, so an enzyme molecule catalyses its reaction many millions of times.

Figure 4.1 shows the digestive system.

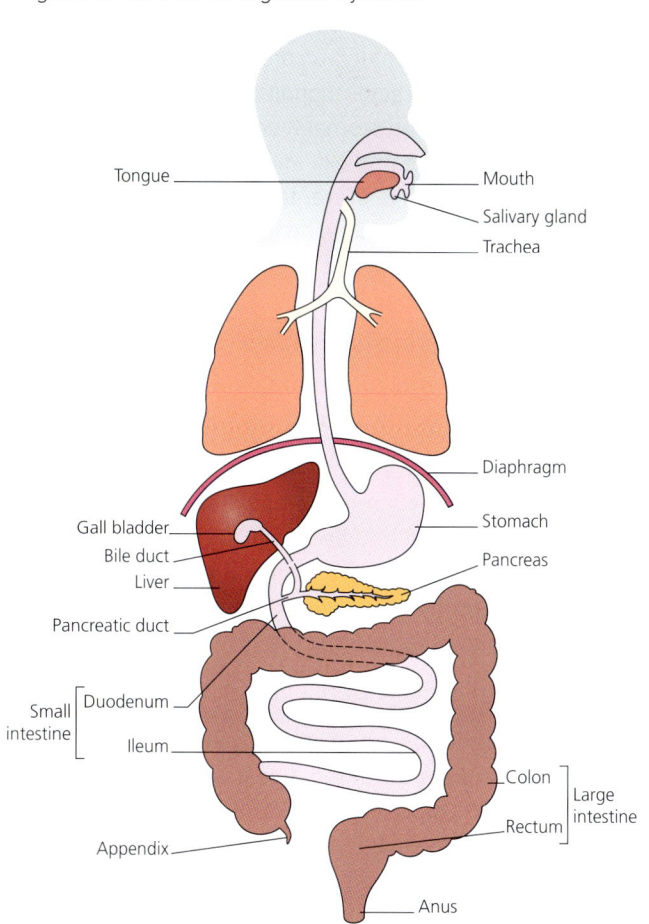

▲ **Figure 4.1** The digestive system

1.2 The process of mechanical and chemical digestion

Chemical digestion

The digestive system makes digestive enzymes in its glands and adds them in with the food. Digestive enzymes catalyse reactions which break down large molecules into smaller, simpler molecules.

Salivary glands in the buccal cavity add amylase, which begins digesting **starch** down to **maltose** molecules. As food reaches the stomach, it is the protein's turn to be broken down. Glands in the stomach wall release the

> ### 🔑 KEY TERMS
>
> **Starch** – An insoluble polysaccharide carbohydrate. Its monomers are glucose molecules. Found in foods made from plants, such as bread and rice.
>
> **Maltose** – Disaccharide composed of two glucose molecules.

enzyme pepsin, which breaks **proteins** down to a mixture of polypeptides, peptides and some amino acids. The glands also add hydrochloric acid because the stomach enzymes work best in acidic conditions. This is quite a harmful mixture, so the stomach seals itself with rings of muscle at its entrance and exit as it churns its contents together. The stomach is prevented from digesting itself by a layer of protective mucus on its inner surface. Also, the pepsin is initially secreted in an inactive form, pepsinogen, which is only converted to pepsin when it comes into contact with the acidic stomach contents.

Once the stomach contents are released into the duodenum, they meet two fluids coming down a side tube called the pancreatic duct.

Bile, from the liver, is a mixture of water, bile salts, bile pigments, cholesterol and salts. It has the double effect of neutralising the stomach acids and **emulsifying** the **lipids**, helping them mix with the watery environment inside the gastrointestinal tract.

Pancreatic fluid is made by the pancreas. It contains:

- trypsin, chymotrypsin and carboxypeptidase, which continue breaking down the digestive products of proteins into dipeptides and amino acids. (The duodenum releases dipeptidase enzyme, which breaks the remaining dipeptides into amino acids.)
- pancreatic amylase, which continues the breakdown of starch to maltose (broken down in turn to glucose by maltase enzyme from the duodenum)
- pancreatic lipase, which begins the digestion of the lipids into glycerol and fatty acids.

🔑 KEY TERMS

Protein – Long chains of amino acids chemically bonded together. Proteins have many different shapes, depending on the sequence of their amino acids.

Emulsification – When the surface tension of a mass of lipid in watery surroundings is broken down, splitting it into smaller droplets which mix more easily with the water.

Lipid – Fats and oils, mostly composed of three fatty acid molecules bonded with one glycerol molecule.

Glycogen – An insoluble polysaccharide carbohydrate, which is chemically similar to starch, made from monomers of glucose. Animals and fungi use glycogen as an energy store.

Assimilation – A molecule becomes an integral part of the body's processes.

Smooth muscle – A type of muscle over which we have no conscious control.

The length of the small intestine gives the enzymes plenty of time to complete digestion.

All along the small intestine, the products of digestion – soluble glucose and amino acids – are being absorbed. They move across the thin barrier of cells which line the inner surface, enter the capillaries on the other side of the barrier and are carried away by the bloodstream. Their first destination is the liver, which processes the products of digestion. Excess glucose is converted to **glycogen** and stored (see section 5.1). Excess amino acids are converted to carbohydrates or lipids, releasing urea (see section 5.1). The products of digestion have been **assimilated** – they are now used by the body cells in their own chemical reactions.

Mechanical digestion

Simply pouring enzymes onto the outside of a piece of food might not be enough to ensure its complete digestion. You need to make sure the food is broken up by movement (another word for this is 'mechanically') into small pieces and the enzymes mixed in so that they touch the outside of each little particle.

Chewing food (mastication) achieves this, mixing in the amylase-containing saliva. Further along the tract, the **smooth muscles** in the stomach wall continually move, mixing the food with enzymes and acid for several hours at a time.

Peristalsis

All along the digestive tract, rhythmic contractions and relaxations of the smooth muscles in the walls move the food continually forwards. Circular muscle at the back of a mass of food contracts, pushing it forwards. Longitudinal muscle around and ahead of the food contracts, shortening the passage in front. These movements, called peristalsis, are organised by the autonomic nervous system (see section 5.1).

GROUP ACTIVITY

Draw it (20 minutes)

You will each be given a short sentence describing a process happening in the digestive system. For example: 'Pepsin is secreted in an inactive form, pepsinogen, which is converted to pepsin when it comes into contact with the acidic stomach contents'.

Using no letters or symbols, you have sixty seconds to make a sketch on the board depicting your sentence. The other members of your group have the sixty seconds to recreate the meaning of the sentence from the sketch, identifying as many technical terms as possible.

1.3 The causes and effects of common digestive disorders

Infection

The stomach and intestines can become infected with harmful bacteria or viruses. The digestive tract responds to infection by making the rate of peristalsis faster. Faeces pass through the large intestine too quickly for water to be absorbed properly, so they are much more watery than usual (diarrhoea).

Vomiting is another symptom of infection. The digestive tract puts peristalsis into reverse, expelling harmful substances through the mouth.

The major cause of bacterial gastroenteritis in developed countries is through eating foods contaminated with *Campylobacter jejuni*, or through ingesting particles of contaminated faeces – for example by eating a salad washed in untreated water. Other harmful species include *Escherichia coli* (*E. coli*), *Salmonella* and *Shigella*. Foods with greater risk of contamination include seafood, chicken, undercooked meat and unpasteurised milk. In Africa and Asia, infection with *Vibrio cholerae* from contaminated water causes the disease cholera.

Viral infections cause about 70% of gastroenteritis in children, mostly caused by rotavirus. Amongst adults, norovirus is the major culprit. Again, infection can be caused by eating contaminated foods, but also by person-to-person contact, so outbreaks can occur where people are in close proximity together for a time; for example, conferences, cruise ships and hotels.

Inflammation

Inflammation is a normal, temporary response to an infection, but it can be damaging when the swelling becomes long term. For instance, in ulcerative colitis the lining of the colon develops small ulcers – breaks in the lining which do not heal – when the immune system mistakenly attacks normal body cells. This causes diarrhoea and abdominal pain.

Irritable bowel syndrome (IBS)

The cause of IBS is still unclear. Problems with digestion lead to either peristalsis in the large intestine becoming faster, or slowing down, causing constipation. Also, the digestive tract becomes much more sensitive, leading to distressing sensations of pain or urgently needing the toilet.

> **KEY TERM**
>
> **Inflammation** – In inflammation, capillaries become more permeable, more fluid and white blood cells than usual escape, and the tissues become swollen.

> **KNOW IT**
>
> 1. List the parts of the digestive system in the order that food passes through them.
> 2. What are the digestive functions of (a) the pancreas (b) the liver?
> 3. Where is the digested food absorbed?
> 4. Which food substances are broken down to (a) glucose (b) amino acids?
> 5. What are the symptoms of gastroenteritis?

LO1 Assessment activities

Below are suggested assessment activities that have been directly linked to the Pass and Merit criteria in LO1 to help with assignment preparation and include top tips on how to achieve best results.

Activity 1 Create a flow chart to show the processes of digestion *P1*

You should present the processes of digestion as a single flow chart, showing how molecules of (a) starch (b) protein (c) lipid (d) the indigestible carbohydrate cellulose are processed by the digestive system. Include, as appropriate:

- location within the system
- chemical digestion, with names of enzymes and products
- physical digestion
- absorption
- assimilation
- egestion.

> **TOP TIP**
>
> ✓ The digestive system could be represented on the flow chart as a straight tube.

Activity 2 Create a guide to explain a common infection *M1*

Being infected with the *Salmonella* bacterium gives the sufferer symptoms of diarrhoea and vomiting. Sometimes the sufferers can be very young. Explain how the infection causes these symptoms, in the form of a guide for worried parents.

> **TOP TIP**
>
> ✓ This book uses a lot of technical terms. These are useful to us because single words can hold a lot of information. This only works when your audience shares the same understanding of the words with you, the author. In this activity, your audience might not recognise the ideas behind technical terms. You will have to explain what is happening without using technical terms, but using commonly used words in clear and informative sentences.

LO2 Understand the role and function of the musculoskeletal systems P2 D1

GETTING STARTED

(2 minutes)

Move your fingers. On the back of your hand and the inside of your wrist, you will see tendons moving. Can you locate the muscles which are actually moving your fingers?

2.1 The components and functions of the musculoskeletal system

The musculoskeletal system has two types of component: the bones of the skeleton and the muscles attached to them. The functions of the skeleton are as follows:

- To support the body, maintaining its shape during movement.
- To protect internal organs. The brain and spinal cord are encased by the skull and vertebral column. The heart and lungs are surrounded by the rib cage.
- To move. Bones in the limbs are arranged as **levers** which move around **joints**. When a lever is pulled close to the joint, the bone moves.

The function of the muscles is to move the body by providing those pulling forces on the bones.

Structure of bone and muscle

We have already seen (in Unit 1 section 3.3) how bone and muscle tissue are related to their functions. Here we will look at the structure of the whole organs. The shafts of the long bones of the limbs are hollow cylinders of compact bone tissue. As well as reducing their own weight, this structure gives them strength in compression, so that they can support the weight of the body.

Where the bone ends meet to make joints, they have a mass of spongy bone tissue with an arched internal structure, which adds to their strength. Bones have outgrowths called *processes*, which are the attachment points for the tendons of muscles.

The bone marrow is fatty tissue found in limb bones and flat bones, such as the ribs and skull. It contains stem cells, which can **differentiate** into cell types which are important to both the skeletal and immune systems:

- Osteoblasts, which lay down the matrix of collagen and minerals which give bones their structure. See Unit 1 section 3.3.
- Osteoclasts, which break down bone tissue in maintenance and repair. They also release calcium ions from the bones when needed to maintain the blood calcium level. See sections 2.2 and 5.1.
- Erythrocytes (red blood cells). See section 3.1.
- Platelets, which aid blood clotting.
- Lymphocytes – white blood cells which run the immune response. Bone marrow produces B cells directly and contributes lymphocytes to the thymus where they become T cells. See section 6.2.
- Macrophages, which engulf and digest invading organisms.

You will see much more about how lymphocytes and macrophages work together in section 6.1.

Striated muscles contract, making the whole muscle shorter and thicker. The muscles in the musculoskeletal system are attached to the process of a bone at both ends by tendons, tough fibres which do not stretch when the muscle pulls on them so they exert the pulling force of the muscles onto the bones.

Types of joint and limitations of movement

As we have seen, the place where bones meet are called joints. There are several types of joint, each allowing movement limited in different ways.

The plates of the skull have no movement at their joints.

The joints between the vertebrae only allow a little twisting and bending but these give the spine its overall flexibility.

Synovial joints have a fuller range of movements. Hinge joints, such as the knee or between the finger bones, allow movement in just one plane, forward and back like

> **KEY TERMS**
>
> **Lever** – A rigid bar with a pivot, used to transmit a force.
>
> **Joint** – The structure where bones meet.
>
> **Differentiate** – Cells become specialised for a particular job, changing in features and abilities.
>
> **Striated muscle** – Muscles which are under conscious control. Viewed under a microscope, they appear striated, meaning 'stripey'.

a book opening and closing. Ball-and-socket joints, for instance at the hip (See Figure 4.2) and shoulder, have a ball on one bone fitting into a depression on another. They allow movement in almost every direction.

Components of a synovial joint

Having a lot of freedom at a joint means the bone could be damaged where the ends move over each other. The synovial joints have features which prevent this (see Figure 4.2). The ends of the bones have caps of smooth cartilage, which minimises friction where they meet. Friction is reduced further by having a layer of lubricating synovial fluid inside the joint, held in by a loose synovial capsule surrounding the joint.

The capsule is held together by ligaments, which prevent too much movement in unwanted directions.

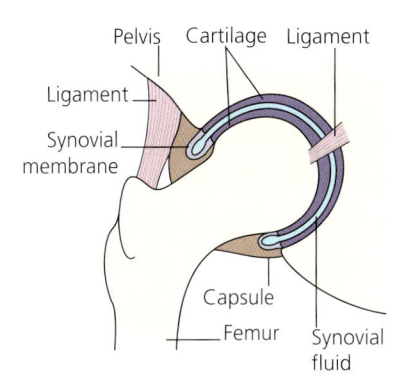

▲ Figure 4.2 The hip: a synovial joint

Muscle action around a joint

Muscles get shorter, but not by much. Attaching close to the joint multiplies the distance moved at the other end of the bone. In Figure 4.3, as the biceps muscle contracts by a small distance, the hand at the far end of the lever moves a larger distance.

Muscles cannot push themselves out to the longer, thinner shape they had before they contracted. For this, a skeletal muscle relies on a second muscle attached to the same side of the joint but pulling the joint in the opposite direction, and at the same time pulling the first muscle longer again.

In Figure 4.3, the biceps muscle is attached to the shoulder blade (scapula) at one end, and the forearm side of the elbow joint at the other. As the biceps contracts, it pulls the elbow joint closed, and the bone in the forearm side of the joint pulls the triceps muscle longer. When the triceps contracts, it opens the joint and pulls the biceps longer again. Skeletal muscles, then, are always arranged in opposing pairs.

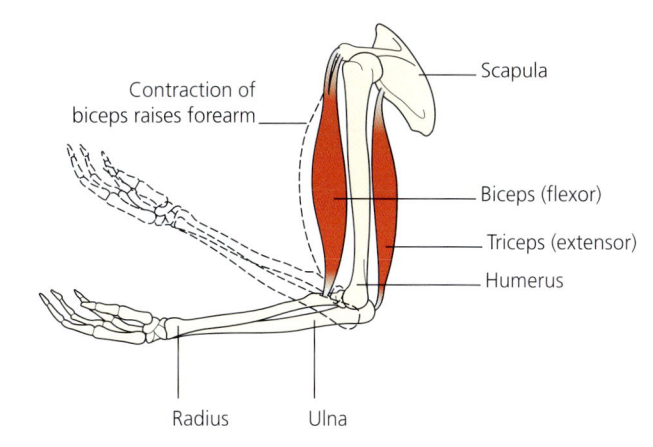

▲ Figure 4.3 Movement at the elbow

PAIRS ACTIVITY

Video diary (30 minutes)

Work in pairs to make a video explanation of how the muscles of the elbow work. No written words are allowed on screen, only spoken explanations. You could use paper cutouts for the muscles and bones, or use your arm to model what happens in bending and straightening the arm.

2.2 Common disorders of the musculoskeletal system

Arthritis

There are two forms of arthritis. In osteoarthritis, due to injury or age, the cartilage becomes thin and the bones themselves rub together. The joints swell and bone spurs grow on the joint surfaces, further increasing friction. Movement at the joint becomes limited and painful.

In rheumatoid arthritis, the body's own immune system attacks the joints, causing inflammation and damage to the synovial capsule, and later to the cartilage and bone tissue. The joint can change shape, making movement of the joint difficult or impossible.

Osteoporosis

With age, the osteoclasts remove more calcium salts than the osteoblasts add. Possibly the bones had a low density due to a calcium-poor diet when young. The bones become less dense, and fragile. Bones can break even under minor stress, particularly wrists, hips and the vertebrae of the spinal column. This is painful and debilitating, so the sufferer becomes less active.

105

> **KNOW IT**
> 1. What are the functions of the skeleton?
> 2. Give one difference between striated and smooth muscle.
> 3. Name two features in a joint which reduce friction.
> 4. Why are bones arranged in pairs at a joint?
> 5. What are the symptoms of arthritis?

LO2 Assessment activities

Below are suggested assessment activities that have been directly linked to the Pass and Distinction criteria in LO2 to help with assignment preparation and include top tips on how to achieve best results.

Activity 1 Prepare flashcards to illustrate the structure of the body *P2*

Prepare postcard-sized illustrated flashcards to describe how the skull, spine and ribs each help maintain the body's structure, and how the bones and muscles of the leg move the body.

Activity 2 Explain the role of stem cells in two different systems *D1*

The bone marrow contains stem cells which can differentiate into very different types of cell. Choose:

- one type of cell that remains in the musculoskeletal system
- one type of cell with a role in the immune system.

Explain the roles that these cells carry out, and why they are important to the functioning of their particular system and to the functioning of the whole body.

> **TOP TIP**
> ✓ You will need to access appropriate resources to fill out the details of the cell types.

LO3 Be able to assess how the cardiovascular system functions in the body *P3 M2*

GETTING STARTED
(5 minutes)

Collect your classmates' ideas on what substances the blood carries around the body.

3.1 The components and function of the cardiovascular system

The function of the cardiovascular system is to transport dissolved substances both to and away from the body's cells.

Blood

Blood is a remarkable multi-purpose fluid. Most of blood's volume is plasma, a yellowish mixture of proteins in water. Water is an excellent solvent, carrying away cells' waste products, such as carbon dioxide and urea, while bringing them glucose and amino acids from the small intestine, along with vitamins and minerals, and keeping them in touch with other parts of the body through chemical messengers called hormones.

The plasma proteins have various jobs: albumen keeps blood pH stable, clotting factors solidify the blood where an injury has happened, antibodies recognise invading microorganisms and start the immune response.

The cells floating in the plasma are mostly erythrocytes – red blood cells – sacs full of the protein haemoglobin. Each haemoglobin has four binding sites for carrying the oxygen that cells need for respiration, making blood able to carry more oxygen than plain water could. Haemoglobin loads up oxygen where it is plentiful, such as in the lungs, but only unloads it in the places it is needed.

The rest of the cells are the various types of white blood cell, which co-ordinate the body's immune response, attacking and engulfing invading microorganisms. (See section 6.2.)

The heart

The heart pumps blood around the system. Its muscular walls press against the blood, increasing its **pressure** and making it flow.

It actually consists of two pumps, one on each side. In both sides, blood enters a small chamber (the **atrium**) at the top while the heart is relaxed. This phase is called diastole. In systole, the heart muscles contract in two stages. The atrial walls contract, pushing blood into the lower chamber (the **ventricle**) to fill it. The ventricular wall contracts, sending blood out of the heart. The heart has valves at its exits, and between the atria and ventricles, to make sure blood only flows in the correct direction when it contracts.

The timings of the heart's contractions are controlled by the **sinoatrial node** in the right atrium wall. The node spontaneously sends out an electrical impulse, one after the next, each of which begins the process of a heartbeat. The node is often called the 'pacemaker' of the heart. The impulse travels across the atrial walls, making them contract first. There is a pause, long enough for the ventricles to fill, as a new impulse travels down conducting fibres in the septum between the ventricles. The impulse emerges at the apex of the heart, making the ventricle walls contract from the bottom upwards, pushing blood up and out of the heart.

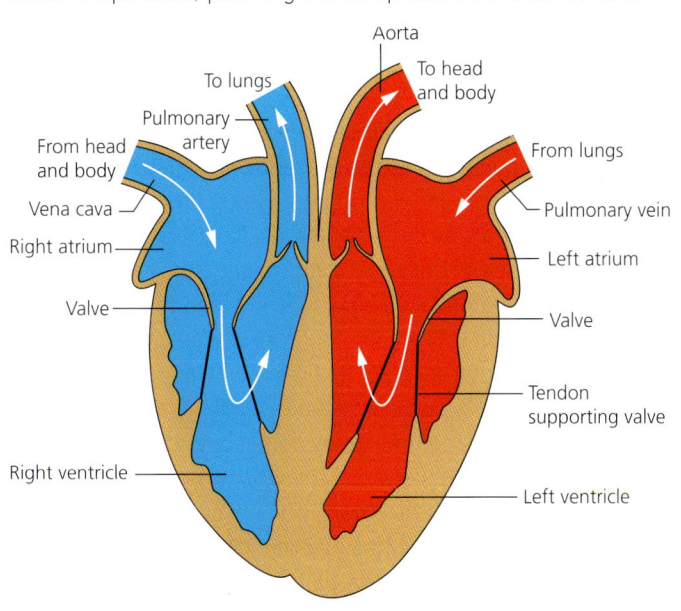

▲ **Figure 4.4** The heart

Types of blood vessels

See figure 4.5. Blood is taken away from the heart in **arteries**. Blood leaving the heart is under high pressure, so artery walls are thick, with **elastic** fibres which stretch and recoil to smooth out the pressure changes

> ### 🔑 KEY TERMS
>
> **Pressure** – Force exerted over an area of surface. For example 10 N/m² is 10 Newtons of force exerted over every 1 m² of surface. An alternative unit, mmHg, is often used in medicine.
>
> **Atrium** – Small chamber on each side at the top of the heart.
>
> **Ventricle** – Large lower chamber at each side of the heart.
>
> **Sinoatrial node** – Specialised muscle cells in the wall of the right atrium.
>
> **Artery** – Vessel which takes blood away from the heart.
>
> **Elastic** – Can stretch and then return to its original size and shape.
>
> **Coronary arteries** – Arteries which supply the heart muscle with blood.
>
> **Capillaries** – Very narrow vessels which supply blood to tissues and which connect arteries to veins.

during the heartbeat. Arteries have muscles in their walls to control the rate of blood flow to different parts of the body, and a smooth lining to let blood flow with a minimum of turbulence.

Blood leaves the heart in the main artery, the aorta. See figure 4.6. Smaller arteries come off the aorta, each taking blood to a separate organ. Two **coronary arteries** let out of the very first part of the aorta, taking blood into the heart muscle, so that the first organ the heart supplies with blood is itself.

Inside each organ, its artery repeatedly splits into smaller and smaller arterioles, then finally into **capillaries**. Arteries are excellent at moving blood around quickly, but their walls are far too thick to let any substances move in or out of the blood; that is the capillaries' job. They are just wide enough for a single erythrocyte to move along them, losing oxygen to the surrounding cells by diffusion on the way. The capillary walls are as thin as possible – only one cell thick – to let substances diffuse in both directions through them. Part of the plasma is forced out through tiny holes between the cells of the capillary walls, to bathe the cells directly as tissue fluid, before returning to the capillaries further along their route, now carrying waste products from the cells.

A harmful stimulus, for example infection, irritants or tissue damage, can cause inflammation, where the capillaries become 'leakier', let out more fluid and white blood cells into the surrounding tissues. This is a normal, healthy part of the body healing and protecting itself (see section 6.1), but if it continues for too long can lead to serious health problems (see sections 1.3, 2.2, 4.3 and 6.3).

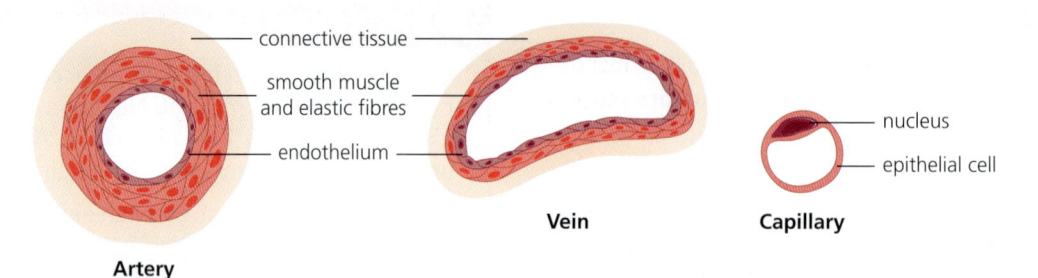

▲ Figure 4.5 Cross sections of an artery, a vein and a capillary

Leaving the organ, the capillaries repeatedly rejoin, forming first venules, then a **vein** – one of the vessels which return to the heart. The vein from each artery joins the major vein, the vena cava, as it takes blood to the heart. Most of the blood's pressure is lost in pushing through the tiny capillaries, so the veins have structures to help return blood to the heart despite the low pressure. The walls are thinner, with fewer elastic fibres, and the space the blood travels through – the lumen – is wider. There are valves so when the blood travels further upwards, it will not then fall back down due to gravity.

> 🔑 **KEY TERMS**
>
> **Vein** – Vessel which takes blood towards the heart.
>
> **Pulmonary** – Involving the lungs.

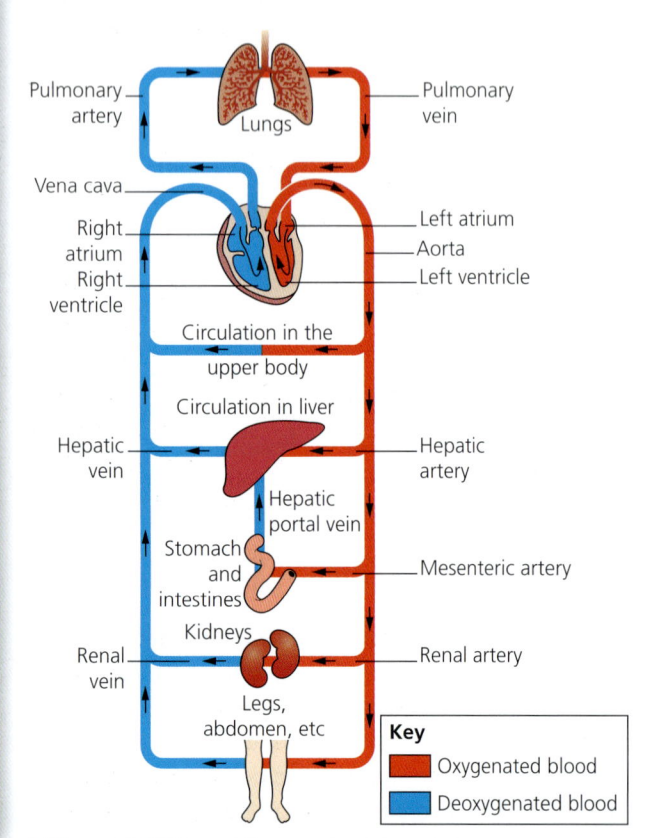

▲ Figure 4.6 The cardiovascular system

Pulmonary circulation

The heart has two separate pumps, one on each side. See figure 4.6.

The right side is the less powerful one, propelling blood through the **pulmonary** circulation – only to the lungs, out via the pulmonary artery, through the capillaries in the lungs and back again in the pulmonary vein. The blood from the lungs, now loaded with oxygen, enters the more powerful left side, which pumps it around the rest of the body. The blood returns to the right side again, having lost some of its oxygen and ready to repeat its route again. If you could follow one erythrocyte, in one circulation, it would pass through the lungs once, the heart twice and just one other organ.

3.2 Monitoring the cardiovascular system in people

Pulse rates

By pushing an artery against a solid structure like a bone, the artery's pulsation as the pressure wave passes through it can be felt through your fingers. This is often done at the wrist, but it can also be felt at the neck, jaw, groin, ankle and at the back of the knee. With training, the rhythm and force of the pulse can give a rough idea of blood pressure and an indication of many possible problems in the cardiovascular system. Its first use is to measure the heart rate; that is, how many times a minute the heart is beating, by counting how many pulses are felt in a timed 60 seconds. The pulse rate at the wrist is not always, though, the same as the heart rate. If a patient's atria are not contracting properly (as in atrial fibrillation), sometimes the ventricles do not entirely fill and that heartbeat may not give a pulse which can be felt at the wrist.

> **PAIRS ACTIVITY**
>
> **Pulse rate (1 day)**
>
> Work in pairs to take each other's pulse rate throughout the day, during different activities. Which activity gave the lowest pulse rate, that is the closest to resting pulse rate? What tips do you have for slowing the pulse rate?

Heart monitors

The activities of the sinoatrial nodes, the conducting fibres and the contracting muscle cells make a complex wave of voltage changes, which passes through the heart and into the surrounding tissues. It reaches the skin, where it can be detected by electrodes stuck to the chest and arms. The pattern of voltage changes (the electrocardiogram or ECG) is displayed on a screen. The whole apparatus is called an electrocardiograph. An ECG can be used to measure the rate and rhythm of heartbeats, the size of the atria and ventricles, and assess damage to the muscles and control system.

Normal sinus rhythm is the rhythmic beating of the healthy heart, shown by Figure 4.7a. The P wave corresponds to the wave of excitation passing through the atria, followed by the QRS and T as the wave passes through the ventricles.

In Figure 4.7b, the waves are normally shaped, but much more frequent. This shows sinus tachycardia, where the atrioventricular node produces impulses at an abnormally rapid rate.

In sinus arrhythmia (Figure 4.7c), again the waves are normally shaped, but they have an irregular rhythm, showing the atrioventricular node is firing irregularly.

In Figure 4.7d, activity in the atria is completely irregular and the P wave is missing. This shows atrial fibrillation, where the atria are beating chaotically.

(a) Normal sinus rhythm
(b) Sinus tachycardia
(c) Sinus arrhythmia
(d) Atrial fibrillation

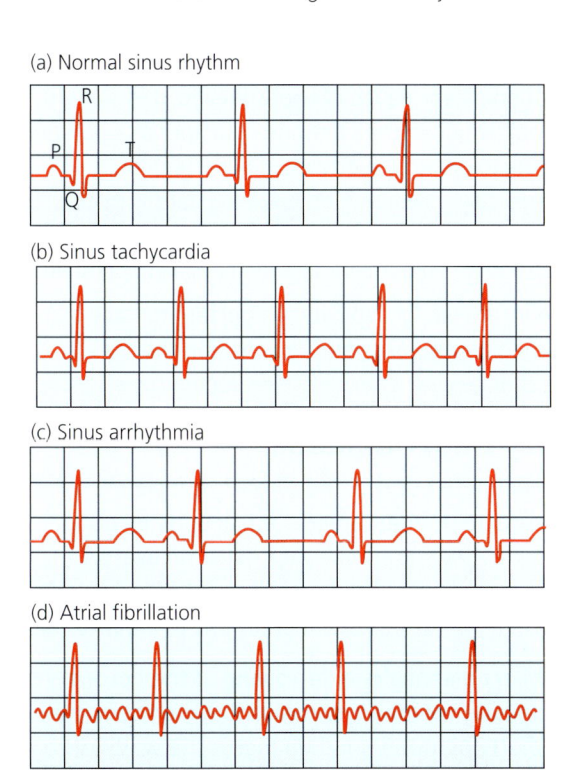

▲ Figure 4.7 Electrocardiograms

Echocardiography uses ultrasound technology to image the heart in motion. It is used to find the size and shape of the chambers, to calculate the **stroke volume**, and to locate internal damage, amongst many other uses.

> 🔑 **KEY TERM**
>
> **Stroke volume** – The volume of blood pushed out by the left ventricle in one heartbeat.

Heart rate during exercise and at rest

The resting heart rate is the number of contractions in one minute when the person is awake, at rest and not under stress – either emotional or physiological stress, such as coping with high or low temperatures.

The normal range depends on a person's age, sex and level of fitness.

An unborn fetus has a resting heart rate of 140–150 beats per minute (bpm). A newborn infant has a rate of 130–140 bpm, gradually falling to 70–75 bpm as an adult. In old age, the range widens to 67–80 bpm.

Women have a slightly higher resting pulse rate than men.

Physical fitness has an effect. A fit person will have larger heart chambers, and so a larger stroke volume. Fewer beats will be needed per minute to supply the resting tissues with oxygen. A resting heart rate of more than 80 bpm indicates poor fitness, 60–69 bpm a good level of fitness and below 50 bpm an outstanding athlete.

During exercise, the muscles are respiring more so need blood – and therefore the oxygen and glucose it carries – delivered to them at a faster rate. The heart rate increases to supply this demand. The exercising heart rate depends on age and the strenuousness of the activity. For instance, a 20 year old with a resting heart rate of 70 could expect to see an increase to 100–120 bpm during a warm up, then to 160–180 bpm in very intense exercise. A 50 year old would see 85–100 bpm for warm up, then 153–170 bpm for very intense exercise. The older heart beats more slowly, so it is wise to scale down what you attempt for each level of exercise as you age.

Pacemakers

If the sinoatrial node is not functioning correctly, or there is a problem with the conducting fibres in the heart, a person may suffer from an irregular or overly slow heartbeat. The node can be supplemented with an artificial electronic pacemaker. This is a small electrical device surgically implanted under the skin of the chest. Inside, there is a small computer, battery and pulse generator. Externally, there are one or more wires, which enter the heart through a vein, ending in electrodes in the walls of the heart chambers.

Sensors in the package monitor the heart's electrical activity. When the heart misses a beat or is beating too slowly, the pulse generator sends electrical impulses to the electrodes, stimulating contractions. When the heart is beating normally, the pacemaker will not send out impulses.

Artificial pacemakers have sensors which detect a person's motion and breathing rate, so that the computer can respond with more frequent impulses, raising the heart rate to match the person's level of exercise.

The battery will last five to ten years, after which the body of the pacemaker is replaced in a minor operation.

Defibrillators

Fibrillation is when the contractions of the heart are unsynchronised, irregular and rapid. The sinoatrial node cannot control muscles which are in this state. A fibrillating heart will be less effective at pushing blood forwards.

A defibrillator is a device which passes a controlled electric current through the heart, 'resetting' the heartbeat, and allowing the sinoatrial node to re-establish its control over the cardiac muscles.

There are several designs of defibrillator.

An implantable cardioverter defibrillator (ICD) is implanted in the chest, similar to a pacemaker. They are fitted to people at risk of a heart attack at some point in the future, in order to prevent one happening. Once it senses the heart beating abnormally, it automatically defibrillates it.

Manual external defibrillators work to apply the current through pads across the heart and need specialist training to operate.

Automatic external defibrillators (AEDs) are designed to be used by non experts. Placed on the chest, the onboard computer will monitor heart activity, then advises the user whether to continue with defibrillation.

Comparison of different populations

The cardiovascular system differs in function depending on a number of factors including age. An ageing cardiovascular system is linked to a decline in the cardioprotective systems. 50% of heart failure diagnoses occur in people aged 70 and over. As a person ages the following issues may arise; stiffening of the arteries, increased blood pressure and increase in left ventricular wall thickness and mass. The threshold for the manifestation of disease is lowered in elderly people, and the structural and functional changes can have significant implications on the strength of the cardiovascular system.

> **KEY TERM**
>
> **Fibrillation** – When the contractions of the heart are unsynchronised, irregular and rapid.

GROUP ACTIVITY

Invent (10 minutes)

Electronic devices become cheaper and can carry out a wider range of tasks as time progresses. In your groups, brainstorm extra life-saving functions that could be built in to a new generation of artificial pacemakers and implantable defibrillators.

3.3 Common cardiovascular disorders, their possible causes and symptoms

Hypertension

The blood is constantly under pressure. The pressure is highest when the ventricles are contracting (the systolic pressure) and lowest while the heart is relaxed (the diastolic pressure). The pressures are written as systolic/diastolic. Normal resting blood pressure is within the range 100–140 /60–90 mmHg.

A reading over 140/90 mmHg indicates high blood pressure, known in medical terms as hypertension but often called 'the silent killer' because it has no obvious symptoms. Left untreated, it greatly increases the risk of heart failure, heart attack, kidney disease, stroke and dementia.

There is not one clear cause of hypertension. It is multifactorial, which means a range of factors in a person's life contribute to an increased risk of hypertension, including age, being overweight, smoking, family history, a diet with insufficient fruit and vegetables and excess salt, little exercise, high caffeine intake and high alcohol intake.

For people in 'at risk' groups, it is important to have blood pressure checks regularly, and to consider changes in lifestyle to reduce your chance of developing hypertension.

Coronary heart disease

For an individual to survive, their heart muscle must make successive uninterrupted contractions through decades of life, taking no rest after extreme exercise. The muscle needs an unhindered flow of blood through its two coronary arteries (see section 3.1) to achieve this.

Coronary heart disease is a narrowing of the coronary arteries, causing the heart to malfunction or become damaged. Fatty plaques build up inside the walls of the coronary arteries, narrowing their lumens. This makes it more difficult to supply the muscle with the oxygen and nutrients it needs. The heart pumps harder, increasing blood pressure in the coronary arteries and risking damaging them.

The forms of coronary heart disease include the following:

- Angina. This is chest pain brought on by lack of blood to the heart muscle. This does not cause tissues to die.
- Heart attack. Part of the plaque-weakened artery wall bursts, and blood clots around the injury site, blocking the lumen. Heart tissue beyond the blockage is starved of oxygen and dies, disrupting contractions. The symptoms of a heart attack are an increasing band of pain around the arms and chest, profuse sweating, nausea and difficulty breathing. A heart attack is serious – but correct, simple first aid greatly increases its survivability.

Varicose veins

Varicose veins are swollen, darkened and twisted veins – usually blue or dark purple – that usually occur on the legs, but can also occur in the oesophagus, uterus and rectum. The legs are prone to muscle cramps, and feel aching and uncomfortable.

The swelling is due to blood pooling within the veins because their valves have become weakened and are unable to stop the blood falling backwards on its journey up the legs towards the heart (see section 3.1).

KNOW IT
1. Which type of blood vessel carries blood (a) away from the heart (b) through tissues (c) towards the heart?
2. Give three functions of the blood.
3. What does an electrocardiograph measure?

LO3 Assessment activities

Below are suggested assessment activities that have been directly linked to the Pass and Merit criteria in LO3 to help with assignment preparation and include top tips on how to achieve best results.

Activity 1 Measure and record athletes' pulse rates *P3*

Measure the pulse rates of an athlete.

Find the resting pulse rate, the rate during exercise and the rates every minute after exercise until the resting rate is regained.

If you are in a centre where appropriate equipment is available, record the athlete's electrocardiogram at rest under the same conditions. Use the recordings to calculate heart rates.

Explain the changes in rates, by saying what physiological changes are taking place in the body.

TOP TIP
✓ You should undertake a full health and safety evaluation, in line with your centre's policies, before undertaking this activity.

Activity 2 Independent research activity *M2*

Research the symptoms of:
- hypertension
- angina
- heart attack
- varicose veins.

Explain how problems with the cardiovascular system:
- affect the normal functioning of the body
- causes each of the symptoms.

LO4 Be able to assess how the respiratory system functions in the body *P4 M3*

GETTING STARTED
(5 minutes)

Discuss which parts of your body are moving when you breathe in and breathe out.

4.1 The components and function of the respiratory system

Structure of lungs

The **lungs** are a pair of air-filled bags in the **thorax**. Their function is to do **gaseous exchange** between air and the blood. Each lung is surrounded by a pleural cavity, a low pressure space which causes each lung to be pushed out into a fluffy structure which almost entirely fills the **thoracic cavity**.

KEY TERMS

Lungs – A pair of air-filled bags in the thorax.

Thorax – The part of the torso above the diaphragm.

Gaseous exchange – Passing oxygen and carbon dioxide in opposite directions. Oxygen enters the body, while carbon dioxide leaves it.

Thoracic cavity – The chamber of the body protected by the thoracic wall.

111

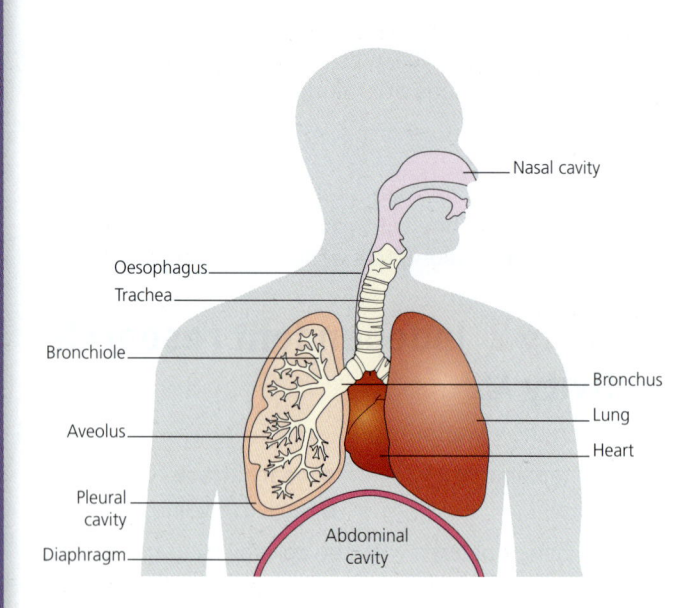

▲ Figure 4.8 The respiratory system

The trachea reaches from the mouth and nose down into the thoracic cavity. It is split into two smaller tubes called bronchi which enter each lung. Once inside, the bronchi branch into smaller and smaller bronchioles. So far the trachea, bronchi and bronchioles have all been held rigidly open with rings of cartilage, but the smallest bronchioles are thinner-walled and end in clusters of rounded air sacs called **alveoli**, where gaseous exchange happens. Carbon dioxide moves from the bloodstream into the air in the alveoli, and in the opposite direction oxygen moves from the air into the bloodstream.

Inspiration and expiration

Constantly breathing in (inspiration) and out (expiration) swaps the air in the alveoli for air from outside, bringing in new oxygen and removing carbon dioxide.

The ribs form the walls of the thorax, surrounding the lungs to the side and above. Two layers of muscles, the internal and external intercostals, run between the ribs. The floor of the thorax is a sheet of muscle called the diaphragm, stretched across the bottom of the rib cage. When the diaphragm is relaxed, it has a domed shape which extends up inside the thoracic cavity.

During inspiration, the diaphragm contracts, pulling itself down into a flatter shape. The external intercostal muscles contract, lifting the ribcage upwards and outwards. The volume inside the thorax, and therefore inside the lungs, increases, which lowers the air pressure in the lungs until it is less than the atmospheric pressure outside the body. The atmospheric pressure pushes air down the trachea and into the lungs, inflating and stretching the elastic alveoli. Air is blown in, not sucked in.

Little work is done during light expiration. The alveoli recoil to their original shape, pushing air out. The diaphragm relaxes, returning to its domed shape, and the rib cage falls down and in, under its own weight. All of this decreases the volume in the thorax, increasing air pressure in the lungs until it is more than atmospheric pressure, and pushes air out through the trachea.

In forced expiration, when we need to expel more air and decrease the volume in the thorax still further, the internal intercostal muscles contract, pulling the ribs inwards. The muscles of the **abdomen** pull down on the rib cage and also force the abdominal organs upwards, pushing the diaphragm further into the thorax.

> 🔑 **KEY TERMS**
>
> **Alveoli** – Where gaseous exchange happens.
> **Abdomen** – The torso below the diaphragm.
> **Concentration** – The number of molecules of a substance within a set volume.

Gaseous exchange

Gaseous exchange happens through diffusion. Diffusion works best when there is a small distance for particles to move across and a large difference in **concentration** between two places.

The alveoli provide a very small distance between blood and air by having thin walls, made from single flat endothelial cells, backed with a network of equally thinly walled capillaries. A molecule of carbon dioxide only has to diffuse across two very slim cells to move from blood plasma to alveolar air. The capillaries are just wide enough to allow single red blood cells to pass through, so oxygen molecules diffuse from the alveolar air into the capillary, and then have only the shortest distance through plasma and into an erythrocyte through its surface membrane.

The lungs maintain a constant concentration difference for both gases. There is always a greater concentration of oxygen in the alveolar air than in the blood, and a greater concentration of carbon dioxide in the blood than in the alveolar air. With inspiration and expiration, the alveoli are always being partly refreshed with new oxygen – rich, carbon dioxide-poor air. The blood is constantly circulating. It is only in contact with the alveolus for seconds, and being replaced with carbon dioxide-rich, oxygen-poor air through the pulmonary circulation.

Having a large surface area means more gas exchange can happen in the same time. The 700 million alveoli per adult pack a huge surface area of about 70 square metres into a pair of lungs.

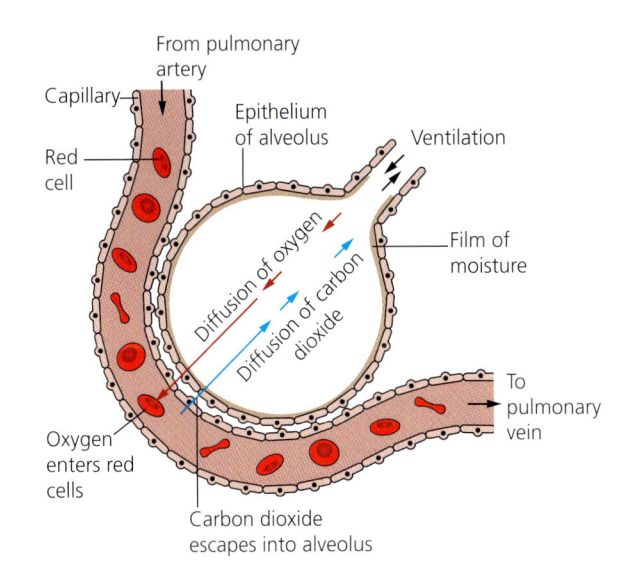

▲ Figure 4.9 Gaseous exchange in the alveolus

4.2 How the respiratory system is monitored in people

Lung volumes and capacities

The volume of air in inspirations and expirations is measured using a spirometer. Older types involved breathing from a fixed volume of gas, which moved a marker pen up and down. (And having your exhaled carbon dioxide chemically absorbed to avoid poisoning yourself!) Modern models are entirely electronic, sensing the speed and direction of airflow in a pipe attached to a mouthpiece.

Tidal volume is the volume of air inspired then expired in a single breath without exerting any extra effort. This is about 0.5 dm^3 for an adult.

Inspiratory reserve volume is the volume of extra air inhaled beyond the tidal volume during a deep breath, fully inflating the lungs.

Expiratory reserve volume is the volume of extra air exhaled beyond the tidal volume during a forced expiration.

Vital capacity is the maximum amount a person can exhale easily from their lungs after their maximum inhalation. It is the same as adding together tidal volume with the inspiratory and expiratory reserve volumes.

Forced vital capacity (FVC) is the same measurement, but done blowing out as fast and hard as possible. A part of this is 'Forced Expiratory Volume in One Second' or FEV1, the maximum volume of air which can be blown out in the first second. The ratio FEV1/FVC is important for showing whether air is being moved easily out of the lungs.

Even after a maximum exhalation, there will still be air left in the lungs, in the bronchi and bronchioles because they are held open by their cartilage. This is the residual volume, which cannot be measured by a spirometer.

Total lung capacity is the volume of air in the lungs after a maximum inhalation. It is the same as vital capacity plus residual volume. Figure 4.10 shows lung capacities, and FEV1 and FVC.

Peak flow

The peak expiratory flow is the maximum speed a person can expire air. It is easier to measure than FEV1/FVC, as it is done with a simple hand-held peak flow meter, which a patient is trained to use themselves at home. Measuring peak flow is important because it shows the amount of obstruction a person is suffering in their bronchi and bronchioles.

Oxygen saturation – oximeter

A spirometer and peak flow meter measure the movement of air in and out of the lungs, but an oximeter indicates how effective the lungs are at getting oxygen into the blood. A clip on the finger or earlobe shines wavelengths of light into the tissue to measure the oxygen saturation in the arterial blood; that is, what percentage of haemoglobin's binding sites are carrying an oxygen molecule.

Comparison of different populations

The lungs are bigger when the overall body size is larger, so vital capacity increases with height. Adult males 150–155 cm tall could be expected to have a vital capacity of 2.9 dm^3, and those who are 175–180 cm tall, 4.3 dm^3. With a generally smaller body size, females' vital capacities are about two thirds those of males of the same age and height. This is reduced a little further in pregnancy, as the uterus pushes the diaphragm further upwards.

Vital capacity decreases with age, from 3.5 dm^3 at 25 years old to 2.85 dm^3 for the average male.

FEV1/FVC for a healthy individual are from 0.75 to 0.85, reducing to the lower end of this range with age.

Smoking is a major challenge to the working of the lungs. Smokers' vital capacities are actually little different from non-smokers. But smoking massively

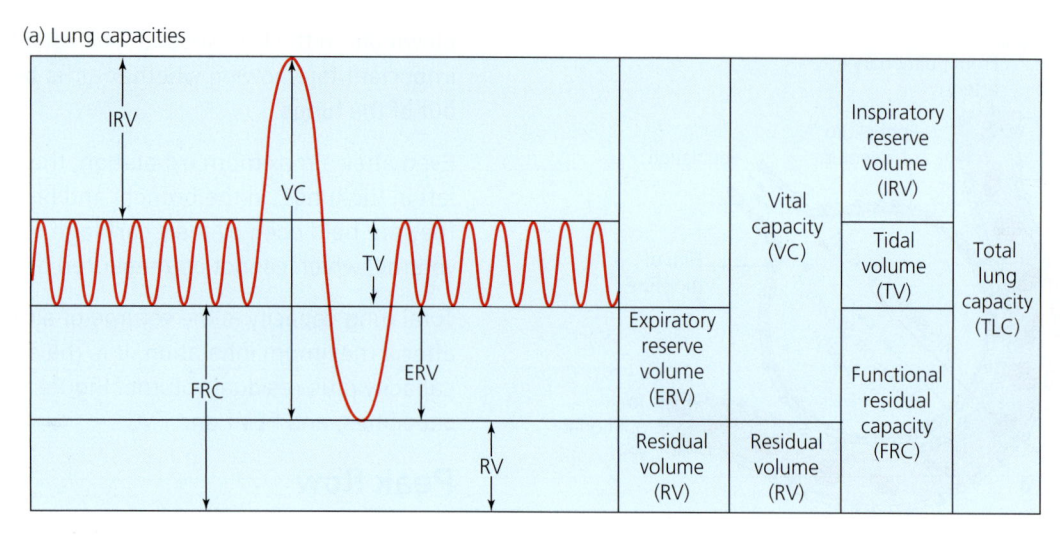

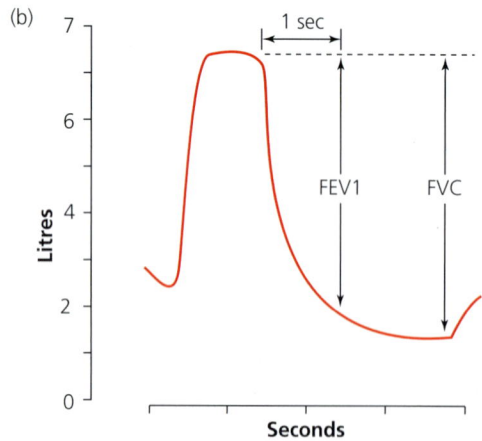

▲ Figure 4.10 (a) Lung capacities (b) FEV1 and FVC

increases the risk of developing lung diseases, which radically change the picture. We will look at one of these, emphysema, in the next section.

4.3 Common disorders of the respiratory system

In an obstructive lung disease, the sufferer cannot expel their normal volume from the lungs. Their total lung capacity is little changed, but the residual volume part of it increases. Crucially, FEV1 has a lower value, reducing FEV1/FVC to less than 0.70 and so showing that the flow of air out of the lungs is being obstructed. The air in the alveoli is not being sufficiently refreshed, the all-important differences in concentration between the alveolar air and blood is reduced, so the rate of gas exchange reduces too. The consequence is the blood has a lower oxygen saturation, so tissues lack the oxygen needed for respiration.

Here are two examples of obstructive lung diseases.

Asthma

An asthma sufferer's smaller bronchioles have become hypersensitive, so that when they come into contact with tobacco smoke, pollen, dust mite faeces and many other 'triggers', they respond by becoming inflamed, producing extra mucus and contracting the smooth muscles in their walls. All these changes make the bronchioles narrower, obstructing air leaving the alveoli. The sufferer experiences shortness of breath, chest tightness, coughing and wheezing.

People can develop asthma at any age. The underlying cause of the bronchioles becoming hypersensitive is not fully understood.

Asthma can be controlled by the sufferer monitoring their own breathing using a peak flow meter, identifying their own triggers, and taking medication, often through inhalers, which can deliver drugs deep into the bronchioles.

Emphysema

Inflammation of the lungs from inhaling tobacco smoke or poor air quality means phagocytes have been leaving the capillaries and patrolling the lung tissue. On the way, they will have digested their way through elastin fibres in the alveolar wall. Not a problem in the short term, but if it continues for too long, the damage done is irreversible and the alveoli lose their ability to expand and recoil. The sufferer faces hard work pushing air out of the lungs on every exhalation. It will take longer to exhale than to inhale, and it will involves much more muscular effort. FEV1/FVC can be reduced to as low as 0.45.

Once started, the effects of emphysema become progressively worse as time passes. There is no cure – but it can be prevented by measures to stop smoking and improvements to air quality.

Cystic fibrosis

Cystic fibrosis is an example of another type of disorder – a restrictive lung disease, where the lungs cannot fully expand. The total lung capacity is reduced but expelling air is not a problem, so FEV1/FVC is within the normal range.

The sufferer has a genetic problem which leads them to make thick mucus in places where it would normally be thin. Mucus is a normal part of the lung's functions, coating the bronchioles, trapping dust and bacteria, constantly on the move as it is transported up out of the airways. This thicker mucus, though, is stationary. It clogs the lungs, bacteria multiply in it and cause frequent infections. The alveoli become scarred in the course of fighting the infections. The sufferer experiences difficulty in breathing and coughing up mucus as the lung-related part of their symptoms.

> ### ❓ THINK ABOUT IT
> **Smokers (20 minutes)**
> Smoking tobacco is a risk factor in developing several lung diseases. Treating patients with these diseases is expensive. The National Health Service only has a limited budget. Prepare a written response to the proposal that smokers should be held accountable for the health risk they put themselves under by being required to pay for a proportion of their treatment if they suffer from a smoking-related disease.

> ### KNOW IT 💡
> 1. Which sets of muscles are used when we breathe in?
> 2. What is meant by 'gas exchange'?
> 3. Which parts of the respiratory system are affected by (a) emphysema (b) asthma?

LO4 Assessment activities

Below are suggested assessment activities that have been directly linked to the Pass and Merit criteria in LO4 to help with assignment preparation and include top tips on how to achieve best results.

Activity 1 Investigate your own lung volume *P4*

If you have access to appropriate equipment, take measurements of your own peak flow, tidal volume, inspiratory reserve volume, expiratory reserve volume, vital capacity and total lung capacity.

These results could be presented graphically, showing their relationship to one another. The illustration could be annotated to show what sorts of activities would need the different sections of lung volume.

> **TOP TIP** ✔
> ✔ You should undertake a full health and safety evaluation, in line with your centre's policies, before undertaking this activity.

Activity 2 Research lung diseases *M3*

Research the lung diseases which have been occupational hazards in the coal mining industry. Use statistics to compare the prevalence of these diseases between coal miners and non-miners.

> **TOP TIPS** ✔
> ✔ Make sure that you use statistics which compare like with like. For instance, data should be for the same gender and same age.
> ✔ If this cannot be done, be aware of how this limits the conclusions you can draw.

4.3 Common disorders of the respiratory system

LO5 Understand how homeostasis maintains balance within the body P5

GETTING STARTED
(5 minutes)

Discuss what your body must keep the same inside itself for its organs to keep working correctly.

5.1 How the following structures and systems maintain homeostasis within the body

The concept of homeostasis

Cells are bathed in **tissue fluid**. The tissue fluid is the 'internal environment' that the cell lives in. To work at their best, they rely on the tissue fluid to provide the right conditions: an optimum temperature, enough (but not too much) oxygen and glucose, the correct concentrations of each of a range of ions, water content and many others.

We have already seen how the body uses its systems to obtain useful substances then deliver them to the cells. (For example, glucose from the digestive system, oxygen from the respiratory system, both delivered by the circulatory system.) Now we will look at how it controls its systems by **homeostasis**: keeping the conditions in the environment around each cell constant, and at the levels that the cell needs to work effectively.

In this Learning Objective, you will see many examples of the body using a control mechanism called **negative feedback** to control an internal condition. In negative feedback, a condition changes from its optimum value. The body takes action which brings back the condition to its optimum value. Regaining the optimum shuts down the action.

The autonomic nervous system

'Autonomic' means 'self-governing'. The conscious part of ourselves has little control over this section of our nervous system. It is a quick-response control system which influences several organs. Its co-ordination centres are in the hypothalamus, in the base of the brain. It has **receptors** and **effectors** scattered throughout the body, linked to the hypothalamus by **neurones**.

The system has two parts, the sympathetic and parasympathetic nervous systems. An organ under autonomic control will have neurones from both the sympathetic and parasympathetic systems attached to it. One system – usually but not always the sympathetic – will activate a response, and the other system will inhibit it.

Both systems are attached to the smooth muscles in places we have already seen – the arteries, the bronchioles of the lungs and the gastrointestinal tract in ways that let small signals have large effects.

The autonomic system has receptors in the arteries of the neck which monitor the pH and pressure of the blood. During exercise, blood pH drops because more respiration is going on, increasing carbon dioxide concentration. In response, the sympathetic system stimulates the sinoatrial node in the heart, increasing the heart rate, supplying cells with more blood and carrying carbon dioxide away to the lungs at a higher rate. This returns the blood pH to its normal levels, causing the sympathetic stimulation to end. Parasympathetic stimulation has the opposite effect, slowing the heart rate in response to the blood pressure becoming too high.

At the same time in the lungs, sympathetic stimulation causes **bronchodilation**, allowing a greater airflow to reach the alveoli. The increased blood flow from

KEY TERMS

Tissue fluid – The internal environment that the cell lives in.

Homeostasis – Keeping the conditions in the environment around each cell constant, and at the levels that the cell needs to work effectively.

Negative feedback – To control an internal condition. In negative feedback, a condition changes from its optimum value.

Autonomic – Means 'self-governing'. The conscious part of ourselves has little control over this section of our nervous system.

Receptor – A cell which converts a change in the environment into an electrical impulse in a neurone. For example, a touch receptor in the skin.

Effector – A muscle, gland or organ which can make a response when it is stimulated by an electrical impulse from a neurone.

Neurone – Very long threadlike cells that carry information from place to place quickly around the body in the form of electrochemical impulses.

Bronchodilation – The smooth muscles in the walls of the bronchioles relax, widening the lumen.

the heart now has more oxygen to carry and more opportunity to lose the carbon dioxide which began this process. Parasympathetic stimulation will have the opposite effect, causing **bronchoconstriction**.

Cells in the core organs only work well within a fairly narrow range of temperatures. Receptors in the **hypothalamus** sense blood temperature rising above normal. The thermoregulatory centre in the hypothalamus responds with sympathetic stimulation to the smooth muscle of selected arterioles in the skin. The muscle relaxes, the arterioles widen (vasodilation) and an increased blood flow is sent to vessels closer to the skin's surface. Excess heat is radiated out to the surroundings and blood temperature begins to fall.

If blood temperature falls, the thermoregulatory centre sends parasympathetic stimulation which contracts the smooth muscle, blood flow is restricted to the deeper vessels in the skin, and the heat is conserved, to be sent back to the core organs.

> **KEY TERMS**
>
> **Bronchoconstriction** – The smooth muscles in the walls of the bronchioles contract, narrowing the lumen.
>
> **Hypothalamus** – A region at the base of the brain.
>
> **Hormones** – Chemical messengers made by glands and carried in the bloodstream.
>
> **Target organ** – The organ which will make a response when it encounters a hormone.
>
> **Chemical receptors** – A molecule in a cell's surface membrane, which will bind with a messenger molecule, such as a hormone. This is not the same as the 'receptor' in the nervous system.

The endocrine system

The endocrine system is slower acting than the nervous system. Organs called endocrine glands respond to changes in the body by secreting chemical messengers called **hormones** into the capillaries passing through them. The hormones are carried all around the body in the bloodstream, but only have effects on their **target organs**. Only the cells of the target organs have **chemical receptors** on their surface membranes for that hormone. Where a hormone binds to a receptor, the cell is stimulated to make a dramatic response.

Pancreas

We have already seen the pancreas secreting enzymes into the duodenum. It has a second function, regulating the concentration of the sugar glucose in the blood. Blood concentrations are important because anything dissolved in the blood will soon be in the tissue fluid, right next to every cell of the body.

Digesting and absorbing food adds a sudden surge of glucose to the bloodstream. Scattered cells in the pancreas detect the increase in blood glucose concentration above normal levels, and secrete the hormone insulin. The blood from the small intestine does not flow immediately back to the heart. Instead, the hepatic portal vein takes it first to the liver (see Figure 4.6). The liver is one of the targets for insulin. The liver cells respond to insulin by taking in glucose from the bloodstream before it can be pumped around the rest of the body. Glucose is very soluble and upsets the water content of cells, so the liver cells convert glucose into insoluble glycogen.

Elsewhere, insulin stimulates fat and muscle cells to take up blood glucose and use it in respiration or use it to make glycogen or lipid.

If the blood glucose concentration falls below normal, the pancreas secretes glucagon, which has the opposite effect to insulin, causing the liver cells to break up the chains of glycogen back into glucose molecules, which they release out into the blood.

Thymus

The thymus, lying just behind the breastbone, produces the hormone thymosin before puberty, stimulating the development of lymphocytes into T cells, an essential part of the immune system.

Pineal gland

Situated inside the brain, the pineal gland secretes melatonin, which has a role in establishing daily rhythms such as the sleep-wake cycle.

Parathyroid

Calcium is essential for neurones and muscles to function. If the blood calcium concentration falls, parathyroid glands secrete parathyroid hormone, which stimulates the osteoclast cells in the bones to break down the calcium salts which surround them, and release calcium into the bloodstream (see section 2.2). Parathyroid hormone also acts on the kidneys, reducing the amount of calcium lost in the urine. A hormone with opposite effects, calcitonin, is secreted by the thyroid. Between them, the two hormones regulate the homeostasis of blood calcium level.

Hypothalamus

Along with its role in the autonomic nervous system, the hypothalamus also controls a hierarchy of glands. It secretes truly minute amounts of releasing factors which control the **pituitary**, a pea-sized gland on a stalk of tissue below the hypothalamus. Some of the pituitary's hormones have direct effects: prolactin causing milk release from mammary tissue, growth hormone dictating when and how quickly we grow. Other pituitary hormones activate larger glands around the body, such as FSH and LH controlling the ovaries. A tiny signal from the hypothalamus cascades into a large-scale effect on the whole body. Here are two more examples.

Thyroid

A thyrotropin-releasing hormone from the hypothalamus stimulates the pituitary to secrete a thyroid-stimulating hormone, which in turn stimulates the thyroid gland in the neck to secrete thyroxine. Thyroxine controls the rate of growth and speeds up cells' rate of **metabolism**.

Adrenal gland

A corticotropin-releasing hormone from the hypothalamus stimulates the pituitary to secrete adrenocorticotropic hormone (thankfully shortened to ACTH), which in turn stimulates the adrenal glands above the kidneys to secrete two types of hormone. It increases the rate of metabolism by making more glucose available.

Mineralocorticoids regulate blood pressure and sodium and potassium levels by controlling which ions are reabsorbed by the kidneys, and how much water is reabsorbed into the blood.

In emergency situations, the adrenal glands release adrenaline which prepares the body to take drastic action (often called 'flight or fight') by overriding normal homeostatic controls. It stimulates the breakdown of glycogen to glucose. It increases the heart rate and blood pressure, dilates the bronchioles, and diverts blood towards the muscles by dilating the arterioles that supply them. Altogether, these changes prepare the muscles for vigorous activity.

🔑 KEY TERMS

Pituitary – A pea-sized gland on a stalk of tissue below the hypothalamus.
Metabolism – The chemical reactions going on inside the body.

🔍 RESEARCH ACTIVITY

Gland failures (20 minutes)

For each of the glands in this section, research what happens if the gland stops functioning. What would be the effects on the whole body?

Structure and functions of the kidneys

The kidneys remove the harmful waste product urea from the blood, along with excess salts and water, forming urine. Removing a waste product that the body has made itself is called **excretion**.

We have two kidneys, one each side of the spine. Each kidney has an outer and inner layer – the cortex and medulla – which contain the microscopic tubules (nephrons) where the urine is formed. Many nephrons pour urine into collecting ducts, joining together in a small chamber called the pelvis, before flowing down a ureter into the bladder for storage.

In each nephron, blood is forced against a microscopic filter, pushing water and everything dissolved in it out of the blood and into the tubule. If this was the whole story, all the body's dissolved nutrients and ions, and all its water too, would soon be lost in the urine. The nephron's job is to reabsorb all the nutrients back into the blood, and just the right amounts of each mineral ion (mineralocorticoid and parathyroid hormones control cells in the tubule here) to ensure the blood keeps the optimum concentration of each one.

The collecting ducts will reabsorb the correct amount of water under the influence of another hormone from the pituitary, **ADH**. When the body is short of water and has to save it for a more important function such as sweating, more ADH is secreted, stimulating the walls of the collecting duct to reabsorb more water.

The poisonous urea is not reabsorbed, so it all stays in the urine.

🔑 KEY TERMS

Excretion – Removal of a waste product made by the body itself, e.g. carbon dioxide.
ADH – A pituitary hormone which stimulates the kidney to reduce the volume of water in the urine.

Role of liver in excretion

The liver is the body's centre for dealing with problem molecules.

When the diet provides the body with more amino acids than it can use in growth and repair, they are used instead as an energy source in respiration. First, though, they must be chemically converted in the liver cells, and this makes ammonia – a highly toxic waste product. The liver cells bond two ammonias with a carbon dioxide molecule, making urea – still harmful, but safe enough to be transported in the plasma then excreted by the kidneys.

Old red blood cells are broken down in the liver. Their haemoglobin yields the compound bilirubin, which the liver excretes in the bile, secreted into the duodenum and lost from the body in faeces.

5.2 Common disorders caused by the inability to maintain homeostasis

Heart failure

The heart cannot pump out enough blood per heartbeat to meet the demands of the body. Several homeostatic systems make short-term changes to cope with the low blood flow – but a malfunctioning heart is a problem homeostasis cannot fix. Each attempt to bring conditions back to normal fails, making a bad situation progressively worse.

The autonomic nervous system's pressure receptors in the arteries of the neck sense the drop in blood pressure, and the sympathetic nervous system responds by causing two changes:

- An increase in heart rate. The ventricular walls – already working hard – contract harder, but soon become fatigued.
- Smooth muscle in the arterioles contracts, trying to keep blood pressure high but at the same time making it difficult for blood to flow through. The heart is now working against a greater resistance from its blood vessels.

The adrenal glands secrete the mineralocorticoid hormone aldosterone to encourage the kidneys to reabsorb more water to keep the blood pressure high, giving the heart yet more work in pumping a greater volume against the greater resistance, so that the flow of blood from the heart decreases yet further. This could lead to total heart failure where the heart stops.

Diabetes

A diabetic person either cannot make insulin (Type 1 **diabetes**), or the target cells do not respond to the hormone (Type 2 diabetes).

Type 1 is caused by the immune system attacking the pancreatic cells that make insulin. Type 2, which is more common, is largely due to lifestyle where obesity, a high sugar diet and lack of exercise are risk factors.

In both cases, as a meal is digested and absorbed, glucose-rich blood flows along the hepatic portal vein, then through and out of the liver without the liver cells being alerted to alter its sugar content. Without a homeostatic mechanism in control, blood glucose concentration rises above normal level (hyperglycaemia).

Later, when the food is all digested and the circulating glucose is being used up, the level drops below normal (hypoglycaemia). Usually, homeostasis would raise the level by breaking down glycogen – but none was laid down earlier, so blood glucose continues to fall.

Hyperglycaemia

The symptoms of being hyperglycaemic are being excessively thirsty, excessively hungry and producing abnormally high volumes of urine. Over time, hyperglycaemia can cause serious problems: damage to nerves, blindness and kidney failure.

A diabetic person would manage their condition by careful diet, exercise, measuring their own blood glucose and injecting calculated amounts of insulin (in type 1 diabetes).

Other hormones have effects on blood glucose, so malfunctions of the pituitary, adrenals and thyroid can also cause hypoglycaemia, as can heart attack and stroke.

Hypoglycaemia

Again, diabetes is not the only cause of hypoglycaemia: other causes include starvation, over-exertion, a sudden large intake of alcohol, disorders of the adrenals or liver. The list is extensive. Neurones are particularly hungry for glucose, so first symptoms involve the nervous system: being unable to concentrate, movements become unco-ordinated and speech slurred. The immediate treatment is straightforward: raise the blood glucose carefully by eating sugars or starchy food, or giving sugar intravenously. Left untreated, with glucose levels dropping further, the situation becomes progressively more serious: loss of consciousness, seizures, coma and death. A diabetic person would manage their condition to prevent a 'hypo' by being aware of their blood glucose and eating the right foods at the right time.

> **KEY TERM**
>
> **Diabetes** – A metabolic disease in which the body has a high level of glucose (blood sugar).

Dehydration

Dehydration is where a person's water intake is less than their water loss, so the water content of the body falls below normal. Potential causes are diseases which cause heavy diarrhoea, vomiting, diabetes, alcohol, sweating through illness, heavy exercise or high temperatures – or simply not taking in enough water.

With body fluids becoming more concentrated, the person will feel thirsty, followed by fatigue and dizziness. Losing more than about 20% of body water will be fatal. Homeostasis through ADH acting on the kidneys will reduce the amount of water lost in the urine, but there are other water losses that homeostasis cannot control. Water is constantly lost from the lungs, so a person becomes more dehydrated with every breath. The body continues sweating to lose excess heat until dehydration is quite advanced, too.

A drink of water treats mild dehydration. More serious dehydration needs oral rehydration therapy – a more effective way of rehydrating the body using a simple but clever solution of salts and sugar.

KNOW IT

1. What is a hormone?
2. What is meant by 'negative feedback'?
3. What effects does ADH have on the body?
4. What is a neurone?

LO5 Assessment activity

Below is a suggested assessment activity that has been directly linked to the Pass criteria in LO5 to help with assignment preparation and includes top tips on how to achieve best results.

Activity 1 Prepare an information booklet P5

Prepare an information booklet:

- outlining the health risks associated with both elevated and depressed blood glucose levels
- giving advice on maintaining a healthy blood sodium concentration.

TOP TIP
✓ Diabetes UK and NHS Choices provide very useful online information.

LO6 Understand the role and function of the immune system P6 D2

GETTING STARTED
(2 minutes)

As a class, list as many different types of infection as you can in two minutes.

6.1 How the innate immune system functions

Our environment is full of tiny organisms. Some of them (**pathogens**) have evolved to live inside the human body, causing damage.

Bacteria are single living cells, much smaller than ours. From their point of view it is full of nutrients and oxygen, warm and mobile – a very suitable home.

From our point of view, once inside, bacteria will use our resources to reproduce, multiplying the problem. They interfere with our body systems, they produce waste products for us to deal with, some of them toxic.

Viruses are quite different. They do not feed, grow, respire or do anything we would consider 'alive'. Once inside a living cell (a 'host'), their genetic material writes itself into the host's DNA. The cell now has only one function: to build more viruses until it bursts. The released viruses infect and destroy more cells.

Pathogens have useful proteins and carbohydrates on their outside surfaces. Human cells have these molecules too, but they are quite different shapes from those on the pathogens. This gives the **immune** system a way of telling 'self' cells from 'non-self' cells, identifying which cells to attack and which to leave alone. **Antigens** are molecules which the immune system recognises as non-self.

KEY TERMS

Pathogen – An infectious agent such as a bacterium, virus, fungus or parasite that causes disease in its host.

Bacteria – Single living cells.

Viruses – These are strands of genetic material surrounded by protein coats.

Immune – Resistant to a particular infection.

Antigen – A molecule which stimulates an immune response and causes the immune system to produce antibodies against it.

The body has several layers of defence.

The first, the innate immune system, responds in the same ways to any non-self antigen. It does not 'learn' from previous encounters.

Surface barriers

To prevent pathogens entering the body in the first place, the skin makes an excellent barrier. The outside layers are dead cells which are constantly being rubbed off and replaced. Something which is already dead cannot be infected. In places like the hair, glands secrete oils into the skin to make it more impermeable. When skin is damaged, blood clots in the wound, preventing pathogens entering.

The skin cannot be a complete barrier. Bacteria landing on the eyeball meet a thin layer of tears containing the enzyme lysozyme which digests their outer layers and swiftly bursts them. Pathogens inhaled into the lungs become stuck in the mucus lining the airways, which contains yet more lysozyme. The mucus is moved to the top of the throat and swallowed into the stomach where it joins pathogens eaten on food, in a bath of acid and digestive enzymes.

Complementary systems

The blood plasma carry over twenty different proteins, together called the **complement**. Complement is activated when one of its proteins binds to an antigen. This activates more molecules, which activate more again. The proteins swiftly organise themselves into assemblies, which puncture pathogens' surface membranes. (see Figure 4.11)

Cellular barriers

There are several types of white blood cell. The macrophages move themselves by a flowing motion. They recognise non-self antigens, then pursue, engulf and digest pathogens by phagocytosis. Macrophages patrol the body, some carried swiftly by the blood, others sliding between cells in the tissues.

One strength of the immune system is that one part will alert another part once an antigen is detected. Activated complement proteins also attract macrophages towards the site of infection and cause inflammation: capillaries become more permeable, letting more antibodies and macrophages out into infected tissues.

Natural killer cells

These **lymphocyte** cells recognise when a body cell is missing some self markers on its surface. This could be caused by the cell becoming cancerous, or by being infected with virus (which by working inside cells would be difficult for the immune system to detect). The cell is too much of a danger to be repaired or saved. The natural killer cell either attracts a macrophage to engulf and digest it, or kills the cell itself.

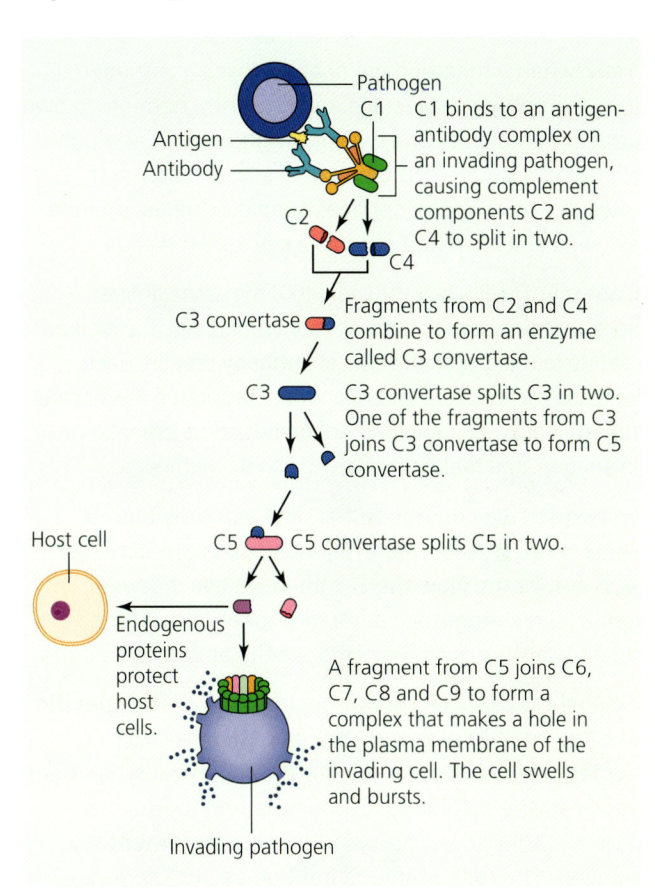

▲ Figure 4.11 Activation of the complement system by bound antibodies

6.2 How the adaptive immune system functions

Lymphocytes

The adaptive immune system 'learns' from each encounter with a pathogen. Next time it meets the same pathogen, it will make a much more rapid and powerful response. The adaptive system is controlled by the lymphocyte family of white blood cells, which arise from stem cells in the bone marrow.

> **KEY TERMS**
>
> **Complement** – A system of plasma proteins that can be activated directly by pathogens.
>
> **Lymphocyte** – A family of white blood cells with large nuclei found in the lymph system. Types of lymphocyte include T cells, B cells and natural killer cells.

The **immune response** is a drastic step. If it does not happen, the infection could kill the body. If it starts when it is not needed, or if it is accidentally directed against body cells, it will be damaging. It is set off by two types of cell working in parallel. Both types must agree before the response is initiated.

Firstly, when a macrophage phagocytoses a pathogen, it saves the antigen molecules and presents them on its own surface membrane. One type of lymphocyte, a T-helper cell, recognises the combination of self and non-self molecules on the macrophage. It rapidly divides, making new copies of itself and produces cytokine molecules.

Meanwhile, B cells have encountered the same antigen. There are many populations or 'clones' of B cell. Each clone is set up to make just one type of **antibody** protein. Each clone has samples of that antibody on its surface membrane. One clone of B cells will have antibodies complementary to the antigen, and that one will bind with the pathogen.

The two stories come together. The similarly bound T-helpers have secreted cytokines which are detected by the B cell clone. Now the B cells have two independent signals, which together stimulate them to rapidly divide and differentiate into B-memory cells and plasma cells.

The plasma cells secrete large amounts of the **specific** antibody into the plasma, and from there easily into infected tissues, thanks to inflammation which has been caused by the cytokines. Pathogens find themselves coated in antibodies, bound to their **complementary** antigens. The tails of bound antibodies stick to one another, clumping pathogens together, immobilising them. Nearby phagocytes target the antibody antigen complexes and engulf the pathogens. The complement system becomes activated, bursting bacteria open. All of this attack is directed solely at the antigen, hopefully minimising damage to healthy body cells.

Immunological memory and the use of vaccination

When the B and T cells divide, some of their offspring became 'memory cells', making those particular B and T clones much larger. The next time the immune system is presented with that same antigen, its response will be much faster because it will not need to spend as much time in selecting and dividing the correct B and T cells. A greater number of plasma cells will be produced from the larger clone, so the concentration of antibodies made in this secondary response will be much greater than in the original primary response. In the primary response, the pathogens were multiplying, leading to symptoms of infection. In the secondary response,

> **KEY TERMS**
>
> **Immune response** – The reaction of the body to something which is not recognised as part of the body itself.
>
> **Antibody** – Proteins made by plasma cells. They have variable regions, which give them an immense range of shapes. Each antibody recognises and binds with a specific shape of antigen.
>
> **Specific** – When a molecule will bind to only one shape of a second molecule.
>
> **Complementary** – The shape of one molecule fits exactly around the shape of a second molecule.
>
> **Vaccination** – A person's immune system is presented with the antigen without the risk of being harmed by the pathogen.

the pathogens have little chance to reproduce, so the person probably does not realise they have even suffered that second infection. They are now immune to the pathogen with that particular antigen.

In **vaccination**, a person's immune system is presented with the antigen without the risk of being harmed by the pathogen. The vaccine contains a weakened, dead or harmless form of pathogen, or the antigen alone. The vaccine is injected, or in some cases taken orally. The system makes the primary immune response, leaving behind populations of T and B memory cells. If the person is later infected by the living, dangerous pathogen, the memory cells are ready to mount the faster, stronger secondary response.

> **CLASSROOM DISCUSSION**
>
> **Getting the green light (20 minutes)**
>
> The immune system battling an infection could be seen as a story. It has heroes, villains, violence, chases, deception and complexity. What if it were adapted as a film? Have a class discussion to cast the movie. Come to an agreement over which actors should play the various molecules and types of cell. Have one section of the class as the studio art department, producing a mock-up billboard poster.

6.3 Disorders of the immune system

Autoimmune diseases

The immune system is quite capable of recognising self-antigens. Cells which do this are regularly destroyed.

In an autoimmune disease, the immune system mounts an immune response against its own cells. It has lost the ability to distinguish that a particular marker is 'self'. As examples, in rheumatoid arthritis the joints are attacked, type 1 diabetes targets cells in the pancreas and myasthenia gravis attacks the junctions between neurones and muscles.

Inflammatory diseases

In inflammation, capillaries become more permeable, more fluid than usual escapes and the tissues become swollen. This is quite normal in combating infection and aiding healing, but damaging if it happens over a long period. We have already seen examples of this in ulcerative colitis, rheumatoid arthritis and emphysema.

Cancer

Damage to the genes controlling growth can lead to a cell growing and dividing uncontrollably, and not carrying out their normal functions. There are many types of cancer, originating from different types of cell. In some types, self markers are changed. The immune system recognises them as non-self and destroys the whole cell with natural killer cells and another member of the T cells, the T-killer cells.

Some cancers have ways of evading the immune system. Their cells are more likely to invade and destroy surrounding healthy tissue.

Immunodeficiency

Immunodeficiency is when parts of the immune system have stopped working.

All parts of the system depend on proteins. Protein-poor diets, as those found in many developing countries, lead to reduced ability to make antibodies, complement and cytokines.

The very young and the old have a lower immune response, and it can also be caused by drug use, alcoholism and obesity.

> **KNOW IT**
> 1. Name two types of pathogen.
> 2. List the ways in which the body resists infections.
> 3. What are the functions of (a) macrophages (b) T-cells (c) B-cells?

L06 Assessment activities

Below are suggested assessment activities that have been directly linked to the Pass and Distinction criteria in L06 to help with assignment preparation and include top tips on how to achieve best results.

Activity 1 Do a table-top simulation on the parts of the immune response *P6*

Present the parts of the immune response as a table-top simulation to be run by several people. Each person plays a component of the immune system, for example as cell types, pathogens, antibodies. Produce plasticine pieces to represent their identity in the simulation. Write a guide for each participant, saying how they would respond to what the other participants are doing. Prepare a board, showing the locations where events are happening.

Make a photographic record of the board and pieces and possibly video the simulation in action.

Activity 2 Show how the immune response is overriden by a vaccine *D2*

Take the participants' guides that were written for Activity 1. Amend them to show how some parts of the immune response have been overridden by administering a vaccine. In the place where a participant's role has been changed, add a taped-hinged flap on a different coloured paper. The flap showing the changed events in a vaccinated person would overlay the events in the primary immune response.

Run the simulation to show how infection is prevented. Write an account of the differences in immune response in cases where a vaccine has been administered and where it has not.

> **TOP TIP**
> ✔ Be prepared to produce guides which have large numbers of instructions. The parts of the immune system are interdependent.

Read about it

Betts, M. (2004) *The Human Body,* Wooden Books

Parker, S. (2013) *The Concise Human Body Book (2nd edition),* Dorling Kindersley

Pickering, W.R. (2009) *AS and A Level Biology Through Diagrams:* Oxford Revision Guides, Oxford

Solomon, E., Martin, C., Martin, D. and Berg, L. (2015) *Biology* (10th edition), Thomson

Unit 05
Genetics

ABOUT THIS UNIT

This unit looks at how characteristics are inherited using the long-established techniques of crossing plants and animals, and in humans, by looking at the inheritance of certain characteristics and clinical conditions.

Advances in DNA technologies have demonstrated the potential of this work in several areas of science, including our understanding of disease and screening for inherited conditions, epidemiology and disease control, forensic science and historical and archaeological investigations. By studying this unit, you will be able to apply techniques used in genetics crosses using mathematical techniques to determine probability of inheritance by producing genetic chromosome maps using data from crosses. You will understand how geneticists are now able to produce detailed chromosome maps using the science of genomics.

LEARNING OUTCOMES

The topics, activities and suggested reading in this unit will help you to:

1 Understand the importance of meiosis
2 Be able to apply techniques used in genetic crosses
3 Understand the techniques of DNA mapping and genomics
4 Understand the impact of an innovation in an application of genomics

How will I be assessed?

You will be assessed through a series of assignments and tasks set and marked by your tutor.

How will I be graded?

You will be graded using the following criteria:

Learning Outcome	Pass	Merit	Distinction
	The assessment criteria are the Pass requirements for this unit.	**To achieve a Merit the evidence must show that, in addition to the Pass criteria, the candidate is able to:**	**To achieve a Distinction the evidence must show that, in addition to the Pass and Merit criteria, the candidate is able to:**
1 Understand the importance of meiosis	**P1** Describe the stages of meiosis	**M1** Explain the importance of meiosis	
2 Be able to apply techniques used in genetic crosses	**P2** Demonstrate the genotypes and phenotypes produced in monohybrid genetic crosses	**M2** Demonstrate that statistics can be used to compare expected and observed data in a cross	**D1** Discuss the limitations of genetic maps of chromosomes
	P3 Demonstrate the genotypes and phenotypes produced in dihybrid genetic crosses involving two non-linked autosomal genes	**M3** Construct a simple genetic map of a chromosome using data on recombinants from dihybrid crosses	
	P4 Use phenotypic ratios to identify gene linkage and epistasis		
3 Understand the techniques of DNA mapping and genomics	**P5** Describe the principles of DNA sequencing	**M4** Evaluate the advantages of next-generation DNA sequencing (NGS)	**D2** Evaluate the significance and limitations of genetic profiling techniques
	P6 Describe the principles of genetic profiling	**M5** Discuss the statistical significance of genetic profiling matches	
4 Understand the impact of an innovation in an application of genomics	**P7** Describe an application of genomics	**M6** Evaluate the impact of the specified genomic application	**D3** Assess the implications and future impact of the specified genomic technique

LO1 Understand the importance of meiosis *P1 M1*

Meiosis is described as a *reduction* division. This means that the cells that are produced have half the number of chromosomes of the parent cells, so a **diploid** cell will produce **haploid** cells by meiosis. This is the process by which **gametes** are formed. In **mitosis** (see Unit 8) the daughter cells are genetically identical to the parent cell, but in meiosis, the daughter cells are genetically different – both to the parent cell and to each other. This genetic difference is highly important, as we will see in the following sections.

GETTING STARTED

The importance of sexual reproduction (5 minutes)

Think about sexual and asexual reproduction. What are the advantages of sexual compared to asexual reproduction?

125

KEY TERMS

Meiosis – Nuclear division that produces four genetically different daughter nuclei.

Diploid – A cell that has two sets of chromosomes, two of each homologous pair.

Haploid – A cell that has only one set of chromosomes, just one of each homologous pair.

Gametes – Specialised sex cells such as egg (female gamete) and sperm (male gamete).

Mitosis – The division of a nucleus to form two genetically identical daughter nuclei.

Chromatids – These are the product of DNA replication, so that after replication a single chromosome consists of two identical DNA strands that are called (sister) chromatids held together by the centromere until the time comes for them to separate. Once separated, they are no longer referred to as chromatids – they are now chromosomes.

Homologous chromosomes – A pair of chromosomes, one inherited from each parent. Homologous chromosomes carry the same genes, but may carry different alleles of those genes.

Independent assortment – This is also known as random assortment or independent segregation and refers to the fact that each daughter nucleus produced in meiosis I receives at random either a maternal or paternal member of each homologous pair, i.e. the daughter nuclei don't contain all maternal or all paternal chromosomes. This is an important source of variation that will be discussed in the section 'The importance of meiosis'.

Recombine/recombination – In the context of meiosis, this means the exchange of genetic material (i.e. alleles) between maternal and paternal chromosomes by formation of chiasmata between homologous chromosomes.

1.1 Meiosis as a reduction division

Stages of meiosis

Meiosis involves two separate nuclear divisions (meiosis I and II). Before the start of meiosis, during interphase (see Unit 8), the chromosomes (DNA) replicate. This means that each chromosome consists of two sister **chromatids**.

Meiosis I

During meiosis I, the **homologous chromosomes** pair up and **independent assortment** occurs. In addition, **recombination** of genetic material between homologous chromosomes also occurs. In this way, the two daughter cells are genetically different to the parent cell.

Meiosis II

In meiosis II the division occurs at right angles to meiosis I. In many ways, the process is similar to that of mitosis.

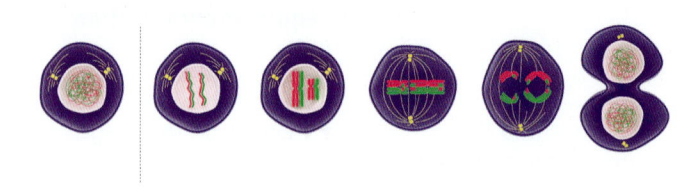

▲ **Figure 5.1** The stages of meiosis I and II

GROUP ACTIVITY

Modelling meiosis (20 minutes)

This will require quite a lot of cutting and colouring, so it is best done in a group so that the work can be shared out! The objective is to explore the role that random assortment plays in generating genetic variation – this will be covered further in the next section.

Make a set of model homologous chromosomes out of two different coloured cards, e.g. red for maternal and blue for paternal. You should make at least four pairs of homologous chromosomes. Each homologous pair should be a different length. You will need multiple copies of each, because you are going to simulate meiosis and, in particular, the random assortment of chromosomes. Each chromosome should consist of two chromatids that can be separated. It is easiest to do this if you use paperclips to hold the chromatids together. Use Figure 5.1 as a guide and run through the stages of meiosis using your model chromosomes. You could take pictures with a digital camera or smartphone to record this.

Then investigate how many different combinations of chromosomes you get in the different possible daughter cells produced at the end of meiosis I and meiosis II. This is why you will need multiple copies of the chromosomes!

If you want to be really adventurous, use stickers (or bits of Post-It notes) to represent recombination and see how many more combinations you will get. Hint: You will soon run out of card and Post-It notes!

The importance of meiosis

By halving the number of chromosomes, from diploid to haploid, gametes are produced (sperm and egg in animals) that can fuse in the process of fertilisation to form a **zygote**, restoring the diploid number of chromosomes. This brings together maternal and paternal **alleles** from two genetically different individuals, which is one source of variation. But the process of meiosis itself is the source of a great deal of variation due to the random assortment of the homologous chromosome pairs and recombination – as you will have found from all the cutting and sticking you did in the group activity.

Another source of variation during meiosis is mutation caused when mistakes in DNA replication cause changes in the sequence of the nucleotides.

KNOW IT

1. The terms *haploid* and *diploid* refer to the number of chromosomes. Think about the amount (or mass) of DNA in the cell instead. Draw a simplified diagram of the stages of meiosis, showing:
   a. the parent cell before DNA replication
   b. the start of prophase I
   c. the two daughter cells after meiosis I
   d. the four daughter cells at the end of meiosis II
2. If we assume that the amount of DNA in a haploid cell is X, label each cell with the amount of DNA it contains.
3. Complete the following passage by filling in the blanks: Mitosis produces two genetically _____ nuclei, whereas meiosis produces _____ nuclei that are genetically _____. The products of meiosis are called _____.
4. What are the two sources of genetic variation in meiosis?

L01 Assessment activities

Below are suggested assessment activities that have been directly linked to the Pass and Merit criteria in LO1 to help with assignment preparation and include top tips on how to achieve best results.

Activity 1 Describe the stages of meiosis *P1*

This could be done by preparing an annotated wallchart showing the stages of meiosis.

Activity 2 Explain the importance of meiosis *M1*

This could be done by adding further annotations to the wallchart for *P1* explaining the importance of meiosis. You may be required to write a detailed account of this subject.

TOP TIPS
- Make sure you understand the order of events in meiosis.
- Make sure you appreciate the importance of halving of chromosome numbers as well as independent assortment of genes and crossing over in gamete formation as sources of genetic variation.

KEY TERMS

Zygote – The diploid cell produced when haploid gametes (sperm and egg) fuse during fertilisation.

Alleles – Different versions of a gene. For example, the gene for seed shape in peas has alleles for smooth and wrinkled peas. Many genes can have more than two alleles.

LO2 Be able to apply techniques used in genetics crosses P2 P3 P4 M2 M3 D1

🔑 KEY TERMS

Gene – A section of DNA that codes for a protein, although some genes code for RNA or regulate other genes.

Phenotype – The characteristics expressed in an organism. In the case of the gene for eye colour, the phenotypes might be blue eyes or brown eyes. However, in many cases the phenotype depends not just on the genes (alleles) that are expressed, but also on environmental factors.

Genotype – The genetic makeup of an individual, specifically the alleles it contains. We usually describe the genotype as the alleles for a particular characteristic present in an individual.

Dominant – An allele that is expressed even if only one copy is present is known as dominant. In other words, the allele is expressed in heterozygous or homozygous individuals.

Recessive – An allele that is only expressed when two identical copies are present, i.e. only in a homozygous individual.

Codominant – Neither allele is dominant or recessive and both alleles contribute to the phenotype.

Heterozygous – When an organism has two different alleles for a particular gene.

GETTING STARTED

Logic puzzles and biochemistry (5 minutes)

Before launching into the rules and methods of genetics, it is worth thinking about how best to approach the subject and to try to understand **genes** and inheritance. Approach A: We can look at how factors are inherited and work out the rules and how to apply them. Approach B: We can try to work out what is happening at the molecular level – the different proteins and enzymes produced and what effect they have on the organism. Approach C: We can use a combination of both methods. Discuss with the group which approach you feel more comfortable with and then make a note of how many prefer each of A, B or C. It will be interesting to see if these numbers have changed after we work through this Learning Outcome.

2.1 Monohybrid inheritance

This refers to inheritance of just one gene or characteristic. Before looking at some examples, we need to be clear about the conventions used in creating genetic diagrams.

Conventions in genetic diagrams

- Start by showing the parental **phenotypes**, then the parental **genotypes** and then the gametes produced by each parent.
- The gene is usually represented by a single letter, with upper case for the **dominant** allele and lower case for the **recessive** allele, e.g. for eye colour, the brown allele (B) is dominant to the blue allele (b). However, this rule is broken so often, it's almost not worth calling it a rule! The important thing is to be able to distinguish between the alleles.
- Some genes have more than two alleles. Although any individual can only have two alleles, there can be more than two alleles in the population. We show this by using an upper case letter for the gene and show the alleles by superscript letters. For example, the alleles for the human blood groups are I^A, I^B and I^O.
- The same convention is sometimes used in **codominant** inheritance.
- Other conventions will become apparent as we go along!

Use of the Punnett square

The Punnett square was devised by Reginald Punnett who was Professor of Biology in the University of Cambridge in the early years of the 20th century. It is the method we use to work out the various combinations of alleles in genetic crosses; in other words, the proportion of different genotypes and the corresponding phenotypes.

Monohybrid inheritance of a normal trait

A simple example is inheritance of eye colour. The allele for brown eyes (B) is dominant to the allele for blue eyes (b). In a family with **heterozygous** brown-eyed parents, for any child they have, there will be a 75% probability of it having brown eyes and a 25% probability of having blue eyes. This is shown in Table 5.1.

Table 5.1 Monohybrid inheritance of eye colour

Parental phenotypes	Mother (brown eyes)		Father (brown eyes)	
Parental genotypes	Bb		Bb	
Gametes	B	b	B	b

We can then use a Punnett square to work out all the possible combinations of alleles and their corresponding phenotypes.

UNIT 5 GENETICS

128

Table 5.2 Punnett square of mono

	Male gametes	
Female gametes	B	b
B	BB Brown eyes	Bb Brown eyes
b	Bb Brown eyes	bb Blue eyes

From the Punnett square we can see the ratio of phenotypes: 3 brown eyes (75%) to 1 blue eyes (25%). Remember that these are really probabilities. A brown-eyed couple with these genotypes will have a 25% chance of having a blue-eyed child. If they have a large enough family, then one quarter will be blue-eyed, but the parents will probably be too tired to count.

In fact, inheritance of eye colour is more complicated than this. There is evidence that as many as 16 genes are involved in determining eye colour in humans.

Monohybrid inheritance of a single gene disorder

We are ready to start thinking about inheritance not just as a logic puzzle, but about the underlying molecular mechanism. First, let's be clear that there are no genes that code for a genetic disease. Genes code for proteins such as enzymes or structural proteins such as muscle or collagen (see Biochemicals in section 4.4 of Unit 1). An altered (mutant) gene might produce an altered or even non-functional protein that leads to the expression of a disease.

Haemoglobin consists of four polypeptide chains, two α chains and two β chains. In the case of sickle cell anaemia, there is an alteration in one amino acid of the β chain. Normal haemoglobin has glutamic acid at position 6 but in sickle-cell anaemia valine is present instead. This changes the structure of the β chain and this leads to the symptoms of the disease.

We can represent the normal allele as H^A and the mutant gene as H^S. A heterozygous individual (H^AH^S) will have one allele (H^S) producing mutant haemoglobin and will have one normal H^A allele, so at least half the haemoglobin will be normal and they will not show symptoms. They will however be a **carrier** of the disease; this is sometimes called 'sickle-cell trait'. If two carriers (heterozygous individuals) have children, there is a 25% chance they will have full sickle-cell disease, because the H^SH^S genotype means that all the individual's haemoglobin is the mutant form and will cause the disease. Table 5.3 illustrates this – notice the similarities with the inheritance of eye colour above.

Table 5.3

Parental phenotypes	Mother: normal (sickle-cell trait)		Father: normal (sickle-cell trait)	
Parental genotypes	H^AH^S		H^AH^S	
Gametes	H^A	H^S	H^A	H^S

	Male gametes	
Female gametes	H^A	H^S
H^A	H^AH^A Normal	H^AH^S Normal (sickle-cell trait)
H^S	H^AH^S Normal (sickle-cell trait)	H^SH^S Sickle-cell anaemia

Monohybrid inheritance of codominant alleles

Codominant alleles are neither dominant nor recessive, but the phenotype depends on the combination of alleles. The human ABO blood groups are an example of co-dominance involving three alleles, I^A, I^B and I^O. Each allele produces different variant of an antigen on the surface of red blood cells. Group A individuals have the A antigen, Group B individuals have the B antigen and AB have both while O have neither.

I^A allele produces the A antigen.

I^B allele produces the B antigen.

I^O allele doesn't produce either antigen.

Incomplete inheritance

This is also known as incomplete dominance or partial dominance. An example is inheritance of flower colour in snapdragons. **Homozygous** plants can have white flowers or red flowers. If a plant with white flowers is crossed with a plant with red flowers, the offspring (which must be heterozygous) have pink flowers. If we assign the genotype C^+C^+ to red flowers and C^-C^- to white flowers, the pink flowers will be C^+C^-.

PAIRS ACTIVITY

Codominance and incomplete inheritance (20 minutes)

Working in pairs, try the following:
- Work out the genotypes associated with the three ABO blood groups.
- In the days before DNA profiling was used in paternity testing, blood groups could be used to at least disprove paternity. Work out the various blood group combinations of mother and supposed father that would disprove paternity.

Sex-linked inheritance

In humans, there are 23 pairs of chromosomes. One pair (number 23) is known as the sex chromosomes; the other 22 pairs are called autosomes. All of the examples we have looked at so far involve autosomal inheritance – the genes are carried on the autosomes. However, the sex chromosomes are different to the autosomes because, in males, one member of the pair (the Y chromosome) is much shorter than the other (the X chromosome). Human males are XY and females are XX. Most of the alleles carried on the X chromosome do not have a corresponding allele on the Y chromosome. This means that a male who carries a recessive allele on his X chromosome will express that phenotype, but in a female the recessive allele is only expressed when homozygous; a heterozygous female is called a carrier.

One of the earliest examples of this being recognised is in the disease haemophilia A. This is caused by a mutation in the gene for a blood clotting factor (factor VIII) carried on the X chromosome. The normal allele (H) is dominant to the mutant allele (h). We can represent the various genotypes as follows:

$X^H X^H$ Normal female

$X^H X^h$ Carrier female

$X^H Y$ Normal male

$X^h Y$ Haemophiliac male

PAIRS ACTIVITY

The descendants of Queen Victoria (15 minutes)

Use a Punnett square to work out the probability of a carrier female and a normal male having a son with haemophilia.

Now look at Figure 5.2. Can you work out any rules that help to identify this as sex-linked inheritance? Why are there no females with the disease? Work out what it would take for a female to inherit haemophilia. Is the British royal family at risk?

KEY TERMS

Carrier – A heterozygous individual in which the recessive allele causes a disease in the homozygote. The carrier does not show symptoms, but can pass the disease on to their children.

Homozygous – When an organism has two identical alleles for a particular gene.

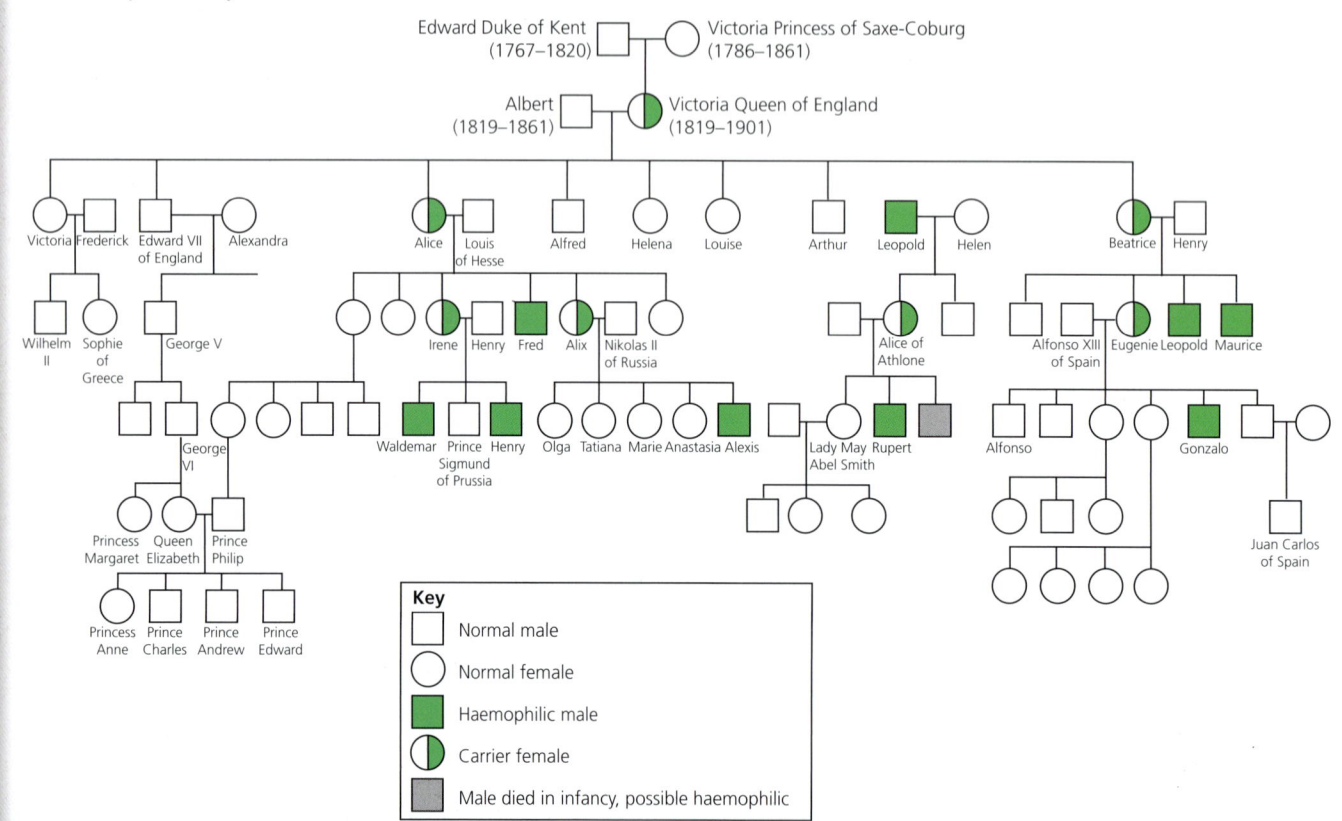

▲ Figure 5.2 Inheritance of haemophilia in the descendants of Queen Victoria

Genetic counselling

Family trees can be used to identify genetic diseases in families and work out the genotypes of couples planning to have children. This allows a genetic counsellor to calculate the risk of any children inheriting the disease. Of course, this only informs the couple of the percentage risk: it can't remove the risk. In some cases of sex-linked inheritance, there will be a 100% chance that any sons will have the disease, so a decision could be taken to abort a male fetus. This is just one of many ethical issues to be considered.

You will see in section 4 of this unit that modern molecular techniques can be used to give a clearer insight into the risk of inherited disorders being passed on, although many ethical issues remain. For information on the implications for drug development see sections 1.2 and 1.4 of Unit 11.

2.2 Dihybrid inheritance

The dihybrid cross – predicting genotype and phenotype ratios

Dihybrid inheritance involves two genes or characteristics. This was studied by Gregor Mendel in his experiments with peas. He crossed two pure-bred strains of peas. One had wrinkled green seeds and the other had round yellow seeds. All the offspring (F_1 generation) had round yellow seeds. From this we can conclude that round (R) is dominant to wrinkled (r) and yellow (G) is dominant to green (g). (In case you are wondering, we are using G for the yellow allele instead of Y to avoid confusion with Y chromosomes.) He allowed the F_1 plants to self-pollinate and collected a large number of seeds. When these seeds were grown, he found the F_2 generation had round yellow, round green, wrinkled yellow and wrinkled green phenotypes in the ratio 9:3:3:1.

We can use a genetic diagram to explain this. The first cross is shown in Table 5.4.

Table 5.4 Dihybrid cross of two pure-bred strains

Parental phenotypes	Round yellow seeds	Wrinkled green seeds
Parental genotypes	RRGG	rrgg
Gametes	RG	rg
Offspring (F_1) genotype	RrGg	
Offspring (F_1) phenotype	Round yellow seeds	

When the F_1 generation self-pollinates, we get a 9:3:3:1 ratio of phenotypes in the F_2 generation: see Table 5.5.

You can use the same approach to work out the possible genotype and phenotype ratios of any cross involving two genes as long as they are autosomal (i.e. not carried on the sex chromosomes) and are on different chromosomes. The significance of the latter point will become clear in the section on gene linkage below.

The test cross

An individual expressing a dominant phenotype may be homozygous or heterozygous. To determine which it is, we can cross it with an individual expressing a recessive phenotype, because this must be homozygous recessive. This works with both monohybrid and dihybrid inheritance. However, it is only practical in organisms where controlled mating / pollination is possible. In human populations we can draw similar conclusions about the genotype of one parent when the other parent shows a recessive phenotype by looking at the phenotypes of the children.

Table 5.5 Dihybrid cross of F_1 generation producing a 9:3:3:1 ratio of phenotypes

Parental phenotypes	Round yellow seeds				Round yellow seeds			
Parental genotypes	RrGg				RrGg			
Gametes	RG	Rg	rG	Rg	RG	Rg	rG	rg

	Male gametes			
Female gametes	**RG**	**Rg**	**rG**	**rg**
RG	RRGG Round, yellow	RRGg Round, yellow	RrGG Round, yellow	RrGg Round, yellow
Rg	RRGg Round, yellow	RRgg Round, green	RrGg Round, yellow	Rrgg Round, green
rG	RrGG Round, yellow	RrGg Round, yellow	rrGG Wrinkled, yellow	rrGg Wrinkled, yellow
rg	RrGg Round, yellow	Rrgg Round, green	rrGg Wrinkled, yellow	rrgg Wrinkled, green

GROUP ACTIVITY

Work out and discuss (20 minutes)

Choose an example of dihybrid inheritance. Use genetic diagrams to work out the ratio of phenotypes of the offspring of a test cross between a homozygous recessive individual and (a) a homozygous dominant or (b) a heterozygous dominant individual. Then, do the same for an example of dihybrid inheritance. Make sure the group has worked out all the possible combinations of the 'unknown' genotypes to be tested before you start constructing genetic diagrams. Once you have worked out the ratios of the phenotypes in the F_1, discuss whether this is a useful test to determine an unknown genotype. How useful would it be in plant and animal breeding? Can it be applied to human genetics?

KEY TERMS

Null hypothesis – In statistics, this is the hypothesis that says there is no effect or no difference between two measurements.

Expected and observed data in a cross

After considering examples from human genetics, we suddenly changed to Gregor Mendel's peas. One reason for this is that crossing the F_1 generation with itself is possible in plants, but not in humans (it would involve brother-sister mating). Even if we carry out the experiments in animals such as mice, the litter sizes are still too small for us to be able to see the ratios clearly. However, with peas we can carry out hundreds of crosses and the large sample sizes make it easier to see the ratios. But we still need to be sure that the numbers we obtain really match the expected ratios. If we expect a ratio of 9:3:3:1, then in a cross that produces 800 offspring, the expected numbers will be 450, 150, 150 and 50. If the observed numbers are similar to the expected numbers, we can assume that any differences are simply due to chance. If the observed numbers are far away from the expected numbers, then we need to re-think our conclusions.

But what counts as 'similar' and 'far away'? To answer this, we use a statistical test called the chi-squared test (χ^2 test). The 'ch' in 'chi' is hard, like the letter 'k' and the 'i' is long, like 'eye'.

The χ^2 test

This uses a formula that looks quite scary, but is actually fairly easy to use:

$$\chi^2 = \sum \frac{(O - E)^2}{E}$$

This says that, to calculate χ^2 we take the difference between the observed value (O) and the expected value (E) and square it, and then divide the answer by the expected value. We then repeat this for every other set of observed/expected values and add them up (that's what the Σ means). Why? Well, the differences could be positive or negative, so we take the square to make them all positive, otherwise positives and negatives could cancel. Dividing by E takes account of the size of the numbers as well as the differences. Taking the sum will take account of the number of comparisons. It's not really necessary we understand exactly why it works; we just need to be able to use it.

Having calculated χ^2, we can then use a statistical table to compare the value we have calculated with the 'critical value'. If our value for χ^2 is larger than the critical value then we reject the **null hypothesis**, meaning it is probable that any difference between our results and the expected results is not due to chance (i.e. it is statistically significant). If our value for χ^2 is smaller than the critical value, then we accept the null hypothesis, meaning the differences are probably due to chance. All this talk of probability and chance will probably make more sense if you follow the worked example below.

Finally, the small print. The χ^2 test can only be used for categorical data – meaning data that we can divide into categories or classes. There also has to be a strong biological theory we can use to predict the expected values. As well as this, the sample size must be relatively large; E must be greater than five for each category (so it probably won't work on inheritance in human families). We have to use raw counts, not percentages or ratios – remember that we divide by E to take account of the size of the numbers; if we have already done this in calculating a percentage, we negate the effect of this. Also, the test doesn't work if there are any zero scores.

If we take the example of the dihybrid cross with peas, we can look at a real example of how to use the χ^2 test. A total of 381 pea plants were produced in the F_2 generation and their phenotypes were as shown in Table 5.6 on the next page. We calculate the expected number based on a 9:3:3:1 ratio (i.e. 9/16, 3/16, 3/16 and 1/16).

Table 5.6 χ^2 analysis of dihybrid cross

Class / category (phenotype)	Observed (O)	Expected (E)	O − E	(O − E)²	$\frac{(O-E)^2}{E}$
Round yellow	216	381 × 9/16	214	2	4	0.019
Round green	79	381 × 3/16	71	8	64	0.901
Wrinkled yellow	65	381 × 3/16	71	−6	36	0.507
Wrinkled green	21	381 × 1/16	24	−3	9	0.375
Total	381			$\chi^2 =$	1.802

Having calculated a value for χ^2 we can now use a table of χ^2 values (Table 5.7) to convert this to a probability (represented by the symbol 'p'). If $p < 0.05$ we reject the null hypothesis because there is a less than 5% probability that the differences between observed and expected values were due to chance. On the other hand, if $p > 0.05$ then we have to accept the null hypothesis because there is more than 5% probability that the differences between observed and expected were due to chance. The $p = 0.05$, or 95% probability level is accepted by statisticians as a cut-off point. We can use Table 5.7 to look up the probability level for our calculated value of χ^2. We have four classes (categories of data) which corresponds to 3 degrees of freedom. If you are using published χ^2 tables you may only see degrees of freedom, which is the number of categories minus 1.

Our calculated value of χ^2 is 1.802, which is smaller than the critical value at 3 degrees of freedom and so we accept the null hypothesis. In fact, our value for χ^2 is less than 2.37 and so we can say that there is a greater than 50% probability that the differences between the experimental values and the expected values are due to chance. Therefore, we can be confident that the ratio is actually 9:3:3:1.

Gene linkage and epistasis

Linkage

We now know that genes are carried on chromosomes. In fact, they are simply sections of the DNA molecule that makes up a chromosome. This is why some genes are always inherited together – they appeared to be 'linked' in inheritance. We now know that they are actually physically linked, because they are carried on the same chromosome and are part of the same DNA molecule. The cross between pea plants producing round yellow seeds and wrinkled green seeds does not show linkage because the gene for seed colour (yellow or green alleles) is on chromosome 1 and the gene for pod texture (smooth or wrinkled alleles) is on chromosome 4.

Table 5.7 Table of χ^2 values

Number of classes	Degrees of freedom	Chi squared							
2	1	0.00	0.10	0.45	1.32	2.71	3.84	5.41	6.64
3	2	0.02	0.58	1.39	2.77	4.61	5.99	7.82	9.21
4	3	0.12	1.21	2.37	4.11	6.25	7.82	9.84	11.35
5	4	0.30	1.92	3.36	5.39	7.78	9.49	11.67	13.28
6	5	0.55	2.67	4.35	6.63	9.24	11.07	13.39	15.09
p =		0.99	0.75	0.50	0.25	0.10	0.05	0.02	0.01
Percentage probability that difference is due to chance		99%	75%	50%	25%	10%	5%	2%	1%
		Accept null hypothesis					Critical value	Reject null hypothesis	

Understanding of the details of meiosis explains another observation. Sometimes the linkage between genes is 'broken' and they are not inherited together. This is due to the crossing over or recombination that occurs during the first meiotic division (see section 1.1). This leads to a different ratio to that expected. Linked genes are inherited as if they were a single gene (monohybrid cross) but recombination leads to genotype (and therefore phenotype) ratios more like those from a dihybrid cross. The farther apart the genes are on the chromosome, the greater the likelihood of recombination and, therefore, the closer the ratio is to that of a dihybrid cross. This has been used to map the genes on chromosomes. Genes on the same chromosome will be linked in inheritance and the greater the degree of linkage, the closer together the genes must be.

Epistasis

Epistasis is a type of interaction between genes that may be on the same or different chromosomes (i.e. it doesn't involve linkage) where an allele of one gene masks the expression of alleles of another gene. Epistasis can be recessive, dominant or the genes can work in a complementary fashion. This can be considered as a logic puzzle, but we can also understand it in terms of biochemical pathways and the fact that genes simply code for proteins.

The key point to remember about epistasis is that it means the phenotype ratio will be different to the usual 9:3:3:1 that you expect in dihybrid crosses.

> **? THINK ABOUT IT**
>
> **Epistasis and linkage (30 minutes)**
>
> Your tutor will provide you with a number of different dihybrid crosses. Use the number of each phenotype to work out the ratio and decide whether this is the normal dihybrid ratio of 9:3:3:1 or if it shows evidence of epistasis or linkage. Work out the expected ratio of phenotypes using a Punnett square. Then compare the expected numbers with the actual numbers using a χ^2 test.

Gene linkage and epistasis are two quite difficult concepts in genetics. However, they are both much easier to understand if we appreciate the fact that genes are carried on chromosomes which produce gametes by the process of meiosis and that genes also code for proteins, which can be enzymes or inhibitors or regulators. The interactions between genes can be quite complex with some characters being controlled by many genes as well as environmental influences. Height in humans is determined about 80% by genes, of which about 50 have been identified so far.

Genetic maps

Simple genetic maps from recombinant data

From our consideration of linkage and how genes may be carried on the same chromosome, we can see how it is possible to construct a map of where the genes are located on the chromosomes. Genes that are linked in inheritance are carried on the same chromosome, so we can assign genes to particular chromosomes. Recombination of genes happens when crossing-over occurs between them and the recombination frequency is proportional to the distance between the genes. This allows us to construct a map showing the relative distances between different genes on a single chromosome.

If you think about it, the maximum recombination frequency must be 50%. This means that half of the time genes linked on the same chromosome are actually separated during gamete formation. This is what you expect when the alleles are carried on different chromosomes, so you can't have more recombination than that. In practice, you seldom see 50% recombination between linked genes – we will look at the reason for this in a moment.

We said that recombination frequency is proportional to the distance between the linked genes, and two units of measure have been used based on this: the map unit (m.u.) and the centimorgan, named after the American geneticist Morgan whose work on locating genes on chromosomes helped to transform the field of genetics. A recombination frequency of 1% equates to either 1 m.u. or 1 centimorgan.

PAIRS ACTIVITY

Map this (15 minutes)

Working in pairs, use the following recombination frequency data to be converted into a chromosome map.

Drosophila have three loci on one chromosome (how do we know they are on the same chromosome?):

1 v (vermilion eyes) versus v+ (red eyes)
2 cv (crossveinless, or absence of a crossvein on the wing) versus cv+ (presence of a crossvein)
3 ct (cut, or snipped wing edges) versus ct+ (smooth wing edges)

A series of genetic crosses were set up and the results of the final test cross are shown in Table 5.8 on the next page. The total number of flies produced from the test cross was 1448. Calculate the recombination frequency for each comparison.

Table 5.8 Recombination frequencies

Gene loci compared	Number of recombinants	Recombinant frequency
v and cv	268	
v and ct	191	
ct and cv	93	

Use the recombination frequency to create a map showing the relative positions of the three loci on the chromosome. Do the numbers add up? Suggest why they might not – this is dealt with in the next paragraph.

Limitations of genetic maps of chromosomes

The process that causes recombination – chiasma formation during meiosis – can also restore linkage between genes. This is because more than one chiasma can form along the length of the chromosome. For two genes that are relatively far apart there is a good chance that two chiasmata will form, restoring linkage. For this reason, accurate mapping using recombination frequencies only works over relatively short distances. Functional maps of whole chromosomes can be made by studying linkage between groups of genes that are relatively close together and then assembling a whole map by putting together the smaller maps.

PAIRS ACTIVITY

Think about it (5 minutes)

How would you go about constructing a map of a whole chromosome given a set of recombination data? What is the importance of overlap in sets of data?

KNOW IT

1. Describe the conclusions of the following statistical tests using the terms *probability* and *chance*:
   a) A genetic cross produced five different phenotypes. A χ^2 test compared the observed and expected numbers of the different phenotypes and gave a value of $\chi^2 = 2.41$ (use Table 5.7).
   b) In another test to compare differences between two sets of data, a value of $p < 0.02$ was obtained.
2. What is the difference between *linkage* and *epistasis*?
3. A man was blood group O and his wife was blood group A. They had a child who was blood group B. The man was suspicious and thought his wife had been unfaithful, so he consulted a divorce lawyer. Use a genetic diagram to help the lawyer construct a case.

LO2 Assessment activities

Below are suggested assessment activities that have been directly linked to the Pass, Merit and Distinction criteria in LO2 to help with assignment preparation and include top tips on how to achieve best results.

Activity 1 Demonstrate the genotypes and phenotypes produced in monohybrid genetic crosses *P2*

You need to be able to demonstrate the genotypes and phenotypes produced in monohybrid crosses. You should also understand the Law of Segregation; that paired genes separate in such a way that each gamete is equally likely to contain either member of the pair.

Activity 2 Demonstrate the genotypes and phenotypes produced in a dihybrid genetic crosses involving two non-linked autosomal genes *P3*

You should appreciate the Law of Independent Assortment, i.e. that allele pairs separate independently during the formation of gametes. Examples should involve two non-linked autosomal genes giving a characteristic F_2 generation 9:3:3:1, as well as examples of 1:1:1:1 ratios in test crosses.

Activity 3 Use phenotypic ratios to identify gene linkage and epistasis *P4*

You should understand that sometimes the F_2 generations do not give the 9:3:3:1 ratio expected; some phenotypes occur more frequently than Mendelian genetics would predict. This leads to an understanding of how genes influence each other, or are physically coupled or linked in some way on the chromosome.

You should also be able to understand epistasis, where crossing two dihybrids produce a modified Mendelian ratio, e.g. instead of a 9:3:3:1 ratio, a 9:7 ratio, 15:1 ratio, etc.

Activity 4 Demonstrate that statistics can be used to compare expected and observed data in a cross M2

You should use the chi-squared test (χ^2 test) to compare expected and observed progeny in a cross. It is important that you can show understanding of statistical significance of differences in data, and probabilities of (apparent) differences having occurred by chance.

Activity 5 Construct a simple genetic map of a chromosome using data on recombinants from dihybrid crosses M3

You should be familiar with DNA recombination during meiosis and its importance in breaking of linkage between genes on the same chromosomes.

You should understand the procedure and the numerical components involved in producing a simple genetic map of a chromosome.

Activity 6 Discuss the limitations of genetic maps of chromosomes D1

Chromosome mapping by counting recombinant phenotypes produces a genetic map of the chromosome and you need to understand the limitations of the technique, e.g. multiple cross-over restoring linkage. You also need to be aware that due to advances in DNA, it is now possible to produce physical maps of the genome.

> **TOP TIPS**
> ✓ Make sure you understand the relationship between genotypes and phenotypes at a molecular level.
> ✓ Make sure you can produce genetic diagrams to illustrate monohybrid inheritance involving:
>   - incomplete dominance
>   - codominance
>   - sex-linkage.
> ✓ Make sure you understand the χ^2 (as in opposite paragraph) test and how to use it.

LO3 Understand the techniques of DNA mapping and genomics P5 P6 M4 M5 D2

> **KEY TERMS**
> **Genome** – The full set of genes contained in an organism. A better definition is that it includes all the genetic material (DNA or RNA) in an organism, which means genes and non-coding sequences of DNA or RNA.
> **Base pairs** – This is the standard unit of size (length really) for DNA. You might come across bp or even kbp (thousands of base pairs). For single stranded DNA the corresponding unit of length is b (base) or kb (1000 bases).

> **GETTING STARTED**
> **The need for sequencing (5 minutes)**
> Throughout the unit so far we have looked at various aspects of genetics. We have learned about the ways we can understand inheritance based on the laws of genetics and an appreciation of the molecular and biochemical mechanisms involved in the expression of genes and alleles.

Now we are going to study genetics at the most fundamental level – the DNA. We have already learned about how the sequence of DNA determines the sequence of proteins. Think about what we could learn about genes if we could just read off the sequence of the DNA or RNA. What would that tell us that we don't already know? Share your thoughts with the group.

Mapping and sequencing the **genome** provides insight into how genes and alleles are organised and how variation arises. It also provides the basis of understanding how gene expression is regulated and gives us a range of powerful tools. These can be used in applications as diverse as personalised medicine, determination of family relationships and forensics.

3.1 Deoxyribonucleic acid (DNA)/whole genome sequencing

One thing to bear in mind is that the human genome is very, very large. Even a single chromosome (one DNA molecule) can contain between 51 million and 259 million **base pairs** – see section 3.2 in Unit 1 for more information. Much of the early work was

> **KEY TERMS**
>
> **Restriction endonuclease** – Sometimes called restriction enzymes, these are enzymes that cut (cleave) DNA at a specific sequence of bases, usually 4–8 base pairs long (the restriction sequence or site for that particularly enzyme). Some restriction endonucleases produce a 'staggered' cut, so that the resulting fragments have 'sticky ends'.
>
> **Genomics** – The study of the genome, its sequence, structure and function.
>
> **Gene technology** – This is the study and manipulation of DNA to provide practical applications.

done using simple organisms with smaller genomes. However, it is still necessary to break down the DNA into manageable pieces – DNA fragments. This is done using **restriction endonucleases**. Having prepared fragments, the sequence of individual fragments can then be determined.

Some restriction endonucleases cut straight across the double helix, while others produce fragments with 'sticky ends', for example the enzyme *Eco*R1:

Recognition sequence   Cut, showing 'sticky ends'

5' GAATTC   5'---G   AATTC---3'

3' CTTAAG   3'---CTTAA   G---5'

These are called 'sticky ends' because they have complementary sequences – this fact is exploited in many areas of **genomics** and **gene technology**.

Another important feature of restriction endonucleases is that we always know the base sequence at the ends of the fragments produced. This proves very useful, as you will see.

The Polymerase Chain Reaction (PCR)

PCR was originally developed as a means of amplifying DNA fragments, i.e. take just a few DNA molecules and replicate them to produce sufficient copies to use

PAIRS ACTIVITY

Draw me (10 minutes)

Work in pairs for this activity. Use the following DNA sequence and make two copies on card, one for each member of the pair:

AGCTTCGATATCCTACCGTAATCGCCCGGGTACGCGCGTAAACCTGGATATCCT

TCGAAGCTATAGGATGGCATTAGCGGGCCCATGCGCGCATTTGGACCTATAGGA

Work out where the following restriction enzymes will cut the sequence:

Table 5.9 Restriction enzyme recognition sequences

Enzyme	Recognition sequence	Cut products
*Sma*I	- - -CCCGGG- - - - - -GGGCCC- - -	- - -CCC GGG- - - - - -GGG CCC- - -
*Eco*RV	- - -GATATC- - - - - -CTATAG- - -	- - -GAT ATC- - - - - -CTA TAG- - -

One member of the pair should choose one enzyme, the other member the other enzyme. Cut your card wherever you see a cleavage site matching your enzyme to produce two or more DNA fragments. Then combine your sets of fragments and mix them up. If the fragments produced were sequenced, could they be overlapped to generate the sequence of the original DNA molecule? Would this method work with a full-length chromosome? If not, how many different restriction endonucleases do you think you would need?

in analysis or gene technology. This is illustrated in Figure 5.3.

Primers will bind specifically to the ends of the DNA single strands (this process is known as **annealing**) through the usual base pairing process. We know the sequence of bases at the ends of our fragments (because we know which restriction endonuclease was used to prepare the fragments), so we know what primer sequence to use.

> 🔑 **KEY TERM**
>
> **Anneal/annealing** — The process of forming a double stranded DNA molecule (dsDNA) from complementary single stranded DNA (ssDNA). The ssDNA is usually produced by heating dsDNA to separate the strands and annealing occurs when the solution is cooled. The process can be used to attach complementary primers to a DNA fragment being prepared for PCR.

The enzyme DNA polymerase is then used to synthesise complementary strands to make a double strand from each single strand. We need to use primers because DNA polymerase needs a short section of double stranded DNA to start making the complementary strand. Also, the enzyme used (Taq Polymerase) comes from a bacterium that lives in hot springs, so it has an unusually high optimal temperature, meaning we have to heat the mixture to 72°C. The result is two new DNA molecules for each one in the original sample. This process can be repeated, so after two cycles the two have become four, which become eight after three cycles, and so on. You can see that you will very quickly generate a large number of identical copies of the original DNA.

The process is carried out in an instrument called a thermocycler, programmed to heat to 95°C, cool to 55°C and then heat back to 72°C, repeated as many times as are required.

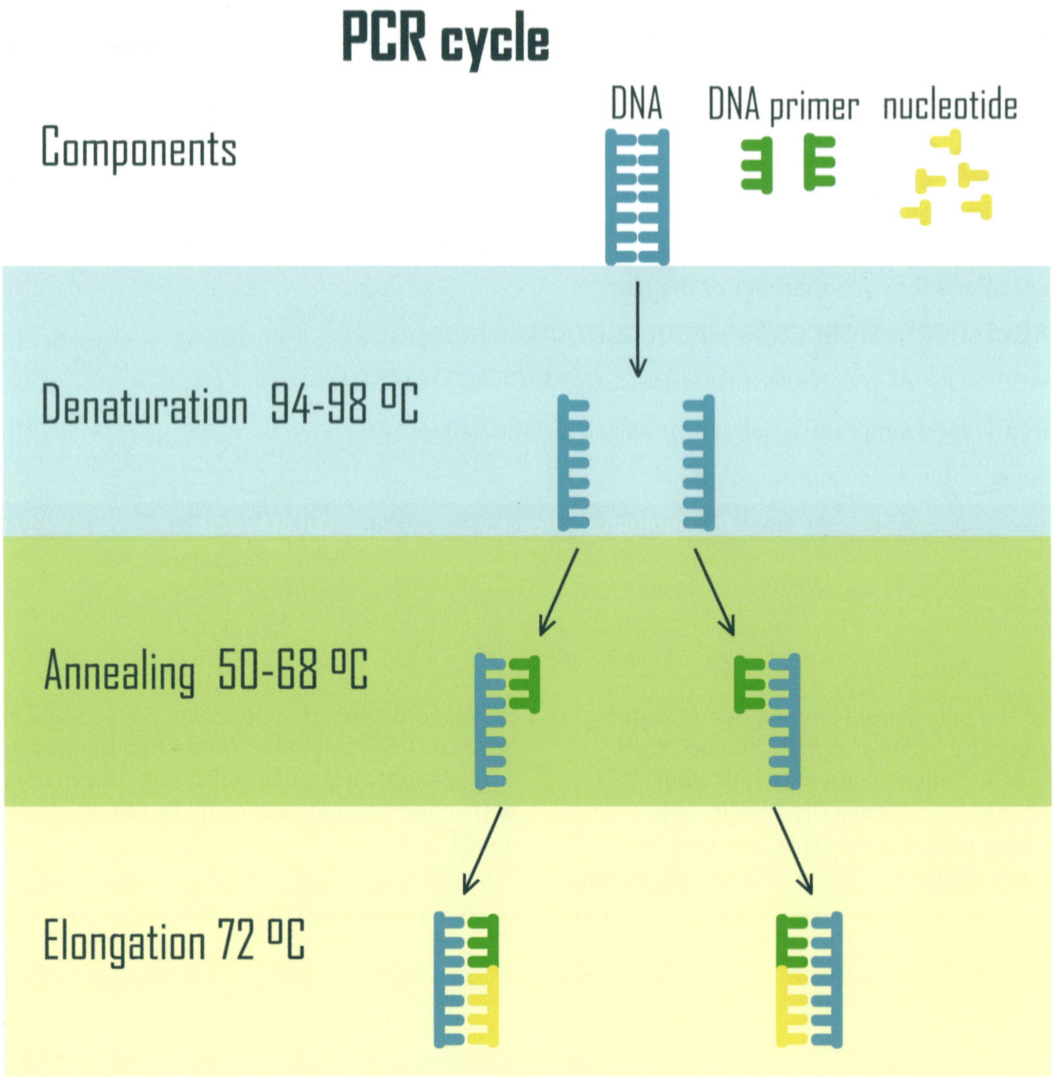

PCR cycle

Components: DNA, DNA primer, nucleotide

Denaturation 94-98 °C

Annealing 50-68 °C

Elongation 72 °C

▲ **Figure 5.3** The Polymerase Chain Reaction showing a single cycle. The amount of DNA doubles after every cycle

The Sanger sequencing process (chain-termination or dideoxy method)

This process was developed by Fred Sanger, working at the MRC Laboratory of Molecular Biology in Cambridge and published in 1977; it earned him his second Nobel Prize in Chemistry in 1980 (the first one was in 1958 for determining the structure of insulin).

Sanger's method is probably easier to understand if we look at a more recent adaptation – known as dye terminator sequencing. In this method, a single incubation is carried out with all four regular nucleotides. In addition, a small proportion of terminator nucleotides (bases) are included. These can be incorporated into the growing chain but the chain cannot then be extended. Each terminator base is labelled with a different fluorescent dye. This means that some chains will be terminated and each fragment will have the colour fluorescence that corresponds to the nucleotide at the end of the chain. Automated gel electrophoresis and fluorescence detection allows the sequence to be read off by computer (see Figure 5.4).

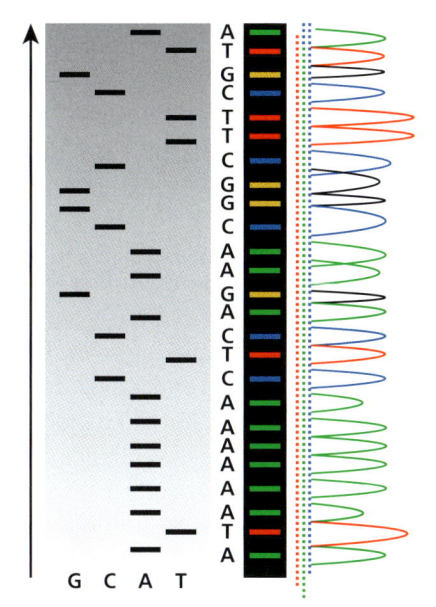

▲ **Figure 5.4** Sanger sequencing by gel electrophoresis (left) or fluorescence (right)

Next-generation DNA sequencing (NGS, or high-throughput sequencing)

Although dye-terminator sequencing allowed automation and, therefore, much quicker progress, it still has limitations. It only works reliably up to about 800–900 bp length fragments and can only sequence a relatively small number of fragments at a time. Next-generation sequencing methods allow for the sequencing of many fragments in parallel and have revolutionised the field of **molecular biology**. The Sanger method made it possible to sequence the entire human genome. However, that was just one representative genome. It is now possible to sequence the genome of individual humans in a relatively short time and for a reasonable cost.

There are various methods of NGS, all of which rely on cleaving the DNA in to short sections.

The Illumina method

In the Illumina method, fragments of 100–150 bp are used. These are attached to a chip rather like a microscope slide and PCR carried so that, where there was previously a single DNA fragment attached to the glass, there will now be a spot with many copies of the same sequence. The slide is then flooded with a mixture containing DNA polymerase, a mixture of all four nucleotides, each labelled with a different fluorescent dye. The nucleotides also have a terminator, a removable chemical group that prevents chain extension, so only one nucleotide is added. This means that the colour of each spot will indicate the base present at that position. An image is taken of the slide and then the terminators are removed. The process is then repeated with an image being taken after each step. Computer analysis allows the sequence of each spot (corresponding to a single DNA fragment) to be read off.

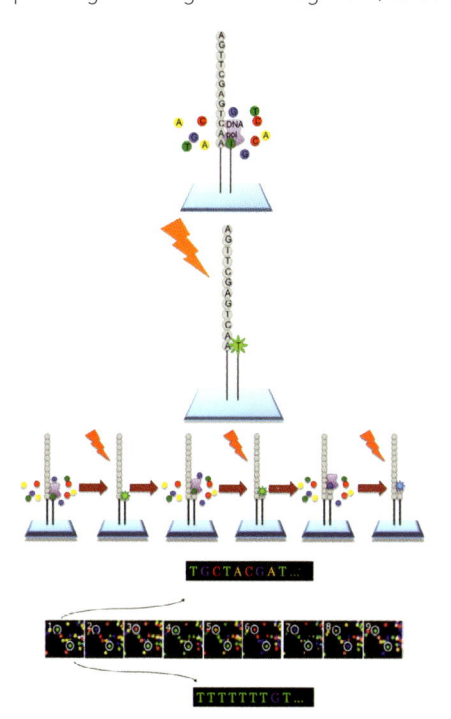

▲ **Figure 5.5** The Illumina process. One nucleotide is added at a time and the colour of the fluorescence identifies the base at each position. The process is repeated until the sequence of the entire fragment can be read off. This happens in parallel for each spot on the slide

◀ **Figure 5.6** 454 sequencing. In this example two T nucleotides are added in the first cycle, no A nucleotides, then one G nucleotides and no C, so the sequence can be read as TTG

> 🔑 **KEY TERMS**
>
> **Molecular biology** — The study of DNA and its application. This includes topics such as DNA sequencing as well as gene technology and molecular genetics – the understanding of genetics at the level of DNA, RNA and proteins.

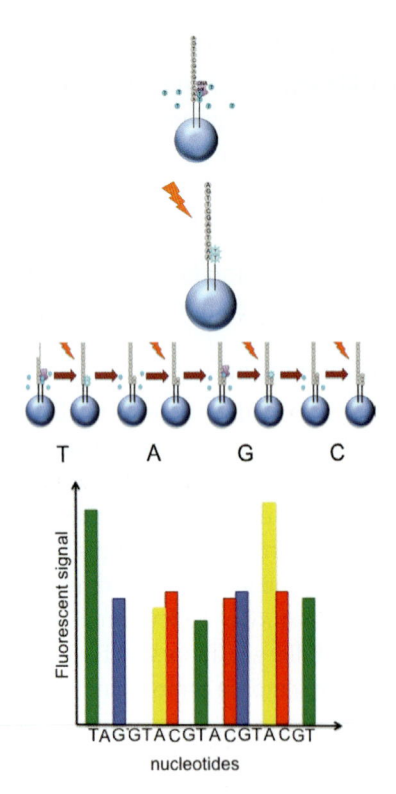

Once the sequence of all the spots is read off, overlap analysis can be used to put them into the correct order.

Roche 454 sequencing

Like Illumina sequencing, the Roche 454 method can sequence many fragments in parallel, but can work with longer fragments (up to 1 kb). Like Illumina sequencing, the method uses adapters to anneal the fragments, but to resin beads, one fragment per bead. PCR then amplifies the number of copies on each bead. The beads are placed in the wells of a chip (slide), one bead per well, together with DNA polymerase. One of the nucleotides is added to the slide and, if this is the next base in the sequence, it will be added to the growing chain and a light signal will be generated. If there are two or more of the same base in the sequence at that point they will all be added and the intensity of the signal will be proportional to the number of bases added. An imaging device records the light emission from each well and so a record can be made of which wells (and therefore beads with a single DNA sequence attached) contained that nucleotide at that position. The nucleotides are washed out and then the next nucleotide added and the process repeated for each of the other nucleotides. This whole process is then repeated until the entire fragment in each well has been sequenced. (see Figure 5.6)

Ion torrent / ion proton sequencing

This method makes use of the fact that, when a nucleotide is added to a DNA chain, a proton (H^+) is released. This is detected by a drop in pH rather than light output, but in other respects the method is very similar to the 454 sequencing method, although it only works with fragments up to about 200 bp.

Advantages of NGS over Sanger sequencing

- Sample size – Sanger sequencing needs one strand for each base being sequenced, because each base is identified by termination of that strand, so many thousands of copies of the DNA are required. In NGS a sequence can be obtained from a single strand.
- Speed – Only one sequence (maximum ~1 kb) can be determined at a time with the Sanger method, whereas NGS is massively parallel. This means that 300 Gb (300 × 10^9 bases) of DNA can be sequenced in a single run on a single chip.
- Accuracy – Repeats are inherent to the NGS methods. This means that each piece of DNA is sequenced multiple times in different overlaps. Any inaccuracies will be removed by analysis of the overlap data.
- Cost – Sequencing the first human genome cost about £300m. Modern Sanger sequencing methods reduce this to about £6m. By contrast, sequencing an individual human genome with Illumina today would cost only £6,000 at the time of writing.

Principles of genetic profiling (PCR of short tandem repeats (STRs))

The technique of genetic profiling was developed by Professor Sir Alec Jeffreys at the University of Leicester in the mid 1980s. The technique was originally called 'genetic fingerprinting' and used restriction endonucleases to cleave DNA samples into different size fragments that could be separated by gel electrophoresis (see section 2.2 in Unit 2). The size of the

Table 5.10 Short tandem repeats (STRs) in the House of Stuart

Allele 1 (5 repeats)	—	GATA	GATA	GATA	GATA	GATA	—				
Allele 2 (6 repeats)	—	GATA	GATA	GATA	GATA	GATA	GATA	—			
Allele 3 (8 repeats)	—	GATA	GATA	GATA	GATA	GATA	GATA	GATA	—		
Allele 4 (10 repeats)	—	GATA	GATA	GATA	GATA	GATA	GATA	GATA	GATA	GATA	—

fragments would depend on the number and location of the restriction sites and the analysis produced a pattern ('fingerprint') that could be used to compare two DNA samples. The technique was soon used for proving family relationships and in forensics for comparing a suspect's DNA to DNA isolated from a crime scene.

The initial method was very time-consuming and used quite large amounts of DNA that had to be radioactively labelled. The method of DNA profiling now used relies on PCR to amplify the DNA of regions of chromosomes known as **short tandem repeats** (STRs) or microsatellites. These STRs are distributed throughout the genome at various specific positions (loci) on the chromosomes. The number of repeats at any one position on the chromosome **(locus)** varies between individuals and is inherited just like regular genes. Table 5.10 shows four alleles of an STR located on the Y chromosome that has been used to trace descendants of the royal house of Stuart. In the table, '—' represents the rest of the DNA sequence before and after the STR.

The method involves use of PCR with sequence specific primers to amplify particular STRs. These DNA fragments produced are then separated by gel electrophoresis, or the similar technique of capillary electrophoresis. The latter method uses less sample and can be automated. Fluorescent dyes are used to label the DNA so it can be detected. The number of repeats at a particular locus represent alleles of that locus. There are relatively few alleles of each locus (each one will be shared by about 5–20% of the population). However, by analysing the DNA at different loci simultaneously, the technique becomes much better able to identify individual DNA.

Statistical probabilities (Second Generation Multiplex Plus SGM*Plus*® and DNA-17)

We need to think about the probability that a match between two DNA samples is due entirely to chance. The two samples might be taken from two individuals

> **KEY TERMS**
>
> **Short tandem repeats** — Also known as a microsatellites, these are sequences that consist of between five and fifty repeats of, usually, four nucleotides.
>
> **Locus** — A specific position on a chromosome, corresponding to a particular gene or section of DNA. The plural is loci.

in a test for paternity or other family relationship. Alternatively, it could be a comparison between DNA from a crime scene and a DNA database, such as the National DNA Database for England & Wales. STRs at loci on different chromosomes, or far apart on the same chromosome, will not be linked. This means that if one locus has a 10% (1 in 10) probability of a match being due to chance and a second locus has a 5% (1 in 20) probability of a match being due to chance, then there is an overall probability of 0.5% (1 in 200) that the match is due to chance – we multiply the probabilities. If we get a match at a third locus that has a probability of 1% (1 in 100) being due to chance, then the probability that the match between the DNA samples being due to chance is now 0.005% (1 in 20,000). You can see that the probability of a match being due to chance becomes very small even when just a few loci are used.

The original method used in forensic science in the UK (in 1994) used just four STR loci and was not very discriminating, i.e. there was a relatively high probability that a match between two samples was due to chance. This was replaced in 1995 with the Second Generation Multiplex (SGM) method that used six STR loci as well as one other to identify gender. This was followed in 1999 by SGM Plus that used ten loci and the gender identifier. This method has a less than one in a trillion probability of a match being due to chance, although a more conservative 'less than one in a billion (1,000,000,000)' probability is quoted in court.

The method has been further enhanced by adding an extra six STR loci to the analysis – DNA-17 has been the standard method throughout the UK since July 2014.

Limitations of genetic profiling techniques

As with all analytical techniques, there are limitations and sources of error. Sample contamination is one source of error. Because PCR amplifies the DNA, i.e. makes multiple copies, it will also amplify any contaminants. It is essential to follow stringent procedures when collecting and processing DNA samples – you don't want to convict the lab technician of murder rather than the suspect! Low Copy Number (LCN) testing is used when there is very little DNA available to sample. In this technique the number of cycles of PCR is increased to amplify the DNA enough to be able to detect. This increases the risk of amplifying contaminants. Also, the PCR method has an inherent level of error, so using more PCR cycles means more overall error.

The DNA-17 technique is much more sensitive than earlier methods, increasing the risk of the results being affected by contamination. However, the greater sensitivity also means that contamination is more easily detected.

Another source of error is the state of the original DNA sample. This can be a particular problem in forensic science with DNA collected from crime scenes. If the DNA is degraded then all the loci might not be present. However, even with only 12 loci present, the probability of a match is still about one in a billion.

> **CLASSROOM DISCUSSION**
>
> ### Ethics and miscarriages of justice (20 minutes)
>
> There are many ethical issues to consider. Select some of the following and discuss them as a group.
>
> - Do juries always understand probability?
> - Could a defence lawyer use errors inherent in any laboratory technique to induce doubt in the minds of the jury?
> - Should DNA profiles be removed from the national database if someone is not convicted of a crime?
> - Should everyone have their DNA profile kept on a national database?
> - Has flawed DNA analysis led to miscarriages of justice?
>
> You could use the discussion as the starting point for project work, e.g. producing a report or information leaflet describing the use of DNA profiling in forensic science.

> **KNOW IT**
>
> 1. DNA mapping, sequencing and profiling requires the use of DNA fragments. State the name of the type of enzyme used to prepare these fragments.
> 2. Describe how the polymerase chain reaction (PCR) is used to amplify (make multiple copies of) a sample of DNA.
> 3. What are the limitations of the PCR method?
> 4. Complete Table 5.11 comparing the different DNA sequencing methods. The first row has been completed for you.
>
> Table 5.11 Comparing different DNA sequencing methods
>
> | Method | DNA fragment length | Suitability for parallel working |
> | --- | --- | --- |
> | Sanger | 800–900 bp | limited |
> | Illumina | | |
> | Roche 454 | | |
> | Ion torrent | | |
>
> 5. Besides suitability for working in parallel, describe how NGS methods have advantages over the Sanger method.
> 6. Describe the limitations of genetic profiling.

LO3 Assessment activities

Below are suggested assessment activities that have been directly linked to the Pass, Merit and Distinction criteria in LO3 to help with assignment preparation and include top tips on how to achieve best results.

Activity 1 Describe the principles of DNA sequencing *P5*

You could describe the principles and techniques involved in DNA sequencing by constructing a timeline of events (display or web page) covering the significant stages.

Activity 2 Describe the principles of genetic profiling *P6*

You need to understand the process of genetic profiling and be able to interpret examples of these.

Activity 3 Evaluate the advantages of next-generation DNA sequencing (NGS) M4

You could research current techniques and evaluate their advantages over other techniques. Your evaluations could be added to the display or web page produced for **P5**.

You could describe these techniques in an information leaflet, e.g. for legal use, which could be supplemented when working on **M5** and **D2**.

Activity 4 Discuss the statistical significance of genetic profiling matches M5

You should understand the probability of a match occurring at the 10/16 loci; because of the number of loci used, the probabilities are multiplied. These probabilities should be discussed in terms of the population of the UK, Europe and the global population.

You could describe these techniques in an information leaflet, e.g. for legal use.

Activity 5 Evaluate the significance and limitations of genetic profiling techniques D2

You need to understand the key points of the significance and limitations of DNA 17 genetic profiling techniques:

- improved discrimination between profiles
- improved sensitivity
- improved comparability between profiles on the national DNA profiles produced in other European Union countries and beyond.

You need to know that there are now a number of companies offering different versions of profiling methodology, and be able to discuss compatibility.

You need to understand errors in amplification using PCR.

You could describe these techniques in an information leaflet, e.g. for legal use.

TOP TIPS
- ✔ Don't confuse DNA sequencing with genetic profiling.
- ✔ Profiling focuses on the analysis of short repeating units of DNA (short tandem repeats, STRs) at specific loci.
- ✔ Sequencing involves the whole genome.

LO4 Understand the impact of an innovation in an application of genomics P7 M6 D3

GETTING STARTED

Science or technology? (5 minutes)

LO3 had a lot of information for you to take in. Genomics is having an increasingly large impact on our everyday lives. As a group, discuss whether you think the basic science (e.g. the development of PCR or the Sanger method) or technological advances (e.g. development of automated sequencing) have been more important.

4.1 Assess the impact and implications of a DNA-sequencing project

DNA sequencing (genomics) has allowed us to understand inheritance, variation, mutation and regulation of gene expression at the molecular level. This has informed our understanding of the basic mechanisms of various disease processes. The technology has also led to the conviction of many criminals that, otherwise, may have escaped justice. You can see to what extent it has had an impact on the whole of modern biology and medicine. You will be working on an assignment to bring all this together. You could consider the Human Genome Project, NHS 100,000 Genomes Project or the US-based ENCODE Project.

Increasing understanding

DNA sequencing has increased our knowledge and understanding at many levels and in many areas of science.

- **Understanding of the genome (genomics) –** We now know more about how the genes are arranged on the chromosomes, giving greater understanding of the molecular basis of genetics and inheritance.
- **Location of genes linked to medical conditions (e.g. cancers); non-medical conditions and complex human traits –** There are many examples, such as the BRCA1 and BRCA2 genes linked to human breast cancer. Knowing the sequence of these genes has allowed development of techniques such as the use of DNA probes (labelled DNA fragments that bind to a target sequence) to identify those at risk of developing the disease.
- **Understanding of functional elements of DNA determined by non-protein-encoding regions –** About 98% of the human genome does not code for proteins and was once thought to be 'junk DNA'. We now know that much of this non-coding DNA has a function, including the following:
  - Coding for tRNA and rRNA.
  - Regulation of gene expression (i.e. protein synthesis).
  - Unknown functions; DNA sequencing allows comparison of the DNA sequences of a wide range of organisms showing that some non-coding sequences are conserved over many millions of years of evolution, so must be important.
- **Ecological relationships –** The science of molecular ecology uses genomics as the basis of understanding subjects such as population genetics, biodiversity, species identification and species relationships at the molecular level.
- **Genetic variation; evolutionary development –** We can now understand variation, which is the basis of evolution, at the level of the DNA sequences in individuals, within species, between species and across millions of years of evolution by comparing organisms that have common ancestors.

Implications of genomics

There are many implications of the technology of DNA sequencing and the field of genomics that has been made possible, in areas as diverse as medicine and ecology. For information on the implications for drug development see sections 1.2 and 1.4 of Unit 11.

- **Medical treatments and personalised medicine –** This is of great potential importance. You can read more about this in the section on pharmacogenetics and pharmacogenomics in Unit 11.
- **Ecological applications –** One aspect of ecology is to study how organisms adapt to their environment. Genomics can be used to study the genetic mechanisms that affect the way in which organisms adapt to their environment.
- **Conservation applications –** Conservation genetics studies the survival of species with small populations that are prone to extinction. Understanding the basis of biodiversity and the way species adapt to changing conditions helps us to make predictions and take action to improve conservation. Conservation genomics takes this science to another level with the ability to use DNA sequences as the basis of understanding the genetics. The European Science Foundation website has some useful resources. Visit www.esf.org and search for ConGenOmics.

Ethical, Legal and Social Implications (ELSI)

In all of these areas, we have to consider the ethical, legal and social implications of genomics and its application. One such implication is that of identifying a strong genetic marker associated with greatly increased mortality, e.g. from cancer, heart disease or inherited disorders such as Huntingdon's disease. Would an individual wish to know they had a 90% chance of premature death? Would it be ethical to withhold this information from them? Would it be right for this information to be made available to an employer or life assurance company?

GROUP ACTIVITY

Discuss it (30 minutes)

Divide into two groups. One group should list all the benefits that have come from DNA sequencing and genomics, the other group should list all the risks or potential harmful outcomes. Take turns to present a case for and against the motion 'DNA sequencing has brought, and will bring in the future, nothing but good for mankind'.

EXTENSION ACTIVITY

Personalised medicine

Genomics (and the related area of pharmacogenomics; see Unit 11) offers the prospect of designing new drugs that are specific to an individual or relatively small group of individuals. You could research this area, possibly in conjunction with Unit 11, to assess the importance of personalised medicine now and how it could develop in the future. This work could form part of your assessment for *D3* (see Assessment activities below). Some points to consider:

- How does the genetic makeup of an individual affect their susceptibility to disease?
- How does the genetic makeup of an individual affect their response to drug treatment?
- How are drugs 'tailored' to an individual?
- What are the social, ethical and economic consequences?

LO4 Assessment activities

Below are suggested assessment activities that have been directly linked to the Pass, Merit and Distinction criteria in LO4 to help with assignment preparation and include top tips on how to achieve best results.

Activity 1 Describe an application of genomics *P7*

You should assess the impact and implications of a DNA-sequencing project, e.g. the Human Genome Project, NHS 100,000 Genomes Project, ENCODE Project or other genome-sequencing application.

You could present your assessment in the form of a report or website.

Activity 2 Evaluate the impact of the specified genomic application *M6*

This could be an application with medical implications, or one involved with systematics and evolutionary development, or conservation. Your tutor will be able to help with suggestions. You could prepare a written evaluation presentation including Ethical, Legal and Social Implications (ELSI) of these initiatives.

Activity 3 Assess the future impact of the specified genomic technique *D3*

The particular impacts here will depend on the application reported on, but could include aspects such as developing personalised medicines as our understanding of the genetic causes of various human conditions are determined, and possible manipulation of the human genome.

> **TOP TIPS**
> - This section focuses on real-life applications of what you have learned about genomics.
> - Make sure you understand the basic technology.
> - Make sure you have covered a range of different applications – there are lots of interesting applications to choose from and they all have a potential to revolutionise treatment of disease.
> - Don't dismiss ethical concerns, but do make sure you appreciate the difference between scientific concerns and ethical concerns.

Read about it

Yourgenome is a website that enables you to find out more about genetics and genomics:

www.yourgenome.org/

The University of Cambridge produces a series of podcasts called 'Naked Genetics', which covers a wide range of topics in genetics:

www.thenakedscientists.com then scroll down to the podcast feeds.

The Clinical Genetics Society website has a section of websites and downloads with links to a number of useful resources:

www.clingensoc.org/information-education/websites-downloads/

145

Unit 06
Control of hazards in the laboratory

ABOUT THIS UNIT

There are risks associated with any sort of lab work and these can be made worse in busy labs when the health and safety of lab workers can be overlooked or forgotten. That is why there are rules and regulations that govern identifying and controlling hazards. It is important that you have a good understanding of not just the rules and regulations, but also the principles of safe working.

This unit covers the most common hazards you might encounter in a typical research lab and will help you maintain a safe working environment. It also links to many other units so that you can apply these skills in the context of the practicals you undertake – and in your future working life.

LEARNING OUTCOMES

The topics, activities and suggested reading in this unit will help you to:

1. Understand the types of hazard that may be encountered in a laboratory
2. Be able to use health and safety procedures to minimise the risk presented by hazards in a laboratory
3. Be able to design a safe functioning laboratory to manage the risk presented by hazards

How will I be assessed?

You will be assessed through a series of assignments and tasks set and marked by your tutor.

How will I be graded?

You will be graded using the following criteria:

Learning Outcome	Pass	Merit	Dinstinction
	The assessment criteria are the Pass requirements for this unit.	To achieve a Merit the evidence must show that, in addition to the Pass criteria, the candidate is able to:	To achieve a Distinction the evidence must show that, in addition to the Pass and Merit criteria, the candidate is able to:
1 Understand the types of hazard that may be encountered in a laboratory	**P1** Describe the types of hazard agents that are found in a laboratory situation	**M1** Explain how disease-causing organisms reproduce and are transmitted	
2 Be able to use health and safety procedures to minimise the risk presented by hazards in a laboratory	**P2** Carry out risk assessments for a laboratory procedure	**M2** Describe how health and safety legislation influences procedures and practices	**D1** Evaluate the effectiveness of current legislation in safe working practices in the control of diseases
	P3 Explain the procedures and practices required to effectively prevent diseases from spreading in a laboratory		**D2** Evaluate the potential impact of poor procedures and practices on individuals and the environment
3 Be able to design a safe functioning laboratory to manage the risk presented by hazards	**P4** Produce a design specification to control risks posed by hazards in a laboratory	**M3** Explain how the design of a laboratory can control the spread of disease in a laboratory	**D3** Analyse how procedures and legislation affects the control of diseases in a laboratory

LO1 Understand the types of hazard that may be encountered in a laboratory *P1 M1*

GETTING STARTED

How bad could it be? (5 minutes)

Think about the different types of hazard you might encounter in a laboratory – and not just the type of laboratory you might have experienced in school or college. Discuss your ideas with the rest of the group and provide a list of at least five examples of hazards and their effects. How bad could things get?

1.1 The types of hazardous agents that may be encountered in the laboratory and the risks that they pose

Chemical hazards were covered briefly in section 1.1 of Unit 2 and we will cover those in more detail in this unit. Before that, we need to think of another type of hazard that you might encounter – biological hazard, or biohazard for short.

There are two main types of biohazard that we need to consider:

- **biological agents**
- biological substances that are **carcinogens**, **mutagens** or **teratogens**.

147

KEY TERMS

Agent – A general term that includes biological agents as well as a chemical substance (which may be a natural product such as a protein).

Biological agent – This term includes bacteria, viruses, fungi and other organisms, as well as toxins that they may produce.

Carcinogen – An agent that is directly involved in causing cancer.

Mutagen – An agent that changes the genetic material (DNA in humans); as many mutations cause cancer, a mutagen may also be a carcinogen.

Teratogen – An agent that can disturb the development of an embryo or fetus. This may cause a birth defect or even spontaneous abortion (miscarriage).

Toxin – A toxic (poisonous) substance produced by a living organism.

Pathogen – An infectious agent such as a bacterium, virus, fungus or parasite that causes disease in its host.

Exposure to either type of biohazard can affect human health and may prove fatal. However, we need to understand the nature of the two types of agent so that we can minimise the risk to ourselves and to others.

Biological agents

You may have come across the term 'biological agent' in the context of biological warfare – the use of biological **toxins** or infectious microorganisms to kill or incapacitate. However, generally, biological agents include various **pathogens**, such as bacteria, viruses, fungi, other microorganisms and their associated toxins. They can be a risk to human health in many ways, ranging from relatively mild allergic reactions to serious medical conditions, even death. These organisms are found in water, soil, plants and animals and because many microbes reproduce rapidly and require minimal resources for survival, they are a potential danger in a wide variety of occupational settings, not just in laboratories.

Biological agents generally cause infections, such as influenza, hepatitis, HIV, tuberculosis (caused by *Mycobacteria*) or Legionnaires' disease (caused by *Legionella*). Some types might cause communicable infections, i.e. they can be passed from person to person. Others will affect only the person infected and cannot be passed on.

Compounds listed as carcinogens, mutagens or teratogens

Most carcinogens, mutagens or teratogens are either organic or inorganic chemicals, although a few are natural products (see Table 6.1).

Table 6.1 Carcinogens, mutagens and teratogens

Type	Example	Description
Carcinogen	Aflatoxins	Some fungi growing on foodstuffs such as peanuts, seeds and grains in storage.
Carcinogen	Viruses	Epstein-Barr virus (EBV), the papillomaviruses that are linked to cervical cancer.
Carcinogen	Bacteria	Their infection causes chronic inflammation and this is known to produce reactive oxygen species that can damage DNA and be a cause of several cancers.
Physical mutagens	Radiation	Gamma-rays, X-rays, etc. damage DNA and cause a range of cancers.
Chemical mutagen	Nitric oxide	Removes amino groups from bases in DNA.
Teratogen	Rubella virus	Causes deformity in human embryos, e.g. deafness, cataracts.
Teratogen	*Herpes simplex* virus, Zika virus	Causes deformity in human embryos, e.g. microcephaly, microphthalmia.

CLASSROOM DISCUSSION

Vaccinations (10 minutes)

Vaccinations are available against many infectious agents, some of which are given routinely to babies and children. Suggest infectious agents that laboratory staff could be vaccinated against and discuss advantages and disadvantages of this approach.

Chemical hazards

Look back at Figure 2.1 in Unit 2. This shows the GHS Hazard Pictograms and summarises the chemical hazards you will encounter. Some of these are obvious and likely to present in most if not all labs – corrosive substances such as acids and alkalis or flammable substances such as solvents. Others might be less common, such as explosive or toxic substances.

However, there are other chemical hazards that you might not immediately think of and some of them will be very common, for example:

- cleaning agents
- disinfectants
- drugs
- anaesthetic gases
- compressed gases.

Physical hazards

These are the types of hazard you might encounter in any working environment – or even at home:

- slips and falls from working in wet locations
- ergonomic hazards of lifting, pushing, pulling and repetitive tasks
- electrical hazards
- acoustic hazards (noise)
- thermal hazards.

1.2 The principles of disease-causing organisms

You have seen that biohazards that are also infectious agents, such as bacteria and viruses, are transmissible – they can be passed from person to person. This means that they can also pose a threat to the health of the general population, not just to the workers in a laboratory exposed to them.

To understand fully the type of risks, we need to understand more about the structures and life cycles of the different types of infectious agents, especially viruses and bacteria. Viruses, unlike all other infectious agents, can only reproduce by infecting a host cell. The host cell could be animal or plant, depending on the virus – viruses are often highly specific in the type of cell they infect. There are even some viruses, called **bacteriophages** ('phage' for short) that infect bacteria.

KEY TERMS

Bacteriophage – A virus that infects bacteria, often shortened to 'phage'.

Reverse transcriptase – A viral enzyme that converts single stranded viral RNA into double stranded cDNA.

Retrovirus – A virus with single stranded RNA that infects a cell and uses its own reverse transcriptase to make a DNA copy that becomes integrated into the host cell genome.

Lysis – The disintegration of the cell caused by rupture of the cell membrane (or cell wall in plant cells).

Lytic cycle – The main method of viral reproduction inside cells. It results in lysis of the cell and release of progeny viruses that go on to infect other cells.

Structure of a virus as nucleic acid, capsid and viral enzymes

All viruses consist of genetic material (nucleic acid – DNA or RNA) and a protein coat or capsid. The nucleic acid may be single stranded or double stranded. The simplest virus contains only enough nucleic acid to code for four proteins whereas the most complex can code for 100–200 proteins. Recently, giant viruses have been found that have genomes that code for more than 2000 proteins.

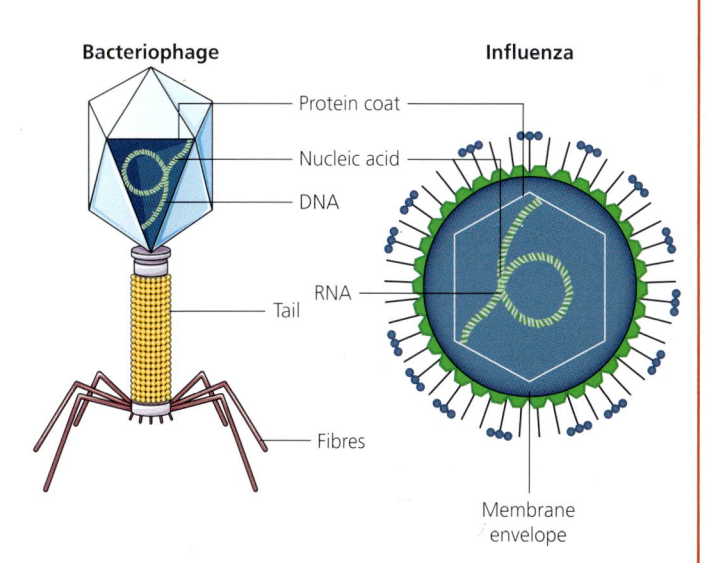

▲ Figure 6.1 Structure of a bacteriophage (left) and animal virus (right). There are many types of virus structure, of which the influenza virus is just one

> **KEY TERMS**
>
> **Genome** – The full set of genes contained in an organism. A better definition is that it includes all the genetic material (DNA or RNA) in an organism, which means genes and non-coding sequences of DNA or RNA.
>
> **Lysogenic cycle** – This involves integration of the viral nucleic acid into the genome of the host cell. The viral nuclei acid is replicated along with the host genome. The term was originally used in the context of bacteriophage infection of bacterial cells – the integrated viral nucleic acid is known as a prophage. The term is now also used to describe the life cycle of retroviruses.

In addition to the genetic material and protein coat, some viruses also contain enzymes. These enzymes can be involved in various stages of the life cycle of the virus:

- Helping the virus enter the host cell.
- Generation of a complementary DNA (cDNA) copy of the viral genome. This is done by an enzyme called a **reverse transcriptase** and happens in **retroviruses** such as HIV. This type of enzyme is now used very widely in molecular biology and DNA technology. Some of these applications are covered in section 3 of Unit 5.
- Integration of the genetic material of the virus into the DNA of the host cell. This occurs in retroviruses and follows on from the production of a cDNA copy of the virus genome.
- Helping the virus to be released from the host cell.

Presence of antigens

In this context, antigens are proteins, glycoproteins, lipoproteins (see section 3.2 of Unit 1 for more information) or other similar molecules that will bind to antibodies produced as part of the immune response to infection. Antigens may be naturally present on the surface of pathogens such as bacteria or viruses. Antigens from pathogens can also be found on the surface of infected cells or cells in the immune system involved in the immune response that fights infection. There is more information about this in section 6 of Unit 4.

Lytic and lysogenic cycles as related to diseases

Once a virus has penetrated a host cell, it uses the cell's own systems to replicate its genetic material (nucleic acid) and synthesise viral proteins used to create new virus particles that are released from the host cell in a process known as **lysis**. The whole process, from penetration of the host cell, through replication and synthesis of new viral proteins to release of the virus is the **lytic cycle** – see Figure 6.2.

Some viruses do not enter the lytic cycle after penetrating the host cell. Instead their genetic material is integrated into the host cell **genome**. This is known as the **lysogenic cycle**. The retrovirus can then enter the lytic cycle, or it can remain incorporated in the host cell genome, sometimes for many years, and is replicated along with the host cell DNA when the cell divides.

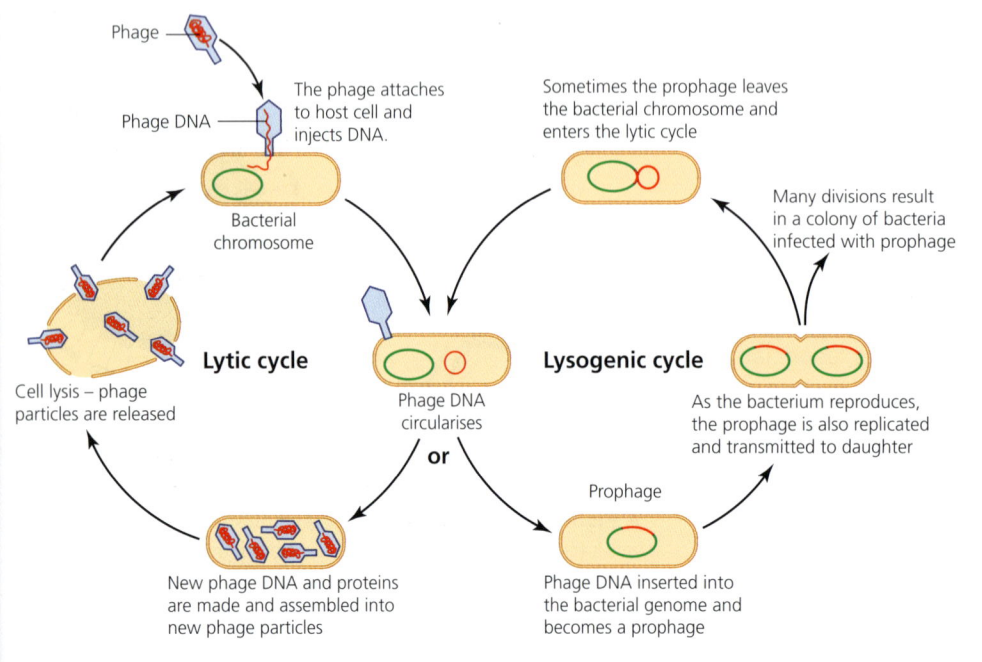

▲ **Figure 6.2** The lytic and lysogenic cycles. The diagram illustrates this using bacteriophage (phage) as an example, but the principles are the same with a retrovirus. The difference is that a retrovirus infects animal cells rather than bacteria. Also, the retrovirus genome is RNA and it uses its own reverse transcriptase to make a cDNA copy that inserts into the host cell genome

> **KEY TERMS**
>
> **Phenotype** – The characteristics expressed in an organism. In the case of the gene for eye colour, the phenotypes might be blue eyes or brown eyes. However, in many cases the phenotype depends not just on the genes (alleles) that are expressed, but also on environmental factors. See section 2 of Unit 5 for more information.

There are two main implications of retroviruses for human disease:

1. When incorporated into the host genome, it can remain inaccessible to the host's immune system; think about why that is the case. There is more information about this in section 6 of Unit 4. Retroviruses, such as HIV or the various herpes viruses, can be difficult, if not impossible, to eliminate completely from the body.
2. Retroviruses can become integrated in the host genome within a gene for one of the many proteins that regulate the cell cycle (see section 3.1 in Unit 8). This could disrupt the normal control systems and cause uncontrolled cell division leading to cancer. That means they are carcinogens or teratogens.

Some bacteriophages (see Section 1.3) have a lysogenic phase during which the bacteriophage can alter the **phenotype** of the infected bacterium. This can be seen in several bacteria that produce toxins only when infected with bacteriophages – examples include diphtheria toxin, cholera toxin and botulinum toxin, all of which cause (often fatal) human disease.

GROUP ACTIVITY

Pathogenicity of viruses (20 minutes)

Pathogenicity means the ways in which viruses are pathogenic. Divide into two groups. One group will consider the lytic cycle, the other the lysogenic cycle. Draw up a chart of the harmful effects that each type of cycle could have on an infected person.

Once you have done this, compare your charts. Which type do you think is more harmful? There may not be a simple answer to this question!

Structure of a bacterial cell and how bacteria reproduce (to include cell wall variation)

All bacteria have a cell envelope – the plasma membrane that regulates entry and exit, and the cell wall that provides strength and structural integrity. The bacterial cell wall consists of peptidoglycan, a polysaccharide made from two types of amino-sugar where the polysaccharide chains are held together by peptide cross-links. This structure gives rigidity to the bacterial cell wall. The two types of bacteria, Gram positive and Gram negative, are classified according to whether they are stained with gram stain (see section 6.5 of Unit 3 for more information about Gram staining).

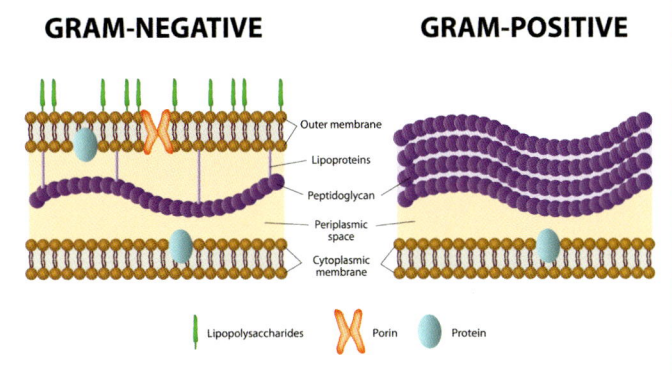

▲ **Figure 6.3** Structure of the bacterial cell wall in Gram positive and Gram negative bacteria. Lipoteichoic acid is a component of the Gram positive cell wall that has similar pathogenicity to endotoxins (lipopolysaccharides)

Gram positive

The cell wall of Gram positive bacteria is composed of a thick layer of peptidoglycan that makes up about 90% of the dry mass of the wall. The cell wall of some Gram positive bacteria can be digested by the enzyme lysozyme. This enzyme is a widespread anti-bacterial found, for example, in tears, where it is responsible for prevention of eye infections – the eye, particularly the cornea, is not protected by the immune system.

Gram negative

The cell wall in Gram negative bacteria consists of only about 10% peptidoglycan and is much thinner than in Gram positive bacteria. Gram negative bacteria have an outer membrane that consists of a phospholipid bilayer (like all biological membranes) – this was covered in section 3.2 of Unit 1. The outer leaflet of the bilayer contains lipopolysaccharides that contribute to the structural integrity of the bacteria and protect it from some types of chemical attack.

1.2 The principles of disease-causing organisms

Plasmids

A plasmid is a small, circular DNA molecule found in bacteria – you could think of it as a mini-chromosome. Plasmids are used to exchange DNA between bacteria – either of the same species, or of different species – during the process known as conjugation (see Figure 6.4). As plasmids often carry genes that give resistance to some types of antibiotic, the exchange of plasmids can have serious consequences in terms of spread of antibiotic resistance.

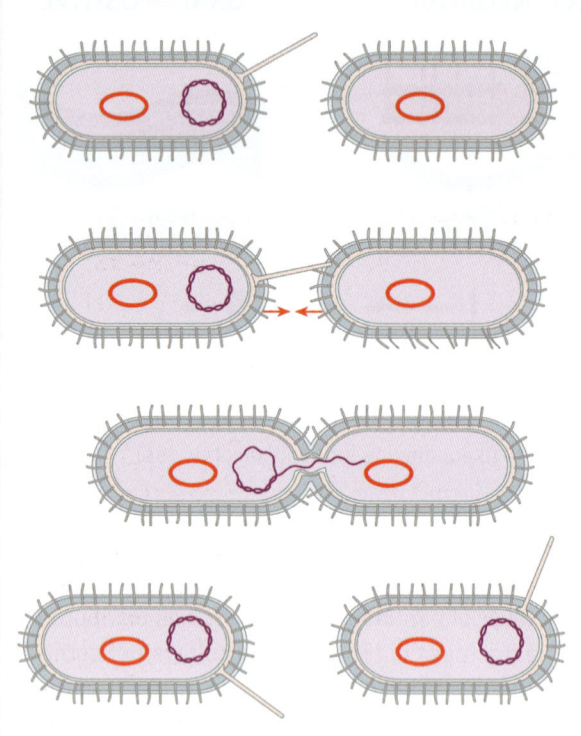

▲ **Figure 6.4** Bacterial conjugation. A section of DNA from a plasmid (circular DNA) in the donor bacterium (left) is passed through an enlarged pilus (hair-like tube) and into the recipient (right). Once transferred, the plasmid genes are incorporated into the recipient's DNA

> 🔑 **KEY TERMS**
>
> **Immunogenic** – Produces an immune response.
> **Toxicity** – The degree to which a substance can poison or harm an organism.
> **Cytokines** – A broad category of small proteins involved in cell signalling. They are released by some cells and act on other cells, often as part of the response to infection.
> **Pyrogenic** – Causes fever.

Endotoxins in cell wall

The lipopolysaccharide (LPS) found in the outer membrane of Gram negative bacteria is also known as endotoxin.

LPS has two main components associated with disease.

1. The polysaccharide component is highly **immunogenic**, i.e. picogram (10^{-12} g) amounts are sufficient to generate an immune response.
2. Lipid A anchors the LPS in the membrane and is responsible for the **toxicity** of Gram negative bacteria such as *Salmonella* and *E. coli*.

When these bacteria are attacked by cells of the immune system, LPS is released into the bloodstream and the Lipid A component causes **cytokine** production, release of histamine and activation of the coagulation (blood clotting) cascade. The result is inflammation, formation of blood clots and, in severe cases, internal bleeding and septic shock.

The presence of pili for adhesion

Pili are protein tubes that extend out from the outer membrane of many Gram negative bacteria, e.g. *E. coli*. They are involved in the attachment of bacteria to surfaces, e.g. to form a biofilm. A biofilm is a population of one or more types of bacteria attached to a surface and embedded in a matrix or slime. The matrix provides a physical barrier that makes the bacteria more resistant to antibiotics or disinfectant chemicals. Pili are also involved in the attachment of bacteria to animal cells during infection.

Capsules to block phagocytosis

Some bacteria, such as *Neisseria meningitides* (also known as *Meningococcus*), have a polysaccharide capsule that protects the bacterium against phagocytosis, which is the process where cells in the immune system engulf and destroy pathogens, such as bacteria and viruses.

Link between toxin production and symptoms

In addition to endotoxins, many bacteria produce exotoxins. These are proteins that are released by the bacteria. Many have two subunits: one facilitates entry of the toxin into the cell, the other has an enzyme activity that is responsible for toxicity. Other toxins form pores in host cell membranes disrupting the normal movement of ions (e.g. Na^+, K^+, Cl^-) across the membrane. **Pyrogenic** exotoxins, such as those produced by staphylococci and streptococci, can cause potent stimulation of the immune system, fever and enhance the septic shock produced by endotoxins.

KEY TERMS

Spore – A small, highly resistant form of bacterial cell produced under adverse conditions. Bacterial spores can resist extreme dryness and many are not killed by freezing, high temperatures or chemical disinfectants. When conditions improve, they can begin to divide again.

Sterilisation – A process that eliminates or kills all forms of life, including viruses and spores.

Binary fission and how quickly a bacterial population can build up given the right conditions

Bacteria reproduce asexually by binary fission. The DNA replicates and the two identical DNA molecules are pulled to opposite poles of the cell. The cell lengthens and a new cell wall begins to grow between the two poles, separating the bacterium into two identical bacteria.

1 DNA replicates — Cell wall, Cytoplasmic membrane, Replicated DNA

2 The cell membrane elongates and DNA molecules separate

3 Cross wall forms and membrane pinches in

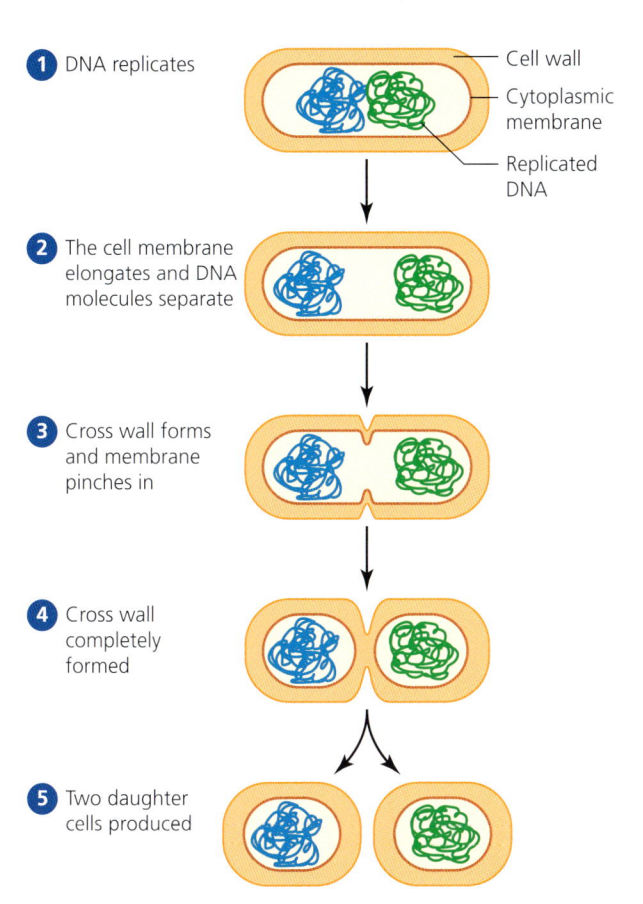

4 Cross wall completely formed

5 Two daughter cells produced

▲ Figure 6.5 Binary fission in bacteria

The process takes about 20 minutes at room temperature, so the size of the population will double every 20 minutes under favourable conditions.

This means that the population will grow exponentially – plotting the logarithm of the cell number against time produces a straight line, which is why this is often referred to as the logarithmic or log phase.

Table 6.2 shows how a bacterial population will double with each generation, from 1 → 2 → 4 → 8 → 16 → 32 cells, and so on.

Many antibiotics, including the penicillins, work by interfering with cell wall synthesis. The weakened cell wall is no longer able to withstand the pressure exerted by the cytoplasm and so the cell bursts. This group of antibiotics, therefore, attack actively dividing bacteria, when new cell wall material is being produced.

PAIRS ACTIVITY

Bacterial population growth (5 minutes)

Divide into pairs. You will need a simple calculator for this exercise (it won't work with a scientific calculator). One person has the calculator, the other counts the generations. If you enter 2 × 2 = on the calculator you will get 4 (I know you don't need a calculator to do that, but bear with me!). That is the number of bacteria produced by one bacterium after two rounds of binary fission. Now press the = key. The answer should be 8. Every time you press the = key it multiplies the answer by two. This represents another generation (i.e. doubling of numbers). The person who doesn't have the calculator is responsible for counting the generations – one generation every time the = key is pressed. After about 34 generations you will exceed the capacity of the calculator and it will display an error. Repeat the process, but this time stop at the last generation before the calculator showed an error. Make a note of the number you got. How many generations was that? Divide that number by 3 (this assumes that the bacteria divide every 20 minutes, which is typical). How many hours did it take to reach the number of bacteria you calculated? How does that make you feel about the ability of bacterial populations to grow?

Conditions affecting bacterial population growth through binary fission

There are various factors that affect bacterial population growth and these are important in helping us understand the ways in which bacterial contamination can become a risk factor (see Table 6.3).

Table 6.2 Exponential growth of bacterial populations

Number of divisions	Number of bacterial cells
0	⊙
1	⊙ ⊙
2	⊙ ⊙ ⊙ ⊙
3	⊙ ⊙ ⊙ ⊙ ⊙ ⊙ ⊙ ⊙
4	⊙ ⊙ ⊙ ⊙ ⊙ ⊙ ⊙ ⊙ ⊙ ⊙ ⊙ ⊙ ⊙ ⊙ ⊙ ⊙
5	⊙ ⊙

Table 6.3 Conditions affecting bacterial growth

Condition	Outcome	Risk
Humidity	Bacteria, like all cells, require water for cellular processes. In very dry conditions the cells become desiccated (dried out).	Many types of bacteria are killed but others will form **spores**. Bacterial spores are extremely resistant and can tolerate long periods (many years) of unfavourable conditions.
Temperature	Increasing temperature increases the rate of chemical reactions (see section 2.3 of Unit 1). Most bacteria can be killed by heat treatment at 72°C for 15 seconds – this is the process used to pasteurise milk.	Bacterial populations will grow more rapidly as the temperature increases. Spore-forming bacteria are more resistant and require live steam (steam under pressure) at 120°C for 20–30 minutes. This is used in the process of **sterilisation** known as autoclaving.
Nutrient source	All bacteria need a source of nutrients. Many require a source of organic matter.	Some bacteria can survive on very basic nutrients such as a carbon source (e.g. a sugar such as glucose) and inorganic salts such as nitrates, phosphates and sulfates.
pH balance	Different bacteria have different pH optima; pH is a measure of acidity or alkalinity – see section 3.1 of Unit 2 for more information about pH.	Most bacteria grow best around neutral pH (i.e. ~pH 7).
Aerobic / anaerobic	Aerobic bacteria, like most animals, require oxygen for growth. For anaerobic bacteria oxygen is toxic. These bacteria are used in anaerobic digesters to produce methane gas from waste organic matter.	Anaerobic bacteria can be responsible for food going bad. Some bacteria can live either with or without oxygen – these are known as facultative anaerobic bacteria.

1.3 How some viruses (bacteriophage) can infect bacterial cells

Some viruses infect bacteria, rather than animal or plant cells. These are known as bacteriophages (phage for short). This was discussed in section 1.2 where we considered the effect of lytic and lysogenic life cycles on human disease. The simplest phage have genomes with as few as four genes, others are much more complex. Bacteriophages have been used as research tools in studying basic molecular biology.

CLASSROOM DISCUSSION

Could it be worse? (15 minutes)

Think about what was discussed at the start of the section (How bad could it be?). Now think about what you have learned – have you changed your view at all? Share your ideas and opinions with the class about just how severe some hazards might be.

KNOW IT

1 What is the difference between the following terms?
  a toxin
  b carcinogen
  c mutagen
  d teratogen
  e pathogen
2 Describe the lytic and lysogenic cycles of viruses.
  a What are the key differences?
  b How do these affect the ways in which these types of viruses cause disease?
3 What is the function of the following in the way pathogens cause disease?
  a bacterial pili
  b reverse transcriptase
  c polysaccharide capsules
  d lipopolysaccharide (LPS)

LO1 Assessment activities

Below are suggested assessment activities that have been directly linked to the Pass and Merit criteria in LO1 to help with assignment preparation and include top tips on how to achieve best results.

Activity 1 Describe the types of hazard agents that are found in a laboratory situation *P1*

Choose a type of laboratory you are familiar with that handles hazardous material. Prepare an information poster, wall chart or presentation that could be used both to educate new laboratory staff and remind experienced staff about the types of hazard that they might encounter in that laboratory. Use illustrations where appropriate to make the poster attractive and informative.

Activity 2 Explain how disease-causing organisms reproduce and are transmitted *M1*

Investigate the methods of reproduction and transmission of infectious agents. Produce a report or presentation explaining each method and use examples of specific organisms to support your reasoning.

TOP TIPS
- Explanations require significant detail and therefore if using a poster or wall chart it is important to note that short bullet points will not meet the requirements.
- This Learning Outcome contains a lot of information about microorganisms that you need to understand in order to be able to assess and minimise risk – subjects that will be covered in LO2.
- In your assessment you should focus on those aspects most relevant to understanding microorganisms as biohazards – you're not studying general microbiology.
- Make sure that you fully document the work that you do so that you can put it into practice in LO3.

LO2 Be able to use health and safety procedures to minimise the risk presented by hazards in a laboratory *P2 P3 M2 D1 D2*

Laboratory health and safety was covered in Unit 2, so this would be a good time to review section 1.1 in that unit. We are now going to focus more on biological hazards – biohazards – because of the particular risks that they pose.

GETTING STARTED

Keeping the lid on risk (5 minutes)

Based on what you have learned about risk assessment and good laboratory practice, think about what might be different when you are dealing with biohazards. Think about risks that might be greater and the control measures that might be needed. Also think about measures that are taken to protect biological cultures (cells, bacteria, viruses) from contamination. Are they sufficient to protect workers in the lab? Share your thoughts with the group.

2.1 Safe working practices in the laboratory when working with hazardous substances

Safe working practices are governed by the Control of Substances Hazardous to Health Regulations (2002), abbreviated to COSHH. These cover safe working and the procedures that must be followed in relation to all laboratory hazards. However, special consideration must be given to biohazards. Biological agents can pose a number of risks, including infection, sensitisation, exposure to an allergen or toxin or other hazardous biological agent. Biological agents can also be hazardous to the environment, a factor that must be considered when disposing of biological wastes.

Hazards in the laboratory (disease risk and hazards)

Infectious agents

Biological infectious agents include microorganisms (viruses, bacteria, fungi, yeasts, etc.), cell cultures and human **endoparasites** (such as tapeworms,

> **KEY TERM**
>
> **Endoparasite** – A parasite that lives insides its host and obtains its nutrition directly from the host.

roundworms and single cell organisms) and have been classified into four hazard groups, 1 to 4, in increasing order of severity of risk.

The Health and Safety Executive (HSE) publishes an approved list of biological agents, giving the classification of all biological agents. This is available for download from the website at www.hse.gov.uk/pubns/misc208.pdf.

Other biohazards

Besides infectious agents, other biohazards are either allergens/sensitising agents or toxins. These can be treated in the same way as the chemical hazards described in Unit 2, although disposal methods or treatment of spillages might be different.

Risk assessment

The principles of risk assessment are the same as those outlined in Unit 2.

The main risk associated with working with biological agents is the potential for infection. There are three main potential routes of infection to be aware of:

- inhalation
- ingestion
- skin penetration.

The HSE publishes advice on assessment of risk and control measures needed when dealing with biohazards; you can download a copy at www.hse.gov.uk/pubns/infection.pdf.

Biohazard symbol

The biohazard symbol (Figure 6.6) is used for labelling of:

- containers, including waste containers
- cupboards and cabinets
- the doors of laboratories

where biohazards may be present.

▲ **Figure 6.6** The international biohazard symbol

COSHH regulations

We saw in Unit 2 how the COSHH regulations control the use of hazardous materials of all kinds in the workplace, including in laboratories. Following on from a risk assessment, we need to put in place appropriate control measures. There are some features of biohazards that require particular types of control measure, but we should still follow the same type of hierarchical approach to control. The hierarchy reflects the fact that eliminating and controlling risk by using physical controls and safeguards (e.g. keeping pathogens isolated) is more dependable than relying solely on systems of work:

- eliminating risks
- controlling risks at source
- minimising risks by designing suitable systems of working.

> **EXTENSION ACTIVITY**
>
> ### Managing risk and your duties under COSHH (1 hour)
>
> You will find a great deal of information about this subject in the HSE publication 'Biological agents: Managing the risks in laboratories and healthcare premises', HSE, May 2005. There is a link to a PDF version that you can download in the 'Read about it' section at the end of this unit.
>
> Research the subject and prepare a short report (bullet points or checklist) that you can refer to in your future work.

Health surveillance and occupational health is another area relevant to labs working with biohazardous materials. Think about the benefits of health surveillance. Should this be a replacement for workers being vigilant about their own state of health and seeking prompt medical attention if they develop early signs of infection?

Immunisation

COSHH requires that if the risk assessment shows there to be a risk of exposure to biological agents for which effective vaccines exist, then these should be offered if the employee is not already immune. The advantages and disadvantages of immunisation/non-immunisation should be explained when making the offer. If immunisation is accepted, the employee should not be charged for it. Immunisation can be a useful supplement to reinforce other control measures but not a replacement for them or a sole protective measure.

RIDDOR regulations

RIDDOR stands for The Reporting of Injuries, Diseases and Dangerous Occurrences Regulations 2013. The following terms apply specifically to RIDDOR.

Accident – An accident is a separate, identifiable, unintended incident, which causes physical injury. This specifically includes acts of violence to people at work.

Work-related – The fact that there is an accident at work premises does not, in itself, mean that the accident is work-related – the work activity itself must contribute to the accident. An accident is 'work-related' if any of the following played a significant role: the way the work was carried out; any machinery, plant, substances or equipment used for the work, or the condition of the site or premises where the accident happened.

Reportable injuries – The following injuries are reportable under RIDDOR when they result from a work-related accident:

- the death of any person
- specified injuries to workers
- injuries to workers which result in their incapacitation for more than seven days
- injuries to non-workers which result in them being taken directly to hospital for treatment, or specified injuries to non-workers which occur on hospital premises.

RIDDOR puts duties on employers, the self-employed and people in control of work premises (the Responsible Person) to report certain serious workplace **accidents**, occupational diseases and specified dangerous occurrences (near misses) as well as reportable injuries.

PAIRS ACTIVITY

The importance of RIDDOR (20 minutes)
Working in pairs, prepare a short briefing paper for management on the implications of RIDDOR – imagine you are working for a company that has just opened its first laboratories and needs guidance on the subject.

The HSE website has a section on RIDDOR that will be helpful. See 'Read about it' at the end of the unit for a link.

Safety of instrumentation including electrical equipment

Besides normal considerations of safe use of equipment and instruments, it is important to be aware when instruments could increase the risk of exposure to biohazards. This could be either during normal operation or in the case of malfunction.

Use of standard operating procedures

When considering the COSHH regulations, we saw how important it was to have good training in good microbiological practice as a way of minimising the risk of exposure, e.g. to a pathogen. The use of standard operating procedures (SOPs) is a part of this. An SOP will be based on a risk assessment and have incorporated into it the necessary control measures required. Following the SOP closely will minimise the risk to the person undertaking the work as well as to others in the laboratory and/or the general public.

GROUP ACTIVITY

Putting it into practice (40 minutes)
Consider what you have learned from this unit and the section in Unit 2 on:

- good laboratory practice, e.g. the principles of assessment and management of risk
- the hierarchical approach to putting in place control measures.

Discuss how these should be designed into the laboratory and work process, the steps that would need to be taken and the different trades, professions and support staff that would need to be consulted or who would have an input.

As a group, draw up an action plan for designing a new laboratory that would work with biological and other hazards. Once you have worked through the next section, it will be interesting to compare this list with what you are about to learn.

KNOW IT

1. What are the four categories of infectious agents? What features of the pathogen determine the category it is placed in?
2. You catch flu from a colleague in the lab. Is this a work-related illness? Justify your answer.
3. What is the hierarchy of control measures recommended by COSHH?
4. What are the three main potential routes of infection?
5. What types of incident must be reported under RIDDOR?

LO2 Assessment activities

Below are suggested assessment activities that have been directly linked to the Pass, Merit and Distinction criteria in LO2 to help with assignment preparation and include top tips on how to achieve best results.

Activity 1 Carry out risk assessments for a laboratory procedure *P2*

Identify a laboratory procedure and carry out the following tasks:

1. Prepare the documentation for the risk assessment including, e.g., procedures, location, date of assessment, name of risk assessor, etc.
2. Carry out a full risk assessment which:
   a) identifies potential hazards
   b) meets all regulatory and legal requirements.

Activity 2 Explain the procedures and practices required to effectively prevent diseases from spreading in a laboratory *P3*

You could use the evidence you prepared in P2 as the starting point and add to it further, explaining for each biohazard the procedures and practices which would be effective in preventing the spread of disease. Alternatively, you could produce a laboratory handbook for new staff, to be used alongside the evidence for P2.

Activity 3 Describe how health and safety legislation influences procedures and practices *M2*

You have been appointed laboratory manager for a new facility that will handle biohazardous material. Prepare a PowerPoint presentation that could be used as a training or induction guide for new lab staff to make them aware of their duties in regard to managing risk. You could use the HSE guide mentioned in the Extension Activity as source material.

Activity 4 Evaluate the effectiveness of current legislation in safe working practices in the control of diseases *D1*

To achieve D1 you could incorporate an evaluation of current legislation into your report for M2 or produce a separate report or presentation. It must cover all relevant legislation and use examples to support your evaluation of the effectiveness of each piece of law.

Activity 5 Evaluate the potential impact of poor procedures and practices on individuals and the environment *D2*

Research examples of where poor procedures and practices have affected individuals and the environment.

You have been asked to prepare a balanced assessment of what the impact would be if proper procedures were not followed in your laboratory. Your intention is to help laboratory staff understand the importance of following the correct procedures and practices, while at the same time avoiding any unnecessary scaremongering.

TOP TIPS
- Link the work in your assessments clearly to what you have learned in this Learning Outcome.
- If you understand the subject matter covered in LO1, you will see how it is the basis of risk assessment and regulations.
- Make sure that you fully document the work that you do, so that you can put it into practice in LO3.

LO3 Be able to design a safe functioning laboratory to manage the risk presented by hazards P4 M3 D3

> **KEY TERM**
>
> **Decontamination** – Cleaning an object or substance so as to remove microorganisms or other hazardous materials.

GETTING STARTED

Sketch it (5 minutes)

Make a sketch of the layout of a laboratory that will handle potentially infectious materials and other biohazards up to Level 3 (see Table 6.4). Indicate the safety and control features that you think should be included. Share your sketch with the group and keep it to refer to later – how close do you think yours will be to a recommended design?

3.1 Design of the area of work

Initial planning

Prevention is better than cure is an old saying that applies to illness, but it also applies to the control measures needed to minimise risk. In fact, it applies to the whole process of designing a laboratory. It is far better (and cheaper) to anticipate any risks and incorporate the control measures needed into the plan from the outset rather than at a later stage.

The plan needs to start with a full risk assessment of the work that will be done in the laboratory. This will then lead to determining the necessary control measures, including containment facilities if you are dealing with biohazards. There are four levels of containment that correspond to the four hazard groups of infectious agent. Table 6.4 summarises the containment facilities corresponding to the four hazard levels; we will cover the details of these next.

Maximum containment (Biosafety Level 4) facilities are rare and highly specialised. For that reason, we will concentrate here on Biosafety Levels 1–3.

Utilisation of space, technician workspace position in relation to other rooms

The most important thing about space is that there should be enough of it – overcrowding will always increase risk. Surfaces (floor, ceiling and walls) need to be:

- smooth and easy to clean
- impermeable to liquids
- resistant to the chemicals and disinfectants that will be used.

Work surfaces (benchtops) should follow the same rules, but also be resistant to acids, alkalis, organic solvents and (moderate) heat.

Other factors to be considered include:

- providing separate write-up areas adjacent to the laboratory or nearby so that the laboratory itself is used only for laboratory work
- adequate storage in the lab
- handwashing facilities
- space for storage of lab coats (remember that lab coats and gloves should not be worn outside of the lab).

Security, access and containment

Only authorised persons should be allowed access to the laboratory and certainly no children or animals

Table 6.4 Relation of risk groups to biosafety levels, practices and equipment

Risk group	Biosafety level	Laboratory type	Laboratory practices	Safety equipment
1	Basic (Biosafety Level 1)	Basic teaching, research	Good microbiological techniques (GMT)	None; open bench work
2	Basic (Biosafety Level 2)	Primary health services; diagnostic services, research	GMT plus protective clothing, biohazard sign	Open bench plus Biological Safety Cabinet (BSC) for potential aerosols
3	Containment (Biosafety Level 3)	Special diagnostic services, research	As Level 2 plus special clothing, controlled access, directional airflow	BSC and/or other primary devices for all activities
4	Maximum containment (Biosafety Level 4)	Dangerous pathogen units	As Level 3 plus airlock entry, shower exit, special waste disposal	Class III BSC, or positive pressure suits in conjunction with Class II BSCs, double-ended autoclave (through the wall), filtered air

(except laboratory animals used in the work of the lab). Additional security is required where biohazards are present. How much depends on the Risk Group / Biosafety Level (see Table 6.4), but there should be a minimum level of security, such as the following:

- There should be vision panels in the entrance door(s).
- The biohazard symbol (see Figure 6.6) should be displayed on the door of rooms at Biosafety Level 2 and above. This should carry the warning 'BIOHAZARD' and the text 'Admittance to Authorised Personnel Only' as well as the following information:
  - the Biosafety level
  - the responsible person (laboratory manager, etc.)
  - contact to call in case of emergency (daytime and home phone).

A Biosafety Level 3 laboratory needs additional security and containment.

Services and utilities

The laboratory needs a supply of electricity, gas, water and air (ventilation). As well as an electricity supply capable of meeting the load of all the laboratory plant and equipment, an emergency electricity supply such as a stand-by generator to support essential equipment (incubators, biological safety cabinets, freezers, etc.) is advisable. Emergency lighting is also necessary to permit safe exit in the event of a power failure.

It is important that there is no cross-connection between drinking water and laboratory water supplies. Contamination of the public water supply should be prevented by using an anti-backflow device – a valve or other device that prevents potentially contaminated water flowing from the laboratory back to the public water supply.

Storage space

There are specific things to think about depending on the nature of the hazard.

Toxic and flammable substances

These should be stored in cabinets of approved design and properly labelled – see Unit 2 for details.

Radioactive substances

The use of radioactive substances in the laboratory is governed by the Ionising Radiations Regulations 1999 (IRR99). Any organisation working with radioactive substances needs to be registered with the Environment Agency (or local equivalent). The Management of Health and Safety at Work Regulations 1999 (MHSWR, see section 1.1 in Unit 2) governs the general management of risk in the workplace, but IRR99 sets out additional duties covering:

- storage and record keeping
- routine checks
- leak testing
- movement and transport
- incidents (e.g. loss or theft).

More information is available in the HSE publication 'Control of radioactive substances', which can be downloaded from the following link: www.hse.gov.uk/pubns/irp8.pdf

Biological materials

There are particular hazards associated with the storage of infectious materials, usually in a fridge or freezer, particularly during cleaning, defrosting or removal of broken containers.

- Face protection and heavy rubber gloves should be worn during cleaning.
- After cleaning, the inside of the cabinet should be disinfected.
- Containers should always be labelled with the full scientific name of the contents as well as the date stored and name of the person who stored them.
- An inventory should be kept of the contents and any unlabelled or obsolete materials safely disposed of.

Glass ampoules should not be stored in liquid nitrogen because scratches or cracks in the glass could cause them to explode on removal from the liquid nitrogen.

Laboratory equipment

Laboratory equipment should be designed, above all, to prevent or limit contact between the operator and any hazardous or infectious material, and to take account of ease of maintenance, cleaning and **decontamination**. Anything that increases risk of cuts or other injury to the operator (e.g. sharp edges) will add to the risk of infection.

Before using any equipment such as autoclaves or biological safety cabinets (BSCs) for the first time, it is necessary to show that they are performing their function correctly. This is known as validation. It is also necessary to repeat these checks at regular intervals. This is known as recertification.

Furniture

Work surfaces were discussed in space utilisation. Open spaces between and under benches need to be accessible for cleaning. Under-bench cupboards and drawers should provide sufficient space for storage of consumables to avoid clutter on benchtops.

Air management and fume control

If you are planning a new facility, a mechanical ventilation system that provides an inward flow of air without recirculation should be considered.

Air management is more stringent in Biosafety Level 3 laboratories, where the ventilation system must direct airflow into the laboratory. This must be monitored so that laboratory staff can see at all times that proper airflow is being maintained. In addition, the ventilation system in the whole building must be designed so that air from the containment laboratory is not recirculated to other areas in the building. The air can be filtered through high efficiency particulate air (HEPA) filters, reconditioned and recirculated into the containment lab. Alternatively, exhaust air from the laboratory may be discharged through HEPA filters to the outside, well away from occupied buildings or air intakes.

Waste disposal

You need to set up a system for identification and separation of hazardous chemicals, such as chlorinated solvents, acids and other hazardous chemicals, as well as infectious materials (and their containers). Space will be required for storage of waste prior to disposal. See section 3.3 for more information about waste disposal procedures.

Changing facilities

Provision must be made for outer clothing and personal possessions to be kept outside of the laboratory area, particularly when working in a Biosafety Level 3 laboratory that may require facilities for changing out of street clothing and into dedicated laboratory clothing.

Ergonomics and aesthetics

Ergonomics, in the context of this unit, is the study of how a workplace or system can be designed to suit the people who need to use it. Essentially, there is no point in doing a full risk assessment and putting in place appropriate control measures if they make the work process unworkable.

Ergonomics helps us understand human reliability and how to reduce the risk from human error. This means we have to think about the layout of the laboratory – does it help to encourage safe working, or are there any features that make work difficult and so lead to potentially dangerous shortcuts or other practices? We also have to think about the procedures that must be followed. Are they appropriate to the level of risk? Imposing unnecessarily cautious SOPs might lead to a lack of respect for the systems, which could mean that important safety procedures aren't followed.

Finally, we have to consider aesthetics. An attractive and comfortable working environment will mean that laboratory staff are more likely to be happy in their work and, therefore, more motivated to follow the necessary procedures.

PAIRS ACTIVITY

Assembling the team (10 minutes)

Working in pairs, think about all the aspects of planning a laboratory that can safely handle biohazardous material. Who would you need to work with to draw up a plan? Think about all the professions, experts and trades that would need to be consulted. Make a list and share it with the group. Were your lists similar, or were there some differences?

Resources

Computer hardware/software

Various software packages are available for helping with risk assessment, managing risk, record keeping and reporting.

Laboratory equipment

Fume cupboards are essential when handling harmful or toxic chemicals.

Biological Safety Cabinets (BSCs) are essential in any laboratory handling biohazardous materials. Their main function is to protect the worker, the laboratory environment and materials from exposure to aerosols and splashes that may contain infectious agents.

Testing equipment

Various types of testing equipment are available. Equipment such as fume cupboards and BSCs will have built-in devices to measure airflow to make sure they are working correctly. The general ventilation system also needs to be monitored. This includes monitoring airflows and air quality. If HEPA filters are used, these must be regularly checked (at least once every 14 months).

Surfaces can also be swabbed and cultures made to check for microbial contamination. Other testing might require a specialist service provider.

It is also necessary to test seals, filters and other equipment to make sure that they do not leak and cause release of a hazard. The testing procedure will depend on the type of equipment, but could involve pressure testing with a non-hazardous substance before use.

Personal Protective Equipment (PPE)

PPE and clothing may act as a barrier to minimise the risk of exposure to aerosols, splashes and accidental contamination with hazardous or infectious materials. When dealing with biohazards, lab coats should be long-sleeved and back-opening gowns or coveralls will give better protection than a coat. Gloves should be pulled over the wrists of the gown or coat so that there is no skin exposed at the wrist. Aprons can be worn over the lab coat or gown to give additional protection.

The type of eye protection required depends on the nature of the risk. Safety glasses (including prescription safety glasses) are the minimum requirement. Goggles give protection against splashes and impacts, but face shields made of shatterproof plastic will give the greatest protection.

Respirators may be required for high hazard procedures, e.g. cleaning up a spill of toxic or infectious material. Different filters are available for the particular application and/or hazard. To give the best protection, the respirator should be individually fitted to the operator's face, and tested. Surgical face masks are designed to protect the patient from the surgeon, so do not offer any respiratory protection to lab workers.

Disposable gloves are available in latex, vinyl or nitrile – the latter gives the greatest chemical resistance. Used gloves should be discarded with infected laboratory waste. Powdered latex gloves have been associated with the development of dermatitis and immediate hypersensitivity, and are not recommended.

All PPE should be removed before leaving the laboratory.

> **? THINK ABOUT IT**
>
> **The importance of PPE (5 minutes)**
>
> When we put control measures in place, PPE is generally the last topic that we consider. Why is this? Isn't personal protection important? Why are other measures put in place before PPE?

3.3 Procedures

Once the laboratory has been designed and built, a full risk assessment carried out and control measures put in place, it is important that a series of SOPs (see section 2.1) should be drawn up to describe safe working practices for all laboratory tasks. Good staff training together with the use of well written SOPs means there is much less likelihood of accidents in the laboratory causing harm to laboratory staff, the general public or the environment.

Working under supervision

SOPs should always incorporate guidance on working under supervision. Inexperienced workers (those who have just started working in the laboratory) should always be under the close supervision of experienced staff (a senior technician or other senior member of staff). The level of supervision will depend on the nature of the risk and the experience of the person doing the work. Therefore, the level of supervision might progress from the supervisor being next to the person at all times, to being present in the laboratory and checking from time to time, being available within the building to provide advice and guidance, to, eventually, unsupervised working. However, unsupervised does not mean working alone. There should always be someone else present in the laboratory in case of accident.

Hazard recognition

Hazard recognition should be a part of initial training and should be incorporated into continuous in-service training for all laboratory staff. In particular, staff should be made aware of the risks that are commonly encountered – see section 2.1 on risk assessment.

Safety awareness

Safety awareness should be incorporated into initial and continuing staff training, as well as into SOPs and laboratory safety manuals. Laboratory managers or supervisors should take responsibility for training staff in good laboratory techniques, including safety awareness. It helps to keep records of staff training, including having signature pages as a way of ensuring that staff have read and understood the relevant guidelines and procedures. Each organisation should have a biosafety officer who should be able to assist with training and development of training materials and documentation.

Risk assessments

The principles and procedures of risk assessment have been covered in detail. It is enough to say here that this is not a once and forever process. Risk assessment is a continuing activity to ensure that assessments are kept up to date and that any new risks are recognised and controlled.

Incident procedures

Every laboratory working with infectious materials and other biohazards should have safety precautions in place that are appropriate to the level of hazard – we covered this in previous sections. Emergency plans need to include:

- the foreseeable types of incidents, accidents or emergencies that might occur
- the role, responsibilities and authority of individuals during an emergency
- procedures for employees to follow – including regular safety drills and identifying the special needs of any disabled employees
- the safety equipment and PPE to be used
- arrangements for liaison with emergency services
- first-aid facilities, including consideration of whether any treatment needs to be given by a qualified medical practitioner
- procedures for cleaning up and disposal of waste.

Biosafety Level 3 and 4 laboratories need a detailed written contingency plan in the event of any accident that leads to release of infectious materials that could be harmful to laboratory staff or the wider community. Detailed information about this is contained in the WHO Laboratory Biosafety Manual – see 'Read about it' at the end of this unit for a link to download a copy of this.

Accident procedures

In all laboratories, including all levels of Biosafety laboratory, emergency procedures should be in place for dealing with a range of accidents, including:

- puncture wounds, cuts and abrasions
- ingestion (swallowing) of potentially hazardous or infectious material
- release of a potentially infectious aerosol (outside of a BSC)
- broken containers and spilled infectious substances
- breakage of tubes containing potentially infectious material in centrifuges with or without sealable buckets (the procedures are different).

Fire precautions

As well as the usual fire precautions in any laboratory, fire and other emergency services need to be involved in the development of emergency preparedness plans. For example, they need to be told which rooms contain hazardous or potentially infectious materials. A site visit will help them become familiar with the layout of the laboratory and its contents.

Storage and security

Storage facilities should be maintained so that stocks are not likely to slide, collapse or fall. They should be kept free from accumulations of rubbish, unwanted materials and any trip, fire or explosion hazards. Freezers/fridges and storage areas should be lockable.

Staff must be made aware of the need to maintain security of the laboratory by not allowing unauthorised access.

Use of fume cupboards and fume control

While a BSC will protect against biohazards, including infectious materials, it should not be used for handling of other hazardous substances. Unless a procedure involves biohazardous material, hazardous chemicals should always be handled in a fume cupboard.

Airflow is important in a fume cupboard where the sash must not be opened further than the maximum indicated so as to maintain adequate air flow. When not in use the sash should be closed to act as a shield in case of explosion. Remember that there are limits to the effectiveness of fume extraction. During your COSHH assessment you need to decide whether the fume cupboard gives adequate protection. For example, handling of larger (e.g.>1 g) quantities of a highly hazardous substance might be more appropriate in a glove box.

Fume cupboards and other forms of fume extraction need to be examined and tested, at least every 14 months.

Disinfection

A basic knowledge of disinfection and sterilisation is essential to ensure safety in a laboratory handling biohazardous materials. Heavily soiled items cannot be effectively disinfected or sterilised but must be cleaned first, as dirt, soil and organic matter can interfere with the action of a disinfectant.

It is important to appreciate the following differences:

- A bacteriostat or antimicrobial will stop the growth of microorganisms but not necessarily kill them.
- A disinfectant or germicide will kill microorganisms but not necessarily their spores.
- A sporicide will also kill spores.

Table 6.5 Types of germicide

Chemical class	Activity	Application and examples
Chlorine based	Wide spectrum, fast-acting oxidant.	Bleach (sodium hypochlorite), sodium dichloroisocyanurate (NaDCC), chloramines and chlorine dioxide (ClO_2). The latter is probably the most selective, meaning that it is effective at lower doses.
Formaldehyde/ glutaraldehyde	Kills microorganisms and spores.	Relatively slow acting, both are hazardous chemicals.
Alcohols	Active against microorganisms including most viruses, but not against spores.	Ethanol and propan-2-ol (isopropyl alcohol, IPA) are most effective at concentrations of about 70% by volume in water. This can be used on the skin (e.g. an alcohol-based hand rub) as well as on work surfaces.
Hydrogen peroxide and peracids	Strong oxidising agents and broad-spectrum germicides.	Safer than chlorine-based agents to humans and the environment.

The choice of germicide (Table 6.5) will depend on a number of factors, including:

- the type of microorganism
- considerations of potential for harm caused by the germicide to humans and/or the environment.

For example, phenolic compounds were amongst the earliest germicides, but their use is now restricted because of safety concerns.

> **? THINK ABOUT IT**
>
> **Use of disinfectants (10 minutes)**
>
> Now that we have considered the hazards that we might have to deal with, think about the types of disinfectant that might be used for:
>
> - routine cleaning of floors, doors and walls
> - routine cleaning of laboratory work surfaces
> - routine cleaning of BSCs
> - clean-up in the event of a spillage or escape of an infectious agent.
>
> What are the factors that we need to consider in each case? Think about the risks we are trying to control in each case, the effectiveness of the disinfectant and the degree of risk associated with its use. Share your conclusions with the group.

Disposal of waste

The COSHH regulations (see section 1.1 of Unit 2) cover the disposal as well as handling of hazardous substances. You will need to take these into account when developing procedures for storing and disposing of all hazardous waste.

There are particular precautions needed when disposing of biohazard waste.

All infectious materials should be decontaminated, autoclaved or incinerated within the laboratory wherever possible so as to remove the need for transport of biohazard waste. Steam autoclaving is the preferred method for all decontamination processes, unless an incinerator is available on site. If there is no incinerator on site, all contaminated materials should be autoclaved in leak-proof containers before transfer to the incinerator.

Personal Protective Equipment (PPE)

This has been covered in section 3.2 above.

> **GROUP ACTIVITY**
>
> **The elements of a good laboratory (20 minutes)**
>
> You should have kept your sketch plan of a laboratory from the start of this section. Now, working as a class or in small groups, pool features of your original designs and list the features you have learned about in this section to produce a design checklist for a Biosafety Level 2 laboratory.

Legislation

Health and safety at work

The general legislation governing health and safety at work was covered in section 1.1 of Unit 2. This includes the Health and Safety at Work Act 1974 (HSWA 1974) and the Management of Health and Safety at Work Regulations 1999 (MHSWR). Aspects specific to labs handling biohazard material were covered in section 2.1 above.

Codes of practice

All organisations must have a written statement of their health and safety policy. Local health and safety policies may be needed to cover specific risks and working patterns in single departments or individual labs. A code of practice can give further detail of how safe working will be achieved on a day-to-day basis, based on the local risk assessment and SOPs.

Control of toxic and flammable substances

This is covered in section 1.1 of Unit 2.

Microbiological hazards

The Public Health (Control of Disease) Act 1984 requires doctors to notify suspected cases of certain infectious diseases – there is more information on the Public Health England website (www.gov.uk/government/organisations/public-health-england); similar arrangements are in place elsewhere in the UK.

The Specified Animal Pathogens Order 1998 prohibits any person from having in their possession, or introducing into animals, any of the organisms listed in the Schedule to the Order except under the authority of a licence issued by the appropriate minister.

Part 7 of the Anti-terrorism, Crime and Security Act 2001 deals with the security of pathogens and toxins. These regulations are overseen by the Home Office and enforced by the police.

In addition, labs working with genetically modified organisms (GMOs) need to pay attention to the Genetically Modified Organisms (Contained Use) Regulations 2000.

Design specifications

Based on what you have learned in sections 3.1–3.4, you should now be in a position to put together a design for a laboratory, including one that will handle biohazards up to Biosafety Level 3. The extension activity will allow you to put this into practice.

EXTENSION ACTIVITY

Design your own laboratory (1.5 hours)

Working in groups, and using your checklist from the group activity, create a design specification for your laboratory, including:

- siting of laboratory
- structures and fittings
- ergonomics and aesthetics
- health and safety
- space efficiency
- storage.

You must justify your design to the class.

KNOW IT

1. What are the minimum requirements for security, access and containment for the following types of laboratory?
   a Biosafety Level 1
   b Biosafety Level 2
   c Biosafety Level 3
2. Ventilation is of great importance in Biosafety Level 3 laboratories. What are the options for handling waste air from such a lab? What precautions must be taken?
3. What is the importance of changing facilities? How do the requirements differ for different levels of Biosafety?
4. When is testing equipment important?
5. What should be included in an emergency plan for dealing with incidents in a biohazard laboratory?

LO3 Assessment activities

Below are suggested assessment activities that have been directly linked to the Pass, Merit and Distinction criteria in LO3 to help with assignment preparation and include top tips on how to achieve best results.

There is an opportunity for synoptic learning and assessment with many of the other units.

Activity 1 Produce a design specification to control risks posed by biohazards in a laboratory *P4*

Your organisation is in the process of developing a new research facility in biological research. You have been asked to take responsibility for the control of risks arising from storing and using biohazardous substances. You should research the types of biohazards which may be used and the means of controlling such risks. From this you should develop a design specification to ensure that risk arising from the use and storage of biohazards is controlled.

Activity 2 Explain how the design of a laboratory can control the spread of disease in a laboratory *M3*

This may be a supplementary document to your specification for P4, or a separate report or presentation. It must clearly identify elements of laboratory design which can control the spread of disease within a laboratory and explain how the design provides protection.

Activity 3 Analyse how procedures and legislation affect the control of diseases in a laboratory D3

Control of disease in the laboratory is multi-factorial – in other words, it requires many things to be done correctly so as to reduce risk and minimise harm.

Prepare an article for an in-house journal (college, company, etc.). The working title is 'It's not just bureaucracy – how proper procedures and legislation help to keep us safe.'

The intention is to show your co-workers how proper procedures and the various pieces of legislation all work together to control biohazards in the laboratory.

TOP TIPS
- The work that you do in LO1 and LO2 will prepare you for LO3.
- Use what you have learned in the LO3 Group Activity (The elements of a good laboratory).
- The extension activity (Design your own laboratory) gives you the chance to really put your knowledge into practice.
- Make sure that you fully document the work that you do throughout LO1 and LO2, so that you can put it into practice in LO3.

Read about it

Laboratory Biosafety Manual (3rd edition), World Health Organization – This contains practical guidance on biosafety techniques for use in laboratories at all level of risk. There is a great deal of information about risk assessment, design and commissioning of biohazard laboratories, laboratory biosecurity concepts and good microbiological techniques. A PDF copy may be downloaded from the following link:

www.who.int/csr/resources/publications/biosafety/WHO_CDS_CSR_LYO_2004_11/en/

'Biological agents: Managing the risks in laboratories and healthcare premises' (May 2005), HSE. This is available to download from:

www.hse.gov.uk/biosafety/biologagents.pdf

Further information about COSHH and RIDDOR is available from the HSE website:

www.hse.gov.uk/coshh/

www.hse.gov.uk/riddor/

Unit 07
Human nutrition

ABOUT THIS UNIT

The human body requires a range of nutrients to ensure appropriate bodily function and health. The aim of this unit is to provide you with knowledge of the nutrients needed and how the body uses them. You will be able to calculate the energy content of food and compare it with the body's requirements. You will also understand the importance of hydration and supplements in maintaining a healthy body. Finally, the unit links the development of deficiency and disease from a biological and psychological perspective.

LEARNING OUTCOMES

The topics, activities and suggested reading in this unit will help you to:

1 Understand human nutritional requirements in the maintenance of health
2 Be able to calculate nutritional requirements to maintain energy for different levels of activity
3 Understand conditions relating to dietary needs
4 Be able to label food with nutritional information

How will I be assessed?

You will be assessed through a series of assignments and tasks set and marked by your tutor.

How will I be graded?

You will be graded using the following criteria:

Learning Outcome	Pass	Merit	Distinction
	The assessment criteria are the Pass requirements for this unit.	To achieve a Merit the evidence must show that, in addition to the Pass criteria, the candidate is able to:	To achieve a Distinction the evidence must show that, in addition to the Pass and Merit criteria, the candidate is able to:
1 Understand human nutritional requirements in the maintenance of health	**P1** Outline the components of a healthy balanced diet		
2 Be able to calculate nutritional requirements to maintain energy for different levels of activity	**P2** Describe energy intake and expenditure when exercising and at rest	**M1** Calculate the BMR for two comparable groups of people taking different levels of exercise	
3 Understand conditions relating to dietary needs	**P3** Measure the energy content of a carbohydrate food		
	P4 Describe the common diseases and conditions relating to nutrition	**M2** Explain physical impacts of developing a condition relating to malnutrition	**D1** Recommend and justify nutritional requirements for a specific group of people
4 Be able to label food with nutritional information	**P5** Describe the factors that affect people's eating habits		
	P6 Label food products to provide nutritional information to consumers	**M3** Explain how legislation affects food labelling in the UK	**D2** Discuss how food labelling may lead to confusion for consumers and suggest solutions

LO1 Understand human nutritional requirements in the maintenance of health *P1*

GETTING STARTED

Food diary (5 minutes)

Write down the food you have eaten over the past week. Put this to one side for discussions throughout the lesson.

1.1 The sources of nutrients

Nutritional requirements to maintain health refers to the sources and requirements of various nutrients and water required by the body. It also refers to the role nutrients play in growth and repair: the importance of a balanced diet and the variation of requirements in individuals.

1.2 Macronutrients and micronutrients and 1.3 The role of nutrients for growth and repair

You may have heard the term 'You are what you eat'. This refers to the importance of the amount and quality of food we eat, in order to allow our cells and organs to function well and to give the body the energy it requires daily.

There are various sources of nutrients the body needs; these are taken from different food groups. The balance of the nutrients taken from various food groups is important as each one helps the body to function.

1 Macronutrients (macro = 'big'). These bulk energy and are protein, fat and carbohydrates, water and atmospheric oxygen.
2 Micronutrients (micro = 'small'). These are vitamins, trace minerals and organic acids.

Table 7.1 Sources of macronutrients and how they are used within the body

Macronutrients	Example of how the body uses it	Example of food source
Protein	• builds new cells • maintains tissues • development of new proteins	• tuna • chicken • cheese
Fat	• fatty deposits narrow arteries (S) • fat needed to absorb vitamins A, D and E (US)	• fatty meat (S) • butter (S) • coconut cream (S) • olive oil (US) • avocados (US) • almonds (US)
Carbohydrates	• supply energy • fuel for the nervous system • fuel for muscles	• apples • spinach • pasta • potatoes
Water	• lubricates mucus membranes • transports waste and toxins out of the body • digests food	• some in foods • tea/coffee • fruit juice • water
Atmospheric oxygen	• allows metabolism • converts food to energy	• good quality air

S = saturated
US = unsaturated

Proteins

Proteins can be found on the outer and inner membranes of living cells. Here are some of the places where they are used.

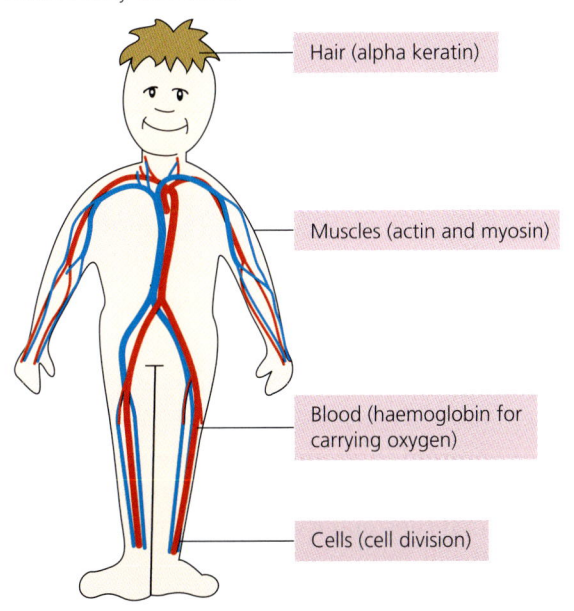

▲ Figure 7.1 Areas in the body where proteins can be found

As you can see, protein has many roles and is important for the health of the body. Proteins are used by the body to build new cells, put together new proteins and to maintain tissues. Worker proteins called **enzymes**, which often need specific vitamins and minerals, help to digest food and pull together or divide molecules to make new cells and chemical substances. Other chemicals such as **neurotransmitters** (nerve cells) send messages backwards and forwards to each other – these also require proteins.

Nucleoproteins are made of amino acids, linked by peptide bonds for forming proteins and **nucleic acids**, which store and transmit genetic information present in every cell.

Essential proteins: there are nine amino acids obtained from food. **Non-essential proteins** are all others manufactured from carbohydrates, fats and other amino acids. To make all the required proteins needed by the body, we require 22 different amino acids.

Table 7.2 Essential and non-essential amino acids

Essential amino acids	Non-essential amino acids
Histidine	Alanine
Isoleucine	Arginine
Leucine	Aspartic acid
Lysine	Methionine
Cysteine	Glutamic acid
Phenlyalanine	Glycine
Thereonine	Hydroxyproline
Tryprophan	Proline
Valine	Serine
	Tyrosine

As well as essential and non-essential proteins, whether a protein is high or low quality is also important. Animal proteins such as meat, eggs, poultry and dairy products are high-quality proteins, as the body can absorb these proteins efficiently. Proteins from plants such as grains, vegetables, fruit, beans or peas, are not as high due to the limited amounts of amino acids. You can, however, provide the missing amino acids in some proteins by combining foods together. For example, pasta is low in the amino acid lysine and isoleucine but milk products have lots of these two amino acids. So, by having pasta and cheese, you are getting the correct amount.

How much protein do we need? Generally, according to the current information from the **Department of Health**, we need approximately 15% of our calories each day to come from protein. This of course may vary depending on individual needs at different stages of life.

169

How do we digest proteins? During **protein synthesis**, proteins combine with **phosphoric acid** to produce **phosphoproteins**, amino acids join with fats to form **lipoproteins** and nucleic acids combine with proteins to create nucleoproteins. Once protein synthesis has been completed, this is then converted to **glucose** and used for energy. The **nitrogen** left is **metabolised** by the liver and is converted to **urea**.

The amount of protein required is dependent on weight and lifestyle. For example, if you do not exercise and lead a sedentary lifestyle, you multiply your body weight by 0.8 g (this is the amount required per kg). If however you are an athlete, you multiply your body weight by 1.4–1.8 g per kg, as this would be the required amount of protein for this level of activity.

Carbohydrates

Table 7.3 Carbohydrates and sugars

Type of carbohydrate	Sugars
Fructose (simple carbohydrate)	Monosaccharide, one sugar
Glucose (simple carbohydrate)	Monosaccharide, one sugar
Galactose (simple carbohydrate)	Disaccharide, two sugars linked
Maltose (simple carbohydrate)	Disaccharide, two sugars linked
Sucrose (simple carbohydrate)	Disaccharide, two sugars linked
Lactose (simple carbohydrate)	Disaccharide, two sugars linked
Raffinose (complex carbohydrates)	Polysaccharide < two unit sugar together
Stachyose (complex carbohydrates – tetrasaccharide)	Polysaccharide < two unit sugar together
Starch (complex carbohydrates – polymeric)	Polysaccharide < two unit sugar together
Cellulose (complex carbohydrates – organic compound)	Polysaccharide < two unit sugar together
Hemicellulose (complex carbohydrates – heteropolymers)	Polysaccharide < two unit sugar together
Pectin (complex carbohydrates – heteropolysaccharide)	Polysaccharide < two unit sugar together

KEY TERMS

Enzymes – A protein which acts as a catalyst. Each one makes a particular chemical reaction happen at the temperatures found in the body, and at a much faster rate. Enzymes are not used up, so an enzyme molecule catalyses its reaction many millions of times.

Neurotransmitters – Chemicals that transmit signals between nerve cells.

Nucleoprotein – A substance made up of a nucleic acid and a simple protein.

Amino acids – These are building blocks of proteins.

Nucleic acid – A group of complex polymers found in living cells.

Essential proteins – Proteins that the body cannot produce itself.

Non-essential proteins – Proteins that the body can produce itself.

Department of Health – This is the part of the government responsible for the policy on health.

Protein synthesis – How individual cells construct proteins.

Phosphoric acid – This is a mineral acid.

Phosphoprotein – A protein modified by the attachment of at least one phosphate group.

Lipoprotein – A biochemical of proteins and lipids.

Glucose – A type of sugar.

Nitrogen – A chemical element found in all proteins.

Metabolised – The breaking down of substances and their reorganisation into another form.

Urea – A chemical compound produced during metabolism.

Dietary fibre consists of non-starch polysaccharides. What makes dietary fibre different from other carbohydrates is that the digestive enzymes cannot break the bonds that hold its sugar together.

Any glucose remaining in the cells is converted into glycogen and stored in the liver and muscles. If there is an excess once the body has used what it needs, the rest will be converted to fat. However, if the body does not have enough glucose, it will take it from glycogen in fatty tissue. If that runs out, it will burn its own protein tissue (muscle).

KEY TERMS

Cholesterol – A lipid molecule synthesised by animal cells.

Bacteria – Single living cells

Biochemical – Relating to the chemical composition of a biological substance.

Myelin – This is a fatty substance which forms a sheath around the axon of some nerves.

Cell membrane – This separates the interior of individual cells from their environment.

Carbohydrates can also:

1 assist in the body's absorption of calcium
2 lower blood pressure and **cholesterol**
3 provide food for **bacteria** in the intestinal tract.

The amount of carbohydrates required daily is worked out by multiplying the amount of kilojoules you require (this will vary depending on age, sex and exercise) by 55, which is the approximate percentage of carbohydrates that should come from the diet.

GROUP ACTIVITY

A day's diet (40 minutes)

In a group, discuss and write down what a typical day's diet consists of. Include food, fluid and types of human activity. Put the foods that you write down into the correct food group section. Record what sources of nutrients you get from each of the foods.

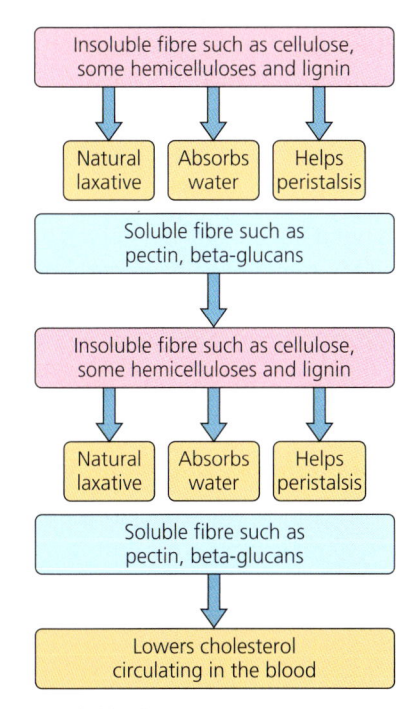

▲ Figure 7.2 Insoluble fibre

Fat

Fats are needed to manufacture **biochemical** substance and to make tissue, so it is important to the body in the correct amounts. Fat has a number of jobs to do in the body.

1 It provides padding for your skin.
2 It acts as an insulation to ensure the body does not lose too much heat.
3 It provides a source of energy.
4 It gives the body shape.

Internally:

1 It sits around the organs to protect them from any impact or injury.
2 Fat is around **myelin** (nerve cells) to ensure the smooth communication between cells.
3 It is around every **cell membrane**.

Working out how much fat we should take from food can be calculated as shown below:

- 5021 kilojoules diet – recommended fat content: 30 percent of 5021 is 1506.3 kilojoules; at 38 kilojoules per fat-gram, this means we should eat no more than 40 grams of total fat, per day. A maximum of 13 grams may be saturated fat.
- 6694 kilojoules diet – recommended fat content: 30 percent of 6694 is 2008.2 kilojoules at 38 kilojoules per fat-gram, this means we should eat no more than 53 grams of total fat, per day. A maximum of 17 grams may be saturated fat.
- 8368 kilojoules diet – recommended fat content: 30 percent of 8368 is 2510.4 kilojoules; at 38 kilojoules per fat-gram, this means we should eat no more than 66 grams of total fat, per day. A maximum of 22 grams may be saturated fat.

Digestion

Digestion takes place in two phases: firstly in the mouth and stomach. The teeth break down the food into pieces and the stomach performs a churning action to break food into smaller pieces; it also releases some acids and enzymes (e.g. peptin) to start the chemical process. Chemical digestion takes place in the digestive tract where enzymes, hydrochloric acid and bile dissolve the food, releasing the nutrients into the body.

Table 7.4 shows the phases of digestion and Figure 7.3 shows the journey of food through the digestive system.

Table 7.4 Phases of digestion

Eyes and nose	They stimulate receptor cell on olfactory nerves and communicate with the brain to send messages to the mouth and digestive tract – salivation begins.
Mouth	Chewing and swallowing break up parts of fibre to allow digestive enzymes to the nutrients inside. Saliva production moistens food with the tongue pushing it to the back of the mouth. Starts the digestion of complex carbohydrates, breaking starch molecules into simple sugars.
Stomach	Peristalsis takes place and breaks food down into smaller particles; cells in stomach wall secrete stomach juices – enzymes, hydrochloric acid and mucus. The stomach juices are acidic; the enzymes in saliva break complex carbohydrates apart into simple sugars. The stomach blends the contents to a mass called chyme; this spills past the stomach into the intestine.
Small intestine	Once chyme is in the small intestine, pancreatic and intestinal enzymes finish the digestion of proteins into amino acids, and then digest fat and polysaccharides. Bile enables fats to mix with water. Alkaline pancreatic juices make the chyme less acidic. Intestinal alcohol dehydrogenase digests alcohol not previously absorbed into the blood. Peristaltic contractions move the food down through the tube so your body can absorb sugars, amino acids, fatty acids, vitamins and minerals into cells in the intestinal wall. Order broken down: Carbohydrates separate quickly into single sugar units Proteins Fats Water-soluble vitamins and minerals Amino acids, sugars, vitamins C and B, minerals and trace elements are carried through the blood to the liver, processed and sent out to the rest of the body. Fatty acids, cholesterol and fat-soluble vitamins go into the lymph system, then are passed into the blood, processed by the liver and sent out to other body cells.
Large intestine	Food moves into the top of the large intestine (colon) then absorbs water from the mixture, then into faeces (indigestible material from food). Some bacteria help to do jobs like break down nitrogen-containing molecules.

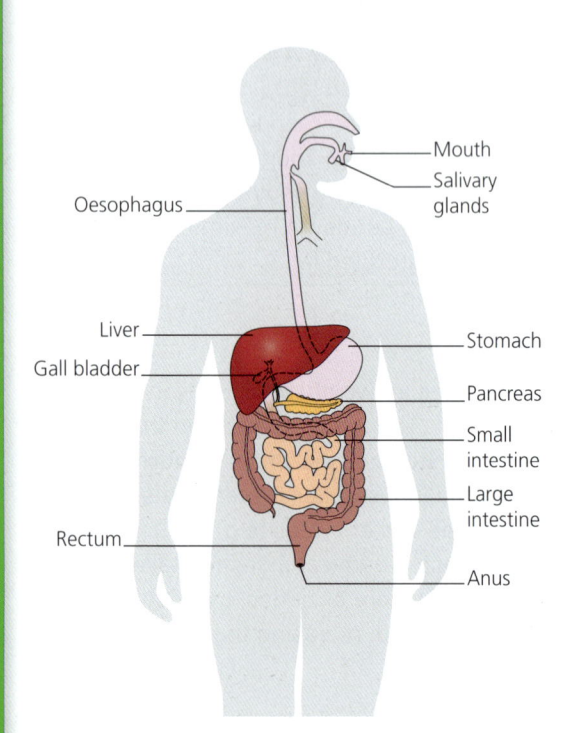

▲ **Figure 7.3** The journey of food through the digestive system

EXTENSION ACTIVITY

(40 minutes)

Individually, research the following diets: Atkins, Weight Watchers and SlimFast. As a group, discuss the advantages and disadvantages of each.

Vitamins, minerals and organic acids

Vitamins contain carbon, hydrogen and oxygen. They are needed to build bones, glands, nerves and blood. They also assist in metabolising fat, carbohydrates and proteins. Being deficient in some vitamins can cause illness, so they are crucial to good health. Table 7.5 on the next page shows the roles of the vitamins needed by the body and the food sources that provide them.

Table 7.5 Vitamins and their role in the body

Vitamin	Positive role in the body	Examples of food source
Vitamin A (fat soluble)	• immune system • eyesight	• cod liver oil • carrots • leafy vegetables
Vitamin K (fat soluble)	• helps blood to clot	• leafy vegetables (e.g. broccoli)
Vitamin E (fat soluble)	• protects cells from damage (antioxidants)	• leafy vegetables • hazelnuts • sunflower oil
Vitamin D (fat soluble)	• healthy bones • helps to process calcium	• cheese • milk
Vitamin C (dissolves in water)	• helps keep immune system strong • helps collagen	• tomatoes • peppers • citrus fruits
Vitamin B (some) – B1, B2, B6, B12 (dissolve in water)		
Vitamin B1 (thiamine)	• helps to process carbohydrates	• whole grains (pasta, bread)
Vitamin B2 (riboflavin)	• helps to make red blood cells • converts food into energy	• chicken • almonds • milk
Vitamin B3 (niacin)	• helps digestion	• fish • wholegrain
Vitamin B6 (pyridoxine)	• supports the nervous system • breaks down proteins and sugar	• jacket potatoes • eggs • bananas
Folate	• promotes cell division • prevents birth defects • promotes heart health	• lentils • dark leafy vegetables
Vitamin B12 (cyanocobalamin)	• helps to make red blood cells	• salmon • soybeans
Biotin	• production of fatty acids • strengthens hair and nails	• carrots • walnuts • strawberries
Pantothenic acid	• used in synthesis of Coenzyme A, transporting carbon atoms within the cell • biosynthesis of cholesterol and fatty acids	• avocado • meat

Organic acids are used in food preservation; this will be discussed later in the unit.

> **? THINK ABOUT IT**
>
> With the high consumption of junk food by some individuals, it seems unlikely that this kind of food is providing the required nutrients for our bodies. Review the nutritional value of food at a popular fast food restaurant of your choice and compare it to the recommended nutrients for good health.

Minerals

Trace elements are also required, albeit in small amounts, for a healthy body. Table 7.6 shows the role of minerals in the body.

> **KEY TERMS**
>
> **Electrolytes** – Substances that produce electrical conducting solutions when dissolved in fluid (for example water or blood).

Table 7.6 Minerals and their role in the body

Mineral	Example of job in the body	Examples of food source
Trace element		
Iodine	• helps to make thyroid hormone	• sea fish
Selenium	• helps with immune system	• fish • Brazil nuts
Copper	• helps in production of red and white blood cells	• nuts
Manganese	• helps to make enzymes	• nuts • bread
Chromium	• influences insulin	• varied balanced diet
Molybdenum	• making genetic material	• oats • tinned vegetables
Major minerals		
Calcium	• strong bones • blood clotting	• milk • cheese
Phosphorous	• healthy teeth and bones	• red meat • oats
Magnesium	• turns food into energy	• leafy vegetables • fish
Iron	• production of red blood cells • carries oxygen around the body	• liver • beans
Zinc	• helps to heal wounds	• shellfish • cheese
Sulphur	• helps make new tissue	• naturally in most foods

1.4 Fluid intake

Our bodies are made up of approximately 60% water. So what does water do? As a solvent, water dissolves other substances and carries things such as blood cells and nutrients around the body. Without sufficient water, organs cannot function effectively. From this it is clear how important water is to the body and its function.

Other vital roles of water are as follows.

- It digests food.
- It enables metabolism.
- It transports waste and toxins out of the body.
- It lubricates mucus membranes.
- It assists in electrical signalling between cells and prevents over-heating.
- It assists with temperature regulation.

Electrolytes are substances that produce an electrically conducting solution when dissolved in a polar solvent, such as water: the body uses electrolytes to regulate the fluid balance. Cells use electrolytes to carry electrical impulses across themselves to other cells. It is important to keep electrolytes balanced; if this balance is disrupted such as during heavy exercise, it is important to replace them. Drinks such as sport energy drinks can help to restore the balance.

For example, when you are thirsty (due to a lack of fluid), electrolytes are increased, especially sodium. If the body does not have enough fluid, certain electrolytes such as sodium will increase, making you thirsty. This is the message the body is sending – telling you to drink more.

Because we do not store water in our bodies, it is important to ensure our intake is adequate.

Where the body loses water per day:

1. urine – approximately 1.5 dm^3
2. faeces – approximately 150 cm^3 per bowel movement
3. breathing – approximately 400 cm^3
4. sweating – approximately 500 cm^3 or 0.5 dm^3.

Where the body gains water:

1. created during digestion and metabolism
2. food intake
3. from drinks.

Water alone is not the only fluid that can keep the body hydrated, but there are things to bear in mind when consuming other fluids. Fruit drinks, milk, squashes, tea and coffee count towards fluid intake, but strong coffee (and alcohol) act as **diuretics**, which make you urinate more often, so these are not as re-hydrating. The daily recommended amount of water (or other fluid) is 6–8 glasses; this is to replace normal water loss.

1.5 The importance and principles of a balanced diet

There is always new research or news about what food is good for you and what is not, as well as information about diets and exercise, so sometimes it is difficult to know exactly what to eat! The NHS cites:

- plenty of fruit and vegetables
- plenty of starchy foods, such as bread, rice, potatoes and pasta
- some meat, fish, eggs, beans and other non-dairy sources of protein
- some milk and dairy foods
- just a small amount of food and drinks that are high in fat and/or sugar.

Figure 7.4 shows the Eatwell Guide, which illustrates what the **Food Standards Agency** think a healthy diet should consist of.

> **RESEARCH ACTIVITY**
>
> We have seen how important a balanced diet is to achieving good health. Put together a diet plan for the following:
>
> 1. a child of eight years, who goes to intense dancing lessons three times a week
> 2. a male adult aged 22 years, who has an office job, doesn't exercise and has diabetes
>
> Refer to the recommended daily calories intake, resting energy expenditure (REE), BMR (see Key Terms in LO2) and food groups.

> **KEY TERMS**
>
> **Diuretic** – Any substance that promotes the production of urine.
>
> **Food Standards Agency** – Responsible for protecting public health in relation to food.
>
> **BMR** – Basal Metabolic Rates; minimum rate of energy expenditure per unit of time by a body at rest.

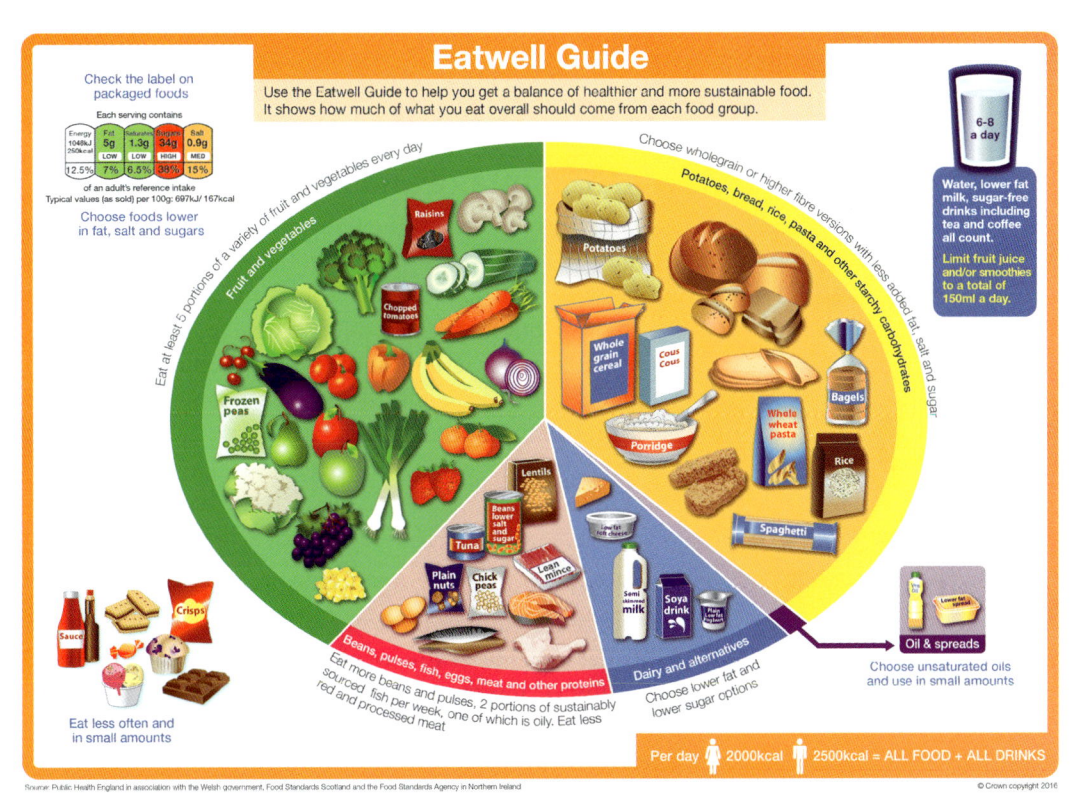

▲ Figure 7.4 What a healthy diet should consist of, according to the Food Standards Agency

1.6 How nutritional requirements may vary in individuals (e.g. calorific requirements, nutrient requirements, allergies/intolerances)

GROUP ACTIVITY

Nutritional variation (40 minutes)

Research the diet of a chosen developing country (for example, a country in Africa). Document the conditions and diseases in relation to nutrition and water, and discuss.

The first step is to be aware of the requirements of individual needs. For example, gender, age, activity level, specific dietary requirements. A balanced diet is represented by the consumption of food taken from all of the food groups, plus the correct amount of water to aid the digestion and nutrients in the body.

One of the things to consider is the portions we consume. If we consume more than we burn, we store it! If we eat a variety of different food groups in moderation according to our requirements, then we should be getting it right.

We do not require food just for energy. The nutrients we get from food also help with growth and repair, as we have previously seen. We will look at the variation in calorific and nutritional requirements later in this unit.

KNOW IT

1. What do macronutrients and micronutrients consist of?
2. Name one food source and job in the body for each of the following: protein, carbohydrate and fat.
3. What role do oxygen and water play in the health of the body?
4. Name two important vitamins and minerals, the role they have in the body and a food source.

LO1 Assessment activity

Below is a suggested assessment activity that has been directly linked to the Pass criteria in LO1 to help with assignment preparation and includes top tips on how to achieve best results.

Activity 1 Produce an information leaflet P1

Produce an information leaflet for a junior school or health centres about what makes a healthy diet. You could produce a PowerPoint presentation.

TOP TIPS
- ✓ You are expected to outline the components of a healthy balanced diet, so not a great amount of detail is required, but you will be expected to know the components.
- ✓ Make sure you understand the importance of macronutrients and micronutrients.

LO2 Be able to calculate nutritional requirements to maintain energy for different levels of activity P2 P3 M1

GETTING STARTED

Eating too much or too little (5 minutes)

Discuss how we determine how much to eat. How do we know we are eating enough or too much?

2.1 How to calculate energy intake and output

Energy intake and energy expenditure

Energy is required for the body to survive; we must feed our cells in order for them to help our organs and tissues to function. Energy requirements vary from one person to another, depending on age, size, sex and physical activity level. Energy intake is usually measured in joules (J) or kilojoules (kJ). Some people are more familiar with calories (kcal).

– 1 kilojoule (kJ) = 1,000 joules

– 1 kilocalorie (kcal) = 1,000 calories, or 1 calorie.

In order to calculate your recommended energy intake, there are a number of useful websites which can do this for you. For example: https://nutritiondata.self.com/tools. You can enter your sex, age, height and daily activity and it will calculate the recommended amount for you, give it a go. You can calculate energy expenditure using basal metabolic rate (BMR), as described below.

Basal Metabolic Rate (BMR)

Even when resting the body requires energy to function. For example, breathing, digestion, making new tissue. This required energy is known as the Basal Metabolic Rate (**BMR**) and accounts for around 60% of the calories you require daily. To calculate your BMR, you need to know your weight and height and the recommended BMR for you. To find out what your daily energy requirements are and the energy you use for various activities, there is a calculation. So, you can find out just how many calories are required to support your body in whatever activities you are doing.

BMR and energy use depend on:

1 weight
2 height
3 sex
4 age

Calculation:

Mifflin St. Jeor Equation

Males: BMR = 10 × weight (kg) + 6.25 × height (cm) − 5 × age (years) + 5

Females: BMR = 10 × weight (kg) + 6.25 × height (cm) − 5 × age (years) −161

Kcals to kJ conversion: 1kcal = 4.184kJ

Aerobic versus anaerobic

Aerobic exercise is a good way to burn energy. This kind of activity uses a great deal of oxygen by generating energy through the use of oxygen, gets the heart pumping and boosts the metabolism, by causing the body to burn energy quickly. Therefore, it is a good way to promote heart health and increase fat burning. Good aerobic exercises are running, cycling, swimming – anything that increases the heart rate over a certain period of time.

Anaerobic exercise is the opposite. It does require oxygen, but the intensity of this kind of exercise forces muscles to work without oxygen, which it can do for short durations of intense activity. In the beginning, the body will burn energy from the body's reserve fat; after that the toning of muscles becomes the basis of the exercise. Good anaerobic exercises, such as weight training, can be done only over short periods of time because of their intensity.

Table 7.7 The average energy expenditure for a person weighing 60kg (kJ per hour)

Activity level	Activity	Average energy expenditure
Resting	• watching TV • reading • sleeping	approximately 293kJ
Light	• painting • walking (slowly)	approximately 502kJ
Moderate	• golfing • cleaning windows	approximately 962kJ
Heavy	• brisk walking • slow jogging • steady cycling	approximately 1255kJ
Very heavy	• football • tennis • skiing	approximately 2092kJ

2.2 How the amount of calories required varies according to level of activity

Estimated Average Requirements (EAR)

According to the **Scientific Advisory Committee on Nutrition (SACN)**, the new Estimated Average Requirements (EAR) are as follows:

The EAR values for teenage boys (from 15 years old), girls (from 11 years old) and adults should be increased by up to 16%, based on increased activity levels. This is based on whether an individual is active, less active or more active.

Table 7.8 Example of how to work out the **resting energy expenditure (REE)**

Age and gender	REE – The multiplier and addition is part of the equation to work out resting energy expenditure	Example of weight	Calculation REE
Males			
10–17	(17.7 × weight kg) + 657	89 kg	2232
18–29	(15.1 × weight kg) + 692	94 kg	2111
30–59	(11.5 × weight kg) + 873	100 kg	2023
60–74	(11.9 × weight kg) + 700	75 kg	1593
Females			
10–17	(13.4 × weight kg) + 692	70 kg	1630
18–29	(14.8 × weight kg) + 487	65 kg	1449
30–59	(8.3 × weight kg) + 846	55 kg	1303
60–74	(9.2 × weight kg) + 687	50 kg	1147

? THINK ABOUT IT

If an individual is over their recommended body weight and leads a sedentary life, but maintains their current weight, what impact would a little exercise have? Work out approximately how many kJ a 60 kg person would burn during an hour's aerobics class.

PAIRS ACTIVITY

BMR calculations (20 minutes)

Work out the BMR of the person you're working with.

The amount of heat produced by metabolising food is measured in kilojoules (SI: the standard unit for energy).

1 gram of protein has approximately 29 kJ
1 gram of carbohydrates has approximately 17 kJ
1 gram of fat has approximately 38 kJ
1 gram of dietary fibre has approximately 13 kJ
1 gram of water has 0 kJ

Table 7.9 shows a broad example of the approximate kilojoules requirements for men and women aged 19–59.

Table 7.9 Approximate KJ requirement

Weight (kg)	Sex	Age	Less active	More active
60	M	19–29	9305	11 410
65	F	19–29	8104	9205
70	M	30–59	9795	11 908
70	F	30–59	8305	9506

As you can see, the amount of kilojoules required are based on sex, weight, age and how active you are. There are other things that can affect the body's ability to use energy. **Thyroid** problems can affect the energy usage as can certain medications. But, in order to remain roughly the same weight, we should only eat the kilojoules needed, unless we intend to increase activity – that is to say, we must not take in more than we use up, if we want to stay the same weight.

GROUP ACTIVITY

Diet variation (40 minutes)

The amount of kilojoules required depends on various factors, such as the age of a person – a baby's requirements are very different to an elderly person, for example. Body weight and the amount of exercise are key factors, as the amount of kilojoules you take in will be stored if not burned off through activity. Research various government bodies (we have discussed bodies such as the NHS) to find out what the estimated average kilojoules requirements for energy are for adults based on age, weight and activity level. Look at the weight range, say from 60–80 kg, an age from 18–60 years and record the activity levels. You will then have a complete table to use for working out other calculations based on individual requirements.

🔑 KEY TERM

Scientific Advisory Committee on Nutrition (SACN) – This is a committee that advises the government on nutrition on public health issues.

Resting Energy Expenditures (REE) – The energy that your resting body uses to maintain bodily functions.

Thyroid – The thyroid gland controls the rate of use of energy and protein synthesis.

Calorimeter – An object for measuring the heat of chemical reactions.

2.3 Measuring the energy content of food, e.g. burning a suitable carbohydrate food, use of calorimeter

Calorimeters

Calorimeters measure the heat that is released or absorbed in a chemical reaction. So, if for example if you wanted to know the amount of calories there are in certain foods, you could calculate this using a calorimeter.

To make your own calorimeter, you will need:

- polystyrene cup and lid
- thermometer
- thermometer stand
- stirrer
- measuring cylinder
- balance to weigh out food.

Process:

1. Measure the solution, pour into calorimeter.
2. Put lid on.
3. Insert the thermometer in a hole in the lid, ensuring the bulb does not touch the bottom of the cup. Attach thermometer to the stand.
4. Mix the contents of the calorimeter with the stirring mechanism.
5. Note the change in temperature as the chemical reaction.

KNOW IT

1. What information can you find out by calculating the BMR?
2. The amount of heat produced by metabolising food is measured in what unit?
3. What is the difference between aerobic and anaerobic?
4. What information does a calorimeter give you?

L02 Assessment activity

Below is a suggested assessment activity that has been directly linked to the Pass and Merit criteria in LO2 to help with assignment preparation and includes top tips on how to achieve best results.

Activity 1 Calculating appropriate food and fluid intake P2 P3 M1

Interview other learners/staff and calculate how much food and fluid intake is appropriate for each individual. Use their data to analyse consumption and expenditure of food, comparing different groups.

TOP TIPS
- Here, to gain a **Pass** the requirement is to describe energy intake and expenditure when exercising and at rest, which means showing the differences.
- To gain a **Merit** requires some research into comparable groups of people taking different levels of exercise. It also requires calculations, using the BMR formula.
- Remember to work in kilojoules, not kilocalories!
- Remember there are differences in the intake of kilojoules depending on age, gender, state of health and activity.

LO3 Understand conditions relating to dietary needs P4 P5 M2 D1

GETTING STARTED
Diet variation (5 minutes)
Discuss the variation in diets across a range of people, for example, diabetics and children.

3.1 Variations in dietary needs

There are situations, conditions and different stages in life that require variations in dietary needs. Let's begin with different age groups. The dietary need for babies is obviously very different to say a teenage or adult, so let's start at the beginning of life!

Pregnancy

The Department of Health recommends pregnant women take **folic acid** up to the twelfth week of pregnancy. Folic acid is important for reducing birth defects such as spina bifida, a neural tube defect. Iron helps cells to make haemoglobin for the woman and their baby, so this is also important as the body tends to absorb more iron. Some women may require slightly more vitamin D.

What to avoid or reduce during pregnancy:

1. Large doses of vitamin A and multivitamins – Why? Vitamin A is associated with miscarriage and congenital malfunctions.
2. Certain fish – Why? To minimise absorption of mercury as this can cause problems with the babies' nervous system.

KEY TERMS

Folic acid – A B vitamin, synthetically produced; can be found in some foods and supplements.

Salmonella – A group of bacteria that can cause food poisoning.

E. coli 0157 – Bacterium that lives in the gut of animals; the toxins it produces are harmful to humans.

Campylobacter – A bacterial foodborne disease.

Listeria – A type of bacteria from contaminated food.

Still birth – Defined as fetal death after 20 weeks of gestation.

Miscarriage – A natural death of an embryo or fetus.

Osteoporosis – A condition where bones become brittle, normally as a result of lack of vitamin D, lack of calcium or hormonal issues.

Malnutrition – A condition from the result of not eating enough and/or not getting the required nutrients.

Dementia – A brain disorder that causes a decrease in the ability to think.

Diverticular disease – A digestive disease affecting the large intestine.

3. Alcohol consumption – Why? Unborn babies cannot process alcohol and consumption can affect development.
4. Caffeine consumption – Why? Can cause low birth rates and miscarriage.

Pregnant women should avoid paté, uncooked/ undercooked meat and soft cheese such as Camembert or Brie. Raw eggs and food containing partially cooked meat present a risk of **Salmonella**. The usual advice of washing hands after handling raw meat is important to the mother and baby to avoid food poisoning germs such as **E. coli 0157**, *Salmonella* and ***Campylobacter***. Food poisoning bacteria, such as **Listeria,** can cause illness in a newborn baby, **still births** and **miscarriage**.

A well-balanced diet will provide what is required for the mother and baby with the help of folic acid and vitamin D. When breastfeeding, the mother should continue with vitamin D and a balanced diet to ensure she stays well and the baby stays well-nourished from the milk.

GROUP ACTIVITY

Nutritional restriction (40 minutes)

Research the recommended allowance for tuna, alcohol and caffeine for a pregnant woman.

Babies' needs

Due to how quickly babies grow in the first year, they require a lot more nutrients to ensure they grow well. This is provided by formula or breast milk, which contains all the correct nutrients for this stage.

- 6 months: cow milk (to maintain iron); solid food (puréed), e.g. fruit, vegetables, meat
- 6–9 months: as above, now introduce bread, eggs, cereal – food will be more lumpy and child given finger food
- 9–12 months: as above – food may be chopped and minced
- 12 months and over: as above – more of what the family is eating and a variety of finger foods (e.g. carrot sticks, cucumber).

Of course, water, usually cooled boiled water, is important at the later stage; fruit juices are not recommended because of the sugar content, as this can cause problems at an early stage for tooth decay.

School age children and teenagers

At these stages, children go through a major growth spurt including the bones accumulating in density, so vitamin D is important to reduce the risk of **osteoporosis** in later life. Calcium is required to allow the vitamin D to absorb it. Iron is also important for making blood and lean muscle.

Adults

The approximate guidelines for the key nutrients and energy required for adults are presented in Table 7.10.

Table 7.10 Approximate guidelines for key nutrients and energy for adults

Nutrient	Women (g)	Men (g)
Energy	8368kJ	10460kJ
Fat	70	95
Saturated fat	20	30
Sugar	50	70
Salt	6	6
Macronutrient	**Kilojoules per gram (g)**	
Protein	17	
Fat	38	
Salt	16	

You can see from Table 7.10 how many kilojoules are required and you have read how important macronutrients are and their function in the body. However, there are times when we need to take in more kilojoules and reduce them: we will come onto this in the next section.

Older adults

Although the body has stopped growing at this stage, the body tissue is being renewed, so a good balanced diet is vital to an elderly person. Did you know that **malnutrition** is common in a proportion of older adults? Calcium and vitamin D are important as bones can become brittle in the elderly. Age-related conditions such as osteoporosis and **dementia** have been linked to low levels of folic acid. Older adults often have below the recommended levels of daily fibre; this can cause constipation and **diverticular disease**.

Prisons, hospitals, care homes and schools

Prisons, care homes, schools and hospitals provide food to the public. The quality and amounts of food may be different. For example, being in prison, hospital or care homes may reduce the amount of activity a person engages in; this would require a change in the amount of food consumed. The quality of the macronutrients and nutrients may be compromised, affecting the body's consumption. Illness may affect the amount of food and water we consume, affecting the body by losing too much weight.

Activity levels

We know that we require kilojoules from food to ensure the body and its organs function properly, and we are also aware that if we consume more than we use, our body will store the extra calories as body fat and we gain weight. To allow us to consume more calories without storing them, that is to say, increasing our weight, we need to burn the excess kilojoules.

This equation is representative of what happens when we consume more energy than we use up – we put on weight! To lose weight we need to consume less or increase our activity.

$$E (in) > E (out) = +W$$

$$E (in) < E (out) = -W$$

So, what do we do? We increase our daily activity either to lose weight based on the recommended kilojoules per day or we may choose to eat more but burn the excess with extra activity. An average age man requires around 6911 kilojoules just to maintain bodily systems. More kilojoules are required on top of that to accommodate activity levels during the day.

Table 7.11 Activity levels and kilojoules required

Activity level	Example of activity	How many more kilojoules required
Mild	housework, gardening	11 067
Moderate	brisk walking	11 757
Heavy	playing a sport	14 523

Gender

For both males and females, to maintain a stable weight, the kilojoule consumption can be higher as long as the amount of activity uses them up. To lose weight, the intake would stay at the recommended amount and the extra activity would burn up more than we take in, so the body will turn to stored energy for burning. This then leads to weight loss. Of course if you were unable to increase your activity that much, it would require taking in fewer kilojoules than the required daily amount. The body will still use up stored energy the same to ensure the day-to-day function of the bodily system and any further activity you engage in.

Specific dietary needs

There are reasons for deviating from the required intake of food, usually when the body has some difficulty in tolerating certain food types.

For example:

- **Diabetes**: Reducing the amount of sugar intake and ensuring a good balanced diet.
- **Lactose intolerance**: Using an alternative to dairy products.
- Nut allergy: Finding an alternative protein source.

3.2 Factors affecting eating habits

Eating disorders

Eating disorders not only affect the physical body but also a person's psychological well-being.

Anorexia nervosa is a condition that is characterised by weight loss due to suppressing the urge to eat. The reasons for this behaviour are often due to the fear of getting fat. Girls in particular experience themselves as being fat even when they are thin: this is the result of a distorted body image. People who suffer from anorexia may be 15% or more below the normal or expected weight for their height and age.

Bulimia is another eating disorder, which has the following characteristics:

- a persistent preoccupation and compulsion with eating
- self-perception of being too fat with an excessive fear of fatness.

To counteract the binging, an individual will often use a range of methods such as self-induced vomiting, purging, the use of appetite and suppressants, and alternating periods of starvation.

Obesity and compulsive eating

Although obesity itself is not classed as an eating disorder, obese people can still have characteristics of bulimia. Similarly, compulsive eating seems to apply to some eating patterns within bulimia.

We have seen how important macronutrients and nutrients are to the body in the previous sections. Having an eating disorder can seriously damage the body by causing malnutrition.

> **? THINK ABOUT IT**
>
> If you look in magazines and the media, it often shows skinny females as models. What influence do you think this has on young girls?

It is not just eating disorders that can affect our eating habits. A great deal of individuals' lifestyles are packed with appointments, work commitments, and childcare. This can be a contributing factor to either missing meals, eating on the go, or eating low nutritional food such as chocolate, crisps and fast food.

Given what we have learnt about the effects of food on our body, we can see how, when we develop poor eating habits, our physiology changes. These changes include eating disorders, obesity and disease. Public health organisations such as the **British Nutrition Foundation** work towards educating people on the importance of a balanced diet. Schools have introduced more nutritious food and developed information to educate children from an early age.

> **RESEARCH ACTIVITY**
>
> ### Balanced diet (2 hours)
> As a group, design a questionnaire on lifestyle. This should include questions such as:
>
> 1 Do individuals eat food from all the food groups daily?
> 2 Do they know what they should be eating and drinking per day?
> 3 Do they exercise?
> 4 Do they know where to find information about nutrition?
>
> Question around twenty people.
>
> Write up your findings individually.
>
> Work out percentages for individuals who eat a balanced diet and the percentages for the ones who knew where to go for information.
>
> Individually make recommendations on how to educate the public on lifestyle choices.
>
> With all of this information about healthy eating, the NHS statistics still show an increase in obesity, particularly in younger people. For children aged 4–5 years during 2011–12, 9.5% were obese! Discuss this with the group and come up with three possible reasons why this is the case.

Lifestyle and education

Lifestyle choices and education can have a huge impact on eating habits, this can be positive or negative. Population studies show a clear link between social classes and food intake. Low income groups may find it difficult to achieve a balanced diet, through lack of money and lack of understanding. For example, if a person isn't informed about the requirement of specific nutrients and the harmful effects of foods such as fast foods or processed foods, they will not be able to make an informed decision on their food intake. People with lower incomes are limited to what they can afford; it may be cheaper for them to buy a burger from a fast food restaurant than a bag of vegetables or fruit.

Diabetes

There are different types of diabetes but the one most prominent and related to eating habits is Type 2 diabetes. This is caused by insufficient blood insulin concentration. Diabetes can develop for three reasons: insulin production can be impaired, insulin resistance has developed or certain essential medicines affect the metabolism. In any event, diet must be managed to maintain health.

Although there is research showing a genetic link to Type 2 diabetes, insulin resistance can develop from eating too much sugar. We will look at the problems caused by an imbalance of food intake in the next section.

Food allergies and intolerances

A food allergy is a response from your immune system to substances that are usually harmless. The body is fighting against specific proteins in food. Some common allergic reactions to food are:

- itching/rashes
- hives
- headaches/migraine
- asthma
- swelling of face, lips, tongue, eyelids, feet and hands
- diarrhoea
- nausea/vomiting
- sneezing/coughing
- breathing difficulties
- anaphylactic shock.

With food allergies, symptoms can come on quickly or there may be a delayed reaction. No allergic food intolerance is reactions to food from a number of causes but is not a result of an **immunological response**. This is often a result of **histamine** (this is released by the immune system providing symptoms of itching) release and can occur after eating a lot of a specific food; spicy food, for example. **Allergens**, such as an antigen, set off an allergic reaction, and food allergies such as nut allergies can be serious and cause some of the more acute symptoms such as **anaphylactic shock**, whereas food intolerance can often be managed either by tolerating in small amounts or cooking differently, or by having an alternative that does not cause nutrition deficiency.

KEY TERMS

Diabetes – A metabolic disease in which the body has a high level of glucose.

Lactose intolerance – When the body cannot easily digest lactose.

Anorexia nervosa – An eating disorder, where an individual keeps their body weight as low as possible.

Bulimia – A mental health and eating disorder. Individuals binge eat and often use laxatives or forced vomiting to get rid of excess food.

British Nutrition Foundation – A registered charity that provides impartial evidence-based information on food and nutrition.

Immunological response – A process that protects the body against disease and substances which can cause it harm.

Histamine – Apart from being part of the body's immune response, it also regulates physiological function in the gut.

Allergen – A type of antigen that causes an immune response called an allergic reaction.

The foods most likely to cause allergic reactions and intolerance are listed in Table 7.12.

Table 7.12 Examples of foods that cause allergic reactions and food intolerance

Allergic reactions	Intolerance reactions
eggs	wheat
peanuts	fruit
milk	chocolate
soya bean based food	red wine
fish	caffeine-based drinks
shellfish	soya products
fruit	cheese
herbs and spices	yeast extract
seeds	

3.3 Problems caused by imbalances in food intake

We have seen in the previous section how important vitamins and minerals are. We will now look at the complications of certain vitamin and mineral deficiencies. There are many supplements available in the shops but is it not better to get these naturally? Taking steps to ensure the uptake of essential vitamins and minerals in the diet can help the body get what it needs from food.

- Well stored food – once food such as fruit starts to go stale it soon loses its goodness.
- Cooking – steaming and stir frying vegetables, for example, helps food to retain more vitamins whereas boiling can 'boil them away'!
- Frozen – because vegetables and fruit are frozen soon after picking the goodness remains in the frozen state.

CLASSROOM DISCUSSION
(20 minutes)
Discuss how the lack of correct nutrients can affect a baby's growth. Review research on the areas of the body that are affected.

Table 7.13 shows where vitamin deficiencies impact on the body.

Table 7.13 The impact of vitamin deficiencies on the body

Vitamin	Example of sign of deficiency
Vitamin A	nerve damage, reduced resistance to respiratory infections
Vitamin D	osteomalacia (soft bones)
Vitamin E	inability to absorb fat
Vitamin K	blood clotting problems
Vitamin C	scurvy – tooth loss, bruising, swollen joints
B1	nausea, inability to concentrate
B2	cracked lips, burning eyes, skin rashes, anaemia
B6	anaemia, upset stomach, nerve damage
B12	destruction of red blood cells
Folate	anaemia

You can overdose on vitamins and this can be dangerous. A document called 'Safe upper levels for vitamins and minerals' was produced by the **Food Standards** Agency expert group in 2003. This document stated the maximum dose for each where toxicity and other side-effects can occur.

Table 7.14 shows the side-effects of too much of a particular vitamin.

Table 7.14 Side effects of too much of a range of vitamins

Vitamin	Side-effect
D	kidney stones, calcium lumps
A	liver damage, damage to unborn babies
Niacin	liver tissue damage
B6	nerve damage (arms, legs, fingers, toes)

We have seen in previous sections where people such as pregnant women take supplements to help during pregnancy. There are other examples of where extra vitamins may be required and we will look at these now.

> **KEY TERMS**
>
> **Anaphylactic shock** – A serious allergic reaction which can result in death.
>
> **Atherosclerosis** – A condition where arteries become clogged up by fatty substances known as plaque or atheroma.
>
> **Cardiovascular disease (CVD)** – Disease involving the heart or blood vessels.

Further examples where lifestyle and medication can affect the uptake of vitamins

Vegans avoid food from animals and can lack vitamins C, D and B12. Smoking takes chemicals into the body; this can affect vitamin C levels. Medication can alter the effectiveness of vitamins and people may require a vitamin B supplement.

Water plays an important role in assisting the body in the absorption of vitamins and minerals. Dehydration results in a loss of water and electrolytes (including potassium, sodium, and chloride). Organs such as the brain, nervous system, heart and kidneys cannot function without enough minerals and water.

Cholesterol is measured by the amounts of HDL (high density lipoprotein, termed good cholesterol) and LDL (low density lipoprotein, termed bad cholesterol). HDL removes LDL from the body and takes it to the liver for repossessing. There is lots of research on the roles and effects these have on the body. High saturated fat diets are associated with cardiovascular disease, increasing fats in the blood that can lead to **atherosclerosis**. Unsaturated fat, however, is beneficial as it protects the heart against disease, so you can see how the balance is important. Some research shows that HDL and fibre may even protect against some cancers such as colon cancer and shows how high LDL cholesterol may be linked to colon cancer. However, some research suggests it may be a marker that may produce cancer-causing substances in the rectum.

Recommendations (World Heart Federation) suggest following these target optima totals:

Cholesterol: less than 90mg/dl

LDL: 54mg/dl or less

HDL: 22mg/dl or more in females and 18mg/dl in males

Obesity is often a lifestyle choice, eating too much and exercising too little. As fat and sugar consumed from diet are stored as fat, if they are not used it is understandable how weight increases so quickly. These fats and sugars often come from sources such as:

- processed/fast food
- sugary drinks
- alcohol.

Comfort eating is behavioural and sometimes a contributing factor to obesity. If activity levels are low the calories do not get used up. The Department of Health recommends 150 minutes moderate aerobic activity per week.

Type 2 diabetes risk can be increased when the diet is poor and uncontrolled glucose levels can cause damage to blood vessels. People with diabetes are more likely to develop **cardiovascular disease** so it is important to control the diet well. Controlling blood glucose can lower the risk of cardiovascular disease (CVD) by 42%. The risk factors of developing type 2 diabetes are:

1. being overweight and consuming large amounts of sugars
2. age – as you get older it sometimes develops
3. family history.

You can see why some individuals have shortened lifespans, because of the damage done to bodily systems.

> **KNOW IT**
>
> 1. What supplement and vitamin should a pregnant woman have, alongside a well-balanced diet?
> 2. What vitamin and mineral are associated with good bone health?
> 3. What kind of conditions are anorexia and bulimia nervosa?
> 4. Name four common symptoms of food allergies.
> 5. Name one sign of deficiency for the following:
>   a vitamin K
>   b vitamin B6

LO3 Assessment activity

Below is a suggested assessment activity that has been directly linked to the Pass, Merit and Distinction criteria in LO3 to help with assignment preparation and includes top tips on how to achieve best results.

Activity 1 Report on a nutritional related disease/condition P4 P5 M2 D1

Research and report the statistics on a chosen nutritional related disease/condition. You could review case studies and could also produce a leaflet showing common conditions/diseases associated with nutrients.

TOP TIPS
- For a **Pass**, only a description of the common disease and conditions relating to nutrition are required and the factors that affect people's eating habits, so no great detail is required.
- For a **Merit**, an explanation of the physical impacts of developing a condition relating to malnutrition is required. It will also require some research on the impact of a lack of vitamins and minerals on the body and how it affects it physically.
- For a **Distinction**, research, understanding and reflection is required to recommend (on the basis of the research) and justify (based on the various requirements depending on medical, age requirements, for example) nutritional requirements for a specific group of people.

LO4 Be able to label food with nutritional information P6 M3 D2

KEY TERMS
Norovirus – A virus causing nausea, vomiting and diarrhoea, passed on through contaminated food and water.
NHS Choices – An NHS website with information and advice on health.
Preservatives – These are added to food to lengthen the shelf life.

GETTING STARTED
(5 minutes)
Discuss the implications of having no guidelines on nutrition.

4.1 The sources of labelling of food products to indicate nutritional information

The Food Standards Agency works with local authorities to ensure food law is enforced. They work with farms to ensure good hygiene and ensure the food fed to animals will not cause health risks to humans on consumption. They ensure food establishments follow the food hygiene rating scheme and also provide information about their foods, such as allergy information on food sold unpacked. They also conduct food research on foodborne illnesses such as **norovirus**, listeria and campylobacter, ensuring the risks are reduced through the food chain.

Organisations such as **NHS Choices** give information about healthy eating and recipes. The British Nutrition Foundation also provides information on nutrients and healthy eating.

PAIRS ACTIVITY
(40 minutes)
Research what the requirements are for food establishments, such as cafés and takeaways. For food establishments such as cafés, takeaways, etc. research the food hygiene requirements and discuss the findings in a group.

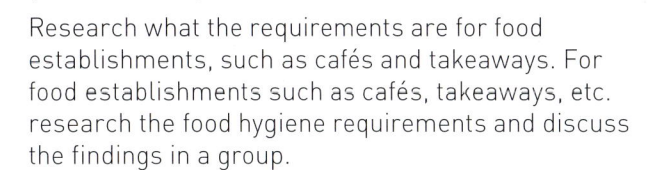

Food preservation, preservatives and labelling

Preservatives
Various acids and other preservation methods are used to preserve food and prevent growth of yeast and bacteria. Here are some methods:

- Drying – Fruit, for example, is dried naturally in the sunlight.
- Cooling – Slowing the growth of microorganisms that cause food to rot.

185

- Salting – Draws moisture from meat.
- Freezing – Inactivates enzymes, which stops the food from deteriorating.
- Heating – Kills microorganisms.

Organic food production

Organic food is free from additives, synthetic pesticides and chemical fertilisers. It is not processed using radiation, solvents or synthetic food additives; for example, the nitrogen levels are lowered in leafy vegetables. This makes organic food desirable as it reduces the risk of health issues relating to processing and additives.

There is an advantage to eating organic food, as it is higher in nutrients due to it not being exposed to chemicals. **Preservatives** help to keep us from getting ill from food poisoning for example, but the preservatives themselves also have an impact on the food.

Figure 7.5 is a typical example of a food label you will see on food packaging.

Nutrition information			
Typical values	Per 100g	Per 30g serving	% based on GDA for women
Energy	2325 kJ / 559 kcal	698 kJ / 168 kcal	8.4%
Protein	0.2g	trace	trace
Carbohydrate of which **sugars**	46.3g / 46.3g	13.9g / 13.9g	6.0% / 15.4%
Fat of which **saturates**	38.8g / 25.4g	11.6g / 7.6g	16.6% / 38.0%
Fibre	nil	nil	nil
Salt of which sodium	trace / trace	trace / trace	trace / trace

▲ Figure 7.5 Example of a food label you will see on food packaging

> **? THINK ABOUT IT**
>
> How many people do you think look at packaging? Think about how much time this would take, as well as working out nutritional values on fresh items that have no information.

Figure 7.6 shows the 'traffic light system', highlighting whether food is low, medium or high in the food groups. This can help people's choices.

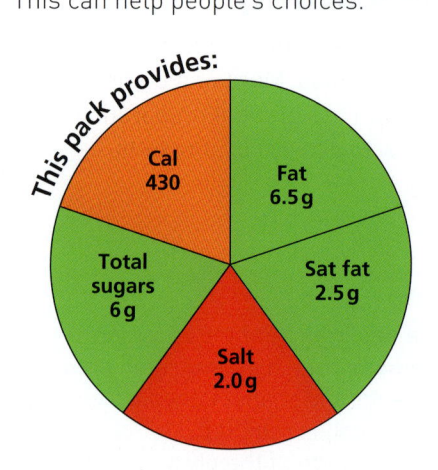

▲ Figure 7.6 Traffic light system

This information is provided per 100 grams. These labels are not only there to give you information of the content but also the amount of salt, total fat, saturated fat and sugars there are. As the food packet states the information per 100 grams, once you know the guidelines of what is classed as high and low amounts, you can make a decision whether to buy it based on its nutritional value (see Table 7.15).

The list of ingredients is also stated on packaging, with the main ingredient listed first. So if you see that the first few ingredients are high in fat then you know that the food is a 'high fat food'. Labels also have to include what additives have been added and state any ingredients that some people might be allergic to, for example nuts or eggs.

4.2 Nutritional requirements

You have seen throughout the unit the effect and importance of nutrients to the body. We have touched on some of the public health sources, such as the NHS. Others include:

- Heart UK
- British Dietetic Association.

Table 7.15 Fats, sugars and salt per 100g

Total fat	Saturated fat	Sugars	Salt
High: more than 17.5 g of fat	**High:** more than 5 g saturated fat	**High:** more than 22.5 g of total sugars	**High:** more than 1.5 g of salt (or 0.6 g sodium)
Low: 3 g of fat or less	**Low:** 1.5 g saturated fat or less	**Low:** 5 g of total sugars or less	**Low:** 0.3 g of salt or less (or 0.1 g sodium)

4.3 Dietary Reference Values (DRV)

This is the European Food Safety Authority's DRV:

- The intake of total carbohydrates – including carbohydrates from starchy foods such as potatoes and pasta, and from simple carbohydrates such as sugars – should range from 45 to 60% of the total energy intake for both adults and children.
- For sugars there is good evidence that frequent consumption of foods high in sugars increases the risk of tooth decay. Data also show links between high intakes of sugars in the form of sugar-sweetened beverages and weight gain. The Panel however found there was insufficient evidence to set an upper limit for sugars. This is because the possible health effects are mainly related to patterns of food consumption – i.e. the types of foods consumed and how often they are consumed – rather than a relation to the total intake of sugars itself. Evidence regarding patterns of consumption of sugar-containing foods should be considered by policy makers when making nutrition recommendations and developing food-based dietary guidelines at national level.
- A daily intake of 25 grams of dietary fibre is adequate for normal bowel function in adults. In addition, evidence in adults shows there are health benefits associated with higher intakes of dietary fibre (e.g. reduced risk of heart disease, type 2 diabetes and weight maintenance).
- Evidence is still inconclusive on the role of the glycaemic index and glycaemic load in maintaining weight and preventing diet-related diseases.
- Intakes of fats should range between 20 to 35% of the total energy intake, with different values given for infants and young children taking into account their specific developmental needs.
- There is good evidence that higher intakes of saturated fats and trans fats lead to increased blood cholesterol levels, which may contribute to development of heart disease. Limiting the intake of saturated and trans fats, with replacement by mono- and poly-unsaturated fatty acids, should be considered by policy makers when making nutrient recommendations and developing food-based dietary guidelines at national level.
- A daily intake of 250 mg of long-chain omega-3 fatty acids for adults may reduce the risk of heart disease.
- For water, a daily intake of 2.0 litres is considered adequate for women and 2.5 litres for men.

KNOW IT

1. What agency works with local authorities to ensure food law is enforced?
2. Name the methods used to preserve food.
3. Give two reasons why people become obese.
4. What is the 'traffic light system'?
5. What does DRV stand for?

LO4 Assessment activity

Here is a suggested assessment activity that has been directly linked to the Pass, Merit and Distinction criteria in LO4 to help with assignment preparation and includes top tips on how to achieve best results.

Activity 1 Research food labelling
P6 M3 D2

Research food labelling in supermarkets and products you use at home. Ask groups of people about their understanding of food labelling.

TOP TIPS

- For a **Pass**, you have to describe the factors that affect people's eating habits, so no great detail is required here.
- For a **Merit**, an explanation on how legislation affects food labelling in the UK is required, so there will be some research required to know the legislators and their requirements.
- For **Distinction**, create a recipe and calculate the nutrients contained, then make a label to show this information. Discuss how food labelling may lead to confusion for consumers and suggest solutions – this will require more in-depth research on food labelling and give an account that addresses a range of ideas and arguments.

Read about it

British Nutrition Foundation:
www.nutrition.org.uk
National Eating Disorders Association:
www.nationaleatingdisorders.org
NHS Choices:
www.nhs.uk/pages/home.aspx
Heart UK:
www.heartuk.org.uk
British Dietetic Association:
www.bda.uk.com
Self Nutrition Data:
http://nutritiondata.self.com/

Unit 08
Cell biology

ABOUT THIS UNIT

Cell biology is the study of the structure and function of cells. Cytology is the study of the structure of cells as seen in microscopes to reveal details of cell structure, their fine structure or ultrastructure, and how parts of the cell function and interact with each other.

Molecular biology and immunology have broadened the field of cell biology. You will also gain an appreciation of the rapidly developing area of cell biology called cell signalling – how cells communicate with each other – and signal transduction – the ways in which signals received by cells are transmitted within the cell leading to a response.

Cell biology is important to the developmental biologist, particularly the processes of cell division and differentiation, with genes being expressed in some specialist cells, but not in others. The cells involved in differentiation are called stem cells and they have the potential to produce the different types of cells required. Stem cell-based therapies offer cell replacement in regenerative medicine.

LEARNING OUTCOMES

The topics, activities and suggested reading in this unit will help you to:

1. Understand the functions of the plasma membrane and endomembrane systems
2. Be able to use cytological techniques
3. Understand the cell cycle and the importance of mitosis
4. Understand the process and significance of differentiation
5. Understand the potential of stem cells in medical therapies

How will I be assessed?

You will be assessed through a series of assignments and tasks set and marked by your tutor.

How will I be graded?

You will be graded using the following criteria:

Learning Outcome	Pass	Merit	Distinction
	The assessment criteria are the Pass requirements for this unit.	To achieve a Merit the evidence must show that, in addition to the Pass criteria, the candidate is able to:	To achieve a Distinction the evidence must show that, in addition to the Pass and Merit criteria, the candidate is able to:
1 Understand the functions of the plasma membrane and endomembrane systems	**P1** Describe the functions of the plasma membrane	**M1** Explain the dynamic nature of endomembrane systems in a eukaryotic cell	**D1** Explain the role of cell signalling pathways
2 Be able to use key cytological techniques	**P2** Demonstrate the use of microscopical and differential staining techniques		**D2** Demonstrate the use of a technique used to extend the limits of light microscopy
	P3 Demonstrate the use of an appropriate cell counting technique		
3 Understand the cell cycle and the importance of mitosis	**P4** Describe the cell cycle, stages of mitosis and cytokinesis	**M2** Explain the importance of mitosis	**D3** Assess the importance of mechanisms that arrest cell division
4 Understand the process and significance of differentiation	**P5** Describe the process of differentiation in the embryo in producing specialised cells	**M3** Explain the role of gene expression in differentiation	
5 Understand the potential of stem cells in medical therapies	**P6** Describe uses of stem cells in medicine	**M4** Explain the laboratory techniques used in stem cell therapy.	

LO1 Understand the functions of the plasma membrane and endomembrane systems *P1 M1 D1*

The basics of the plasma membrane and cell organelles were covered in section 3.2 of Unit 1. We now need to look at these in more detail to understand their role in the function of the cell. This means studying the way in which substances move into and out of cells, the functions of the various membrane-bound organelles and the way in which the activity and function of the cell is controlled at the molecular level. A good appreciation of this will help in understanding the remaining sections of this unit.

GETTING STARTED

Border controls in the cell (5 minutes)

Think about what was covered in Unit 1 about the structure of the cell membrane, particularly the phospholipid bilayer. As a group, discuss what that means in terms of what types of substances are able to enter or leave the cell – and what types of substance need to enter or leave the cell. Can you see a need for specific mechanisms for some types of substance?

1.1 The movement of substances into and out of the cell

> **🔑 KEY TERMS**
>
> **Pumps** – A protein complex embedded in a membrane that uses energy from ATP to move specific substances across the membrane.
>
> **Macromolecule** – A very large molecule, containing thousands or more atoms. They are often, but not always, polymers and the term is used particularly to describe large biological molecules.
>
> **Vesicle** – A structure inside the cell consisting of liquid enclosed by a membrane. Different types of vesicle are formed in different ways or have different functions, such as secretory vesicles and phagocytic vesicles.

Mechanisms

Passive movement by diffusion and facilitated diffusion

Any substance will diffuse from a region of high concentration to a region of low concentration. Drop a crystal of copper sulfate into a beaker of water and watch the blue colour spread. It is also true in biological systems, as long as there is no barrier to movement. However, the phospholipid bilayer with its hydrophobic core can be a barrier.

- Non-polar molecules (e.g. carbon dioxide, oxygen, steroid hormones, lipids or fat-soluble vitamins) can 'dissolve' in the phospholipid bilayer and can diffuse across the membrane.
- Polar molecules, such as water, glucose and ions such as Na^+ or Cl^- that cannot easily diffuse across the membrane, rely on facilitated diffusion.
- Facilitated diffusion usually involves a carrier protein in the membrane (e.g. the glucose transporter protein in the plasma membrane of most cells) or an ion or water channel.
- The rules of diffusion still apply – from high concentration to low concentration.

Active transport

- Active transport is a type of facilitated diffusion – it requires a transport protein.
- Active transport can move substances *against* a concentration gradient.
- This requires energy, usually from ATP.
- Such systems are often referred to as **pumps**.

Endocytosis and exocytosis

- Some materials, like **macromolecules** or bacteria, are too large to pass through a membrane by facilitated diffusion.
- These enter or leave the cell by the process of endocytosis or exocytosis (see Figure 8.1).
- Endocytosis involves the plasma membrane engulfing the material and then fusing to form a **vesicle**.
- Exocytosis is the process by which cells secrete proteins (enzymes, hormones) and other molecules.
- The membrane of a secretory vesicle (see next section) fuses with the plasma membrane and then empties its contents to the outside of the cell.

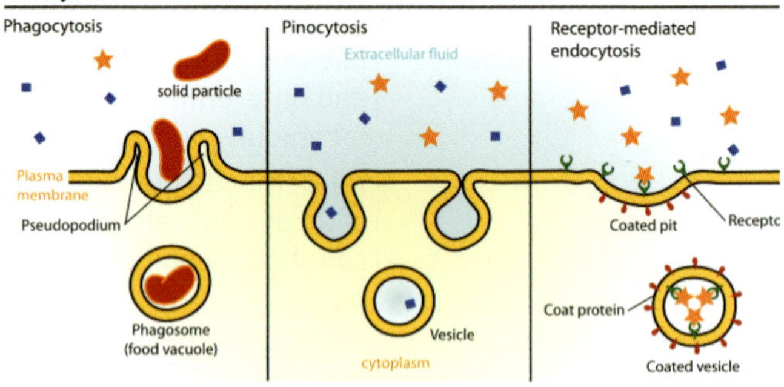

▲ **Figure 8.1** Endocytosis involves the plasma membrane engulfing the material and forming a vesicle. Exocytosis is essentially this process in reverse

Endomembrane systems in eukaryotic cells

A good way of understanding the role of different membranes and membrane-bound structures (e.g. vesicles) is to consider the route taken by proteins that are secreted ('exported') from the cell via the rough endoplasmic reticulum (RER), Golgi apparatus and secretory vesicles (see Figure 8.2 on the next page). Proteins made on the RER are packaged into transport vesicles which move to the face of the Golgi nearest the nucleus and fuse with it. The proteins move through the Golgi where enzymes anchored to the membrane carry out various post-translational modifications, such as adding carbohydrate. Following this processing, the proteins reach the face nearest the plasma membrane where they are packaged into vesicles that bud off the Golgi. Secretory vesicles, containing the processed and packaged secretory proteins, then fuse with the plasma membrane and are released by exocytosis in response to extracellular signals (regulated secretion). Transport vesicles are also budded off the Golgi and these contain soluble proteins as well as cell membrane proteins contained within the vesicle membrane. These fuse with the plasma membrane (unregulated fusion) and release their contents by exocytosis (constitutive secretion). The vesicle membrane, containing embedded proteins, becomes part of the cell membrane.

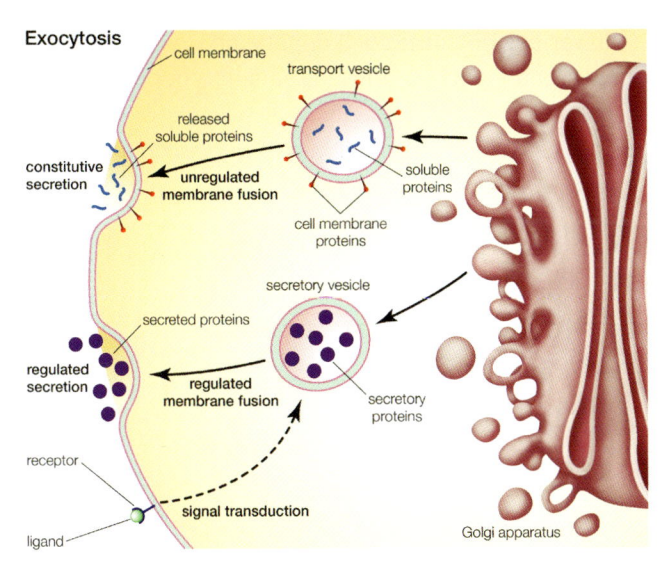

▲ **Figure 8.2** The route taken by a secreted protein

Not all proteins processed and packaged by the Golgi are secreted. Some of the vesicles that bud off the Golgi become lysosomes. These contain digestive enzymes such as proteases. One function of lysosomes is to fuse with phagocytic vesicles and digest their contents. This is the process by which phagocytes engulf and destroy pathogens, such as bacteria (see section 6.1 in Unit 4). Another important function of lysosomes is the breakdown of worn-out cellular components, such as proteins.

The function of the Golgi in processing and packaging of secreted proteins is only one example of the dynamic nature of the cell. Other examples include the role of the smooth endoplasmic reticulum in the synthesis of lipids and their incorporation into membranes. Images obtained with an electron microscope (see Unit 2, section 4.2) show great detail of these various components, but a single image only represents a snapshot – a single point in time – and so we must not make the mistake of thinking that all these organelles remain fixed.

Cell signalling (chemical signalling; electrical signalling)

When we study **cell signalling** we are studying the ways in which cells communicate with each other. This applies to communication between single cell organisms as well as to communication between cells of a multicellular organism. It covers all the events between a signal being generated (e.g. secretion of glucagon) and the response to that signal (increase of blood glucose concentration) – see section 5.1 in Unit 4 for more information about this. To avoid confusion, we need to distinguish between the extracellular and intracellular **cell signalling pathways** – we usually refer to the latter as **signal transduction** pathways (see Key Terms for more information).

We also need to understand the difference between chemical signalling and electrical signalling. In chemical signalling a chemical (hormone, neurotransmitter, growth factor, etc.) is produced in one cell and acts on another. In electrical signalling, electrical impulses transmit the signal, such as in nerve cells or the heart.

🔑 KEY TERMS

Cell signalling – The mechanisms of communication between single-cell organisms or between cells in a multicellular organism.

Cell signalling pathways – The role of hormones, growth factors and other regulatory substances in communication between cells. The term is also used to apply to the various biochemical events inside a cell that follow binding of a signalling molecule to the cell surface.

Signal transduction – The events inside the cell following binding of a signalling molecule to the cell surface.

Ligand – A general term for a substance, such as a hormone, growth factor or other regulator, that binds to a receptor.

Second messenger – The intra-cellular signalling molecule produced when a ligand (the 'first messenger') binds to its receptor. There are a number of second messengers such as cAMP (cyclic adenosine monophosphate), cGMP (cyclic guanosine monophosphate), inositol phosphate, or even inorganic ions such as Ca^{2+}.

Kinase – An enzyme that phosphorylates (adds a phosphate group to) another protein, usually causing that protein to become activated. Many of the components of signalling pathways are kinases – you can usually recognise them because their abbreviated name contains at least one letter K.

Phosphatase – An enzyme that removes a phosphate group from another protein, usually reducing its activity. Many signal transduction pathways involve both kinases and phosphatases working in opposition.

Chemical signalling

- A **ligand** binds to a receptor on the plasma membrane.
- This activates an enzyme on the inner face of the membrane that produces a **second messenger**.
- Signal transduction pathways connect the signal received by the receptor with the process that brings about a response, such as activation of enzymes, expression of genes, differentiation of the cell, regulation of mitosis and cell division amongst others.
- Most pathways include one or more **kinase** enzymes, but these often work in opposition with **phosphatases**. The various outcomes will depend on the balance between the two activities in the different pathways.

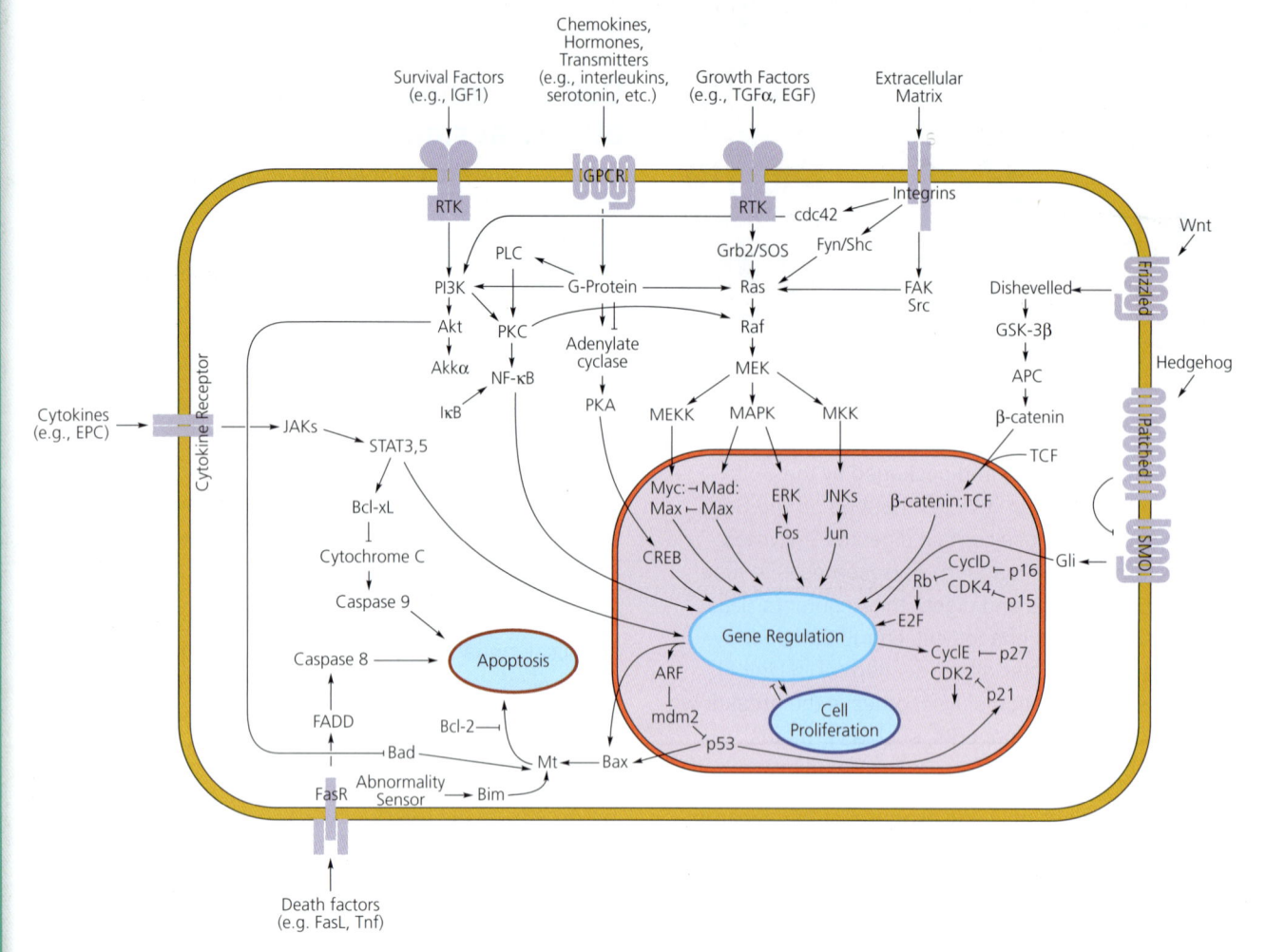

▲ **Figure 8.3** Signal transduction pathways illustrating how binding of various ligands to their receptors leads to a cascade of reactions resulting in switching on or off of genes, cell proliferation (mitosis) or apoptosis (programmed cell death)

GROUP ACTIVITY

Signal this (30 minutes)

Choose a signal transduction pathway – one taken from Figure 8.3, or one of your own choosing – and make a wallchart to show:

- the original stimulus
- the receptor on the surface of the target cell
- some of the components of the signal transduction pathway (don't get carried away – some pathways are hugely complex!)
- the response that is triggered by the pathway
- ways in which this pathway might interact with other pathways or processes; again, don't get carried away – some scientists spend their entire careers working out these interactions.

You will find Wikipedia a useful source of information, but don't be put off by some of the detail of the information. You are just trying to get a feeling for, and demonstrate the complexity of, the processes involved.

Electrical signalling

Nerve communication involves electrical as well as chemical signalling. Neurones (nerve cells) have Na$^+$/K$^+$ pumps and ion channels in their membranes that maintain a potential difference (voltage) across the membrane known as the resting potential. The inside of the neurone is slightly more negative than the outside (about -60mV). A nerve impulse is transmitted when changes in the ion channels cause depolarisation of the membrane – in other words the inside now becomes positive relative to the outside. This is known as an action potential and its transmission along the membrane of the neurone constitutes a nerve impulse.

Chemical signalling occurs when the nerve impulse reaches the end of the neurone, where it makes a connection (synapse) with the next neurone or neurones in the nerve pathway. Chemical neurotransmitters are released and bind to the membrane of the next neurone and cause depolarisation of the membrane, which starts a nerve impulse in that neurone.

A similar mechanism allows electrical signals to be transmitted from cell to cell in the heart initiating contraction of heart muscle.

KNOW IT

1. Make a list of the similarities and differences between diffusion, facilitated diffusion and active transport.
2. What would you expect to see a lot of in an electron micrograph of a cell that was carrying out a lot of active transport?
3. What is the difference between cell signalling and signal transduction?
4. Define the following terms:
   a. signal transduction
   b. ligand
   c. second messenger
   d. phosphatase

L01 Assessment activities

Below are suggested assessment activities that have been directly linked to the Pass, Merit and Distinction criteria in LO1 to help with assignment preparation and includes top tips on how to achieve best results.

Activity 1 Describe the functions of the plasma membrane *P1*

You need to consider the functions of the plasma membrane in more detail than was covered in Unit 1, in particular those related to transport of substances into and out of the cell. You could prepare annotated wallcharts, PowerPoint or other multimedia presentations illustrating the function of the plasma membrane:

- The nature of the substance to be transported.
- How the membrane functions as a partially permeable barrier.
- Passive movement of substances – diffusion and facilitated diffusion.
- Active transport.
- Endocytosis and exocytosis.

TOP TIP
✓ Make sure you understand the difference between active and passive, simple and facilitated.

Activity 2 Explain the dynamic nature of endomembrane systems in a eukaryotic cell *M1*

You need to show an understanding of the functions of the endomembrane system, in the production of proteins and lipids, and their modification, packaging, transport and secretion. You also need to appreciate that the system as observed with electron microscopy is not static, but a constantly changing, dynamic system. You could demonstrate this using a series of diagrams, a series of models or an animation.

TOP TIP
✓ It helps if you can visualise the path taken by a protein from synthesis to secretion – prepare an annotated diagram to help you remember.

Activity 3 Explain the role of cell signalling pathways *D1*

You need to be able to give an overview of cell signalling pathways, for instance in a wall display, explaining their various roles; look at the big picture, and don't become focused on detail.

TOP TIP
✓ Remember that cell signalling pathways can be general (found in all cells) or specific to certain cells. This specificity explains why, for example, some hormones have one effect on one target tissue and a different effect on another.

LO2 Be able to use cytological techniques P2 P3 D2

GETTING STARTED

Looking at cells (5 minutes)
Think about the different types of structure you might want to look at when studying cells – how small are they? How big do they need to appear for you to be able to make out important features? What range of sizes do you need to be able to look at? Would it help if you could stain or label different parts of the cell?

> **KEY TERMS**
>
> **Cytology** – The study of the structure, function and chemistry of cells.
>
> **Histology** – The study of the anatomy of cells and tissues using microscopy.
>
> **Inclusions** – Non-living components of cells such as glycogen granules, lipid droplets, crystals and pigments.

The basic techniques and methods of **cytology** have been in use for many years. However, more recent advances in technology, such as confocal microscopy and high content imaging, have transformed our ability to study biological events at the cellular level.

2.1 How to use a light microscope to examine cells

Unit 2 section 4.1 covered the use of a light microscope to study stained tissue sections as well as the limitations of light microscopy, and then section 4.2 went on to look at how electron microscopy could be used to show finer detail (cell ultrastructure).

Limits of light microscopy
The resolution of any microscope depends on the wavelength used. For light microscopy, this means the maximum resolution is about 0.2 µm (2×10^{-5} m) and the maximum magnification about ×1500. As a result, most organelles cannot be seen in the light microscope. Those that can, such as the nucleus, Golgi and mitochondria, cannot be studied in detail because their size puts them close to the limit.

Microscopical and differential staining techniques
Staining is used in microscopy to increase contrast and make it easier to see cells because unstained cells are almost transparent, making it very difficult to see them. Staining can also identify particular structures, such as nuclei (haematoxylin), or chemicals, such as starch grains (iodine). Some stains can be used to distinguish between live and dead cells, or to identify cells undergoing apoptosis (programmed cell death).

Individual cells, such as bacteria, cells in culture or blood cells can be stained without further treatment, although it is usual to fix the cells to preserve them. Tissue samples, however, are too thick to be viewed in the microscope and must be sectioned, i.e. cut into very thin slices. The procedure involves fixing (e.g. with formaldehyde) followed by processing to remove water and replace it with a medium, such as paraffin wax, that will support the tissue during sectioning. The whole tissue sample is then embedded in a support (e.g. agar, paraffin wax) so that it can be sectioned and mounted on a microscope slide. Once mounted, the section can then be stained.

Bacteria are divided into Gram positive and Gram negative depending on whether or not they stain with the Gram stain – see section 6.5 of Unit 3 for more details.

Staining of organelles, inclusions or secretions
Haemotoxylin and eosin (H&E) staining is widely used in **histology** to study tissue sections. Haemotoxylin stains nuclei blue while eosin stains cytoplasm, collagen and other extra-cellular substances pink or red. This makes it much easier to distinguish individual cells. When looking at stained sections it is important to remember that the cell membrane cannot be seen with the light microscope, but its position can be inferred as the edge of the stained cytoplasm.

Some other stains that you might encounter will stain different types of cell or tissue, **inclusions** or even specific molecules. These are shown in Table 8.1.

Table 8.1 Some useful stains

Stain	Target / function
Alcian blue	Mucopolysaccharides produced by mucus-secreting cells.
Masson's trichrome	Blood, smooth and striated muscle and mucus.
Carmine	Intensely red dye used to stain glycogen.
Ethidium bromide	Stains DNA with a red-orange fluorescence. The cell membrane is normally impermeable to ethidium bromide, but becomes permeable in the late stages of apoptosis, so ethidium bromide is useful as a marker for cells undergoing apoptosis.

Immunohistochemistry (IHC)

IHC is a very powerful technique that uses antibodies to identify specific antigens in tissues or on cells. Strictly speaking, when used with cells (e.g. growing in culture) rather than tissue sections we should refer to immunocytochemistry (ICC).

> **KEY TERMS**
>
> **Primary antibody** – An antibody raised against a specific target antigen; primary antibodies can be polyclonal or monoclonal. Monoclonal antibodies are more specific and show much less cross-reactivity.
>
> **Secondary antibody** – An antibody raised against the antibody type of the species used to make the primary antibody. If we raise a primary antibody in rabbits against the human cell-surface protein Fas (rabbit anti-human Fas) and it is an IgG, then the secondary antibody could be raised in sheep against rabbit IgG (sheep anti-rabbit IgG).

Having bound an antibody to the antigen in the tissue section or cells, we need to be able to visualise (view) it. This can be done in several ways. A chromophore (coloured dye) or fluorophore can be attached to the antibody and the colour observed in a light microscope or fluorescence microscope. Alternatively, the antibody can be conjugated with an enzyme that will convert a chromogenic or fluorogenic substrate into a coloured or fluorescent product. In the direct method, the label or enzyme is attached directly to the **primary antibody**. In the indirect method, the label or enzyme is attached to a **secondary antibody**. This can give greater sensitivity, because several secondary antibodies may bind to a single primary antibody, increasing the level of signal.

IHC can be used to identify and locate specific antigens (proteins or glycoproteins) in tissues and so finds many applications in basic research in cell biology, as well as in drug development (see Unit 11) and in diagnosis. There are many cell-surface proteins associated with particular diseases that can be identified using IHC. Intracellular markers can also be identified in this way if the cell membrane is made permeable, e.g. by use of detergents. Similar techniques can be used in ICC with cells in culture.

Extending the limits of light microscopy

We saw above how the resolving power of light microscopes is limited by the wavelength of visible light. Also, conventional light microscopy usually requires thin sections that have been fixed, so is unsuitable for use with living tissues. We will now consider ways in which these limitations can be overcome.

Oil immersion

Oil immersion microscopy uses an oil with a high refractive index between the slide cover slip and the objective lens to increase the resolving power of the microscope. The refractive index of the cover slip, oil and objective lens are all very similar and this increases the amount of light from the specimen that enters the objective lens.

Confocal microscopy

Confocal microscopy represents a major advance in our ability to study living systems at the cellular level. The technique is a development of fluorescence microscopy where the whole sample is illuminated. A fluorescence detector or camera then forms an image, but this includes a large amount of out-of-focus background. Confocal microscopy uses a light source focused on the same point as the imaging lens used to observe the sample, thereby eliminating the out-of-focus signal. This means that it is possible to focus on a point deep within a thick section (50μm or more), overcoming the limitation of conventional light microscopy to use only thin sections. In Figure 8.4 below, a thick section of fluorescently stained human medulla in widefield fluorescence exhibits a large amount of glare from fluorescent structures above and below the focal plane (a). When imaged with a laser scanning confocal microscope (d), the medulla thick section reveals a significant degree of structural detail. Likewise, widefield fluorescence imaging of whole rabbit muscle fibres stained with fluorescein produce blurred images (b) lacking in detail, while the same specimen field (e) reveals a highly striated topography in confocal microscopy. Autofluorescence in a sunflower pollen grain produces an indistinct outline of the basic external morphology (c), but yields no indication of the internal structure. In contrast, a thin optical section of the same grain (f) acquired with confocal techniques displays a dramatic difference between the particle core and the surrounding envelope.

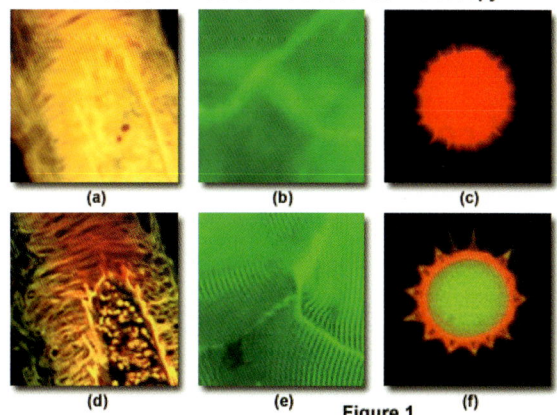

▲ Figure 8.4 Confocal and widefield fluorescence microscopy

The method can also be used with living cells as well as fixed sections. This means that processes can be observed within living cells in real time. Confocal microscopy relies upon a laser light source to provide fluorescence excitation. This is scanned across the sample and then the resultant fluorescence is detected by an objective lens following the same scanning path. The image is then assembled by a computer and displayed on a monitor. It is even possible to assemble a 3-D image of an object by scanning layers through the sample.

GROUP ACTIVITY

Microscopic art gallery (30 minutes)

Use Google images with the search term 'cell staining' and make a selection of different types of light micrograph of sections or whole cells. Annotate them where you can see interesting features or good examples of the use of staining. Try to get a range of different types of staining – use of traditional microscopic stains as well as more recent fluorescence staining and IHC or ICC. Use these to make a wall display.

High content imaging

High content imaging (HCI) makes it possible to study the effect of substances such as peptides, proteins, nucleic acids or potential drug molecules on cells. Effects could include change of shape or appearance of the cells or changes in the production of specific cellular products, such as proteins. The cells are incubated with the substance being investigated and then examined, often using fluorescent tags that label specific cellular components or products (usually proteins). Examination can be via conventional microscopy although it is normally done using confocal microscopy. An enormous quantity of data is collected via a digital imaging system. Powerful image analysis software is then used to process the data and provide a meaningful image.

The technique can be extended to study many samples in parallel, making it an extremely powerful tool for use in drug development (see Unit 11).

2.2 Cell counting techniques

Sometimes it is necessary to be able to count the number of cells in a sample. For example:

- the number of blood cells (blood count) in a patient's blood sample
- the number of viable sperm cells in order to assess fertility
- the number of yeast cells for use in brewing
- the number of cells in a culture, when monitoring cell growth or sub-culturing.

The original method used a haemocytometer, although there are now various automated methods that have much higher throughput.

Haemocytometer

This consists of a thick glass microscope slide with a rectangular indentation that creates a chamber. This chamber is then engraved with very fine lines (modern haemocytometers use laser-etched lines) creating squares and rectangles of different areas, the largest being 1.0 × 1.0 mm and the smallest 0.05 × 0.05 mm. The cover slip is 0.1 mm above the grid, and so the volume of each square or rectangle can be calculated.

The number of cells in several different areas is then counted and a mean obtained. This is then used to calculate the number of cells per unit volume and then multiplied by any dilution factor to obtain the number of cells in the original sample.

Coulter counter

The Coulter principle relies on the fact that particles moving in an electrical field cause disturbances to that field. In a Coulter counter, cells in an electrolyte solution are drawn through an aperture. As each cell passes through, it disturbs the electrical field so that a count can be made. However, the size of the disturbance to the field depends on the size (volume) of the cell, so this can also be recorded. This allows the number and size distribution of the cells in a sample to be determined.

The technique has been extended to continuous flow systems, such as the flow cytometer. This uses a flow cell with electrodes at either end instead of an aperture. Not only does this allow for greater throughput, but it can also be used in conjunction with fluorescently labelled antibodies and a fluorescence detector to count cells carrying one or more antigens. In Fluorescence Activated Cell Sorting (FACS), cells are not just counted but can be sorted and collected.

> **KNOW IT**
>
> 1. What precautions would you need to take to ensure accurate and reliable results when using a haemocytometer?
> 2. A 100 cm^3 sample of yeast was taken from a fermentation tank. 10 cm^3 was taken from this and made up to exactly 250 cm^3 with water. The diluted sample was studied using a haemocytometer. The smallest squares measured 0.05 × 0.05 mm. Ten squares were selected and were found to contain 3, 9, 15, 2, 10, 8, 35, 8, 5 and 11 yeast cells respectively. Calculate the yeast cell density in the fermentation tank.

L02 Assessment activities

Below are suggested assessment activities that have been directly linked to the Pass and Distinction criteria in LO2 to help with assignment preparation and includes top tips on how to achieve best results.

Activity 1 Demonstrate the use of microscopical and differential staining techniques P2

In Unit 1, you will have examined some animal and plant cells. You will need to use staining techniques to identify cell organelles, inclusions or secretions. A worthwhile activity might be to prepare permanent slides by fixing, dehydrating, embedding and sectioning before staining, then dehydrating and mounting. Your assessment will be largely practical and so you should keep laboratory notebooks to provide evidence of your work.

Activity 2 Demonstrate the use an appropriate cell counting technique P3

You need to be able to demonstrate skills in cell counting using a haemocytometer. Some brewers use this technique with a drop of methylene blue solution to assess yeast cell viability (viable cells will be colourless, dead cells, blue). Your tutor will help by providing an optimum concentration of yeast cells to give a countable number of cells (25–100) per large square (1 mm × 1 mm).

Activity 3 Demonstrate the use of a technique used to extend the limits of light microscopy D2

You could start by using oil immersion techniques to observe cells under high power, overcoming the limitations explained in P3. You could then go on to research a range of microscopical techniques, including different optical systems and confocal microscopy, which will extend the limits of microscopy. These could be compared with the advantages conveyed by electron microscopy (Unit 2). As well as practical assessment of oil immersion techniques, you could produce a presentation (paper-based or electronic) showing what you have achieved using light and oil immersion microscopy and how this can be extended using techniques such as confocal microscopy or high content imaging.

TOP TIPS
- Light microscopy is very powerful, but has its limitations – make sure you understand what they are.
- Make sure you understand the principles of staining: to make cells and tissues more visible by increasing contrast as well as identifying specific molecules, cell components, cell types or tissues.
- Make sure you understand not just how to count cells, but also the reasons why it is important.

LO3 Understand the cell cycle and the importance of mitosis P4 M2 D3

GETTING STARTED

Growing up and growing old (5 minutes)

A fertilised egg develops into an embryo and then a fetus; a baby grows into an adult; throughout adulthood, cells and tissues wear out and need to be replaced. During all these processes cells must divide and multiply. Discuss what is important in cell division to maintain the integrity of the developing organism and what might happen if things go wrong.

Mitosis is the process whereby nuclei divide to form two genetically identical nuclei. Strictly speaking, cell division (cytokinesis) is a separate event that usually, but not always, follows on from nuclear division. However, mitosis is only one small part of the 'circle of life' – the cell cycle – that cells undergo. The rate at which the cell cycle takes place determines whether a cell is rapidly dividing or perhaps not dividing at all. There are critical control points – checkpoints – in the cycle that regulate the process. These are involved not just in the regulation of normal cells but also underlie the development of cancer cells and are important targets for the development of anti-cancer therapies.

3.1 Mitosis as the cell division leading to growth and repair

Mitosis produces genetically identical cells and so is the basis of multiplication in single-celled organisms and of growth in organisms as diverse as plants and humans. It is also the process by which old and/or worn-out cells are replaced.

The cell cycle

Figure 8.5 summarises the parts of the cell cycle as well as showing the details of mitosis. Many animal cells continue to divide throughout life and so keep repeating this pattern. Other cells enter the G_0 resting phase and become quiescent either permanently, e.g. nerve cells, or temporarily (e.g. some liver and stomach cells).

Interphase (G_1, S and G_2)

The cell cycle can be divided into mitosis and interphase, i.e. the cell is either dividing (mitosis) or doing everything else that cells do (interphase). During mitosis, all the usual biochemical activities of the cell slow down significantly. Interphase can be divided into three stages: G_1, S and G_2.

- **G_1 phase** – G for 'gap' or 'growth' follows on from mitosis. This phase can vary significantly in length, even in different cells within the same species. All of the normal biochemical activities of the cell resume their normal rate, and protein synthesis and organelle production or replication occurs.
- **S phase** – S is for 'synthesis' and during S phase the DNA is replicated. Each chromosome becomes a pair of identical chromatids.
- **G_2 phase** – During the gap between DNA replication and mitosis the cell continues to grow and prepare for mitosis.

Stages of mitosis

The stages of mitosis (see Figure 8.5) are listed in Table 8.2.

Cytokinesis

Mitosis ends with the formation of the new nuclear envelopes around the two daughter nuclei. It is usually

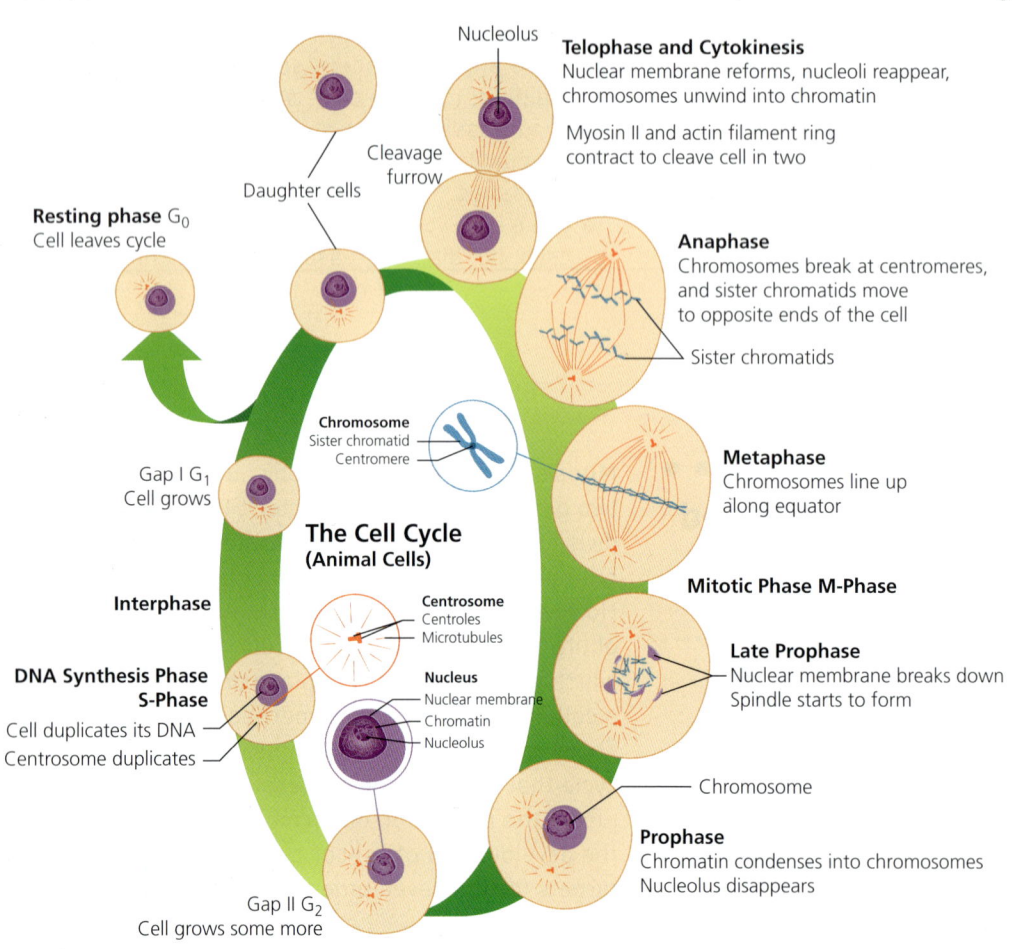

▲ **Figure 8.5** The cell cycle in animal cells. The cell cycle in plant cells follows a similar pattern, although there are differences of detail

Table 8.2 Stages of mitosis

Stage	Observations
Prophase	Chromosomes condense and become visible. DNA is coiled around histone proteins – this is known as chromatin. As the chromosomes condense, the chromatin becomes coiled and super-coiled with other proteins, forming a highly packed structure. Towards the end of prophase (prometaphase), the nuclear envelope breaks down and the centrosomes move to opposite sides (poles) of the cell, and a spindle of microtubules forms.
Metaphase	Chromosomes become aligned along the metaphase plate (equator). The chromosomes become attached to the spindle fibres by their centromeres.
Anaphase	Centromeres split and the spindle fibres shorten so that one chromatid of each pair is pulled towards opposite poles.
Telophase	Chromatids reach the two poles and nuclear envelopes reform around each set of chromosomes to form two identical daughter nuclei.

followed immediately by cytokinesis, or cell separation. The cell membrane becomes pinched in and eventually the membranes fuse and two separate cells are formed.

In some cell types, cytokinesis does not follow mitosis. This happens in some liver cells that form multi-nucleate cells.

In plant cells a cell plate forms, starting at the centre of the cell and spreading outwards until it reaches the cell membrane at the equator. The cell plate fuses with the cell membrane at which point two daughter cells have formed. Cell wall material is then laid down to complete the process.

The importance of mitosis

Mitosis produces genetically identical cells and so it is the basis of asexual reproduction, such as that found in simple single-cell organisms or more complex multicellular organisms such as greenfly that reproduce asexually when conditions are favourable. In contrast, meiosis (see section 1.1 in Unit 5) halves the chromosome number to form gametes that then fuse during fertilisation to form a diploid zygote.

Mitosis is also the way in which cells divide when an organism grows or repairs damaged or worn-out tissues. Accurate DNA replication is important to ensure that the daughter cells are genetically identical to the original cell and there are several enzyme systems that help to ensure this. Nevertheless, errors do arise over many generations of division. This leads to a gradual accumulation of mutations in individual cells that is one aspect of the ageing process.

Mitosis is also important in clonal expansion – the process whereby T cells and B cells specific for a particular pathogen undergo many rounds of cell division to defend against infection – see section 6.2 in Unit 4.

Mechanisms that arrest cell division (checkpoints and clinical techniques)

Progression through the cell cycle is regulated by two classes of proteins, the cyclins and the cyclin dependent kinases (CDKs). The 2001 Nobel Prize in Physiology was awarded to American scientist Lee Hartwell and British scientists Sir Paul Nurse and Sir Tim Hunt for their work in discovering the importance of these proteins. These proteins are found in all eukaryotes and the basic process is the same – much of the early work was done using yeast. More complex organisms may have more elaborate control systems in addition to these.

CDKs and cyclins form heterodimers, functional proteins consisting of two different polypeptide chains. When the active cyclin-CDK dimer is formed, phosphorylation of various proteins occurs and leads to the cell cycle progressing.

Cell cycle checkpoints (Table 8.3) regulate the cell cycle and can be a way of preventing the passing on of damaged DNA to daughter cells.

GROUP ACTIVITY

Compare this (20 minutes)

Divide into small groups. Each group should download a diagram of the plant cell cycle, e.g. from https://en.wikipedia.org/wiki/Cell_cycle and compare it with the animal cell cycle, and make a list of all the similarities and differences. You could also prepare an onion root tip squash to observe under the microscope.

Table 8.3 Cell cycle checkpoints

Checkpoint	Process
G_1/S	The 'decision point' for the cell to proceed through the cell cycle and undergo mitosis. Cells can delay progression and remain in G_1 (e.g. in yeast if food is scarce) or even enter G_0 (quiescence or senescence). Some differentiated cells (see section 4.1), such as heart muscle cells or nerve cells, remain in G_0 for the whole of their functional life.
G_2/M	Ensures that the cell correctly replicated the DNA during S phase and that it is now ready to undergo mitosis.
Mitotic	Occurs at the point in metaphase where the chromosomes are aligned at the equator and are under tension from the spindle fibres. Once this tension is sensed, anaphase can begin.

If DNA damage is detected at either the G_1/S or the G_2/M checkpoints, the cell cycle is arrested until the damage is repaired. There are various DNA repair enzymes that do this. If DNA repair does not occur, the cell undergoes apoptosis (programmed cell death) so that defective DNA is not passed on to the daughter cells.

Cell cycle and cancer

Understanding the cell cycle has improved our understanding of the way in which various cancers develop. In addition, the cell cycle has been a target for anti-cancer therapies.

Control of the cell cycle involves activators, such as the cyclin / CDK enzymes, as well as inhibitors, such as the tumour suppressor p53. Defects in this regulation mechanism can lead to cancer. For example, mutation in p53 can lead to cells dividing uncontrollably.

Cancer cells are generally more rapidly dividing than normal cells, so many anti-cancer drugs target mitosis. For example, paclitaxel (better known by its trademark Taxol®) and the vinca alkaloids (e.g. vincristine and vinblastine) disrupt spindle action in mitosis by targeting the cytoskeletal protein tubulin.

The cyclin / CDK enzymes have also been a target for cancer therapy because they control progression through the cell cycle. Figure 8.6 shows the cell cycle together with a selection of the anti-cancer drugs, already in use or in development, that target various components.

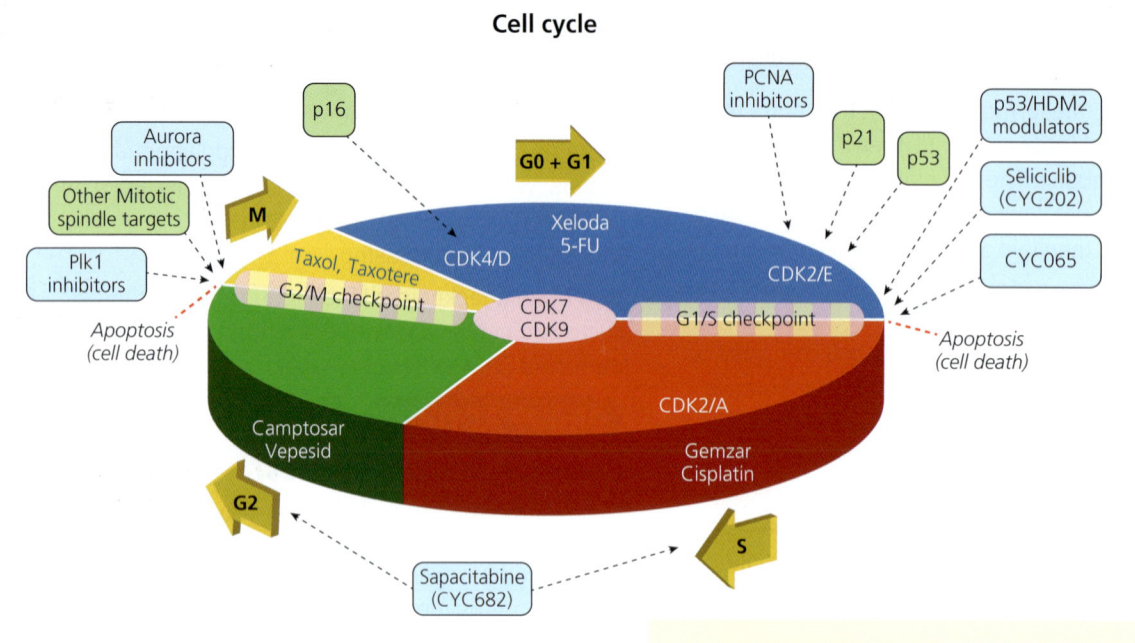

▲ **Figure 8.6** Anti-cancer drugs that target stages in the cell cycle

KNOW IT

1. List the stages of mitosis in the correct order.
2. Complete the following passage by filling in the gaps: Mitosis is a type of nuclear division that produces genetically _____ cells. It is important for making new cells needed for _____ as well as _____ of old cells. Mitosis is also important in _____ reproduction.
3. What regulates progression through the cell cycle?
4. Why are checkpoints important in the cell cycle?
5. What can happen if there are defects in the control of the cell cycle?

L03 Assessment activities

Below are suggested assessment activities that have been directly linked to the Pass, Merit and Distinction criteria in LO3 to help with assignment preparation and includes top tips on how to achieve best results.

Activity 1 Describe the cell cycle, stages of mitosis and cytokinesis P4

You should be able to show your understanding of the processes on the workbench, assembling poppet beads as chromosomes (different colours can be added to illustrate genes), string as spindle fibres, and a board marker to illustrate degradation and then re-formation of the nucleus. You could also draw and animate sequences to illustrate the process, photographing what you do to support the assessment.

Activity 2 Explain the importance of mitosis M2

You should understand the importance of mitosis in producing genetically identical cells for asexual reproduction, growth and repair. This links with LO1 in Unit 5, which covers the importance of meiosis in halving the chromosome number in gamete formation and leading to genetic variation in sexual reproduction.

TOP TIPS
- Make sure you understand the need to replicate the DNA before mitosis so as to produce genetically identical cells.
- Make sure you appreciate the differences between meiosis and mitosis, particularly as both begin with replication of the DNA, but whereas mitosis maintains the chromosome number, in meiosis it is halved.

Activity 3 Assess the importance of mechanisms that arrest cell division D3

You should consider a cell's use of checkpoints during the cell cycle and be able to discuss the timing of checkpoints in mitosis and how these have been linked with cancer. Some medications can introduce stopping points, and you should assess how these are important in many cancer therapies. Toxicity, tumour types affected and their unpredictability are all considerations and you could examine how new strategies are being researched. Environmental factors – either natural or induced by chemicals or radiation – can also be important in cell cycle arrest in certain organisms, leaving the organism in a type of suspended animation until conditions become more favourable.

You might be required to produce a formal report or web pages to address this assessment criterion.

TOP TIPS
- Make sure you address the importance of arrest of cell division at all three stages.
- Make sure you appreciate the importance of maintaining the integrity of the genome by mechanisms of checking (proofreading) the DNA following replication.
- Try to consider the implications of failures in any of these checkpoints.

LO4 Understand the process and significance of differentiation P5 M3

GETTING STARTED
Name that cell game (10 minutes)
Divide into groups and take it in turns to give the name of a type of cell. Your tutor will write these on the board. If a group can't come up with a name, they can pass on one turn.

In multicellular organisms, different cells become specialised to fulfil different functions. This is controlled by regulation of gene expression. Every somatic cell (i.e. excluding the gametes) contains in its nucleus the whole genome of the organism. How the cell functions is determined by which of the many genes are expressed.

4.1 Cell differentiation

The process by which a cell changes from one cell type to another is known as cell differentiation. The process can be understood by considering how a zygote (fertilised egg cell) develops into an embryo and how one single cell gives rise to not just many cells, but many different types of cell.

KEY TERMS

Stem cells – Non-specialised cells that can give rise to one or more types of specialised (differentiated) cell.

Embryonic stem cells (ESCs) – Cells in the embryo that can give rise to all the different types of cells in the adult.

Adult (somatic) stem cells – Cells in the adult that divide and differentiate to maintain and/or repair the tissues. The term somatic stem cell is often used instead, because adult stem cells are also found in children! In this context, 'adult' simply distinguishes this type of stem cell from embryonic stem cells.

Totipotent stem cells – Differentiate into embryonic and extraembryonic (placenta) cell types. Such cells can construct a complete, viable organism. The zygote is, by definition, totipotent as are the cells produced by the first few divisions of the zygote.

Pluripotent stem cells – Differentiate into all embryonic but not extraembryonic cell types.

Multipotent stem cells – Differentiate into a number of different cell types, e.g. adult bone marrow cells.

Unipotent stem cells – Differentiate into just one type of cell type, e.g. skin cells.

Undifferentiated cells (stem cells)

Stem cells are undifferentiated cells that can divide and differentiate to form specialised cells. The human embryo contains **embryonic stem cells** that can give rise to the more than 200 cell types of the adult human body; these are **pluripotent stem cells**. In the adult (or child), most cells have already differentiated. However, **adult stem cells** persist and are responsible for cell turnover, e.g. production of red blood cells from bone marrow cells; these are **multipotent stem cells**. These bone marrow cells also differentiate to form the various types of lymphocytes involved in the immune response (see section 6 in Unit 4). Most epithelial cells need to be replaced throughout the life of an organism. Examples include skin, lungs, cornea and intestine. In each case, **unipotent stem cells** are responsible for replacement of worn out cells.

Most differentiated plant cells lose the ability to divide (undergo mitosis). Plant stem cells are known as meristematic cells; the meristems (where they are located) include the cambium of vascular bundles (that give rise to xylem and phloem cells) and the growing tips of shoot and root (that give rise to cells of the stem and root respectively).

Gene expression and repression

Totipotent stem cells, e.g. the zygote or cells of the early embryo, can give rise to every cell type. Different cell types produce some proteins but not others. For example, some pancreatic cells need to produce digestive enzymes and others produce insulin or glucagon, whereas muscle cells need to produce large amounts of the muscle proteins actin and myosin. Development and differentiation of these cells requires that some genes are expressed ('switched on') whilst others are repressed ('switched off'), because all somatic cells retain the full amount of DNA – no genes are lost from cells.

The expression or repression of genes during embryo development and differentiation of different cell types is controlled by a range of growth factors and other signalling factors. This process is under the control of the homeobox genes; you could research how these genes regulate the development of organisms from primitive worms through to humans. These act through a variety of intracellular signalling pathways and can result in modification of the DNA itself (e.g. by methylation) or of the histones that the DNA is wound around in the chromatin structure. Both of these can lead to expression or repression of genes. The changes can be passed on to daughter cells, although they do not affect the germ line (i.e. the sex cells) and so they are known as epigenetic.

PAIRS ACTIVITY

The development tree (30 minutes)

Divide into pairs and prepare a chart of development of an embryo into an adult human. Start with the **zygote** (totipotent) and make a 'family tree' of the different types of stem cells that originate from either the embryo or from other stem cells. For each type of stem cell, include the other stem cells or types of tissue that they differentiate into.

KNOW IT

1. For the following list of cells, state whether each is totipotent, multipotent, unipotent or differentiated:
   - haemopoeitic (bone marrow) cell
   - zygote
   - cambium
   - blast cell
   - nerve cell
2. Fill in the blanks in this passage: Development of an adult organism requires cells to multiply by _____ so that all the cells produced are _____ identical. The process of forming specialised cells is known as _____ and requires that various _____ are switched on or off. This can lead to modification of the _____ and these changes are passed onto daughter cells. These changes are known as _____ because they do not affect the germ line.

LO4 Assessment activities

Below are suggested assessment activities that have been directly linked to the Pass and Merit criteria in LO4 to help with assignment preparation and includes top tips on how to achieve best results.

Activity 1 Describe the process of differentiation in the embryo in producing specialised cells P5

You need to be able to distinguish between totipotent, pluripotent, multipotent and unipotent stem cells. You could research some of the different types of stem cells and produce a map of differentiation during development of the embryo showing the various types of stem cells, labelled as to whether they are totipotent, etc.

Activity 2 Explain the role of gene expression and repression in differentiation M3

You need to appreciate that the changes that occur during differentiation are the result of regulated gene expression – the switching on or off of different genes.

The process by which the information in the DNA is transcribed into messenger RNA for protein synthesis is dealt with in detail in Unit 1. You need to understand how cell signalling is involved in control of differentiation and how the changes are known as epigenetic. You could research one aspect of differentiation, e.g. haematopoiesis, and prepare illustrations of how cell signalling and regulation of gene expression control the process.

TOP TIPS
- Don't include too much detail – this subject can be extremely complicated!
- Concentrate on illustrating the principles involved.

LO5 Understand the potential of stem cells in medical therapies P6 M4

GETTING STARTED
Potential for stem cell therapy (5 minutes)

Think about the various types of diseases you know about. Discuss the types of disease that might be treated by stem cell therapy and any that might not be suitable. What about trauma?

Stem cells offer the opportunity to treat or prevent a range of diseases. Our understanding of the nature of stem cells has led to the development of the field of **regenerative medicine**.

KEY TERMS
Regenerative medicine – The replacement, regeneration or re-engineering of human cells, tissues or organs in order to treat disease.

Pluripotent stem cells – Originate from totipotent cells and can differentiate into nearly all types of cell.

5.1 Stem cells, their function and potential use in modern medicine

Stem cells are the focus of research in regenerative medicine, aimed at treating a wide range of diseases. Some of these are the result of trauma (injury) whereas others are degenerative diseases.

Origin of stem cells for research and therapies

One of the most controversial aspects of the use of stem cells in research and therapy is the origin of the stem cells, originally from early stage embryos, although more recently it has become possible to use adult stem cells.

Embryonic (ESCs)

Because ESCs are **pluripotent**, they can potentially be used to treat a wide range of degenerative and other diseases and have a range of research applications. They are obtained from the blastocyst stage of the embryo (4–5 days post-fertilisation in human embryos). The cells are grown in cell culture dishes for several months, during which time they are transferred to a number of fresh culture dishes (sub-culturing) several times. This ensures that a stable cell line is established that is capable of growing and self-renewing over a long period of time. The cells have to be checked regularly to ensure that they are not differentiating.

Adult stem cells

Adult stem cells can be obtained from tissues such as bone marrow. Most bone marrow stem cells are haemopoietic, i.e. they give rise to all the types of blood cells. However, a small percentage are mesenchymal (or skeletal) and give rise to bone, cartilage and fat cells. During the 1990s it was accepted that brain tissue contained stem cells that could give rise to the three main types of brain cell. However, it is very difficult to grow adult stem cells in culture. This makes them harder to use in research and more difficult to obtain the large quantities that would be needed for use in therapy.

Induced pluripotent stem cells

Induced pluripotent stem cells (iPSCs) are adult cells that have been reprogrammed (dedifferentiated) to express genes for transcription factors (these control expression of other genes) and make them become pluripotent like ESCs. These iPSCs have the ethical advantage of not being derived from embryos. Also, it is possible, in theory at least, to prepare iPSCs from a patient's own cells – each patient could have their own cell line. This would overcome issues such as rejection that can occur with donated stem cells.

However, there are other risks associated with the use of iPSCs in therapy because of the need to use viruses to reprogram the adult cells – in the original technique used, this introduced oncogenes (cancer-causing genes). Methods are being developed to produce iPSCs in ways that reduce or eliminate these risks. Nevertheless, iPSCs are very useful tools for use in disease modelling and drug discovery (see Unit 11).

Stem cell research

Stem cell research includes using them to study development and differentiation in human and non-human embryos. However, the main focus of stem cell research is into their use in therapy, particularly in the area of regenerative medicine.

Model cells

Stem cells are used as model systems to study the way in which undifferentiated stem cells become the differentiated cells that form the tissues and organs of the body. They can be used to investigate the genetic and molecular events that control this process.

Disease development

Some cancers and birth defects are caused by abnormal cell division or differentiation, so gaining a better understanding of what controls these events will enable us to develop treatments or methods of prevention.

Drug testing

Cultured cells have been used for many years in developing and testing new drugs. For example, cancer cell lines have been used to test new anti-cancer drugs. The availability of pluripotent stem cells means that, in theory, cell lines of all possible different tissues can be produced and used to test new drugs. However, to make a valid comparison between different drugs, the cell lines used must be identical to each other and to the target tissue for the particular class of drugs. This means that we need a good understanding of the way in which stem cells differentiate into the different cell types. We do not yet have this knowledge for all cell types.

Stem cell therapies

Adult stem cells have been used in therapy for many years: bone marrow transplant uses haematopoietic stem cells from the patient's own bone marrow (autologous transplantation) or from a donor's bone marrow (allogenic transplantation) to treat cancers of the bone marrow such as leukaemia. The procedure normally involves destruction of the patient's own immune system (including the cancer cells) through chemotherapy and/or radiotherapy after which the transplanted bone marrow cells divide and differentiate to replace the destroyed cells. More than 26,000 patients are treated in this way in Europe every year.

The treatment has various risks, particularly graft-versus-host disease (rejection) in allogenic transplantation. However, it is now widely used and it is also being used to treat other potentially fatal conditions such as some autoimmune diseases.

Since the 1970s, skin stem cells have been used to grow skin grafts for patients with severe burns on very large areas of the body. However, this is highly specialised treatment available in only a few clinics. It is usually reserved for patients with life-threatening burns and it is not a perfect solution: the new skin has no hair follicles or sweat glands.

A new stem cell based treatment to repair the cornea after injury has recently been given conditional approval in Europe.

Besides these, all other stem cell therapies remain experimental.

Figure 8.7 illustrates some of the other potential uses of stem cells in therapy.

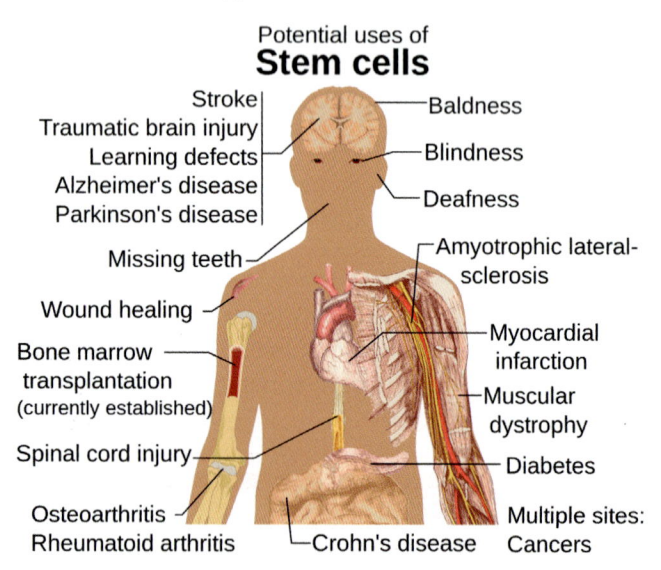

▲ **Figure 8.7** Diseases and conditions where stem cell treatment is promising or emerging. Bone marrow transplantation is the only widely accepted use of stem cells, although approval has been given to a small number of other treatments in the last few years

In order to be successful, stem cell therapy must function appropriately for the rest of the recipient's life and not cause any harm to the recipient.

Scientific issues

Working with stem cells is not easy! Maintaining cells in culture in the lab is now fairly routine, but not trivial. Growing cells on a larger scale presents more challenges. Converting adult cells to iPSCs without use of viruses or other techniques that might cause unwanted side-effects is still a work in progress. Finally, whether there will be any long-term harmful effects of the therapeutic use of stem cells remains unclear. A lot of work is being directed towards addressing these problems in animal systems, such as mouse models of human degenerative diseases.

THINK ABOUT IT

Parkinson's disease

Parkinson's disease (PD) involves destruction of nerve cells in the brain. Stem cell therapy is a potential cure for PD. One option is fetal stem cells. You could research stem cell therapy for PD (see 'Read about it' for some places to start) and think about the following issues:

- The disadvantages of using fetal tissue – not the ethical issues; we will consider those in the next section.
- How could stem cell therapy be of benefit?
- What types of stem cells could be used?
- What risks might be associated with stem cell therapy?

Moral and ethical issues

ESCs are obtained from human embryos that have been formed by in vitro fertilisation (IVF). Spare embryos that are not required for implantation may be donated for use in research. This requires informed consent on the part of the donor. More recently, in the UK, the law has been changed to allow researchers to create embryos (using spare donated eggs from IVF) specifically for research purposes; the law says that these cannot be implanted in the womb, so can never give rise to new individuals.

The key moral and ethical concept is the status of the embryo. The UK consensus, enshrined in the Human Fertilisation and Embryology Act in 1990, is that the embryo does have moral rights but not to the same extent as a living person. This interpretation is challenged by some groups, such as the Catholic Church, whose teaching is that new life begins at the point of conception.

This implies that a fetus at any stage of development, including the early stage embryo, should hold full human rights. Others, including some other faiths, consider that the status of a fetus changes as it develops, for example as its nervous system appears.

The rights of the embryo / fetus are balanced against the potentially very great benefits that could be delivered by research and therapies using ESCs. Some people argue that the use of ESCs is not justified because alternatives, such as adult stem cells or iPSCs, exist. Many scientists argue that research on ESCs is needed to work out the basic biological mechanisms and help decide which type of stem cells might be of most therapeutic benefit.

GROUP ACTIVITY

The ethical dilemma (30 minutes)

Consider the ethical dilemma when we are forced to choose between the two moral principles:

- the duty to prevent or alleviate suffering
- the duty to respect the value of human life.

There are different moral arguments put forward:

1 The embryo has full moral status from fertilisation onwards.
2 There is a cut-off point at 14 days after fertilisation.
3 The moral status of the embryo increases as it develops.
4 The embryo has no moral status at all.

Discuss these different points of view, but keep your discussion calm and reasoned! Don't expect to come to any conclusions in half an hour – these are topics that have been debated for years and will be debated for many more years.

Clinical studies

Besides bone marrow transplantation, all other therapeutic applications of stem cells remain highly experimental and tightly controlled. For this reason, they are normally only used in clinical studies that involve small numbers of patients, often with serious diseases that could not be treated in any other way.

Here are some current and recent clinical studies involving stem cells:

- Use of ESCs differentiated into β-cells (insulin producing cells) to treat diabetes.
- The British company ReNeuron is developing stem cell therapies, initially for treatment of strokes.
- Several studies are looking at the use of stem cells in the treatment of motor neurone disease (MND), in particular its most common form amyotrophic lateral sclerosis (ALS). Remember the ice bucket challenge? That was to raise money for ALS research.

EXTENSION ACTIVITY

Potential and risks of stem cell therapies (2 to 4 hours)

Prepare a presentation, poster or website that describes the therapeutic potential of stem cell therapy in treatment of a named disease or other medical condition (e.g. injury, degenerative disease). Use the online resources listed under 'Read about it' as a starting point for your research.

L05 Assessment activities

Below are suggested assessment activities that have been directly linked to the Pass and Merit criteria in L05 to help with assignment preparation and include top tips on how to achieve best results.

Activity 1 Describe uses of stem cells in medicine *P6*

Stem cell therapies are not new; in 1968 doctors performed the first successful bone marrow transplant. Stem cell transplants have been used in the treatment of various cancers (leukaemia, lymphoma, myeloma) and following chemotherapy or radiotherapy for many years.

Figure 8.7 could be used as a starting point for an illustrated essay, poster or electronic presentation demonstrating the use of stem cells in medicine – both those currently in use as well as those being developed.

Activity 2 Prepare a report on stem cells used in therapy *M4*

You could prepare a report that builds on your work for P6. Outline the types of stem cells used in therapy and the ways in which they are obtained. For each, you should discuss associated risks and benefits as well as the relevant ethical issues.

TOP TIPS
- Make sure you understand the differences between the various types of stem cells.
- Make sure you understand the normal function of the different types of stem cells. This will help you understand their potential as well as limitations in stem cell therapy.
- Ethical concerns must be taken seriously. Make sure you understand the difference between ethical concerns and the scientific concerns about the risks of stem cell therapy.

Read about it

This is such a fast-moving field of modern biology and medicine that any publications rapidly become out of date, so the internet is one of the best sources of up-to-date information.

National Institutes of Health (NIH)

The NIH is part of the US government Department of Health and Human Services and is the largest biomedical research agency in the world. They have a whole section of their website devoted to the subject of stem cells:

https://stemcells.nih.gov/

If you want to go directly to some basic information about stem cells, start here:

http://stemcells.nih.gov/info/basics/pages/basics1.aspx

There are also pages on stem cell research and clinical trials.

The EuroStemCell organisation

This organisation is based in Edinburgh, UK, and exists 'to help European citizens make sense of stem cells'. Their website www.eurostemcell.org is an excellent source of information.

Medical Research Council Centre for Regenerative Medicine (CRM)

Scientists and clinicians at CRM study stem cells, disease and tissue repair to advance human health. Their website is www.crm.ed.ac.uk/

The UK Stem Cell Foundation

A UK charity set up in 2005 to focus specifically on supporting research into the use of stem cells in treating a wide range of conditions and diseases. The website contains useful information about stem cells and stem cell research, as well as many links to other useful sources.

www.ukscf.org/

Unit 11
Drug development

ABOUT THIS UNIT

New medicines are needed to provide more effective treatments for existing diseases, as well as meeting the changing needs of ageing populations. The pharmaceutical industry devotes a significant proportion of its large budget to drug development. Drug development describes the process whereby new medicines are developed, from basic research ('drug discovery') through the various pre-clinical (laboratory tests) and clinical phases (tests in humans) of development and trialling before the drug can be approved for sale. This is a truly multi-disciplinary endeavour involving a wide range of specialists, including synthetic organic chemists (medicinal chemistry), analytical chemists, biochemists, pharmacologists, pharmaceutical (formulation) chemists and many others in the pre-clinical stages, as well as clinicians and statisticians in the clinical trials phase.

In this unit you will learn some of the practical preparative and analytical skills used in the drug development process, as well as gain an appreciation of the wider context in which you might work.

LEARNING OUTCOMES

The topics, activities and suggested reading in this unit will help you to:

1 Understand principle drug discovery and development processes
2 Understand the range of techniques used in drug production
3 Be able to carry out a basic extraction, synthesis, isolation and purification of a simple drug or pharmaceutical
4 Understand the importance of product formulation and dosage form
5 Understand the importance of planning clinical trials when introducing new drugs

How will I be assessed?

You will be assessed through a series of assignments, practicals and tasks set and marked by your tutor.

How will I be graded?

You will be graded using the following criteria:

Learning Outcome	Pass	Merit	Distinction
	The assessment criteria are the Pass requirements for this unit.	To achieve a Merit the evidence must show that, in addition to the Pass criteria, the candidate is able to:	To achieve a Distinction the evidence must show that, in addition to the Pass and Merit criteria, the candidate is able to:
1 Understand principle drug discovery and development processes	**P1** Identify the initial stages of a drug development programme	**M1** Describe modelling techniques used in drug development	**D1** Evaluate pharmacogenomics approaches to drug development
2 Understand the range of techniques used in drug production	**P2** Outline techniques used in drug production		
	P3 Describe analytical techniques used in assessing the purity of products or intermediates	**M2** Explain how analytical techniques are used in assessing the purity of products or intermediates	
3 Be able to carry out a basic extraction, synthesis, isolation and purification of a simple drug or pharmaceutical	**P4** Demonstrate the use of a procedure to extract a pharmaceutical from a plant, animal or microbial source	**M3** Describe the chemistry involved in the extraction of a pharmaceutical from a plant, animal or microbial source	
4 Understand the importance of product formulation and dosage form	**P5** Identify the basic principles of pharmaceutics		**D2** Evaluate the formulations of named drugs for oral, subcutaneous and intravenous administration
	P6 Identify methods used to increase stability of medicines		
5 Understand the importance of planning clinical trials when introducing new drugs	**P7** List the steps in developing a new drug including an appreciation of timescales, costs and clinical trials	**M4** Interpret the results of clinical trial data	**D3** Evaluate the validity and reliability of a clinical trial and justify their conclusions based on an understanding of data

L01 Understand principle drug discovery and development processes *P1 M1 D1*

Most pharmaceutical companies will focus on specific disease areas or types, such as cardiovascular disease, cancer, respiratory diseases (e.g. asthma) or neurological disorders (e.g. depression or dementia). Traditionally, most drugs have been synthetic organic chemicals that can be taken as a tablet, although there have been important exceptions such as insulin (given as an injection). However, many recent drugs, particularly anti-cancer drugs, have been based on monoclonal antibodies or enzymes.

GETTING STARTED

Know me (5 minutes)

Think about all the medicines you or members of your family have taken. Now think about how your life or theirs would be affected if those medicines did not exist. Share your thoughts with the group.

1.1 Categories of drugs

> **KEY TERMS**
>
> **Drug** – A chemical or biological compound used to treat disease. Most drugs are available only with a doctor's prescription and are strictly regulated. Don't confuse pharmaceutical drugs with recreational drugs.
>
> **Biological/biologics** – An enzyme, protein, peptide or antibody (see section 6.2 in Unit 4) used in this context as a drug.
>
> **Medicinal chemistry** – The process where organic chemists make analogues hoping for greater activity.
>
> **Analogues** – A series of compounds with similar chemical structures.

Drugs can be categorised into two main groups:

- Those that are relatively small organic chemicals.
- Those that are **biological** substances, generally much larger and more complex, and beyond the capability of synthetic chemistry.

Synthetic (small molecule) drugs

Many synthetic drugs began as what we would now call herbal remedies. The best known is probably Aspirin (acetyl salicylic acid), which was first isolated from willow bark in 1763 and first synthesised in 1897. Aspirin has given rise to a whole family of drugs – the non-steroidal anti-inflammatory drugs (NSAIDs) that act as painkillers, anti-inflammatories (reduce inflammation) and anti-pyretics (reduce fever). This illustrates an important principle – new drugs are often developments of older drugs rather than being completely new classes of drug.

Many small molecule drugs originate from natural compounds isolated from microorganisms such as fungi – penicillin is an example. Once isolated, purified and characterised, the drug is then produced synthetically and can be the starting point for a programme of development through **medicinal chemistry**. All of the currently available antibiotics were developed in this way.

Small molecule drugs have the advantage of being easy to administer in tablet form and can be absorbed through the gut. For many years the pharmaceutical industry focused almost exclusively on small molecules, with the exception of products such as vaccines (see section 6.2 in Unit 4) or insulin. This is now changing.

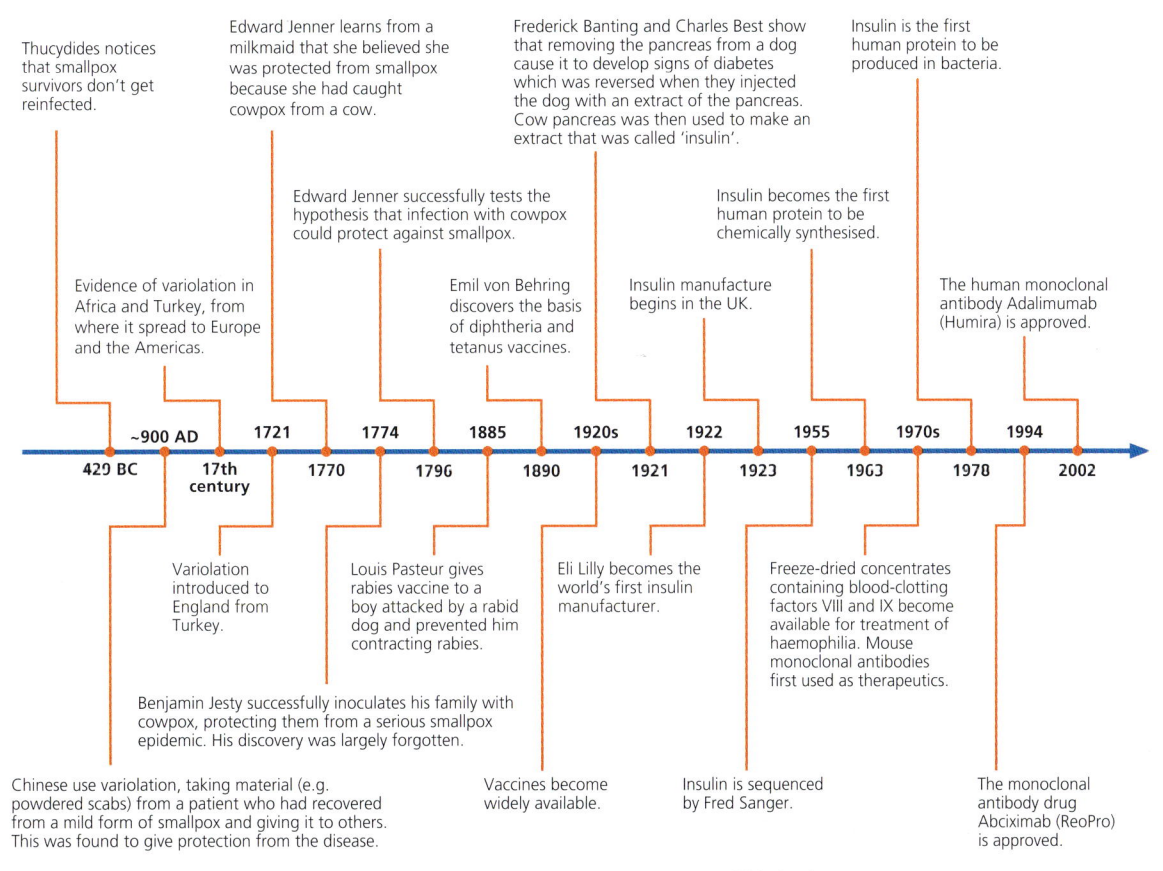

▲ **Figure 11.1** Timeline of key developments in the therapeutic uses of biologics

Timeline events:

- **429 BC** – Thucydides notices that smallpox survivors don't get reinfected.
- **~900 AD** – Chinese use variolation, taking material (e.g. powdered scabs) from a patient who had recovered from a mild form of smallpox and giving it to others. This was found to give protection from the disease.
- **17th century** – Evidence of variolation in Africa and Turkey, from where it spread to Europe and the Americas.
- **1721** – Variolation introduced to England from Turkey.
- **1770** – Benjamin Jesty successfully inoculates his family with cowpox, protecting them from a serious smallpox epidemic. His discovery was largely forgotten.
- **1774** – Edward Jenner learns from a milkmaid that she believed she was protected from smallpox because she had caught cowpox from a cow.
- **1796** – Edward Jenner successfully tests the hypothesis that infection with cowpox could protect against smallpox.
- **1885** – Louis Pasteur gives rabies vaccine to a boy attacked by a rabid dog and prevented him contracting rabies.
- **1890** – Emil von Behring discovers the basis of diphtheria and tetanus vaccines.
- **1920s** – Frederick Banting and Charles Best show that removing the pancreas from a dog cause it to develop signs of diabetes which was reversed when they injected the dog with an extract of the pancreas. Cow pancreas was then used to make an extract that was called 'insulin'.
- **1921** – Vaccines become widely available.
- **1922** – Eli Lilly becomes the world's first insulin manufacturer.
- **1923** – Insulin manufacture begins in the UK.
- **1955** – Insulin becomes the first human protein to be chemically synthesised.
- **1963** – Insulin is sequenced by Fred Sanger.
- **1970s** – Freeze-dried concentrates containing blood-clotting factors VIII and IX become available for treatment of haemophilia. Mouse monoclonal antibodies first used as therapeutics.
- **1978** – Insulin is the first human protein to be produced in bacteria.
- **1994** – The monoclonal antibody drug Abciximab (ReoPro) is approved.
- **2002** – The human monoclonal antibody Adalimumab (Humira) is approved.

Biological (biologics)

The earliest **biologics** were vaccines and antitoxins. The understanding that exposure to a disease can lead to resistance dates back to the ancient Greeks, but we think of the science of vaccination beginning in the 18th century with Edward Jenner (see Figure 11.1).

Most biological drugs are proteins. Some biologics can be given orally, for example therapeutic enzymes intended to replace or supplement digestive enzymes. However, most biologics cannot be given orally as the drug is usually digested in the stomach rather than being absorbed. Therefore, most have to be given by injection. This is acceptable in a vaccine that is only given once or twice. Other drugs where injection is generally acceptable include life-saving drugs like insulin or anti-cancer drugs.

In 2014, seven out of the top ten best-selling drugs were biologics (given by injection). These included an insulin **analogue** as well as monoclonal antibodies or recombinant proteins aimed at treatment of cancer and/or autoimmune disease.

Other biologics in widespread use in medicine include blood-clotting factors (e.g. Factor VIII used in treatment of haemophilia) and the various thrombolytic enzymes used to treat heart attacks and strokes.

The common theme is that biologics are used in situations where the need for administration by injection is acceptable. There is still a challenge in developing ways to deliver biological drugs by 'traditional' routes such as oral (by mouth) or inhaled.

1.2 Research stages of drug discovery programme to identify new drugs

> 🔑 **KEY TERMS**
>
> **Drug target** – This is usually a protein (enzyme or receptor) or gene implicated in the disease that a new drug aims to treat.
>
> **Unmet medical need** – A medical condition for which there is, as yet, no available treatment.

Before the drug discovery and development process can even begin, it is necessary to have an understanding of the disease being treated. This will require input from many different scientists including biochemists, cell biologists, physiologists, pharmacologists and others. This work is carried out in universities and research institutes, as well as in pharmaceutical companies.

However, developing a new drug does not just start in the lab. Before then, the need for a new product has to be established – we are talking, in the main, about the pharmaceutical industry, a multi-trillion pound or dollar industry worldwide that is answerable to its shareholders. This means that there are many factors – medical, scientific, strategic and regulatory – that influence the direction that a drug discovery programme will take. This is illustrated in Figure 11.2.

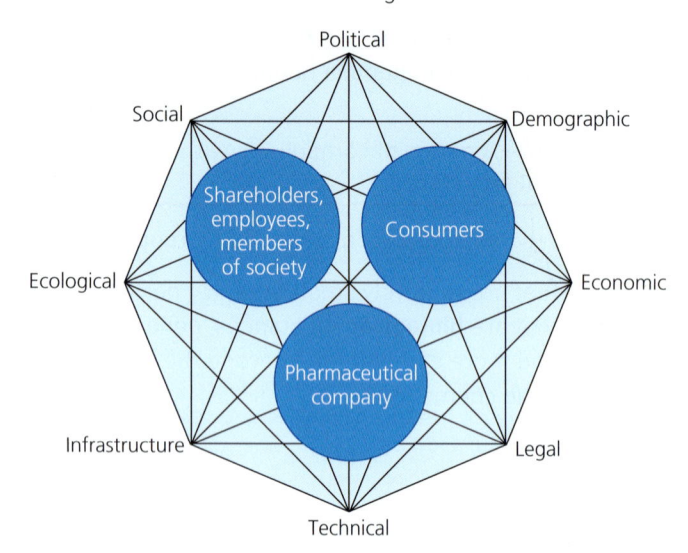

▲ **Figure 11.2** The complex network of influences that have to be considered when embarking on a drug discovery programme. The pharmaceutical company (or drug discovery company) interacts with consumers (see text for explanation of who these are) as well as all those who have an interest in the company (e.g. shareholders, employees) or its wider impact on society

Consumer target

Here, the consumer of a new drug isn't necessarily the patient and target doesn't mean the same as **drug target** (usually a protein). A more general meaning of target could include the therapeutic area (e.g. diseases of the central nervous system) or a particular disease (e.g. Alzheimer's disease). However, in this context, the term 'consumer target' means 'who will buy the drug?' That, in itself, is complicated – does it mean the patient, the doctor who prescribes the drug, or the organisation such as the NHS or insurance company that pays? All of these must be taken account of in developing a strategy for drug development.

Effectiveness of existing products

Except in the case of an **unmet medical need**, any new drug will have to compete with existing treatments.

These could be similar types of drugs, different classes of drugs or even non-drug treatments. An example of the latter might be a new drug to treat a mental illness where the existing treatment is Cognitive Behaviour Therapy. Any new drug is going to be tested against the best currently prescribed drug (see LO5 for details of the clinical trials process). If existing products are perfectly adequate in treating a disease, there is no point in embarking on a multi-million (more likely multi-billion) pound or dollar programme of drug discovery to produce yet another **me-too drug**.

Legislative requirements

The pharmaceutical industry is tightly regulated in most developed countries. This means that a new drug needs to demonstrate safety and **efficacy**. However, many countries now consider the cost-effectiveness of a drug as part of the approval process. In the UK this is the responsibility of the National Institute for Health and Care Excellence (NICE). There is no point developing a drug that will pass regulatory approval on the basis of safety and efficacy if it doesn't meet the criteria for prescription on the NHS or insurance reimbursement.

New and emerging technologies and materials

DNA technology offers the prospect of developing gene therapy to treat and correct inherited disorders such as cystic fibrosis. The Human Genome Project (see section 4 of Unit 5) has made possible the field of personalised medicine. Developments in the technology of diagnosis such as MRI or CT scanning all have an impact on treatment of disease that must be considered when developing new drugs. For example, will advances in personalised medicine mean that drugs become targeted at a smaller group of patients, reducing the potential sales?

New materials, particularly biomaterials (based on peptides or proteins), make novel formulations possible. For example, sustained release formulations, where the drug is injected or implanted once and then released over a prolonged period, can replace the need for daily injections.

The use of technology in therapy is covered by the field of medical devices or MedTech (medical technology). Developments in such technologies can provide opportunities for new drugs to work in tandem with a new medical device. This can bring advantages for the patient, but not necessarily for the pharmaceutical company. A technological solution may make a drug, developed at the cost of many millions, redundant.

Environmental pressures

The pharmaceutical industry operates in a business environment where ethical issues and social responsibility have to be accommodated. Everything from the environmental impact of chemical and pharmaceutical manufacturing processes to the social impact of pricing policy are subject to close public and government scrutiny.

1.3 Modelling techniques in drug development

> **CLASSROOM DISCUSSION**
>
> ### Science and society (20 minutes)
>
> Figure 11.2 illustrates the complex web of influences that determine a company's strategy for developing new drugs. Think about the effects this has on development of new drugs and discuss the following questions:
>
> - To what extent is drug discovery driven by commercial considerations rather than science?
> - Do the strategic goals of pharmaceutical companies match the broader needs of society for development of new drugs?

> **KEY TERMS**
>
> **Pre-clinical phase** – This starts with identifying a **hit compound** and finishes when a potential drug is ready to be tested in humans.
>
> **Drug screening** – Use of a model system that aims to identify compounds (hits) with the necessary activity against the drug target.
>
> **Lead optimisation** – (Lead is pronounced the same as a dog lead, not like lead the metal.) The process of modifying a hit compound, through several rounds of screening different analogues, to produce one or two compounds with high activity.
>
> **Me-too drug** – A drug that is very similar to one or more existing drugs that treat the same disease.
>
> **Efficacy** – A measure of how well a drug works at its required dose.
>
> **Hit compound** – A substance with promising characteristics or activity worthy of further development, sometimes just called 'hit'.

The pre-clinical phase of drug development

Before looking at modelling techniques, we need to consider the **pre-clinical phase** of the drug development process in more detail to understand how these techniques contribute to the overall project.

Target validation

This is the first step in finding new drugs. Before using a drug target in a development programme, we must be sure that the drug target really is instrumental in the disease. This is a crucial part of the process and there are many ways of validating drug targets. Some of them involve computer modelling techniques. Others use structure-activity studies (see later in this section). Proteomics (the study of the protein make-up of the cell) can also be used to look at patterns of expression of different proteins and correlate this with development of the disease.

The use of gene knockouts has become one of the most useful methods of target validation. This uses gene technology methods to prevent expression of a gene for the potential drug target. This is mostly done in mice, although zebrafish are becoming more widely used.

> **CASE STUDY**
>
> **P2X7 knockout mice**
>
> The P2X7 ion channel was thought to have a role in the pain associated with neuropathy (a disease of the nervous system) and inflammation. Mice lacking the P2X7 ion channel protein did not exhibit the type of pain associated with neuropathy or inflammation, although other responses to pain were normal. This indicated that the P2X7 ion channel protein was a valid drug target for development of new analgesic (pain-killing) drugs.
>
> If you are interested to learn more about knockout mice and their role in medical research and drug development, the US National Institutes of Health (NIH) has an online fact sheet at www.genome.gov/12514551 and Science Education has an article on the subject at www.nature.com/scitable/topicpage/scientists-can-analyze-gene-function-by-deleting-6526138.

Drug discovery

Once a valid drug target has been developed, **drug screening** can begin. This involves testing a large number of compounds (a compound library that could consist of tens or even hundreds of thousands of compounds) against the drug target. Automated or robotic systems are often used to handle such large numbers of compounds in a process known as high throughput screening.

Compounds that show activity in the screen are known as hits and are the starting point for further work.

Drug design and optimisation

Most drugs work by binding to a target. It is possible to use computer modelling to study this process and predict ways in which the interaction between the drug and the target can be optimised. This is the basis of rational drug design and although using computer modelling can speed up the process, it still involves many rounds or iterations, also known as **lead optimisation**. This is discussed in more detail in the next section.

Quantitative structure-activity relationships (QSAR)

The relationship between the structure of a molecule and its properties has been known for many years. Trends in the boiling points of the members of the alkane homologous series (see section 4.1 in Unit 1) are a simple example of this.

Structure-activity relationships (SARs) have long been at the heart of the drug discovery and development process, which can be summarised as follows:

- Screen a library of compounds against a drug target.
- Identify compounds with activity against the target.
- Compare the structure and activity of these compounds to identify relationships.
- Construct a model for drug action.
- Use this information to make analogues of these to enhance the desired characteristics (this is where medicinal chemistry is important).
- Screen these analogues for activity.
- Refine the model and design new, hopefully more active, analogues.
- These steps are repeated, until a compound with the desired combination of characteristics is found.

Quantitative structure-activity relationships (QSARs) use statistical methods to refine this approach of correlating chemical structure with pharmacological activity. Having built a QSAR model, it can then be used to predict the activity of a range of chemicals.

QSAR methods are used at all stages of the drug design and optimisation stages. QSAR models can also provide insights into the mechanism of action of a drug and suggest new design strategies and synthetic targets. QSAR models can also be used to predict physicochemical and pharmacokinetic properties (important in formulation and determining bioavailability – see section 4.1) and toxicological properties (important in determining safety). This makes it possible to weed out at an early stage any compounds that might have poor physicochemical, pharmacokinetic or toxicological profiles – it is much cheaper to do this, because late stage failures are very expensive (see section 5.1 for more information about this).

Computational techniques

QSAR methods rely heavily on computers to perform the necessary calculations and generate appropriate models. As well as the statistical methods of QSAR based on chemical structure, 3D-QSAR is a complementary approach that uses computers to generate molecular models of the drug (small molecule) and then to compute the energy of interaction between the drug and the target. Most drug targets are proteins, so a structure of the target derived from protein crystallography or nuclear magnetic resonance spectroscopy (NMR) is required for this approach. These methods allow computer modelling of the binding of the drug to the target. This can be used to optimise the structure of the drug molecule and virtual screening methods now exist where molecular modelling is used to identify a shortlist of candidates for in-vitro screening against the actual target.

1.4 Pharmacogenomics approach to drug development

> **KEY TERMS**
>
> **Pharmacogenetics** – The study of how individual genes affect a person's response to drugs.
>
> **Pharmacogenomics** – The study of how a person's entire genetic makeup (genome) affects their response to drugs.
>
> **Polymorphism** – The existence of two or more alleles of a gene, i.e. variants of a particular DNA sequence. Generally refers to the existence of two or more forms of a drug target (usually a protein).
>
> **Single nucleotide polymorphism (SNP)** – A change in a single base in a DNA sequence, e.g. C replaces G. This means that the two variants are alleles.

Section 4.1 in Unit 5 discussed the impact of DNA sequencing and the human genome project. One major impact of this technology has been the development of **pharmacogenetics** and **pharmacogenomics**. Most people in a population will respond well to a drug, but others will respond less well or not at all, and some will experience side-effects (adverse reactions).

Pharmacogenomics helps us understand and hopefully address these issues.

Genomic profiling of different diseases and identification of targets

The β2 adrenergic receptor is the drug target for bronchodilator drugs such as salbutamol, used in the treatment of asthma. Some individuals have reduced expression of these receptors and respond unpredictably to salbutamol. This is an example of a **polymorphism**, where different variants of drug targets exist in the population (i.e. there are different alleles). Drug targets that show this type of polymorphism are probably better avoided.

However, variation in response to a drug is usually due to the interaction between many genes. For this reason, pharmacogenomics can be more useful, as it can use information from the whole genome rather than just individual genes to predict responsiveness. **Single nucleotide polymorphisms (SNPs)** underlie differences in susceptibility to disease.

A single SNP might be the cause of genetic disease. Sickle-cell disease (see section 2.1 in Unit 5) is caused by a single base substitution in the β-globin gene. However, more complex diseases are often the result of interaction between several SNPs. To understand this we need to prepare SNP maps that show the location of all the SNPs present in a single individual.

A pharmacogenomics study can be used to compare SNP maps and gene expression between normal and affected individuals. This is known as a genome-wide association study and can identify the genetic factors associated with the disease and provide new drug targets to study. Those that could be potential future drug targets are known as 'drugable'.

Identification of sub-groups of patients that will benefit from targeted therapy

Tamoxifen is a drug that was commonly prescribed to women with breast cancers that carry the receptor for the hormone oestrogen (ER+). However, 65% of women taking Tamoxifen developed resistance. There are several causes of resistance, but one group of women were found to have a mutation in the CYP2D6 gene that codes for the enzyme that metabolises Tamoxifen and so were not able to convert it to the active form of the drug.

Other types of breast cancer are characterised by the presence of extra copies of the HER2 gene and so produce higher than normal quantities of the growth-promoting protein HER2/neu that stimulates tumour growth (this is known as a HER2 positive tumour). The drug Herceptin® (trastuzumab) is a monoclonal antibody that targets the HER2 receptor. This blocks binding of the HER2/neu protein to the receptor and so growth of the HER2 positive tumour is prevented. There is no benefit to patients who do not have the over-expression of HER2/neu in being treated with Herceptin®. As Herceptin® – like most monoclonal antibody drugs – is very expensive, it is not given to patients with HER2 negative tumours.

> **KNOW IT**
>
> 1 Describe the differences between the following pairs of terms:
>   a hit compound and drug target
>   b polymorphism and single nucleotide polymorphism
>   c activity and efficacy
>   d pharmacogenetics and pharmacogenomics (a tricky one!)
> 2 What is meant by an 'unmet medical need'?
> 3 Why is it important to carry out structure-activity relationship (SAR) studies?

L01 Assessment activities

Below are suggested assessment activities that have been directly linked to the Pass, Merit and Distinction criteria in LO1 to help with assignment preparation and include top tips on how to achieve best results.

Activity 1 Identify the initial stages of a drug development programme *P1*

Consider the various factors that have to be addressed when designing and planning the development of a new drug. A new start-up company, Unigentech Ltd., is seeking funding to develop a promising new class of drugs that have been developed in a university. You have been asked to prepare a report or presentation for potential investors in the company. This report should summarise the various issues that Unigentech will face in the early stages of its work before they have a product that can enter clinical trials.

> **TOP TIPS**
> ✓ Bear in mind that you are writing for investors, who may not be familiar with scientific terminology, so don't use too much of it, or where you do, be sure to explain it.
> ✓ Be sure you give a balanced overview – don't focus too much on one or two issues.

Activity 2 Describe modelling techniques used in drug development *M1*

Modelling is central to the process of drug development. You will need to display an understanding of the techniques used, such as QSARs and computational methods, during the pre-clinical stage of drug development. Unigentech's investors are still interested, but they have commissioned a consultant to look into the detail of how their money will be spent. You have been asked to prepare a list of the techniques that Unigentech will use, with justifications, so that their consultant can evaluate this and prepare a budget for equipment purchase.

Activity 3 Evaluate pharmacogenomics approaches to drug development *D1*

Pharmacogenomics opens up the possibility of using genomic information to transform the drug development process. Unigentech believes that its expertise in genomics will make it quicker and cheaper to develop drugs up to the clinical trial stage. Prepare a report or PowerPoint presentation making Unigentech's case.

There is opportunity for links with the extension activity and assessment activities for LO4 in Unit 5 here.

LO2 Understand the range of techniques used in drug production
P2 P3 M2

A wide range of chemical and physical methods are used in drug production, depending on whether the drug in question is a natural extract or synthetic chemical, small molecule or biological.

GETTING STARTED

Types of drugs (5 minutes)

Divide into pairs and make a list of the drugs that you know about – one member of each pair should consider synthetic chemical (small molecule) drugs, the other should consider natural product or biological drugs.

KEY TERMS

Active pharmaceutical ingredient (API) – The actual active drug substance contained in the medicine.

Natural product – A drug substance extracted from or produced by an animal, plant or microorganism.

Process development – The application of chemistry to the scale up of new synthetic processes from the laboratory, through pilot plant to full scale commercial manufacture. It brings together disciplines such as synthetic organic chemistry, process technology and chemical engineering.

Good manufacturing practice (GMP) – The principles of GMP include following clearly defined standard procedures that are validated to ensure consistency and compliance with specifications. Detailed record keeping of all steps in manufacture, including documentation of quality of raw materials and intermediates, is essential. The intention of GMP is to reduce the risk to consumers and patients, particularly in ways that might not be detected through analysis of the final product.

Production of a medicine also includes the formulation stage, covered in LO4. For now we will consider just the synthesis of the **active pharmaceutical ingredient (API)**.

2.1 Techniques used in drug production

Extraction

Most **natural product** APIs are obtained by extraction. For example, the anti-cancer drug paclitaxel (Taxol®) is extracted from the bark of the Pacific willow tree. Others are produced by microorganisms such as fungi (e.g. yeasts) or bacteria. The first widely used antibiotic, penicillin, was originally extracted from a fermentation broth of *Penicillium* fungi. Bacterial and fungal fermentation broths are still useful sources of novel compounds for screening programmes and subsequent lead generation.

Traditional extraction methods rely on solubilising the material (plant, animal or culture of microorganisms) in a solvent or mixture of solvents. Extraction methods include:

- maceration – soaking a solid in a solvent
- digestion with proteases, cellulases or other enzymes
- distillation and steam distillation
- cold pressing.

When designing an extraction process, solvent selection is important and we need to consider some or all of the following:

- selectivity
- recoverability
- viscosity
- surface tension
- safety.

Organic solvents, particularly chlorinated hydrocarbons, have many desirable features, explaining their widespread use in the past. However, they also have associated safety and environmental disadvantages. Supercritical fluids are highly compressed gases that combine the properties of gases and liquids. Some of these, particularly supercritical CO_2, are becoming more widely used as solvents for extraction.

All natural products will be mixtures with other compounds from the source, often very complex mixtures. Having obtained a crude extract, further isolation and purification is required. This can involve some of the methods you will have studied in Unit 2 and is covered in the section 'Purification' below.

Synthesis

Although biologics are of growing importance, most APIs are still small molecules (organic chemicals), although they can be quite complex structures. This would probably be a good time to review what you learned in LO4 of Unit 1 (Understand the principles of carbon chemistry) because all of that, and more, will be encountered in the synthesis of APIs.

The chemistry might be the same as that carried out in a chemistry lab, but the scale of operation brings its own set of problems that have to be addressed. The chemical process used to manufacture an API will have started out as a laboratory synthesis making perhaps a few grams of product, but it is unlikely that this will be appropriate for multi-tonne synthesis – hence the need for **process development**.

▲ **Figure 11.3** Pharmaceutical factory. Technician monitoring reactions in a factory producing chemicals for the manufacture of drugs in the pharmaceutical industry

Some of the factors that need to be considered when designing a chemical process include the following:

- Availability and cost of starting materials.
- Overall yield needs to be maximised, which often requires reducing the number of steps compared to a laboratory synthesis.
- Intermediates need to be purified and characterised, partly to meet the requirements of **Good Manufacturing Practice (GMP)**. However, expensive purification techniques should be avoided where possible.
- The environmental impact of all processes, including use and disposal of solvents and waste products, must be assessed and minimised where possible.
- Time is money! It can be more economical to accept a lower yield if the reaction happens faster – especially if unreacted starting materials can be reused.

As well as addressing these issues, there is now a growing emphasis on Quality by Design (QbD), where all the processes involved in pharmaceutical manufacture (not just synthesis of the API) are designed to ensure quality of the final product and avoid or reduce any risks associated with the process.

Synthesis of biologics has its own challenges:

- They are manufactured in living systems.
- It requires large-scale fermentation or tissue culture.
- They are mostly large, complex molecules or mixtures – that makes it difficult, if not impossible, to fully characterise and analyse.
- Minor changes in the raw materials used, or the conditions for growth of the organisms, or harvesting and purification of the products could lead to potentially harmful changes in the API.

Purification

Purification involves separation of the desired products from any impurities, so the methods covered in LO2 of Unit 2 will be applicable here – bearing in mind the effect of scale.

Precipitation and crystallisation are two methods that scale well. Precipitation is often one of the early steps used in purifying a product obtained by extraction, e.g. a natural product, or a biologic made by fermentation. Crystallisation is used to both separate and purify intermediates and final products. Re-crystallisation is often used at the final step.

Solvent extraction is also widely used as a purification technique. This depends on the product being more soluble in one solvent (e.g. water) and the impurities being more soluble in another, immiscible solvent (e.g. a non-polar organic solvent).

Chromatographic methods are particularly useful in purification of biologics, but can also be used in purification of small molecule drugs. Methods used include the following:

- Gel filtration (size exclusion chromatography). Molecules are separated on the basis of their size.
- Ion exchange. This separates molecules based on their charge.
- Affinity chromatography. This makes use of a specific interaction between the substances being purified and the solid phase.

- High performance liquid chromatography (HPLC) and the related **fast protein liquid chromatography (FPLC)**. HPLC was covered in section 2.2 of Unit 2, although application on the large scale brings a whole series of issues that have to be addressed.

2.2 Analytical techniques in assessing the purity of products or intermediates

> **KEY TERM**
>
> **Fast protein liquid chromatography (FPLC)** – A type of HPLC (high performance liquid chromatography – see section 2.2 in Unit 2) that uses separation methods and materials more suitable for use with proteins than small molecules.

The requirements of GMP mean that it is necessary to analyse and characterise all intermediates as well as the final product. Once a synthetic method is established in production, strict adherence to a standard procedure should ensure that it is only necessary to analyse the final product. Nevertheless, every batch of API must be analysed using a range of different techniques. Some of these will include techniques covered in LO2 of Unit 2. However, there are others that are particularly useful in confirming the identity and measuring the purity of APIs.

Infra-red (IR) spectroscopy

When chemical bonds absorb energy in the infra-red region of the electromagnetic spectrum, it causes them to bend, stretch and vibrate. This absorption is measured in an IR spectrometer (the technique is known as IR spectroscopy). Each type of bond will absorb a particular frequency of IR radiation corresponding to the energy of that vibration – see Figure 11.4.

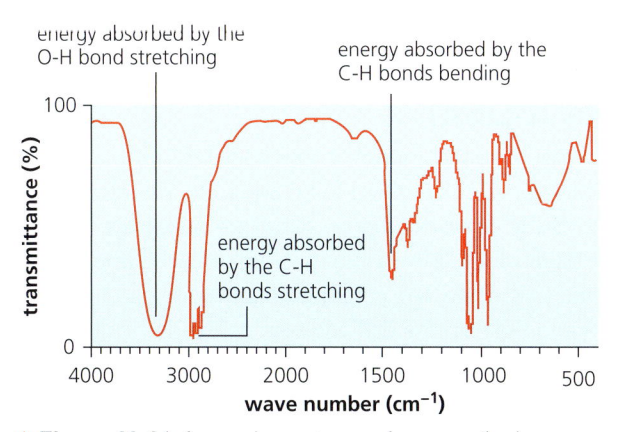

▲ Figure 11.4 Infra-red spectrum of propan-1-ol

IR spectroscopy gives a great deal of information about the types of functional group present in a compound, such as acids (-COOH), carbonyls (-C=O), alcohols (-OH) etc. Tables are available (search for 'IR correlation table') that help with interpretation of the spectrum. There is also a region of the spectrum, known as the fingerprint region, that is characteristic of each particular compound. This can be compared with a spectrum database to help identify a compound.

Low resolution and high resolution proton nuclear magnetic resonance spectroscopy (^{1}H-NMR)

NMR is a very powerful technique for characterising small molecule drugs. It can be used to confirm that we have made the compound that we set out to make and also to characterise any impurities; both are important in meeting the requirements of GMP.

In NMR, a powerful super-conducting electromagnet is used to generate a very strong magnetic field into which a sample is placed. There are various types of NMR, but proton NMR (^{1}H-NMR) is the most widely used. The protons in the sample can be aligned either with the field or against the field. Energy is required to make the protons flip from one alignment to the other. An NMR spectrum represents the energy absorbed by the protons when they do this.

The exact amount of energy absorbed by each type of proton will depend on the local environment of that proton. This allows us to work out what functional groups are present in the molecule, because different types of proton will appear in different regions of the spectrum. The area of peaks is proportional to the number of protons in each position, so we can tell the difference between – CH-, –CH$_2$- and –CH$_3$, for example – see Figure 11.5.

Protons in a molecule interact with each other. In high resolution NMR this can be seen as splitting of peaks into 2, 3, 4 or more. This allows us to work out which groups are next to each other. From this we can usually work out the whole structure of a sample.

> **EXTENSION ACTIVITY**
>
> **SpectraSchool (1 hour)**
>
> IR and NMR can appear quite daunting and it is possible that you will never need to use them in your work. However, if you like the challenge of solving brainteasers, you might find the Royal Society of Chemistry's SpectraSchool website interesting. It is available at:
>
> www.rsc.org/learn-chemistry/collections/spectroscopy

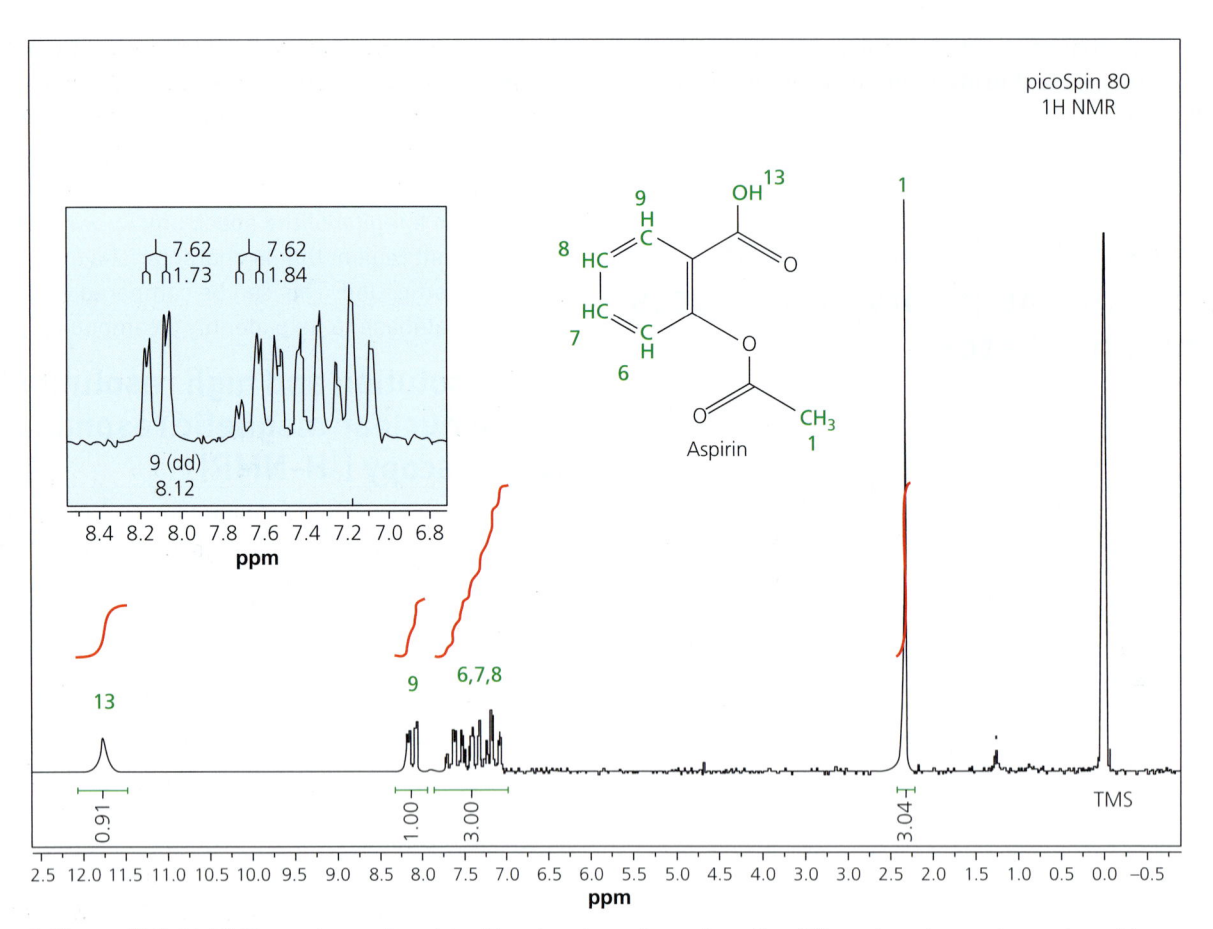

▲ **Figure 11.5** ¹H-NMR spectrum of aspirin. The structure shows how the different peaks can be assigned to the different protons in the molecule. The sloping green trace indicates the area of each peak or set of peaks

You will find a lot of information about different types of spectroscopy, but it is probably best to stick to IR, NMR and mass spectrometry (MS). If you are not put off by the explanations of the principles of spectroscopy, you could test your understanding with some of the interactive spectra. You could produce a short portfolio of spectra that you have interpreted, annotated with how you worked them out.

Biochemical and immunological screening

Biochemical and immunological screening methods are used widely, both as part of the drug discovery and development process and in assessing the purity of intermediates and products in the production process.

Colorimetric methods

Colorimetric methods can be used to measure protein concentration. Various stains will bind to proteins and the intensity of the colour can be used to calculate the protein concentration by using a standard curve prepared with a series of solutions of known concentration. The use of standard curves was covered in section 2.2 of Unit 2.

Other colorimetric methods can be used to measure enzyme activity. An enzyme might produce a product that will react with a reagent to produce a coloured compound. Alternatively, a model substrate could be used that released a coloured compound when cleaved by the enzyme. The enzyme reaction can be followed by the rate at which the colour appears.

Enzyme-linked immunosorbent assay (ELISA)

In enzyme-linked immunosorbent assay (ELISA), an enzyme is covalently bound to an antibody. The enzyme catalyses the conversion of a colourless substrate to a coloured product. If the antigen is present, the antibody will bind to it. Then, the colourless substrate can be added and the enzyme will catalyse conversion to the coloured product. The appearance of the colour indicates the presence of the antigen. The amount of colour indicates the amount of antigen presence. The technique is highly sensitive and can detect less than 1 ng (10^{-9} g) of protein antigen.

> **KEY TERMS**
>
> **Microtitre plate** – A plastic plate with 96 wells arranged in an 8 × 12 layout used for ELISA and other types of assay.
> **Primary antibody** – An antibody raised against a specific target antigen.
> **Secondary antibody** – An antibody that binds to the detection antibody.
> **Capture antibody** – An antibody bound to a well in a microtitre plate used to capture an antigen added to the well.
> **Detection antibody** – An antibody that is specific for a target antigen and will bind to it once captured on a plate.

An ELISA is done in the well of a **microtitre plate**. It starts with either an antigen or antibody attached to the well. Anything that binds specifically to the antigen or antibody will remain in the well and everything else can be washed away.

Direct ELISA is the simplest form and can be used to measure the amount of an antigen in a sample. The antigen is used to coat the well and then a **primary antibody** is added. However, the primary antibody has to have an enzyme attached to it so that the colour can be developed. This means that every primary antibody used must be labelled in this way.

Indirect ELISA is more useful, because it uses a **secondary antibody** labelled with the enzyme. The secondary antibody binds to the primary antibody, after which substrate is added and the colour develops. Secondary antibodies bind specifically to the constant region that is common to many different types of primary antibody, so just a few different labelled secondary antibodies are needed.

A sandwich ELISA allows us to measure the amount of antigen in a sample. In this case a **capture antibody** is used to coat the well. The sample is added to the well and the antigen binds to the capture antibody. Everything else is washed away. Then we add a **detection antibody** that also binds to the antigen. The amount of detection antibody (and hence the amount of antigen) can be measured by adding an enzyme-labelled secondary antibody that binds to the detection antibody.

These methods are illustrated in Figure 11.6.

As you can see, ELISA is a very powerful tool for measuring quantities of biological compounds. As such it is widely used in assessing the purity of biologicals during the production process as well as testing for contaminants.

ELISA can also be used for ligand binding assays. In such assays, the ligand is used to coat the well and then the corresponding receptor can be captured and measured. Alternatively, the receptor can be used to coat the plate and the ligand captured. If the receptor is a drug target, then the method can be used to screen compounds for binding to the target. This has become an enormously powerful tool in the drug discovery and development process, particularly as it can be automated.

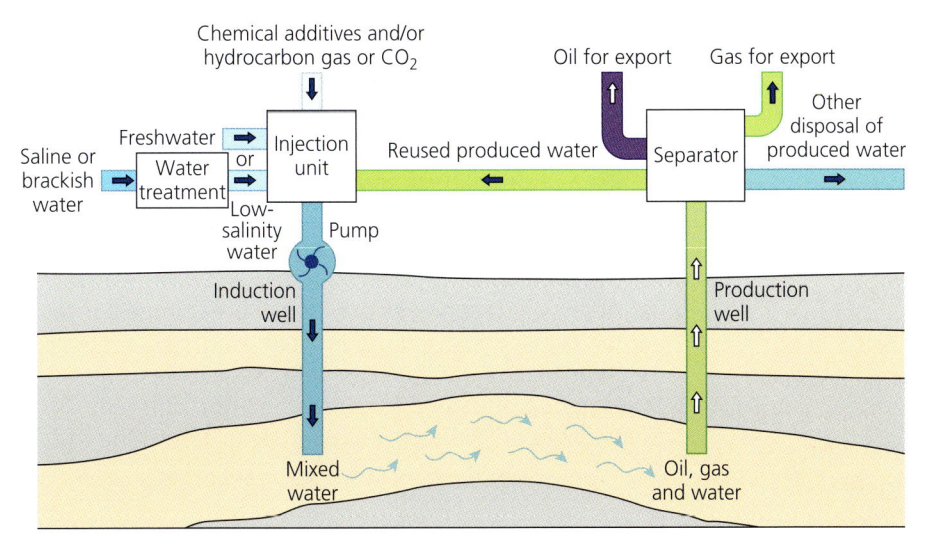

◀ **Figure 11.6** Direct, indirect and sandwich ELISA. After addition of each component (antigen, primary antibody, secondary antibody, etc.) the wells are washed to remove anything that is not specifically bound. The enzyme (E) converts an added substrate to a coloured or fluorescent product that can be measured

> **? THINK ABOUT IT**
>
> **Sandwich ELISAs (5 minutes)**
>
> In a sandwich ELISA, it is important that the capture antibody and the secondary antibody recognise (bind to) different parts of the antigen – why is that?

KNOW IT

1. How well do you know your abbreviations? Define the following:
   - GMP
   - IR
   - NMR
   - ELISA
   - HPLC
   - QbD

2. What is the difference between a capture antibody and a secondary antibody in ELISA?
3. What are the important factors to consider when developing a new chemical process for synthesis of a drug?

EXTENSION ACTIVITY

Virtual tours (30–60 minutes)

Chemistry on the large scale looks very different to what you would see in a school or college laboratory. Take virtual tours of synthetic and analytical laboratories and a chemical pilot plant to get an idea of how things work in the pharmaceutical industry. You can find these at www.abpischools.org.uk. Search for 'Lab & pilot plant tours'.

This online resource is published by the Association of the British Pharmaceutical Industry (ABPI).

LO2 Assessment activities

Below are suggested assessment activities that have been directly linked to the Pass and Merit criteria in LO2 to help with assignment preparation and include top tips on how to achieve best results.

Activity 1 Outline techniques used in drug production P2

The management of start-up company Unigentech have asked you to produce a summary (written report or PowerPoint presentation) of the techniques that will need to be used to put any potential products into production. At this stage it is not clear whether the final product will be a biologic or a small molecule drug. You could choose one or other, or even compare the techniques used in production of both types of product – this might influence their decision as to what path to take.

Activity 2 Describe analytical techniques in assessing the purity of products or intermediates P3*

(*Synoptic assessment – Unit 2 Laboratory techniques)

The Unigentech management has realised that it will need access to a range of analytical techniques throughout the development process. Prepare a report for them outlining the techniques that will have to be used. You could support this assessment with information (e.g. laboratory notebooks) about analytical methods you have used in the course of Unit 2.

Activity 3 Explain how analytical techniques are used in assessing the purity of products or intermediates M2

Expand on your report for P3 to include information about how the biochemical and immunochemical methods, including colorimetric methods and ELISA, that you have described will be used to assess purity of products or intermediates during the drug development stage.

TOP TIPS ✓

- ✔ You need to be familiar with the techniques used in drug production; if you are developing new drugs, they will have to go into production eventually.
- ✔ You should understand that extraction and purification techniques can be used to obtain drugs from natural sources. However, they are also important in the synthesis of drugs both in the course of a development programme and in the production process.
- ✔ Analytical techniques are also of great importance in development and production.
- ✔ Analytical techniques are seldom used in isolation. You need to understand how different techniques can be used together to determine the structure and purity of a drug.

LO3 Be able to carry out a basic extraction, synthesis, isolation and purification of a simple

3.1 Production processes of a simple drug or pharmaceutical

You now need to put into practice the techniques you learned in sections 2.1 and 2.2, depending on the facilities that you have available.

EXTENSION ACTIVITY

Production of a pharmaceutical (2–4 hours)

Working in pairs, you will produce a pharmaceutical compound. Your tutor will provide you with one or more options appropriate to the facilities you have available.

You might extract a compound such as salicylic acid from willow bark. This has been used for centuries as a folk remedy, but can also be converted to acetylsalicylic acid (Aspirin). Alternatively, you could synthesise Aspirin chemically.

You should document your work in your laboratory notebook. This should include details of the extraction methods used or chemical synthesis route and methods. You need to describe the methods you used to isolate and purify the product.

LO3 Assessment activities

Below are suggested assessment activities that have been directly linked to the Pass and Merit criteria in LO3 to help with assignment preparation and include top tips on how to achieve best results.

Activity 1 Demonstrate the use of a procedure to extract a pharmaceutical from plant, animal or microbial source P4*

(*Synoptic assessment – Unit 2 Laboratory techniques)

Extraction of drugs or lead compounds from natural sources is still of great importance in finding new sources of drugs, and some medicines are still extracted from natural sources. Your tutor will provide you with a procedure that is appropriate to the facilities you have available – this could be the one used for the Extension activity above. Carry out the extraction and use your laboratory notebook to document it. This task could also be incorporated into a practical assessment for Unit 2.

Activity 2 Describe the chemistry involved in the extraction of a pharmaceutical from plant, animal or microbial source M3

Use the extraction that you undertook for P4 and identify the chemical principles involved at each stage. Think about how the method could be improved and suggest stages where problems might have been encountered.

TOP TIPS
- ✔ If you haven't already done so, make sure that you learn good habits in keeping a detailed laboratory notebook.
- ✔ Think about what you are doing at each step in a procedure and why you are doing it.
- ✔ Think about sources of error and potential areas for improvement whenever you undertake practical work.

LO4 Understand the importance of product formulation and dosage form P5 P6 D2

GETTING STARTED

Formulations (5 minutes)

Discuss the different types of medicines you have encountered and think of the advantages and disadvantages associated with tablets, liquids, injections, inhalers, etc. Do any patterns emerge?

> 🔑 **KEY TERMS**
>
> **Formulation** – The process by which different chemical substances, including the active drug (API) are combined to produce a medicine. Formulation is a verb, i.e. 'to formulate a medicine', but you will often see it used as a noun, as in 'sustained release formulation'.
>
> **Enteric administration** – A route for drug delivery that involves absorption through the intestines.
>
> **Parenteral administration** – A route for drug delivery that involves injection, inhalation or absorption through the skin, i.e. a route other than the intestines.
>
> **Pharmacokinetics** – The study of how a drug is absorbed by the body, how it reaches the target tissues, how it is metabolised (broken down) and excreted.
>
> **Hygrophobic** – Intolerant of moisture.
>
> **Hygroscopic** – Hygroscopic substances absorb water molecules from the surroundings. In extreme cases this leads to them dissolving in the water they absorb.

4.1 The basic principles of pharmaceutics

Pharmaceutics is the study of the relationships between the **formulation** of a drug, its method or route of delivery, subsequent metabolism in the body and the clinical response to the drug.

Route of administering

In section 1.1 we divided drugs into small molecules and biologics and made the observation that small molecules could usually be given as a tablet, whereas biologics generally needed to be injected. Of course, that is a simplification. There are several routes of administration which generally fall into one of two classes, either **enteric** (via the intestines) or **parenteral** (via other routes).

Enteric routes are principally:

- oral (via the mouth)
- rectal (via the rectum).

Parenteral routes include:

- topical (usually via the skin)
- inhalation
- injection.

Bioavailability

Bioavailability is a measure of the rate and extent to which a drug reaches its site of action. A **pharmacokinetic** study can be done to obtain a plot of the concentration of the drug in blood plasma against time. Relative bioavailability compares the bioavailability of two formulations of the same drug. Think about the factors that might affect bioavailability.

4.2 Solubility and stability of pharmaceutical formulation

Solubility influences the bioavailability of a drug, but it also determines the formulation of the drug. In solid (tablet or capsule) formulations, the particle size influences solubility as well as the inherent solubility of the drug active ingredient. In addition, **hygrophobic** or **hygroscopic** solids should be protected from moisture.

The formulation will normally have been finalised by the time phase 3 clinical trials are reached (see section 5.1).

Expiration

All medicines have to be labelled with an expiration date. The expiration date is the last date that the manufacturer guarantees the full **potency** and safety of a medicine. This is determined following stability testing, usually under accelerated conditions such as elevated temperature. Factors that affect stability include:

- susceptibility to oxidation
- susceptibility to hydrolysis.

Hydrolysis can be controlled in solid formulations (e.g. tablets or powders) by using **excipients** with a low water content or using less hygroscopic salts. Hydrolysis is a particular problem in liquid formulations and it might be necessary for them to be stored at 4°C rather than at room temperature. The susceptibility of semi-solid formulations such as ointments or creams can be controlled by changing the nature of the base, usually an emulsion, to reduce the water content.

Shelf life

Shelf life is not the same as expiration date, although the terms are often used interchangeably. Shelf life relates to the quality of a drug over time, whereas the expiration date relates to both the quality and safety of the drug.

Factors influencing expiration also influence shelf life. As well as the excipients used in the formulation, these factors are the packaging and storage conditions.

Packaging, labelling and storage

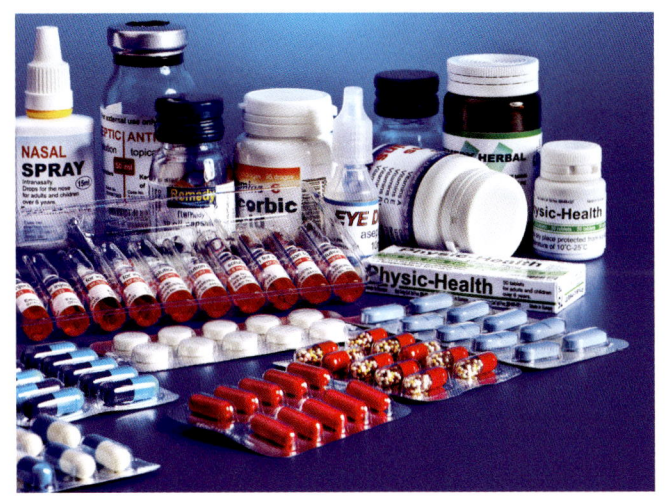

▲ Figure 11.7 A selection of different types of package used for medicines. Note the use of blister packaging as well as child-proof closures amongst others

Bulk APIs might be stored in fibre drums, but the packaging of a formulated medicine is more complex and depends on the nature of the formulation. Tablet or capsule formulations are the easiest to package, with liquids requiring more or less specialised packaging depending on the nature of the medicine. Consideration must be given to the material used, particularly when packaging liquids. Some drugs will bind strongly to glass or certain plastics, thereby reducing the concentration of active substance after a period of storage.

Part of the approval process for new medicines (see section 5.1) covers the labelling as well as the packaging of medicines, both of which are tightly regulated. In the UK this is governed by the Medicines and Healthcare products Regulatory Agency (MHRA) under UK and EU legislation.

PAIRS ACTIVITY
Packaging of medicines (20 minutes)
Working in pairs, research the features and advantages of the following types of packaging:

- blister packs
- child or tamper-proof packaging
- metered dose inhalers
- liquid formulations
- injectables.

Use the following publication to help in your research: www.gov.uk/government/publications/best-practice-in-the-labelling-and-packaging-of-medicines

Labelling
The Human Medicines Regulation 2012 – Part 13 covers the information and warnings that must be included on the label or in a Patient Information Leaflet (PIL) included in the product's carton.

Storage
Expiration dates and shelf lives assume storage under ideal conditions of temperature, humidity, light levels, etc. so it is important that these factors are controlled. Some medicines should be stored in a locked room, cabinet or safe, or in some other form of access-controlled facility.

KEY TERMS
Excipient – A component of a formulation other than the active ingredient. Excipients include bulking agents (tablets) or diluents (liquid medicines) as well as other substances that assist in manufacture or delivery of the drug.

Potency – Compounds with high potency (or activity) can be given in lower doses, which helps to reduce side-effects.

THINK ABOUT IT
Labelling of medicines (20 minutes)
Although there are regulations about labelling of medicines and the information that must appear on packages, there are instances of quite different medicines having similar packaging. This could be because the manufacturer uses colour and design elements to emphasise their brand rather than distinguishing between the different products. Think about the potential risks and produce a poster to draw attention to the issue. You could take some inspiration from the EZDrugID website (www.ezdrugid.org).

CLASSROOM DISCUSSION
Expiration dates (15 minutes)
Expiration dates are usually about two years from the date of manufacture or packaging. How would this date be determined? What tests would the manufacturer need to carry out?

There is evidence that many drugs are still safe and active after the expiration date. Why doesn't the manufacturer put a longer expiration date on the product? It may not be as simple as them wanting to make more profit! Discuss this as a class.

> **KNOW IT**
> 1. Routes of administration of a drug fall into two classes. What are they?
> 2. What factors affect the bioavailability of a drug?
> 3. What factors will affect the stability of a drug once packaged?
> 4. What body governs the approval process for medicines in the UK?
> 5. What regulations cover the information and warnings on drug labels in the UK?

LO4 Assessment activities

Below are suggested assessment activities that have been directly linked to the Pass and Distinction criteria in LO4 to help with assignment preparation and include top tips on how to achieve best results.

> **TOP TIPS**
> - Make sure that you link the types of formulation to factors such as route of administration and bioavailability.
> - Leave space on your wallchart to include information to meet the P6 activity.

Activity 1 Identify the basic principles of pharmaceutics *P5*

Development of an active ingredient (API) is only the start. Prepare a wall chart illustrating the different types of formulation and identifying the various advantages and disadvantages of each.

Activity 2 Identify methods used to increase stability of medicines *P6*

Use the wallchart produced for *P5* and add notes about the stability issues associated with each type of formulation. Then add information about the ways in which methods of formulation, packaging and storage can help increase stability. This might require a separate chart.

Activity 3 Evaluate the formulations of named drugs for oral, subcutaneous and intravenous administration *D2*

To achieve *D2* you will need to incorporate a well-informed evaluation of different formulations of drugs, pointing out the advantages and limitations of each of these criteria in the work you produce for *P5* and *P6*.

LO5 Understand the importance of planning clinical trials when introducing new drugs *P7 M4 D3*

GETTING STARTED

The drug development team (5 minutes)

Working as a group, make a list of all the different specialists (scientists / doctors as well as non-scientists) that will be involved in the development of a typical drug. Hint: The list is longer than you think! Keep a record of this list to use in the group activity.

> **KEY TERMS**
>
> **Clinical phase** – This involves testing of the drug in humans; there are three phases of clinical testing before the drug is approved for use as a medicine.
>
> **Validity** – The extent to which the variable being measured actually does measure the underlying trait being investigated. It is the ability to provide correct answers to the questions being asked.
>
> **Blinding** – This refers to the fact that the subjects don't know if they are receiving the placebo or the drug being tested. A double-blind trial is one where neither the subjects nor the medical staff know and so avoids bias.

> **KEY TERMS**
>
> **Reliability** – The reproducibility of a measurement when repeated at random in the same subject or specimen. It is often confused with validity. Reliability can be assessed as the proportion of all variation in a clinical trial that is not due to errors in measurement; it can be estimated from replicate measurements.
>
> **Placebo** – A medically ineffectual treatment used as a control in clinical trials. A placebo should represent the drug being tested as closely as possible, with the exception of the active ingredient.

5.1 Planning the development of new drugs

Development of a new drug can take ten years from identification of a lead compound to approval for use as a prescription medicine. Not only is it a long process, it is also very expensive. In 2010, it was estimated that the cost of developing a new drug averaged £1.15 billion and it is rising every year. There are several reasons for the high cost:

- The need for rigorous testing to ensure safety and efficacy.
- Most potential new drugs don't make it past the pre-clinical phase.
- Of those that do, many don't make it through clinical trials.

The timescales and costs of developing a new drug

Figure 11.8 illustrates the timeline, cost and number of candidates at each stage in the drug development process.

Clinical trials

The **clinical phase** of drug development is when the small number of drug candidates that emerge from the pre-clinical phase are finally tested in humans. The first phase will use healthy volunteers, but Phase 2 and 3 studies will be carried out in patients who have the disease that the drug is designed to treat (see Table 11.1).

Validity and reliability of clinical trials

The **validity** of a clinical trial depends on a number of factors:

- selection of participants
- control groups
- randomisation
- **blinding** and masking.

The **reliability** of clinical trials depends on the size of the study. Clinical trials should ideally be large and multi-centre to ensure reliability and reproducibility of the results and the statistical validity of any conclusions.

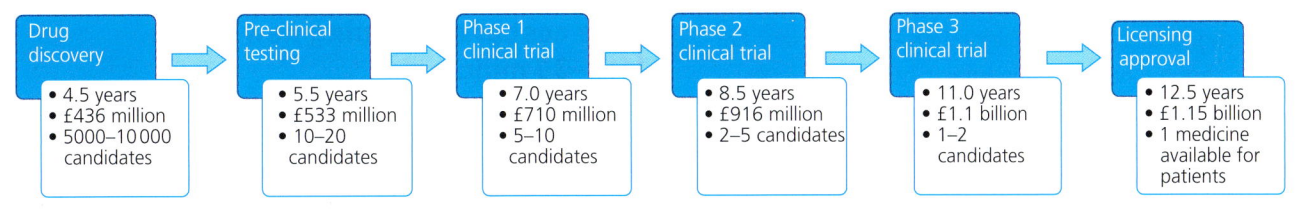

Drug discovery	Pre-clinical testing	Phase 1 clinical trial	Phase 2 clinical trial	Phase 3 clinical trial	Licensing approval
• 4.5 years • £436 million • 5000–10 000 candidates	• 5.5 years • £533 million • 10–20 candidates	• 7.0 years • £710 million • 5–10 candidates	• 8.5 years • £916 million • 2–5 candidates	• 11.0 years • £1.1 billion • 1–2 candidates	• 12.5 years • £1.15 billion • 1 medicine available for patients

▲ **Figure 11.8** Timeline of the drug development process. Each box shows the cumulative time and cost, as well as the number of candidate compounds until one medicine is available for use in patients

Table 11.1 Stages of clinical trials

Phase 1	To test the safety of the drug candidate. Usually done with healthy volunteers, typically 20–100. To reduce the risk, these start with low doses of the drug. The volunteers are closely monitored for side-effects and, if all is well, the dose will be gradually increased. This allows the doctors running the trial to judge what the best dose of the drug will be. This is known as a dose-ranging study.
Phase 2	Those drug candidates that show acceptably low levels of side-effects will then move into Phase 2. This is where the drug will be tested on larger groups of 100–300 patients with the disease to see if it has any biological activity. The early part of Phase 2 aims to identify the dosing requirement, i.e. how much drug should be given. The later part of Phase 2 is designed to study efficacy, i.e. how well the drug works at the required dose. The drug will usually be tested against a **placebo**.
Phase 3	Usually only about 20% of the drugs tested in Phase 2 trials will show sufficient promise to move into Phase 3 trials. These are large (300–3,000 patients depending on the disease) and carried out in many hospitals, often spread over several countries. The purpose of these trials is to compare the new drug to the currently accepted best treatment, and so the new drug will be compared to an existing drug rather than the placebo.

Successful clinical trials share similar characteristics:

- They are simple in concept and tailored to the patient group.
- They address questions of clinical relevance where genuine uncertainties exist.
- They avoid unnecessarily complex/restrictive entry criteria to ensure the results can be applied generally.
- They avoid unnecessarily complex data requirements.
- They ensure the most appropriate choice of control.
- They ensure lack of bias in allocation to test or control groups.
- They ensure robust blinding.

Statistical interpretation of the results

Statistical analysis of the results of a clinical trial is necessary to assess whether the differences in outcome in the test and control groups is significant or if it is solely due to chance. However, statistical analysis is not something that is simply bolted on to the end of a clinical trial. Appreciation of statistics is required at all stages:

- designing the trial
- choice of an appropriate outcome
- providing justification of the sample size
- choice of appropriate randomisation methodology
- drawing up a statistical analysis plan
- handling and structuring collected data.

Licensing approval

Not all drug candidates make it through Phase 3 trials. For example, rare but harmful side-effects might only show up when the drug is used on large numbers of patients. Assuming all goes well, one compound out of the original 5,000–10,000 (or more) will finally reach the stage of approval for sale as a prescription medicine. The regulatory authority for each country or territory is responsible for giving approval. In the European Union (EU) this responsibility lies with national authorities, such as the MHRA in the UK. The European Medicines Agency (EMA) is responsible for co-ordinating standards of approval across the EU. In the USA the Federal Drug Agency (FDA) is responsible for licensing approval. Huge amounts of data will normally be submitted in support of the licensing application and sometimes further research will be required before approval is given. Finally, after an average of 12.5 years and a cost of £1.15 billion, the drug will be approved to treat patients.

GROUP ACTIVITY

The drug development process (20 minutes)

In the Getting Started activity you made a list of all the specialists that you thought would be involved in the drug development process and your tutor may have added to the list. You are going to make posters for each of the stages in the drug development process. Divide into groups, one for each of all the stages in the drug development process (use Figure 11.8 as a guide). Each group should work on one poster. Show the specialists that would be involved in each stage and note, briefly, what their role would be. Present your results to the entire class.

PAIRS ACTIVITY

Taking sides (5–10 minutes)

This is a preparation for the Classroom Discussion. Working in pairs, prepare two lists: one will support the argument 'The cost of new drugs is too high' whereas the other will support the opposing position, 'The cost of new drugs is acceptable considering the cost of development'. Once you have made the two lists, decide which member of each pair will support which argument.

CLASSROOM DISCUSSION

The cost of developing new drugs (15–30 minutes)

You should be able to form two groups based on the work done in the pairs activity. One group should take the position 'The cost of new drugs is too high' whilst the other group takes the position, 'The cost of new drugs is acceptable considering the cost of development'. Take a few minutes to assemble the arguments prepared during the Pairs Activity and then use the remaining time to debate the issue. Try to keep your discussion factual and reasoned!

L05 Assessment activities

Below are suggested assessment activities that have been directly linked to the Pass, Merit and Distinction criteria in L05 to help with assignment preparation and include top tips on how to achieve best results.

Activity 1 List the steps in developing a new drug including an appreciation of timescales, costs and clinical trials *P7*

You need to demonstrate an understanding of the 'big picture', putting what you have learned into a wider context. You should consider the implications for society as a whole, as well as the need for effective regulation of new medicines. You could produce a wallchart showing the steps and timescales, either of a generalised drug, or one that you have researched. Annotate this with information about the costs of each stage and the reasons behind the whole process. Include information about what can go wrong (e.g. failure of a drug to meet safety or efficacy requirements) as well as the regulatory aspects of the process that should ensure new medicines are safe and effective.

Activity 2 Interpret the results of clinical trial data *M4*

Activity 3 Evaluate the validity and reliability of a clinical trial and justify their conclusions based on an understanding of data *D3*

Your tutor will provide you with sample data from a clinical trial. Prepare an assessment of this. To obtain Merit, you should be able to demonstrate the ability to interpret the clinical trial data to show what the conclusions are. For Distinction, you have to show an insight into the essential features of a valid and reliable clinical trial and the reasons why the double-blind crossover study is considered the 'gold standard'. An understanding of statistical methods will inform your interpretation – particularly the ability to identify statistical significance and use it to draw valid conclusions.

> **TOP TIPS**
> ✔ Make sure you have an understanding of the overall costs and timescales of drug development, as well as the order in which the steps are carried out and why that order is important.
> ✔ Make sure you can put into practice what you learned in section 4, 5 and 7 of Unit 3. To interpret and evaluate clinical trials, you need to be able to:
> - analyse and evaluate the quality of data
> - draw justified conclusions from data
> - record, report on and review scientific analyses.

Read about it

Goldacre, B. (2008) *Bad Science*, Fourth Estate

Ben Goldacre is a British doctor and science writer who has made his name unpicking the evidence behind claims made by drug companies, politicians, journalists and others. This book is an entertaining read as well as containing much useful information about the importance of evidence-based science, particularly in the field of medical research and clinical trials.

Here are some useful online resources:

www.abpischools.org.uk

An excellent range of online resources for schools and colleges published by the Association of the British Pharmaceutical Industry (ABPI). This includes virtual tours of chemical laboratories.

The Royal Society of Chemistry website (www.rsc.org) has a wide range of educational resources.

www.rsc.org/learn-chemistry is a good starting point.

www.rsc.org/learn-chemistry/collections/spectroscopy gives access to the SpectraSchool website. This covers all types of spectroscopy, but would be particularly useful in practising interpretation of IR and 1H-NMR spectra.

Unit 13
Environmental surveying

ABOUT THIS UNIT

Understanding environments is essential in many roles and industries, from conservation of vulnerable habitats or sites of special scientific interest, to built environment planning, industrial development, and mining and oil and gas activities. In order for this to be achieved effectively, a sound, evidence-based, scientific approach is required so that meaningful outcomes of studies are obtained, and informed recommendations enacted.

In this unit, you will take a practice-based approach to understanding, evaluating, interpreting and reporting on environmental information and data, drawing on your learning in fundamental science, and your practical and analytical skills. You will examine various types of environment and how they interact. You will then apply relevant scientific methods and analytical techniques in the field and the laboratory. You will survey different environments and information in the scientific literature to address scientific questions relevant to you, and report your findings and recommendations.

LEARNING OUTCOMES

The topics, activities and suggested reading in this unit will help you to:

1. Understand environmental impacts of human activity and natural processes
2. Understand environmental surveying
3. Be able to use field and laboratory techniques to conduct environmental investigations
4. Be able to analyse and present environmental survey findings

How will I be assessed?

You will be assessed through a series of assignments and tasks set and marked by your tutor.

How will I be graded?

You will be graded using the following criteria:

Learning Outcome You will:	Pass The assessment criteria are the Pass requirements for this unit.	Merit To achieve a Merit the evidence must show that, in addition to the Pass criteria, the candidate is able to:	Distinction To achieve a Distinction the evidence must show that, in addition to the Pass and Merit criteria, the candidate is able to:
1 Understand environmental impacts of human activity and natural processes	**P1** Describe how a human activity has impacted on the environment		
	P2 Describe how a natural process can impact on the environment		
2 Understand environmental surveying	**P3** Describe the purpose of environmental surveying for natural and human environments		
3 Be able to use field and laboratory techniques to conduct an environmental investigation	**P4** Conduct safety assessments of field activity and laboratory activities	**M1** Explain the factors which affect the choice of field and experimental techniques	**D1** Analyse results from the investigation
	P5 Carry out an environmental investigation, to include field and laboratory work which produces both qualitative and quantitative data		
	P6 Demonstrate the keeping of a field and laboratory log		
4 Be able to present environmental survey findings	**P7** Use statistical techniques to analyse experimental data	**M2** Explain causes of experimental uncertainty	**D2** Discuss the validity of the outcomes of the investigation, in terms of their statistical significance, and showing an awareness of correlation versus causality
	P8 Produce a scientific report on an environmental investigation	**M3** Recommend actions to positively affect the environmental implications of environmental investigation	

229

LO1 Understand environmental impacts of human activity and natural processes *P1 P2*

GETTING STARTED

What 'environment' means to you (10 minutes)

The term 'environment' can mean many things. To get started in this unit and learning outcome, think about what 'environment' means to you. Is it just the natural environment, or should you include your own 'human' environment? To what extent should you think locally or globally? Can you look at 'environments' in isolation, or do you have to think about how they connect?

KEY TERMS

Human activity – These are actions by humans which affect environments. These can be immediate, as in land clearance for agriculture or building, or as a consequence of activity (for example, mining activities might affect water and air quality). Human activities need not be negative, but can be positive as well, for example where action is taken to conserve or repair environments.

Natural processes – In contrast to human activity, natural processes arise as a result of non-human activity and events. This can include weather events, earthquakes, volcanoes and water transport and action (e.g. erosion, flooding) both at the surface and through underlying rocks.

Environments – This is a very broad term, but for the purposes of this unit, we can understand it as all natural and human habitats on Earth (air, oceans and land environments) as well as the sub-surface environment of underlying geology.

Greenhouse effect – the warming of Earth's surface and the air above it, caused by gases in the air that trap energy from the sun. These heat-trapping gases are called greenhouse gases. The most common greenhouse gases are water vapor, carbon dioxide, and methane.

Environmental impacts are the outcomes of both **human activity** and **natural processes**. **Environments** can experience negative impacts, as in oil spills or volcanic eruption, or positive impacts where human action works to preserve an environment, or new land becomes available for colonisation by life.

1.1 Increased human activity

The environmental impacts of human activities are varied, both in their cause, the degree of permanence and the level of severity. In general though, most impacts can be attributed to:

- human population growth
- human industrial activity
- human agricultural activity.

In this section, we will look at how each of these categories affects the environment, especially in the context of water quality and availability.

Human population growth

For much of human history, population growth has been limited by availability of food, lack of control of disease, and vulnerability to natural events. In consequence, the total global population was only about 5 million individuals up to about 5000 years before the **Current Era** (for BCE see Key Terms).

KEY TERMS

Current Era (CE) and Before Current Era (BCE) – These are the preferred ways to express dating for historical purposes. They are equivalent to the older terms 'Anno Domini' (AD) and 'Before Christ' (BC). As examples, the Great Pyramid of Giza was built in the decades around 2570 BCE, the Roman Empire annexed Britain in 43 CE and this book was written in 2016 CE.

Table 13.1 The top ten urbanised countries (information compiled by the World Bank, 2014)

Rank/Country	Percentage urban
1 Bahrain	100%
1 Malta	100%
1 Monaco	100%
1 Nauru	100%
1 Qatar	100%
1 Singapore	100%
7 Belgium	99%
7 Puerto Rico	99%
9 Guadeloupe	98%
9 Kuwait	98%

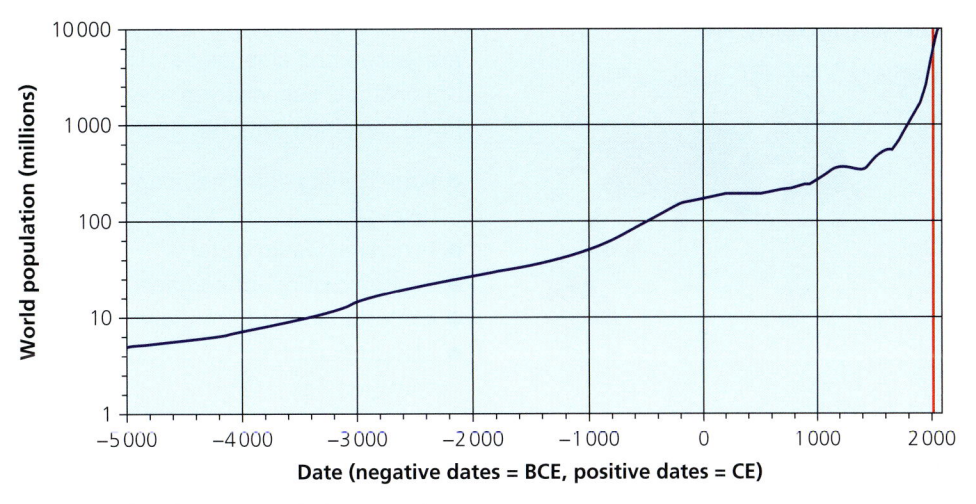

▲ Figure 13.1 Human population growth

Table 13.2 The top ten least urbanised countries (information compiled by the World Bank, 2014)

Rank/Country	Percentage urban
1 Burundi	10%
2 Papua New Guinea	13%
3 Trinidad and Tobago	14%
4 Sri Lanka	15%
4 Liechtenstein	15%
6 Malawi	16%
7 South Sudan	17%
7 Rwanda	17%
7 Nepal	17%
7 Ethiopia	17%

Impacts of human population growth – water use

Like all animals, humans require resources such as food, water, air and shelter to survive. In this section, we will look at human water consumption globally, and, importantly, examine the impact this has on water availability and on the transport and disposal of waste.

PAIRS ACTIVITY

How much water do you use? (10 minutes)

Spend a few minutes thinking about how much water you use per day. This will include drinking, cooking, washing, cleaning and using the toilet. You may be surprised at the result!

The figure you've obtained is fairly typical of a developed country; in fact, the UK has a relatively low figure, compared to similar countries such as France and Germany (see Figure 13.2).

From an environmental point of view, arguably a more useful measure is the water stress, which is measured as the percentage of total withdrawals of water to the total renewable supply (that is, rainfall or other precipitation supplying rivers, lakes and aquifers). The water stress by country is shown in Figure 13.3; a striking point is that water stress doesn't necessarily follow total populations. So, while we might not be too surprised to see that India (with a population in excess of 1 billion) suffers water stress, the fact that the mid-Saharan countries are not stressed, whilst Australia is, does need some thinking about.

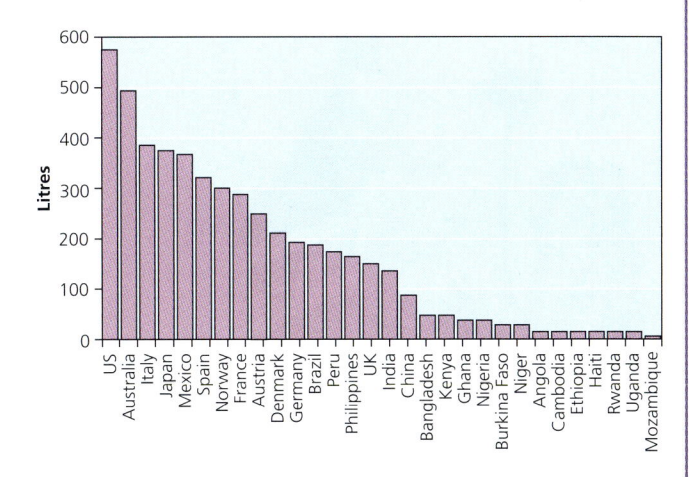

▲ Figure 13.2 Average daily water consumption per person by country

1.1 Increased human activity

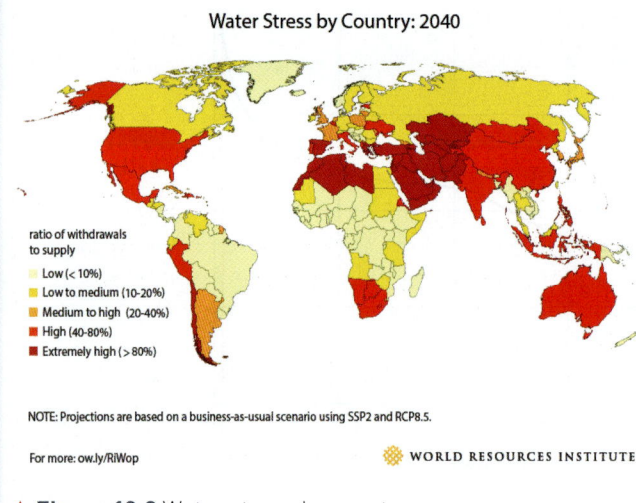

Figure 13.3 Water stress by country

Following extraction and use, waste water has to be processed and disposed of. This is an issue of great importance, especially in areas of high population and lower development. The main points of concern are:

- vulnerability of human populations to water-borne disease
- toxicity of waste water
- evolution of polluting gases and materials
- contamination of groundwater supplies
- contamination of downstream supplies for other communities
- damage to aquatic habitats and environments
- damage to wetland and estuarine habitats and environments.

CASE STUDY

The London sewer system

The issue of waste water came to prominence in the mid-19th century, with serious pollution of the River Thames in London due to sewage from the rising population and pollution from the various industries (such as leather tanning, food production and engineering). The issue was highlighted by Michael Faraday in 1855, who wrote:

"Near the bridges the feculence rolled up in clouds so dense that they were visible at the surface, even in water of this kind. ... The smell was very bad, and common to the whole of the water; it was the same as that which now comes up from the gully-holes in the streets; the whole river was for the time a real sewer."

Other impacts of the pollution of the Thames were:

- outbreaks of cholera and other water-borne diseases
- destruction of the Thames ecosystem, such that only bacteria and microorganisms which thrive on waste products remained.

Initial solutions to the problem included coating the curtains of the Houses of Parliament with lime to neutralise the smell, as well as spreading similar materials on the river itself, at great cost. These interventions though merely masked the problem.

A permanent solution took the form of an extensive underground sewer system, designed by the engineer Joseph Bazalgette, which, through a sequence of gravity-fed and pumped stages, discharged waste into the Thames beyond London (Figure 13.5).

Although improvements were evident, lack of regulation of industrial discharges meant the water quality remained poor. Further, damage to infrastructure due to bombing in the Second World War led to a decline in quality and the Thames was again biologically dead in the 1950s. Since this time, additional environmental regulation for waste treatment, transport and water use has restored the Thames to an environmentally clean state, supporting a thriving ecosystem.

Figure 13.4 A satirical cartoon of Michael Faraday studying the Thames

> **? THINK ABOUT IT**
>
> **Cleaning the Thames – unintended impacts? (30 minutes)**
>
> Whilst Bazalgette's system undoubtedly cleaned up the Thames through London, what problems can you envisage from a natural environment point of view?
>
> Points you might like to consider are:
>
> - Some tributaries of the Thames were incorporated into the system.
> - The waste was, at least initially, untreated.
> - Barges were used to transfer waste from storage areas to the North Sea to be dumped, untreated.
> - As a tidal river, water can flow back up the Thames as the tide comes in.

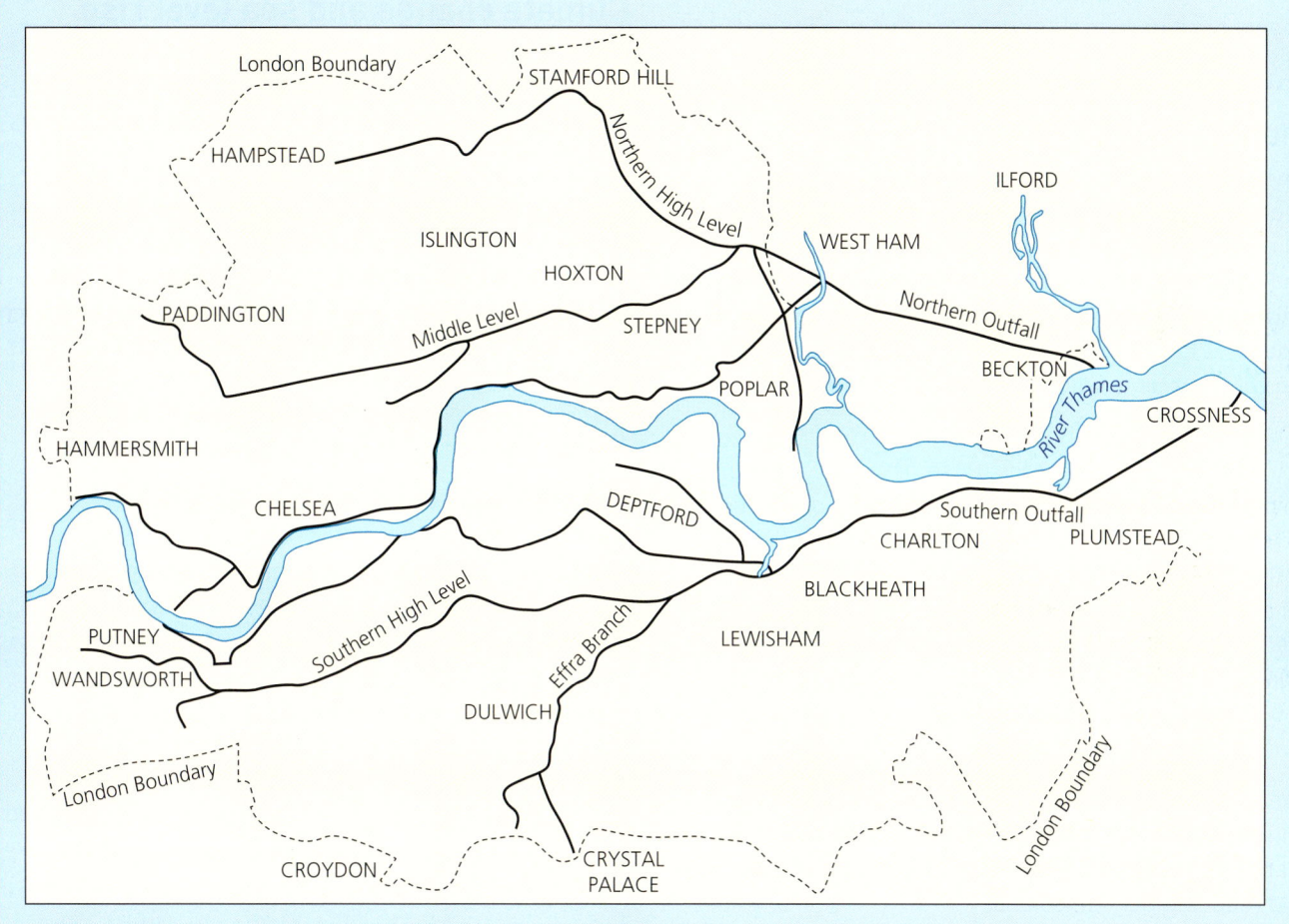

▲ Figure 13.5 The main sewers of London, as designed by Bazalgette

Impacts of human industry – water use

The impact of industrial activity has, at a local level, aligned closely (but not universally) with areas of high population growth and water usage. This is evident to a large extent in Figures 13.2 and 13.3, where regions such as northern Europe and east Asia have both high populations and either developed or developing industrial economies, together with moderate to high water stress.

> **GROUP ACTIVITY**
>
> **Water in industry (10 minutes)**
>
> In small groups, think about the purposes to which water might be put in one of the following industries:
>
> - mining
> - steel production
> - oil and gas production
> - construction.
>
> Consider also to what extent water is consumed (i.e. extracted and discharged back into the environment) or re-treated and recycled in industrial processes.
>
> Report back to your class your conclusions for your chosen industry.

Industrial and global transport

Since the industrial revolution in Britain, we have needed to transport goods by road, rail, air and sea all of which have an environmental impact such as air and noise pollution, loss of habitat to build the roads, railways and airports. Pollution is often worse near industrial towns but increasingly factories are built in the countryside causing increased traffic loads.

As populations increase, so does the demand for housing and cars for personal use. This puts pressure on green field sites that would normally be protected by legislation. Roads are sometimes built across national parks and can cut through sensitive habitats. An example of this was the Newbury bypass built in the 1990's which saw protesters camping for years in woodland to prevent the machinery moving in. In total 360 acres of land and 120 acres of woodland including mature trees were removed.

These days manufacturers and house builders try to minimise their carbon footprint by using locally sourced materials, using less fuel, more efficient power sources and reducing CO_2 emissions.

Deforestation, soil erosion and desertification

We often think about rainforests when we talk about deforestation but we forget that Britain was once covered in woodland. Humans have removed trees for agriculture, to graze animals and grow crops and the wood has a vast range of uses to us.

When trees are removed from large areas, they do not naturally grow back. Sometimes new types of vegetation move in such as moorland, but sometimes, without the roots and covers of the trees, the soil is washed away. This is known as soil erosion and can be a major problem for agriculture, causing farmers to add more fertilizers and chemicals to the soil to increase productivity. This in turn can lead to pollution of water courses and poisoning of wildlife.

Desertification occurs on a global scale usually in drier climates, it is caused by overuse of the land such as grazing and also global warming. Without top soil, plants cannot take root and are washed away by heavy rain or wind. The top soil is never replaced as decomposing organic matter is the main component of soil.

Climate change and sea level rise

It is now known that human activity is causing climate change and not just global warming, but may lead to increased rainfall leading to flooding and more storms.

We know that the planet has warmed by nearly 1°C. This is largely caused by the huge amount of CO_2 that we release into the atmosphere from industry, our homes, transport systems, burning fossil fuels, and farming. As the Earth warms, the impacts can fuel each other and cause irreversible changes. Polar ice reflects sunlight away from the earth, without it more heat is absorbed, thawing permafrost which then releases methane further increasing the **greenhouse effect**.

Climate change will impact on farming, making it harder to grow certain crops and may increase the number of crop pests and diseases. A temperature rise will increase melting of ice sheets in Greenland and Antarctica which will lead to sea level rise and cause the sea to become more acidic which kills coral and krill, destroying food chains.

Water use in industry

It is important to consider the volume of water required for a particular process and the consequent effects of extraction and discharge.

The oil and gas industry is a major user of water globally. At the same time, oil and gas production also yields water from the reservoirs as **produced water**, which requires proper handling to prevent contamination of surface and sub-surface water. As an example, Figure 13.6 illustrates the use and flow of water in the **secondary recovery** phase of a well. The purpose of the water here is to drive oil from the reservoir into the production well, and is typically used after **primary recovery** once reservoir pressure has dropped to a point where it is insufficient to drive the oil to the surface.

KEY TERMS

Produced water – Water is invariably trapped within the pore spaces of rocks when they are formed, along with the hydrocarbon reservoir. Produced water is water extracted at the same time as, and along with, the hydrocarbon resource. Produced water is normally saline with a high temperature by nature of its long residence time in the rocks and its depth. It is also heavily contaminated with both free and dissolved hydrocarbons, and may contain chemicals used in the extraction process, heavy metals, and naturally occurring radioactive material.

Secondary recovery – As the processes driving oil and gas to the surface become less effective over time (for example, the reservoir pressure falls as oil and gas are extracted) production moves into the secondary phase. Additional methods are now needed to drive oil and gas to the surface.

Primary recovery – In the first phase of production, natural processes within the well are sufficient to drive oil and gas to the surface. This can simply be pressure in the well, but can also include flows from other parts of the reservoir and the release of dissolved gases (by analogy, think about how a fizzy drink behaves when the bottle or can is rapidly opened).

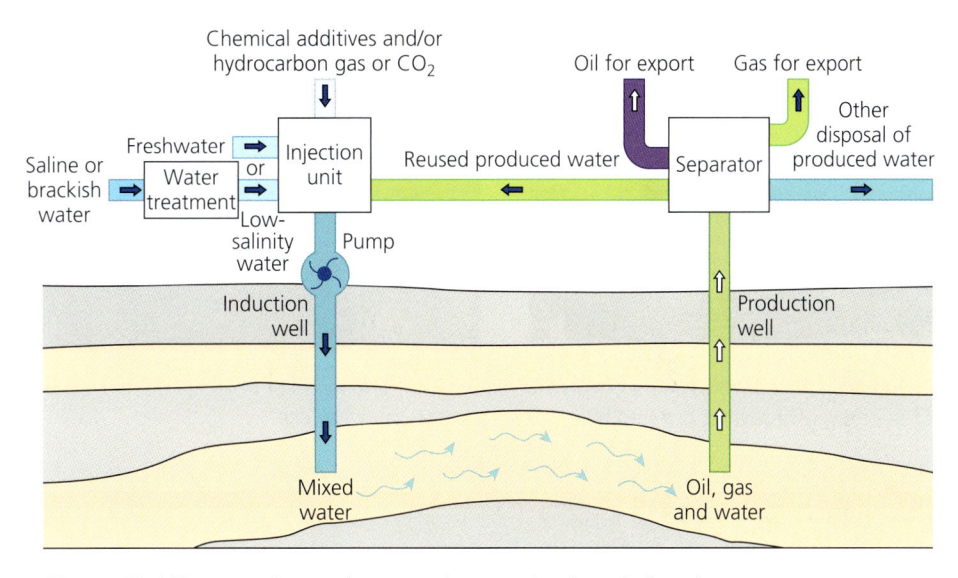

▲ Figure 13.6 The use of water in secondary production of oil and gas

From an environmental perspective, the key questions in a secondary recovery process are:

- How much water is needed to be extracted from sources?
- How much water can be reused?
- How much needs to be discharged?
- How should discharged water be treated?

The first two questions are related to the question of water stress we looked at earlier. Clearly, if an operation can reuse water efficiently, local water stress will be reduced. In regard to discharge, a general principle is that the amount to be discharged be minimised, and that contaminants must be removed as far as possible.

A good example of practice in these areas is the Alberta oil industry, which is subject to oversight from the Alberta Energy Regulator (AER). In particular, the document 'Water and Oil: An Overview of the Use of Water for Enhanced Oil Recovery in Alberta' (www.aer.ca/documents/applications/WA_WaterOil_UseOfWaterForEnhancedOilRecovery.pdf) discusses trends in water use over a 30-year period, how this has related to local water sources, cycles and uses. It is notable that, of the total water used in secondary recovery, about 83% is produced water, with only 17% of water required from other sources.

1.2 Environmental impacts of natural processes

As with human activities, the environmental impacts of natural processes are varied in their type, reach (local or global), scale of impact, timescale and consequences. Typical examples are:

- **volcanic** and **seismic activity**
- extreme weather and oceanic circulation
- erosion and water-borne processes.

In this section, we will look at these in turn, their impacts on both the natural and human environments, and how, where possible, the effects may be controlled.

Volcanic and seismic activity

Of all the natural processes, volcanic and seismic events are probably the most dramatic, and amongst the most damaging. However, they are also a vital part of the geological cycle by which new environments can be created. These two factors are illustrated in Figure 13.7, showing both the destructive and creative power of volcanic eruptions.

▲ **Figure 13.7** The destructive and creative aspects of volcanic activity. (a) The ruins of Pompeii destroyed by an eruption of Vesuvius in 79 CE and (b) Surtsey, off Iceland, created by a volcanic eruption in 1963 CE

🔑 KEY TERMS

Volcano – There are many types of volcano, but all of them result from deep fractures in the upper layers of the Earth, allowing material such as lava and volcanic ash to reach the surface, along with often toxic gases such as sulfur dioxide and hydrogen fluoride. Common types include:

- **Fissure vents**, where eruptions occur along a linear crack, rather than at a single point. Examples include Laki and Holuhraun, both in Iceland.
- **Shield volcanoes**, formed by eruptions of lava which flows long distances before solidifying. This gives these volcanoes their characteristic gentle slopes and low profiles. Examples include the volcanoes of the Hawaiian Islands.
- **Stratovolcanoes**, which form hills or mountains of alternating layers of lava and other materials such as ash. These are perhaps the most dramatic-appearing volcano type, and are responsible for many events damaging to human environments. Examples include Vesuvius in Italy and Fuji in Japan.

Seismic activity/Earthquake – This is a rapid release of energy, typically deep in the Earth's surface, resulting in surface movements from imperceptible to extremely violent. They are most common in areas where sections of the Earth's crust are moving relative to one another, especially around the Pacific rim ('The Ring of Fire'), Southern Europe and the Middle East, and the Himalayas. An earthquake results when stresses built up by sections of crust being unable to move smoothly are suddenly released; the movement may be vertical, horizontal or a combination.

The events shown in Figure 13.7 were largely local, restricted to the areas around the volcanic eruptions. However, global consequences are possible, especially in the case of explosive eruptions where significant amounts of material are ejected into the upper atmosphere. In historical times, there have been several such events, notably the Laki eruption of 1783–4 CE in Iceland which led to famine, destruction of crops and livestock, severe health issues and raised mortality, as well as global effects. Also, the Tambora eruption of 1815 CE on the island of Sumbawa in the Dutch East Indies, which was rated as a VEI-7 (Volcanic Explosivity Index – see below); a major event, which led to the so-called 'year without a summer' in 1816 CE.

Volcanic Explosivity Index (VEI)

The VEI is a scale designed to provide a measure of the relative sizes of volcanic events, including the amount of material produced and the height to which this is ejected. The scale runs from VEI-0, which covers volcanoes which erupt almost constantly at a low level (for example, Kilauea in Hawaii) through the eruptions of Vesuvius in 79 CE, Mount St Helens in 1980 CE and Eyjafjallajökull in 2010 CE (all around VEI-4 and VEI-5) to globally significant eruptions such as Tambora in 1815 CE (VEI-7) and mega-volcanoes such as Toba (about 74,000 BCE) at VEI-8, which has been credited with generating a period of sustained global cooling, with a possible link to human and some animal species near extinction before recovery of populations from very small gene pools.

Table 13.3 The Volcanic Explosivity Index (VEI)

VEI	Amount of material ejected / km^3	Description	Plume height / km	Material in lower atmosphere	Material in upper atmosphere	Examples
0	< 0.00001	Effusive	< 0.1	negligible	none	Kilauea, Hawaii
1	> 0.00001	Gentle	Up to 1	minor	none	Stromboli Island, Italy
2	> 0.001	Explosive	1–5	moderate	none	Sinabung, Indonesia (2010 CE)
3	> 0.01	Catastrophic	3–15	substantial	possible	Nabro, Eritrea (2011 CE)
4	> 0.1	Cataclysmic	> 10	substantial	definite	Eyjafjallajökull, Iceland (2010 CE)
5	> 1	Paroxysmic	> 10	substantial	significant	Vesuvius, Italy (79 CE)
6	> 10	Colossal	> 20	substantial	substantial	Krakatoa, Indonesia (1883 CE), Pinatubo, Phillipines (1991 CE)
7	> 100	Super-colossal	> 20	substantial	substantial	Tambora, Indonesia (1815 CE)
8	> 1000	Mega-colossal	> 20	vast	vast	Toba, Indonesia (74,000 BCE)

CLASSROOM DISCUSSION

What affects the impact of an eruption? (10 minutes)

The VEI scale is very useful in describing many eruptions, but there is some information it doesn't include. Discuss as a class the missing information and how important you think it is.

To help you with this discussion, you might like to think more about the Laki eruption. This was rated VEI-6, so would have ejected about ten times less material than Tambora, yet the effect was at least as devastating.

In particular, think about the timescale of the eruption. The major Tambora eruption essentially lasted about three months (April to July 1815 CE, though some activity lasted for several years) whereas the Laki eruption lasted from June 1783 to February 1784, with eruptions of the associated Grimsvötn volcano continuing to 1785.

It is also notable that many of the human and livestock fatalities were as a result of poisoning, notably fluoriosis caused by excess fluorine in the water, and that gas emissions (notably about 120 million tonnes of sulfur dioxide) were produced – similar to Tambora.

Often associated with volcanic events, **earthquakes** and other seismic processes have major environmental impacts. As with volcanoes, the magnitude of earthquakes is frequently represented by a numerical scale – the most common being the Richter Scale.

Richter Scale

The Richter Scale was devised by the seismologists Charles Richter and Beno Gutenberg in 1935, for the study of earthquakes in California. In its original form, it was based on a measure of movement of the ground, but for many purposes we can think of the Richter Scale in terms of the amount of energy released.

An important point to note is that the Richter Scale is not linear. Rather, each level of magnitude corresponds to about 32 times increase in energy release. As an example, we will compare the energy released in an earthquake which occurred in Rutland in England in 2015, with a magnitude of about 4, with the Christchurch, New Zealand earthquake of 2011, which had a magnitude of about 6.

Starting from magnitude 4, going to magnitude 5 would mean an increase of 32 times the energy. Similarly, going from 5 to 6 means another factor of 32. We therefore need to multiply these two factors, giving 32 × 32, which is about 1000.

We can therefore say that the Christchurch, New Zealand 2011 earthquake was 1000 times more intense than Rutland 2015.

From a human perspective, earthquakes greater than about 4 on the Richter Scale are frequently damaging to the human environment. For example, the Christchurch, New Zealand earthquake in 2011 was magnitude 6, and killed 185 people, and damage to buildings and city infrastructure – primarily because of its proximity to the city. Other notable examples include the 2004 Indian Ocean earthquake between Sumatra and the Andaman Islands at magnitude 9, resulting in the 2004 Boxing Day **tsunami** event. This resulted in the death, injury or displacement of around two million people in countries and islands around the Indian Ocean, and caused significant problems for coastal economies, industries and agriculture for many years afterward.

Arguably more serious was the 2011 Tohoku earthquake off the city of Sendai in northern Japan, also at magnitude 9 which actually moved the main island of Japan 2.4 m east, shifted the Earth on its axis by between 10 cm and 25 cm and triggered a severe tsunami event, destroying much of the east coast of northern Japan. In particular, the city of Sendai was almost destroyed with the tsunami travelling up to 10 km inland.

The seriousness of this event was exemplified by the damage caused to the Fukushima nuclear reactor complex, resulting in a cooling system failure, followed by a nuclear reactor meltdown and the release of radioactive materials. In December 2011, Tokyo Electric Power Company (Tepco) said 45 metric tons of radioactive water had been released into the ocean.

EXTENSION ACTIVITY

(30 minutes)

We've discussed the immediate aftermaths of two serious earthquake and tsunami events – the 2004 Indian Ocean tsunami and the 2011 Tohoku earthquake. It's important though to look at the longer-term effects, especially on the environments affected.

Research these longer-term effects, from either a human or natural environment perspective; as suggestions, you might like to choose one of the following:

- the destruction or creation of new habitats
- the effects on surrounding environments due to human displacement
- levels of ocean and land contamination (especially in the case of the Fukushima reactors.

Compare how the two areas affected have recovered.

Take a look at the historical and geological record – how frequently do such events occur?

Extreme weather and oceanic circulation

The Earth's climate system is dependent on the circulation of oceans. Oceans absorb heat from the sun in summer and redistribute the heat via the movement of oceanic currents and slowly release the heat during winter. The sea surface temperature drives anomalies in winds, weather and climate across the planet.

Many weather patterns are predictable, such as El Niño and La Niña, and we are getting much better at identifying changes well in advance using data models based on the collection of over 40 years of ocean data. Satellites are also used to monitor movement of weather patterns across the globe. This is very useful if extreme weather such as storms, floods, droughts or heat waves are imminent.

Erosion and water-borne processes

About 70% of the Earth's surface is covered by water. Even without the man-made effects of pollution, water has shaped the surface of the earth for billions of years. Most dramatically, glaciers have shaped mountain ranges and left lakes that still exist today. Rivers have carved gorges such as the Grand Canyon and formed huge land masses such as the Nile Delta.

Water is corrosive, dissolving the minerals in rocks, changing the pH of rivers and lakes which in turn affects the type of plants and animals living there.

> **KEY TERM**
>
> **Tsunami** – A tsunami (from the Japanese 'harbour wave') is a surface water wave caused by the displacement of large volumes of water, often due to rapid movements of the sea bed in an earthquake, though other events such as meteor impacts and landslides can have the same effect. Tsunamis are sometimes called 'tidal waves', but it is important to be clear that they have nothing to do with the normal tides caused by the sun and moon.

> **KNOW IT**
>
> 1 What human-related factors have the largest environmental impacts?
> 2 What natural events have environmental impacts on (a) local and (b) global scales?
> 3 Do the factors and events in your answers to the previous two questions affect constructively or destructively the human and natural environments? Explain your reasoning.

L01 Assessment activities

Below are suggested assessment activities that have been directly linked to the Pass criteria in L01 to help with assignment preparation and include top tips on how to achieve best results.

Activity 1 Describe how a human activity has impacted on the environment *P1*

Select a human activity in your local area and describe any positive and negative impacts it has had on both the human and natural environments.

Activity 2 Describe how a natural process can impact on the environment *P2*

Search for a natural event or process in your area and describe any positive and negative impacts it has had on both the human and natural environments.

> **TOP TIPS**
>
> ✔ Think about what kind of industries have operated, or are operating, in your area. For example, if you live in an area where mining is or has been carried out, what impacts are due to soil and water pollution or ground movements?
> ✔ Natural events can be as simple as a severe weather event. Perhaps there's an event you remember.

LO2 Understand environmental surveying *P3*

> **GETTING STARTED**
>
> **What science do we need to survey environments?**
>
> An important aspect of environmental surveying is how it connects with other disciplines. As a group, think of five or six other subject areas which environmental surveying might contribute to. Some of these may be surprising.

In the previous section, we looked at a range of environmental impacts caused by both human and natural events and processes. We will now look at the purposes and reasons for environmental surveying so we can understand when and how such a survey should be undertaken, and the kind of questions the survey should ask.

The kind of areas we'll look at are:

- planning developments (e.g. industrial and energy developments)
- infrastructure (e.g. transport, utilities transmission, waste handling, housing)
- health and well-being of the human population (e.g. detection and analysis of pollution, on-site and off-site monitoring of industrial activities)
- natural environment conservation and protection
- protection and conservation of regional environments and ecosystems
- global strategies for environmental management.

Of these, the first three are most concerned with the human environment, with the latter three connected to the natural environment which, of course, directly affects the human.

> **KEY TERMS**
>
> **Planning** – In the context of development and construction, planning is the legal process which ensures the need for a development and the suitability of the site and construction methods. It also ensures that impacts on local people, facilities and the human and natural environment are minimised.
>
> **Site of Special Scientific Interest (SSSI)** – A Site of Special Scientific Interest (Area of Special Scientific Interest in England, Scotland and Wales, or ASSI in Northern Ireland) is a legal designation protecting an area from development or change of use. The SSSI designation may be connected with any aspect of the area – for example, the area may contain a habitat that supports unique or rare species, have important monuments, contain interesting landscapes and geology, or be historically important.
>
> **Climate change** – Changes in the Earth's climate have occurred throughout its history, from high temperatures and little or no ice cover, as at the end of the Cretaceous period 65 million years ago, to extremely cold periods such as the most recent glaciation which ended about 10 000 years ago. The causes of climate change are varied and can be slow, such as those caused by changes in the Earth's orbit, or they can be rapid, due to, for example, massive volcano events or meteor impacts. In recent times, evidence has emerged which links human activities to possible climate change, in large part caused by CO_2 emissions from transport and industry.

2.1 The purpose of environmental surveying

When deciding on construction or development, at small local levels such as housing or road development, to larger regional and national levels such as power stations, rail infrastructures and airport expansions, the **planning** process must include environmental assessments as well as looking at and balancing other factors at a human, economic and political level.

Much of the planning process is based on legal procedures, ensuring all opinions and evidence has been considered. In this section, we will look at how the science of environmental surveying can affect and influence decisions before, during and after a development.

? THINK ABOUT IT

The Hindhead Tunnel

The A3 London to Portsmouth road in southern England is a major route, which has been progressively widened and improved. Up until 2011 CE, the only major section running through the centre of a town or village was the Hindhead and The Devil's Punch Bowl section in Surrey. This led to major delays on an increasingly busy road through the town of Hindhead, and disrupted a **Site of Special Scientific Interest** (**SSSI**).

Hindhead is a village in the south west of Surrey, with many parts in excess of 200 m above sea level. Historically, it was located on the main London to Portsmouth road. Nearby is the Devil's Punch Bowl, a deep hollow in the landscape. It is a SSSI due to the unusual geography and the presence of rare plants, insects and birds.

▲ **Figure 13.8** The Devil's Punch Bowl, Surrey

Plans for re-routing the A3 began in the late 20th century, with the importance of environmental surveying being evident in the process. In all, nine possible routes were proposed, but all were rejected on the grounds of economics, environmental concerns or community opposition. The final decision, to route the road through tunnels, had to balance a number of factors, for which environmental surveys were vital. The main questions to be asked by such surveys include:

What is the impact to nature conservation?
- What loss of natural environments would be caused by the re-routing? How would this balance against the restoration of the old road to natural land, taking the road and its associated pollution and disturbance away from a SSSI?
- How would the changes affect the pattern of land use (e.g. visitors to the area, dog walkers, etc.) and what is the impact of this on wildlife?

What are the impacts on water, soils and geology?
- The area has many springs arising from a natural aquifer, which also supplies local farms and some houses. Can the tunnel be constructed so that the aquifer is unaffected, and soil contamination avoided?

How would noise from the road and construction affect the human and natural environment?
- Can construction noise be minimised, or at least reduced, without compromising the construction schedule?
- What effect will the re-routing have on noise levels through the SSSI and surrounding area?

What effect will the re-routing have on the local communities?
- Both positive (e.g. reducing noise and traffic nuisance and re-connecting communities by removing heavy traffic through small towns and villages) and negative (e.g. reduced 'passing trade' for local businesses) need to be considered.

GROUP ACTIVITY

The Hindhead Tunnel environmental impact (30 minutes)

In small groups, investigate the outcomes of the environmental assessment, and report your group's findings back to the whole class.

Choose one aspect of the assessment (e.g. noise, conservation impact, community impact, etc.) and use the online resources given at the end of this unit, or find your own!

Use the questions raised in the main text as guidelines; can you think of anything else an environmental assessment might ask?

Environmental surveying in regional and global conservation

We've seen in the previous section that environmental surveying at a local level is vital in planning human developments and in minimising their impacts. Equally important is the need to assess environments at a regional level, and, indeed, globally. The kind of assessment here is different in that larger areas need to be considered, and the evaluations need to be continuous and ongoing, rather than one-off interventions. For example, in the local case of the Hindhead Tunnel, a limited evaluation of the conservation impact of moving the road was needed. However, in evaluating the conservation effects of increasing global temperatures, a wider set of data is required, so that impacts in one area can be understood in the context of others.

As an example, we'll look at **climate change**; in particular current and predicted effects of regional and global temperature changes. We can measure such changes based on historical observations and determining differences from a long-term average temperature. Figure 13.9 illustrates this, showing the global temperature anomaly (the difference between current temperature and a long-term average – in this case, the 30-year average between 1951 and 1980) for February 2015 CE.

An important point to notice in Figure 13.9 is that there are large regional variations, with Siberia and the North Polar

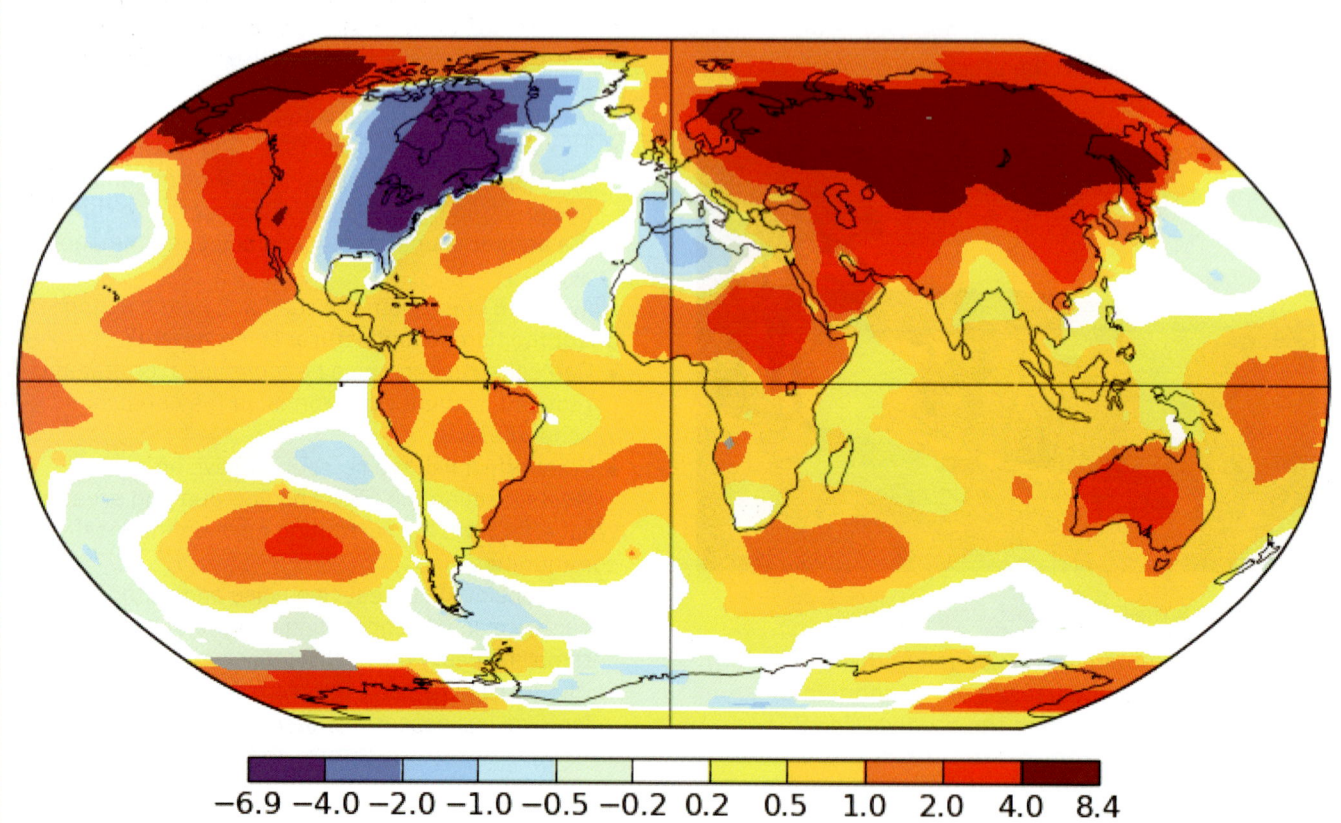

▲ **Figure 13.9** Global temperature anomalies for February 2015 CE, measured against the 1951–1980 30-year average

region in particular experiencing anomalies of between 4 and 8°C, whilst much of North America is experiencing a significant reduction in temperature by up to -8°C.

PAIRS ACTIVITY

Interpreting global temperatures (10 minutes)

Figure 13.9 illustrates the global temperature at a particular time. Think about whether you regard this as representative of changes in global temperature. You may like to look at weather reports from North America for February 2015 in particular.

If you think the figure is not representative, what information do you think you'd need to make a proper assessment?

In the Pairs activity, you thought about how representative one month's data is. You should reach the conclusion that it's not very useful since it contains one-off weather events (such as the cooling of North America). It is of more value to look at average anomalies over a number of years, such as that shown in Figure 13.10, for the period 2000–2009 compared to the 1951–1980 average.

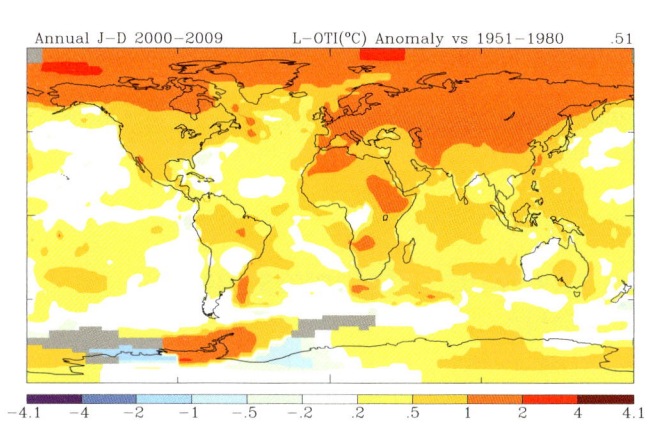

▲ **Figure 13.10** Global temperature anomalies for the period 2000-2009

We can now see that, far from cooling, much of the northern hemisphere, including North America, experienced warming by up to 2°C, with some cooling over parts of Antarctica and stable temperatures over parts of Africa and the Pacific Ocean.

Such global surveys can now be used to model and predict possible future environmental consequences, and inform additional surveys, research programmes and actions. For example, in the context of the polar regions (especially the North Polar region), we can ask:

- Are the anomalies consistent with:
  - natural climate cycles
  - human activities
  - or are there both natural and human contributions?
- What regional and global effects can we observe and predict? For example:
  - Can we observe changes to habitats and ecologies?
  - Can we observe any natural processes which might accelerate or mitigate the changes?
- What are the timescales of any regional and global effects?
  - Are changes on the timescale of centuries, or might events happen more rapidly?
- Will the effects be, overall, harmful or beneficial to human and natural environments, or a mixture?
- What regional or global strategies need to be enacted to address any harmful effects?

GROUP ACTIVITY

Investigating the effects of global temperature (30 minutes)

In small groups, investigate and think about the kind of global and regional environmental surveys you'd need to conduct to address the questions raised by the temperature surveys. Report your group's findings back to the whole class.

Choose one of the questions for your group (e.g. What are the regional and global effects?) and use the online resources given at the end of this unit, or find your own!

Remember, the surveys you suggest can be either field-based (that is, scientists and researchers taking measurements and observations on location) or remote-sensing (such as using satellite data to collect data such as temperatures, sea levels, sea ice thickness, etc.).

KNOW IT

1. What kind of information can environmental surveys provide with regard to (a) human well-being and (b) protecting the natural environment?
2. How do environmental surveys contribute to the planning process?
3. On a global scale, why are long-term studies more useful than one-off observations?

LO2 Assessment activity

Below is a suggested assessment activity that has been directly linked to the Pass criterion in LO2 to help with assignment preparation and includes top tips on how to achieve best results.

Activity 1 Describe the purpose of environmental surveying for natural and human environments *P3*

Select a case study of a development – for example, a new housing development, and describe why an environmental survey would be necessary.

> **TOP TIP**
> ✓ A good source of information will be your local planning office, where all developments, both for the human and natural environments, are assessed.

LO3 Be able to use field and laboratory techniques to conduct an environmental investigation *P4 P5 P6 M1 D1*

> **GETTING STARTED**
>
> **Choosing scientific techniques (10 minutes)**
>
> In this Learning Outcome, you will look at techniques of collecting and analysing environmental data, both in the field and in the laboratory. For each of a field and laboratory investigation, note down two or three advantages and disadvantages of these approaches.

So far, we've looked at how environmental impacts have driven the need for environmental surveys at local, regional and global levels, and we've looked at a few examples. These have ranged from severe pollution events (the 19th century CE River Thames), through natural disasters (tsunamis and earthquakes) to potential consequences of global temperature changes.

In this section, we'll look at some of the techniques used in environmental surveying, and some of the precautions we need to take to ensure our results and conclusions are relevant and valid. You will have encountered the relevant laboratory techniques earlier in this book, specifically in Unit 1 and Unit 2, but equally important are field techniques. These enable us to study environments in situ, with minimal disturbance, and, crucially, without removing any specimens in, for example, an ecological survey.

3.1 Objectives in environmental investigations

To conduct an investigation which has scientifically valid outcomes, we need to apply the scientific method. In outline, we first define our problem based on observations, conduct appropriate tests, draw valid conclusions and refine our knowledge accordingly.

In any environmental investigation or study, a clear idea of the questions being asked is vital, so that the techniques selected can match the data needed. It's also important to define the objectives. This isn't to say you're pre-judging the outcome of the investigation, but rather it reflects the idea that you should specify the general nature of these outcomes. As an illustration, we'll use a hypothetical investigation in this, and other, sections, which will help us understand these, and other, concepts.

An environmental investigation scenario

The village of Woodford Halse in Northamptonshire, England, hosted, until 1965 CE, a major steam railway depot. The land is now woodland, and the river Cherwell runs alongside the site.

After heavy rain, in particular, orange-coloured water can be seen even today, running into the river from the woodland, suggesting pollution is leaching into the river from buried contamination such as old iron and steel scrap and coal, oil and ash waste.

The river supports a thriving ecosystem, including crayfish, so evidently the pollution is not serious. However, it is still important to know what pollutants are present, and in what quantities, to understand and deal with problems further downstream.

▲ Figure 13.11 The River Cherwell in Woodford Halse, Northamptonshire. Note the orange muddy puddle at the bottom right

Our first task is to define a research question and associated objective. One question could be:

- Is there a different concentration of iron in the River Cherwell after it has passed through Woodford Halse, compared to before it enters the village?

The associated objective is then:

- To determine if the concentration of iron in the River Cherwell has increased after it has passed through Woodford Halse, compared to before it enters the village.

These two statements now define our investigation, and help us formulate the next step, a statement of **hypotheses**.

To measure the concentration of iron in the river, we will need to take multiple samples of water from both before the river enters the village, and after it leaves, possibly at different times of day and in different weather conditions. These samples will then need to be analysed in the laboratory to determine their iron content, and the results interpreted statistically. Going back to our question, we can rewrite this as an **alternative hypothesis**:

- There is a difference between the concentration of iron in the river measured before and after the village.

Remembering the definition of the **null hypothesis**, this is now written as:

- There is no difference between the concentration of iron in the river measured before and after the village.

In science, it's good practice to leave nothing to chance. This is especially true when taking measurements from the natural environment, where the outcomes may have important consequences. One procedure we'll need to do is validating any laboratory measurements we take by using **controls**, or samples which have well-known or tightly controlled properties. These can then be used to validate and calibrate our measurements. Also, the validation and calibration process can help us identify and control any **extraneous** or **confounding** **variables** which may arise.

> **CLASSROOM DISCUSSION**
>
> **Thinking about variables (20 minutes)**
>
> Thinking about the types of variables in this investigation, the **independent variable** is the locations at which the samples are taken and the **dependent variable** is the concentrations of iron in those samples.
>
> As a class, think about what (if any) extraneous and confounding variables there might be. Things to consider include:
>
> - the sample collecting procedures
> - sample collecting equipment
> - time of collection
> - ambient conditions (weather, river height and flow, etc.).

3.2 Safe working practices in the field and laboratory

In the units on laboratory and analytical techniques, we saw that a wide range of procedures are adopted within the laboratory to ensure both the safety and welfare of investigators and the integrity and reliability of the investigations. In laboratory-based environmental investigations (for example, studying collected samples), the same procedures and precautions apply, but with additional points to note:

- In normal laboratory investigations, you will be used to looking at, for example, samples which are purely chemical or biological in nature.

- You will therefore be applying safety and hazard control procedures relevant to those investigations.
- For environmental samples, you may well have a mix of sample types. For example, the samples collected from a river to monitor pollution will most likely contain both chemical and biological hazards.
- You will therefore need a higher level of hazard awareness when handling and analysing such samples. For example, although you may be interested in the chemistry of the sample, you will need to assess and apply biological hazard controls.

At some point, most environmental surveys will require field work, which requires a new set of risk assessments and hazard control skills. As an example, we'll look at the risks, hazards and controls which are enacted for the survey being undertaken in Figure 13.12. (see the next page).

For this scenario, we first need to identify the hazards. These include (but are not limited to):

- proximity to a cliff edge (note the sea in the background)
  The main hazards here are the potential for falling or, as may happen especially after a storm, cliff collapse due to erosion.
- exposure to weather, which may include excessive sun exposure or, in poor weather, hypothermia
- possible uneven surface
  Whilst not visible in this image, there is evidence of rabbit activity on this site, with deep holes forming trip hazards.
- biological contamination
  possibly from pathogens in the soil and plant toxicity
- fragile soil and ecosystem
  This is a hazard that you, as an investigator, are bringing to the environment. The soil structure and ecosystem may be easily damaged by the investigation.

We can now assess the level of risk and control measures associated with each of these hazards. These are summarised in Table 13.4, defining levels of risk from 'Trivial', where an event is unlikely, with low severity of harm, to 'Substantial' for highly likely events leading to severe harm.

> **KEY TERMS**
>
> **Hypothesis** – A proposed explanation for a phenomenon or observation. We will also use the term 'hypothesis' when talking about statistical tests, where we're looking at the validity of outcomes based on probability.
>
> **Alternative (or experimental) hypothesis** – Contrary to the null hypothesis, this says that there is an effect or there is a diffference.
>
> **Null hypothesis** – In statistics, this is the hypothesis that says there is no effect or no difference between two measurements.
>
> **Control** – A scientific control is a method for minimising the effect of variables in an experiment. In an environmental context, it could be as simple as using a well-known sample to calibrate or verify the correct working of a procedure, or it might be more complex, such as splitting an area under investigation into sections where interventions take place (the experiment, for example, cultivating a field) and those where no interventions are carried out (the controls, for example, leaving a field fallow).
>
> **Extraneous variables** – These are variables which we are not directly interested in, but still affect the outcome.
>
> **Confounding variables** – These are similar to extraneous variables, but potentially more serious. Whereas the extraneous variable is something external to the investigation, a confounding variable changes along with the variables we are trying to control and measure.
>
> **Variables** – In science, a variable is any quantity which we can measure or control. There are several types of variable we need to think about such as the ones below.
>
> **Independent variable** – A variable (often denoted by x) whose value does not depend on that of another variable or the investigator has no control over. For example, in an investigation to look at carbon dioxide emissions from an industrial site, the independent variable might be time of day.
>
> **Dependent variable** – A variable (often denoted by y) whose value depends on that of another variable. In an experiment, we usually measure the independent variable.

▲ Figure 13.12 A survey of plant species present on part of the South Downs at Birling Gap, Sussex

Table 13.4 Example risk assessment for an environmental field study

Hazard	Severity of harm	Likelihood of harm	Risk level	Control measure(s)
Proximity to cliff, with possible fall hazard or cliff collapse hazard	Extremely harmful	Unlikely	Substantial	Ensure situational awareness and remain alert; do not work alone; define a minimum distance from the edge of the cliff or set up a 'safe zone' for work.
Exposure to weather (hypothermia, over-exposure to sun, etc.)	Harmful	Unlikely	Moderate	Ensure correct clothing is worn for the weather conditions; in hot conditions, ensure water is available.
Uneven surface (rabbit holes)	Slightly harmful	Highly unlikely	Trivial	Ensure situational awareness (in particular, do not write/take notes while walking); use sound footwear.
Biological contamination	Harmful	Highly unlikely	Tolerable	Ensure all cuts or grazes are covered; use disposable gloves if handling specimens.
Damage to fragile ecosystem	Harmful	Likely	Substantial	Do not remove specimens unless necessary; limit the area of the survey as far as possible; do not leave any litter or equipment at the site.

> ### 💬 CLASSROOM DISCUSSION
>
> **Assessing risks and hazards in environmental surveying (10 minutes)**
>
> Assessing risk is a balancing act between ensuring the safety of investigators and environments, and enabling investigations to take place.
>
> - Do you think the risk levels for the hazards in Table 13.4 are sensible?
> - Do you think any are over- or under-estimated?
> - Are there circumstances where one or more of the risks would become intolerable?
> - Can you think of other control measures you'd put in place?

3.3 Data collection and analysis techniques

In earlier sections of this unit, we looked at a number of case studies and scenarios of environmental events and surveys. You may well be asking how such data can be collected and processed, given the complexity of environmental systems and, when thinking about global surveying, the sheer scale.

Where we conduct laboratory analyses of environmental samples, we identified in the previous section a range of additional safety precautions we need to enact to ensure safe working and sample integrity. Apart from this, on the whole, we can draw on our experience from previous laboratory work as far as techniques are concerned, though there are a few specific methods we'll find useful when analysing for certain pollutants.

Chemical tests and indicators for common pollutants

We saw in Unit 2 that we can identify the presence of certain ions in water by using suitable indicators. These then show the presence of a particular ion through a reaction forming either a precipitate or a colour change.

Looking back to Figure 13.11, and our suspicion of iron pollution in the River Cherwell, we can identify the presence of iron in water through addition of sodium hydroxide to a sample. The resulting iron hydroxide is insoluble, and appears as a coloured precipitate (for iron, green or red depending on the oxidation state of the iron).

This reaction is qualitative, in that it tells us that iron is present (and which oxidation state it's in), but not the concentration.

To find the concentration, one simple method is to use titration, which we've encountered before when we investigated acids and bases, but using different reactants involving iron.

An example is the use of potassium permanganate ($KMnO_4$) which forms an intensely violet solution. Reacting this with dissolved iron (specifically Fe^{2+} ions) reduces the permanganate ion ((MnO_4)$^+$) to colourless manganese(II) (Mn^{2+}) ion.

Suppose our contaminated river sample contains both iron(II) and iron(III) ions. We'll use a 50.0 cm³ sample, which we'll treat with zinc metal to reduce all the Fe^{3+} to Fe^{2+}. We'll now suppose that the resulting solution is titrated with 0.0010 M $KMnO_4$, with 15.0 cm³ required to reach the end point.

The stoichiometric relationship of permanganate to Fe(II):

$5\ Fe^{2+} + 8\ H^+ + MnO_4^- \rightarrow 5\ Fe^{3+} + Mn^{2+} + 4\ H_2O$

So 1 mol of MnO_4^- reacts with 5 mols Fe^{2+}

Calculate mols of Fe^{2+} reacted:

Multiply the volume of permanganate solution used by the concentration:

0.0010 mol dm⁻³ × 0.0150 l = 0.000015 mol of MnO_4^-

Multiply this by 5 to obtain the number of mols of Fe^{2+}:

0.000015 mol Mn × 5 mol Fe / 1 mol Mn = 0.000075 mol Fe^{2+}

Determine the molarity of Fe(II) – remember, this represents the total amount of iron:

0.000075 mol / 0.050 dm³ = 0.0015 mol dm⁻³, which corresponds to about 84 mg dm⁻³.

According to World Health Organization (WHO) standards, this would be classed as heavily polluted.

Statistics and interpreting data

In the previous section, we calculated a single value for the concentration of iron present in our river sample, and concluded that it exceeds WHO standards. Whilst this might make a good headline in a news report, scientifically, we should be asking:

- Is this value typical of the river as a whole?
- Is the value repeatable?
- What uncertainties are there in the measurements?

The first depends on our sampling method – if we just sampled one part, maybe where we thought there'd be most pollution, then no – it isn't. Likewise, if we returned to the river after, say, heavy rain, the value from the same place might be very different, so we can't guarantee repeatability. Finally, remember that uncertainties are a part of science – for example, if we're using visual methods such as looking for a colour change in a titration, there may be some variation in your assessment of the end point.

The above tells us that to make sure our results are valid, we need to repeat them. This should involve:

- Making sure we take samples from several parts of the river selected at random, not just parts which are obviously polluted. This makes sure we're not unconsciously biasing our results.
- Analysing each sample several times (making sure we collected enough to do this). This helps us uncover any random uncertainties, such as our observing the end point of a titration.

As an example, let's suppose our measurements at several locations are as in Table 13.5. Note that we've assumed that several measurements for each location were recorded, and that the values in the table are averages of these. We've also realised that our measurement of 84 mg dm^{-3} was from a particular location – the polluted muddy puddle in Figure 13.11, and hence is not representative. Additionally, let's suppose we recorded locations 1–4 downstream of the possible pollution source, and locations 5–8 upstream.

Table 13.5 Iron concentrations at upstream and downstream locations

Location	Concentration of Fe, mg dm^{-3}
	Downstream
1	3.5
2	5.3
3	4.1
4	3.2
	Upstream
5	2.1
6	3.1
7	3.7
8	2.5

Recalling our null hypothesis regarding iron pollution in the river, there is no difference between the concentration of iron in the river measured before and after the village.

We can now think about testing this. At first sight, it may seem obvious that the null hypothesis can be rejected; the downstream values tend to be higher than upstream. However, we have to accept that there is a possibility that the outcome we see could have happened by chance, and we can quantify this through a statistical test.

Many tests have been developed, but a common one for many circumstances is the **Student's t-test.**

KEY TERM

Student's t-test – This is a statistical test developed by the mathematician and chemist William Gosset (who published his work under the pen-name 'Student') in 1908. It's commonly used to test whether the means of two sets of values are equal, and place a probability on the outcome occurring by chance.

The t-test can be evaluated by hand, or, more commonly, using a statistical or spreadsheet computer package or online calculator. A suitable example of an online resource is: www.graphpad.com/quickcalcs/ttest1/.

If we apply the Student's t-test to our data, using the online resource given in the Key Term box, we obtain a probability, or P-value, of about 0.09.

To understand the meaning of this, note that the t-test is, mathematically, asking what the probability is that the means of two sets of numbers are the same; in effect, a null hypothesis. For our example, we've found $P = 0.09$, which means there is a 9% chance of observing a difference as large as you observed even if the two population means are identical, i.e. a 9% probability that our result is due to chance.

GROUP ACTIVITY

Discussion (20 minutes)

In our example of river water samples, we used a statistical test (the t-test) to determine the probability that our result was due to chance.

As a group, discuss:

- Whether you regard the value of 9% as sufficient for you to reject the null hypothesis in this case.
  - You might like to consider the purpose of the investigation – e.g. for legal processes (e.g. pollution prosecution), health reasons (e.g. drinking water) or an academic environmental study.
  - You might also investigate recommended P values for various disciplines or applications.
- What steps you could take to improve (i.e. reduce) the P value.
  - What value do you think would be suitable?

KNOW IT

1. What is a null hypothesis?
2. What types of variable do you have to consider in an experimental investigation?
3. In a field study, why is it important to record observations or take samples from several locations?
4. State three reasons a safety assessment is important in any field or laboratory investigation.
5. Why, scientifically, is it important to conduct relevant statistical analyses?

LO3 Assessment activities

Below are suggested assessment activities that have been directly linked to the Pass, Merit and Distinction criteria in LO3 to help with assignment preparation and include top tips on how to achieve best results.

Activity 1 Conduct safety assessments of field activity and laboratory activities P4

When planning your field and laboratory investigations, record safety assessments of each part in your log book.

Activity 2 Carry out an environmental investigation, to include field and laboratory work which produces both qualitative and quantitative data P5

As a class, you will define an environmental investigation. You should participate fully in the investigation your class defines, both as an individual and as part of a group.

Activity 3 Demonstrate the keeping of a field and laboratory log P6

You should record all of your observations, activities, data and relevant information (such as field study locations) in an experimental log book. This log book in itself will provide the evidence for this assessment.

Activity 4 Explain the factors which affect the choice of field and experimental techniques M1

To achieve this assessment, you could ensure you recorded in your log book your contribution to the discussion of the design of the investigation. In this, you should have noted factors such as choice of locations, sampling protocols and laboratory analysis techniques, and the reasons for their selection.

Activity 5 Analyse and draw conclusions using results from the investigation D1

Using the results you obtained from your investigations, select the methods (eg, statistical methods) you need. Use these methods to analyse your results and record your key outcomes in your log. You can then follow this up by deciding on relevant conclusions which your outcomes support.

TOP TIP
✔ The key to all the assessments in this Learning Outcome is keeping a good log book.

LO4 Be able to analyse and present environmental survey findings P7 P8 M2 M3 D2

Many of the aspects of this Learning Outcome are covered in earlier outcomes as well, especially statistics and analysis of data. We will therefore concentrate in this section on two aspects which are vital for good science:

- Reporting or communicating your results and conclusions.
- Critiquing your outcomes, with a view to assessing the validity of the data and improving your results.

GETTING STARTED

Why do we communicate science (10 minutes)

In science, it isn't enough to collect and analyse data; reporting and presenting your outcomes is vital for the whole scientific process.

Individually, think of three reasons why reporting and presenting scientific outcomes is important, and report back to your group.

4.1 Reporting science

The environmental survey findings which you've carried out as part of this unit comprised both descriptive or narrative elements as well as quantitative results and outcomes. How you'll be preparing and presenting your results for this unit will vary, but two common methods are:

- a scientific report or paper
- an oral or conference presentation.

Which method you choose, and how you present the information, will depend on the target audience. Factors to consider will include:

- the use of language (formal or informal)
- the use of jargon or technical terms
- the types of figures or graphics used to display visual information.

PAIRS ACTIVITY

What is your target audience? (15 minutes)

In pairs, select one of the target audience, and think carefully about the type of reporting which would be most suitable. Why would other types of report be less suitable?

The scientific report

A report of this type is usually prepared as a written, formal paper and would normally be targeted at the scientific community. It's also important to remember that scientific reports are often 'peer-reviewed', meaning they are scrutinised by colleagues in the scientific area of study (usually anonymously) to ensure scientific rigour is maintained.

A typical structure of a scientific report is shown in Table 13.6, together with some guidelines for writing the content.

The scientific presentation

Scientific presentations are often given at conferences or public events, and are considerably less formal than a written report. They are, however, no less important, providing you with a way to communicate your ideas to an audience, and give the audience a chance to think about your work.

There are no hard and fast rules for giving presentations, though there are plenty of guidelines available for what makes a good presentation (and, indeed, what not to do!) Here are a few tips.

- Prepare your material carefully and logically. Give a clear narrative with four parts: (a) Introduction (b) Method (c) Results (d) Conclusion/Summary.
- Don't include too much material. Have one or two key points you want to get over, and concentrate on those. As a rough guide, a single slide in a presentation should take between 1–2 minutes to talk about.
- Likewise, have only a few conclusion points.
- Talk to the audience, not the screen. This can be daunting, especially in a large conference theatre. A good way of addressing this is to find a friend in the audience, and imagine you're just talking to them.
- Design your graphics to be clear. Remember, some of your audience might be at the back of the room, so clarity of slides is of the essence. In particular:
  - Don't use small or 'fancy' fonts for text.
  - Keep graphics simple and bold.
  - Use colour to highlight particular points.
  - Don't be afraid to include humour, if appropriate.
- Avoid too much mathematics.
- Don't treat your slides as 'scripts' for you to read out. Think of the slides as a framework which you build the talk around. Your audience will hear what you say – they don't need to read it too.

Table 13.6 Typical structure of a scientific report

Section	Description
Title	Provide a brief, but informative, title, e.g. 'A field and laboratory investigation into iron pollution in the River Cherwell'.
Abstract	A very brief indication of: • the aim of the report • what you did • what you found • what you concluded. Abstracts are often compiled together, or provided online, so it's important for the abstract to give sufficient information to a potential reader.
Introduction	This section provides the context for the report. It should: • give a scientific context for the investigation, e.g. say briefly what other scientists have found • state why the topic is important or useful • state the primary aims and objectives of the study • explain any abbreviations or special terms.
Method	Here, you set out what you did in sequence. You should: • indicate what materials, techniques or equipment you used • explain the procedures and protocols used • provide sufficient information for the reader to potentially replicate the study.
Results	This is a key section, allowing you to present your findings. You should: • provide a clear narrative of your results and their analysis • include clearly titled and labelled graphs, tables and figures as appropriate.
Discussion	Here, you explain what the results mean. You should: • indicate whether the results were consistent or inconsistent with your expectations • discuss any statistical significance • indicate how the study could be improved or extended.
Conclusion	This need not be a long section. Its purpose is to: • briefly restate the main results and • briefly explain the significance of the findings.
References	It's important that you put your investigation into the context of the scientific field. For that reason, you should provide references to other papers and other work carried out previously. You should give this as a list at the end of the report, but you should also 'cite' the work in the text of your report, e.g. 'Smith *et al.* (2010) found that …'

- Practise. This goes without saying – one or two dry runs will help you get the timings right, and help identify any points you need to be clearer on. You might find it helpful to do this in pairs.
- Finally, remember to be courteous to your audience. Remember to thank the audience for their attention at the end, and answer any questions (even difficult ones) calmly and courteously.

GROUP ACTIVITY

Other methods of communication (20 minutes)

We've looked at two methods of communicating your scientific outcomes. As a group, are there other methods which would be suitable to your investigation?

You might like to think about the following questions:

- Are there target audiences you thought of in the pairs activity which are not addressed by the two methods we've looked at?
- What technological solutions are there to help you reach wider audiences?

4.2 Improving your investigations

A final element to your investigation should be how you would improve or build on it. It may be that our example investigation of river water was very preliminary, so you only had a small number of samples, taken at one time. In this case, an improvement would be to collect more samples, but also collect these at varying times, for example over the course of a year to look at seasonal variations.

GROUP ACTIVITY

Reflect (20 minutes)

Look back through your records for the investigation and think about how you would improve the outcomes. Factors to consider are:

- sampling or analysis protocols
- calibration of equipment
- scientific controls.

Were there any extraneous or confounding variables you didn't consider?

> **KNOW IT**
> 1. Why is it important to communicate your findings?
> 2. Give three possible audiences for your communication.
> 3. Why is it important to reflect on the outcomes of your investigation?

LO4 Assessment activities

Below are suggested assessment activities that have been directly linked to the Pass, Merit and Distinction criteria in LO4 to help with assignment preparation and include top tips on how to achieve best results.

Activity 1 Use statistical techniques to analyse experimental data *P7*

You carried out an analysis in D1 of LO3. In this assessment, now think more about the outcomes, and, in particular, what they mean and their interpretation. This assessment could be contained within P8, where you will prepare a report or other communication.

Activity 2 Produce a scientific report on an environmental investigation *P8*

You will decide as a class whether you will be preparing formal reports, giving presentations or using some other method of communicating your outcomes. You will prepare your communication yourself, using means agreed within your class.

Activity 3 Explain causes of experimental uncertainty *M2*

This will be part of your report or communication. You could provide a list of the uncertainties in your investigation, together with their origins, and their impact on the final result.

Activity 4 Recommend actions to positively affect the environmental implications of environmental investigation *M3*

On completing the main reporting of your investigation, you could provide a table of actions you've decided will affect the implications of your environment. For example, if your investigation has focused on pollution, do your outcomes suggest any actions to alleviate this problem?

Activity 5 Discuss the validity of the outcomes of the investigation, in terms of their statistical significance, and showing an awareness of correlation versus causality *D2*

As a scientist, you need to be able to critique your own work and assess its validity. You could include this assessment as part of your communication, where you discuss not just the mathematical statistical significance, or otherwise, of your results, but also any causal relationships.

> **TOP TIPS**
> - Most of the assessments in this Learning Outcome connect with a well-written report or other communication.
> - Follow guidelines for a good communication.
> - Include references to show your awareness of other work.
> - Be careful in interpreting statistics – remember that statistical significance can mislead.

4.2 Improving your investigations

Read about it

United Nations reports and resources

Water quality
www.unep.org/esm/Waterecosystems/WaterQuality/tabid/794532/Default.aspx

Climate change
www.unep.org/climatechange/

Human population
www.un.org/en/development/desa/population/

Natural disasters (including conflicts)
www.unep.org/disastersandconflicts/

Reports and documents for the Hindhead Tunnel project

General information on the Hindhead Tunnel
https://en.wikipedia.org/wiki/Hindhead_Tunnel

Appraisal Summary Table (AST) for the Hindhead Tunnel
http://webarchive.nationalarchives.gov.uk/20120810121037/http://www.highways.gov.uk/roads/documents/AST_A3_Hindhead.pdf

Post-Opening Project Evaluation (One Year After Study)
http://assets.highways.gov.uk/our-road-network/pope/major-schemes/A3-Hindhead/POPE___A3_HindheadOYA__Final_web_version.pdf

Non-technical summary of the environmental statement
http://webarchive.nationalarchives.gov.uk/20120810121037/http://www.highways.gov.uk/roads/documents/61_nts_may2004.pdf

Online resources for statistical analysis:

Student's t-test calculator
http://graphpad.com/quickcalcs/ttest1.cfm

A guide to choosing statistical methods
www.graphpad.com/downloads/InStat3Mac.pdf

A guide to interpreting statistics
https://researchrundowns.wordpress.com/quantitative-methods/significance-testing/

Common misinterpretations of the t-test
www.graphpad.com/guides/prism/6/statistics/index.htm?common_misinterpretation_of_a_p_value.htm

A more in-depth look at statistical misinterpretation
www.informationweek.com/big-data/big-data-analytics/9-causes-of-data-misinterpretation/d/d-id/1321338

Science reporting

An example of the reasons and procedure for peer-review publishing
www.ncbi.nlm.nih.gov/pmc/articles/PMC3474310/

Guidelines for a good scientific report
https://writing.wisc.edu/Handbook/ScienceReport.html

Guidelines for a good presentation
www.ncbi.nlm.nih.gov/pmc/articles/PMC1857815/

Things to avoid in a presentation
www.mindtools.com/pages/article/presentation-mistakes.htm

Unit 14
Environmental management

ABOUT THIS UNIT

In this unit, you will study the legal and regulatory frameworks underpinning environmental management practice and specific issues of importance. These will include water quality management, managing industrial and natural environments and environmental assessments and reporting.

You will study environmental management, from small-scale, local issues to larger, national and international infrastructure developments, analysing and proposing solutions to key environmental questions in a scientifically and logically sound manner.

You will carry out an environmental survey of a site or sites using environment testing techniques on water, air, soil, diversity of flora and fauna. You will report on your findings to relevant authorities such as land owners or local authorities.

LEARNING OUTCOMES

The topics, activities and suggested reading in this unit will help you to:

1. Understand principal characteristics of environments
2. Be able to identify pollution in the environment
3. Understand how legislation, regulation and agreements impact on managing natural and built environments
4. Understand environmental management assessments
5. Be able to carry out an environmental management study

How will I be assessed?

You will be assessed through a series of assignments and tasks set and marked by your tutor.

How will I be graded?

You will be graded using the following criteria:

Learning Outcome	Pass	Merit	Distinction
	The assessment criteria are the Pass requirements for this unit.	To achieve a Merit the evidence must show that, in addition to the Pass criteria, the candidate is able to:	To achieve a Distinction the evidence must show that, in addition to the Pass and Merit criteria, the candidate is able to:
1 Understand principal characteristics of environments	**P1** Describe principal characteristics of a natural environment		
	P2 Describe a lifecycle of a built environment		
2 Be able to identify pollution in the environment	**P3** Conduct safety assessments of field activity and laboratory activities	**M1** Analyse results from the investigation in P4	
	P4 Carry out an environmental investigation, to include field and laboratory work which produces both qualitative and quantitative data		
3 Understand how legislation, regulation and agreements impact on managing natural and built environments	**P5** Describe how domestic or EU legislation impacts on the management of an environment		
	P6 Describe how natural or built environments are influenced by Supra-national agreements		
4 Understand environmental management assessments	**P7** Describe a case study of the use of one environmental management assessment technique	**M2** Evaluate the use of the environmental management assessment technique in P7 in terms of its advantages, disadvantages and consequences	
5 Be able to carry out an environmental management study	**P8** Provide a report on at least one environmental management case study for a given target audience	**M3** Justify the suitability of the report for the target audience in terms of the management technique, and the level and scope of the content	**D1** Critically reflect on the report, and recommend changes so that it would present the management scenario to other audiences
	P9 Describe how the report is made relevant to the given target audience		

LO1 Understand principal characteristics of environments P1 P2

In the previous unit, we looked at how we investigate environments and the kind of factors we need to think about.

These include:
- the type of environment (natural or human/built)
- the kind of interactions and impacts
- how we investigate environments
- how we report our findings.

In this unit, we'll look at another vital stage of environmental science, that of managing environments, and ensuring outcomes of environmental surveys are valid and enacted.

Biomes are large ecological areas on the earth's surface. Biomes are usually identified by their climate, geology and vegetation. Examples of biomes include:

- deserts
- forests
- grasslands
- tundra
- polar
- aquatic

GETTING STARTED

Characterising your environment (10 minutes)

Firstly, we need to be clear about the characteristics of different environment types (natural and human) and any cycles or processes they possess. This means we need to be able to assess and describe an environment, be it natural or built, at all levels, from a regional perspective (the '**biome**') down to individual small-scale **habitats**. We will then be able to assess and apply suitable **environmental management** techniques, depending on the scale of intervention required.

Thinking about where you live, how would you characterise the environment? Think about this as a hierarchy, from the general biome right down to local habitats. To what extent does the habitat reflect the wider biome, and what kind of interactions are important between the two?

1.1 Principal biomes

KEY TERM

Biome – In contrast to a habitat, which may be a very small region or environmental niche, a biome is a larger region, typically on a continental or oceanic scale, which has a relatively uniform climate and types of plants and animals. There are a number of ways of defining biomes, probably the most comprehensive being that of the World Wildlife Fund (WWF), which recognises fourteen terrestrial biomes, as well as a number of freshwater and marine biomes. For the purposes of this course, we will use a smaller range of definitions, as indicated in Figure 14.1.

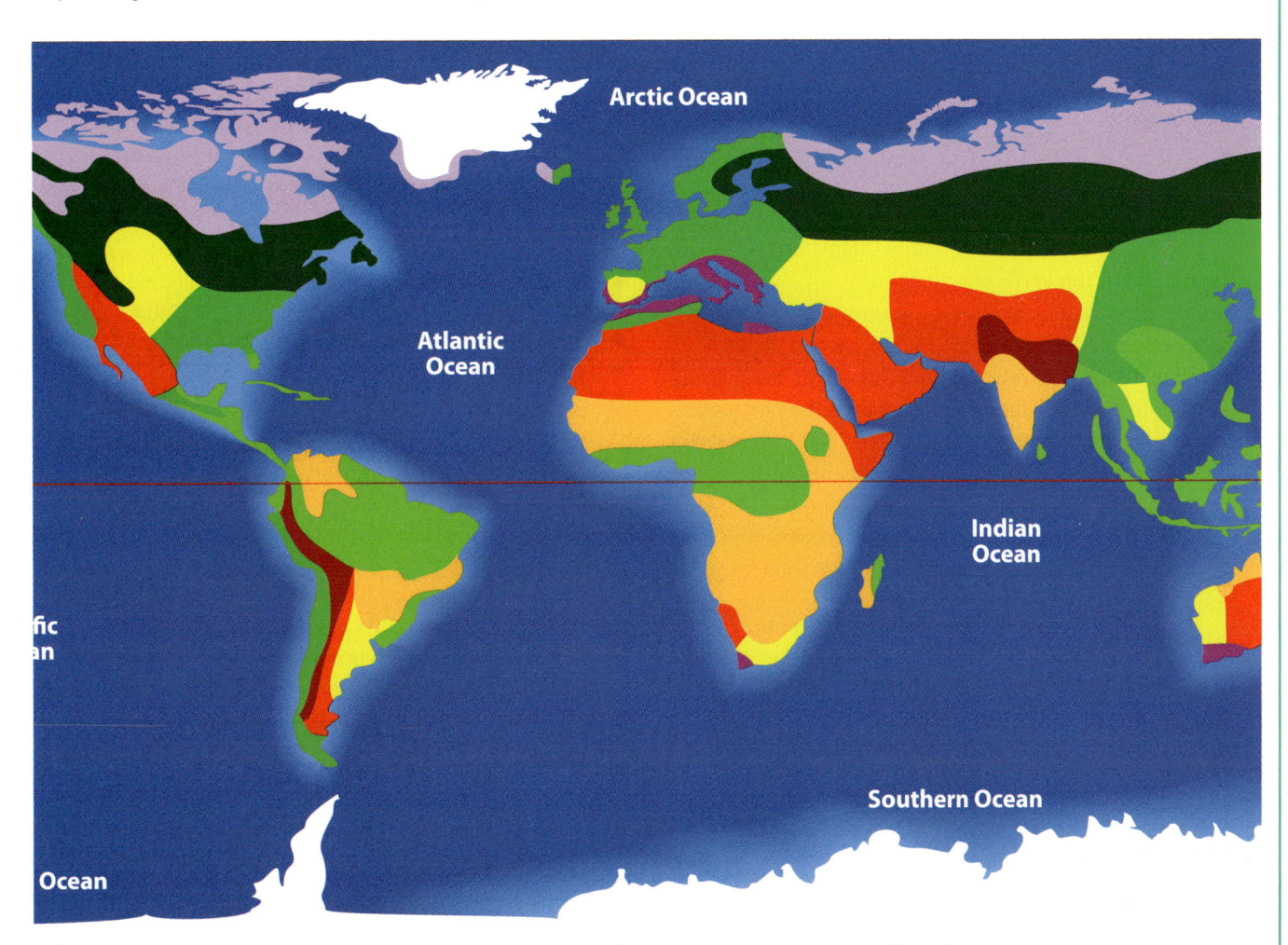

▲ Figure 14.1 The major terrestrial biomes. Note these are further sub-divided in the WWF scheme

257

> **KEY TERMS**
>
> **Habitat** – This is an environmental area inhabited by humans or other organisms. Habitats aren't necessarily natural, or geographical. For example, the habitat of a human might be a city, town or village; that of a flea, the skin of a cat or a dog.
>
> **Environmental management** – In the context of this course, we will define environmental management as the process of managing human-environment relationships. This doesn't mean 'hands-on' looking after an environment, but relates to ideas and development of regulation, policy and practice.

Characteristics of natural environments

As an example, let's consider the central region of Australia, addressing each of the above questions in turn. When thinking about the characteristics of a natural environment, there are a number of key elements to address:

1 What is the biome?
2 What is the geographical location?
3 What are the climatic and seasonal factors?
4 Are there any geological features to consider?
5 What, if any, are the aquatic features of the environment?
6 What organisms exist in the environment?

1 The biome is desert – specifically Desert and Xeric Shrubland in the WWF classification.
2 The geographical location is the southern hemisphere, between about 30°S and 15°S in latitude and about 120°E to 150°E. The region is largely central continental, with some coastal influence especially in the north and south. Elevations vary between 300 and 600 m above sea level in the west to near sea level in the east, with some areas below sea level (the Lake Eyre Basin).
3 The climatic conditions are characterised by high temperatures and low rainfall, though extremes of conditions do occur. Table 14.1 provides information on temperature and rainfall by month.

In terms of seasonal factors, there is a trend of 'wetter' months occurring in the warmer months, with cooler months characterised by lower rainfall.

4 The geology of central Australia is complex, but consists largely of rocks dating to older than 550 million years (pre-Cambrian).
5 As a dry desert, central Australia has no large-scale permanent surface aquatic features. However, a relatively high level of rainfall for a dry desert leads to temporary surface water and oases.
6 The flora is dominated by drought-resistant grasses and shrubs, often with extensive or deep root systems taking advantage of deep ground water or surface rains. In addition, many plants adopt an opportunistic lifecycle, flourishing and setting seed rapidly after rain, with the seed remaining dormant until the next rainfall.

The animal life is characterised by drought-tolerant species, including reptiles, those able to travel long distances in the search for food and water (e.g. red kangaroos) and extreme adaptations, such as Main's Frog, which emerges and reproduces rapidly after rainfall, before absorbing water, burrowing and cocooning itself in a dormant state until the next rains, potentially years away.

> **GROUP ACTIVITY**
>
> ### Investigate (20 minutes)
>
> In groups, investigate a particular geographical location for one of the biomes shown in Figure 14.1, briefly addressing each of the key questions as we have done for the Australian desert. For example, you might like to consider:
>
> - the north coast of Africa
> - the Himalaya mountain range north of India
> - the Amazon region of South America
> - Central Asia.
>
> Be sure to cover as many of the biome types as possible.
>
> On completion, compare the results between your groups. As well as differences, are there any similarities between the various biomes – for example, are there any classes of animals common to all? Are there any common seasonal variations?
>
> In your group activity, you will have looked at the key biomes of Figure 14.1 and identified their key characteristics. In all of this though, we have not yet considered the human dimension; whilst such impacts are relatively light, such as in traditional society in the Australian desert, other human activities, such as those we saw in Unit 13, have led to the arguably new 'biomes' of built environments.

Table 14.1 Climatic conditions by month for the area around Uluru (Ayers Rock) (adapted from data provided by the Australian Government Bureau of Meteorology)

Month	Jan	Feb	Mar	Apr	May	Jun	Jul	Aug	Sep	Oct	Nov	Dec	Year
Record high °C	46.4	45.8	42.9	39.6	35.7	36.4	31.1	35.0	38.7	42.3	45.0	47.0	47
Average high °C	38.6	36.9	34.3	29.9	24.2	20.4	20.5	23.5	28.6	31.9	34.8	36.3	29.99
Average low °C	22.7	22.2	19.1	14.4	9.4	5.6	4.6	6.0	10.7	14.8	18.2	20.7	14.03
Record low °C	12.7	12.5	8.0	1.3	1.3	−1.3	−3.6	−2.2	−1	4.5	6.5	9.9	-3.6
Average rainfall mm	26.4	40.3	35.8	12.7	12.0	19.9	21.2	5.0	8.6	24.3	34.6	41.8	284.6
Average rainy days	4.5	4.5	2.9	2.2	2.8	2.4	2.9	1.4	2.5	4.4	5.7	6.0	42.

1.2 Characteristics of built environments

When considering the characteristics of built environments, we need to think about:

- What is the purpose of the built environment?
- What is its extent?
- How does it interface to natural environments?
- What cycles does it experience?

In looking at the first two of these points, typical purposes and extents in an industrialised country such as the UK, and most European countries, will include:

- Urban. We will define this here as a region of permanent, high human population density, typically covering a large area. We saw some impacts of large urban areas in Unit 13 (e.g. the Greater Tokyo area), but we should also recognise that urban areas can be at much smaller scales, down to smaller towns. The term 'urban' would not normally be applied to small settlements (villages) in industrialised countries, or temporary or traditional settlements of nomadic or indigenous societies.
- Industrial. As a key defining feature of an industrialised society, the industrial environment supports the main economic and production activities of a population. This can represent anything from resource extraction (e.g. mining or quarrying) through to manufacture. Historically, the industrial environment was often embedded within the urban environment; this is, however, now rare, with industrial environments manifesting as radical changes to landscapes (open-cast mining), extensive, dedicated zones or smaller, individual activities.
- Transport networks. As the term suggests, these are built environments facilitating movement of people, goods and services. These may be distributed, as in rail and road networks, or more concentrated and akin to industrial environments, such as sea ports or airports. In these latter cases, we may also include shipping and air transport vehicles as part of the network.

The final two points – interactions and lifecycles – refer to how the built environments connect with the natural environment, and how we can understand and control the development of a built environment from its original inception, through use, to decommissioning and reuse or abandonment of land.

We will look at interactions in more detail later, especially in connection with controlling these interactions and regulations governing this control and assessment.

The life cycle aspect of the built environment involves thinking about its development from initial inception and planning, through use to final demolition, decommissioning and abandonment.

> **KNOW IT**
> 1 What is the distinction between a biome and a habitat?
> 2 When thinking about environmental management, what are we not seeking to do?
> 3 Give three examples of built environments and describe specific examples of each in your local area.

1.3 Principal characteristics of environments

- Geographical location. This can be described globally or on a smaller scale within a country or even within an area within a country. Globally, the environment is affected by the position on the planet in respect of the equator. For example, tropical rainforests are found either side of the equator, desert environments tend to sit on the land masses on the equator and the further you travel to either poles the cooler it gets.

On a smaller scale, for example Britain, it can often be a lot warmer in the south compared to the north. On a more local scale, the environment can change according to the height above sea level, the higher up a mountain, the cooler and more barren it gets.

- Climate. Not to be confused with 'weather' which tends to be a snap-shot of what is happening at a particular moment whereas climate is a record of weather, usually over a 30-year interval. It is measured by assessing the patterns of variation in temperature, humidity, atmospheric pressure, wind, precipitation, atmospheric particle count and other meteorological variables in a given region over long periods of time.
- Seasonality (e.g. temperature ranges, humidity, light). This will vary according to the geographical location of the environment in question and can be described as the changes in the weather over a period of a year. The nearer to the equator or the poles, the less the seasonality there is. In Britain, we have a temperate climate and therefore experience four seasons spring, summer, autumn and winter across the year in fairly even amounts. In winter, days are shorter, temperatures are lower and humidity is high. In summer, we experience longer days, warmer temperatures and drier air.
- Geology and soil. Often forgotten as it is mainly unseen but a vital component of any biome. The geology beneath our feet affects the type of soil above it and therefore the plants that can grow there which form the basis of the food chains living on it. For example, a hard igneous rock, tends to be non-porous and contact with water releases acids which cause the soil above to be acidic, limestone, a soft sedimentary rock, is composed mainly of skeletal fragments of marine organisms such as coral and molluscs so it is mainly formed from different crystal forms of calcium carbonate ($CaCO_3$). Rain water is slightly acidic which erodes limestone causing the soils above to be more alkaline.
- Aquatic (e.g. pond/lakes, stream/river, estuarine, marine). Environments should not be thought of as based on land only. 71% of the Earth's surface is water from large oceans, to lakes and ponds, rivers and streams. Each have their own characteristics and specialist plant and animal species that are adapted to live in each type, some of which can move between more than one type for example estuarine species cope with brackish (slightly salty) water and freshwater from the rivers flowing into them. Deepwater oceans are known to have species living at 5000m, the pressure is enormous but they are adapted by having high internal pressures and no air pockets within their body. They do not survive if brought up to the surface.
- Chemical (e.g. salinity, pH). Some of which has been discussed above i.e. the soil pH or water salinity. This could refer to the pH of rainwater which is often acidic but could also refer to the chemicals found in the air either naturally or from pollutants.
- Atmosphere. This can be defined as a band of gases that envelop the surface of the earth, whereas the term environment refers to all living and non-living things occurring and forming the totality of surrounding conditions. There are four layers within the atmosphere, the Troposphere which is closest to the earth, the Stratosphere which contains the ozone gas, the Mesosphere where meteors from space burn up in this layer and the Thermosphere, the biggest of all the layers of the earth's atmosphere.
- Plant and animal life. All species are affected by all of the environmental characteristics listed above such as geographical location or height above sea level which affects the climate, the geological effect of rocks on the soil type and the seasonal variations in light, temperature and water availability.

1.4 Lifecycles of built environments (e.g. plan, construct, use, decommission/redevelop)

All buildings have a lifecycle. In the past, only the planning and construction stages were considered and historically some buildings were built to last such as medieval castles in Britain while others such as the timber buildings of London in Elizabethan times had a much shorter lifespan. Today, there is legislation in place that requires building planners to think about how the building will be designed, the materials they can use, how energy efficient it is, how long it will last and how it will be decommissioned in the future i.e. how much can be recycled, how much will end up in land-fill sites.

LO1 Assessment activities

Below are suggested assessment activities that have been directly linked to the Pass criteria in LO1 to help with assignment preparation and include top tips on how to achieve best results.

Activity 1 Describe principal characteristics of a natural environment *P1*

You can address this assessment by making a comparison between two selected natural environments. You could present your work either as a tabulated list, highlighting differences or similarities, or as a narrative discussing the various characteristics. The characteristics you could list or discuss include: the biome the environment is part of, climatic conditions, geological and geographic features and typical biodiversity.

Activity 2 Describe a life cycle of a built environment *P2*

We briefly discussed the concept of a life cycle for a built environment. For this assessment, you could investigate a site such as an industrial area, housing area or civic building in your environment over time, and examine how historically land reuse or redevelopment was managed. You could do this using local libraries and museums, contacting local historical or industrial preservation societies or researching local council archives.

TOP TIPS
- When thinking about a natural environment, remember to consider both the biome it belongs to and the habitats it contains.
- Remember a wide range of properties characterise a natural environment, from climate to underlying geology.
- Built environment life cycles can take many forms. Can you identify some of these in your local area?

LO2 Be able to identify pollution in the environment *P3 P4 M1*

An important role of environmental management is the ability to identify pollution in the environment. From an environmental management perspective, we need to think about both human and natural sources of pollution and the design of safe and effective field, laboratory and reporting practices.

GETTING STARTED

Pollution in your environment (15 minutes)

Pollution in the environment isn't necessarily obvious. Think about your local environment, and briefly note down any types of pollution you experience and how they affect the natural and human environments.

2.1 Natural and human-generated pollutants

In the Getting Started activity, you probably thought about particular types of pollution.

It's important though to also think about the way in which pollution is transported, as this has implications for how we can identify and monitor the quantity and effects of pollution.

In this section, we will look at the more common transport routes and think about the types of pollution involved.

Air-borne pollution

Urban pollution

Air-borne pollution is probably the most obvious type in urban or industrial settings, as a result of human activities such as transport or industry. In the urban context, the most important are the traffic-related emissions and particles, as illustrated in Table 14.2.

All are irritants, particularly for those with respiratory issues; the particles especially have the potential to cause long-term problems as they are able to penetrate and accumulate deep in the lungs. The gases are also corrosive, with ozone in particular being responsible for attacking polymers and other materials.

In order to provide guidance for public health purposes, Table 14.3 gives the current levels acceptable. Typically, levels of any one pollutant reaching the 'moderate' bands are considered cause for concern, with implications for health services.

Naturally occurring air-borne pollution

Air-borne pollution can, and does occur naturally, for example, during volcanic eruptions when large quantities of pollutants such as sulfur dioxide can be released (see Unit 13). Pollution events need not be so dramatic however; for example, an issue of concern in some parts of the UK is naturally occurring radon gas accumulating in houses.

Radon

Radon (specifically the isotope ^{222}Rn) is a naturally occurring radioactive gas associated with the decay of the normal low levels of uranium and thorium in many rocks, such as granite, shales and in some cases, limestones.

Radon is continually produced in rocks and soils, making its way to the surface via fractures and groundwater. Normally, concentrations in open air are negligible, but the gas can accumulate in mines or poorly ventilated premises.

Table 14.2 Common urban pollutants

Pollutant	Type	Source	Description
Ozone, O_3	Gas	Photochemical reactions between other pollutants	An irritant gas at ground level, affecting plant and animal life. Also attacks many polymers
Nitrogen dioxide, NO_2	Gas	Traffic exhaust	An irritant gas
Sulfur dioxide, SO_2	Gas	Traffic exhaust	An irritant gas
$PM_{2.5}$	Particles	Traffic exhaust	Dust or soot particles with diameter 2.5 µm or less
PM_{10}	Particles	Traffic exhaust	Dust or soot particles with diameter 10 µm or less

The half life of ^{222}Rn is just under four days, and it is an alpha emitter, meaning its decay pathway is via the emission of a helium nucleus, also known as an alpha particle. It is one of the noble gases (along with helium, neon, argon, krypton and xenon) and is hence chemically very unreactive.

Figure 14.2 shows a map of indoor radon emissions in the UK, measured in units of Bq m^{-3} (bequrels, or decays per second, per cubic metre). There is a clear concentration in the south west of England, coinciding with significant amounts of near-surface granite. As with pollution caused by humans, there are levels above which action concerning radon pollution is advisable. For the UK, this Action Level has been set at 200 Bq m^{-3} for domestic housing by the Health Protection Agency. Whilst radon, in itself, represents a hazard, the decay products include polonium (specifically ^{210}Po), which is also a long lived alpha emitter as well as being highly toxic and capable of being absorbed onto inhalable dust particles.

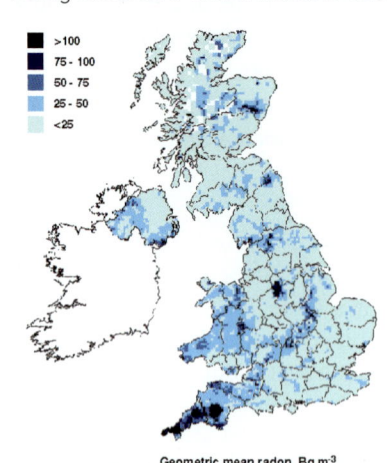

▲ Figure 14.2 Indoor radon concentrations for the UK (DEFRA, 2003)

Table 14.3 UK Air quality bandings applicable until 31st December 2011 (DEFRA, 2017)

Band	Index	Ozone Running 8 hourly or hourly meam* µg m^{-3}	ppb	Nitrogen hourly mean µg m^{-3}	ppb	Sulphur Dioxide 15 minute mean µg m^{-3}	ppb	Carbon Monoxide Running 8 hourly mean mg m^{-3}	ppm	PM_{10} particles Running 24 hour mean µg m^{-3} (Grav. Equiv.)	µg m^{-3} (Ref. Equiv.)
Low											
	1	0–33	0–16	0–95	0–49	0–88	0–32	0–3.8	0.0–3.2	0–21	0–19
	2	34–65	17–32	96–190	50–99	89–176	33–66	3.9–7.6	3.3–6.6	22–42	20–40
	3	66–99	33–49	191–286	100–149	177–265	67–99	7.7–11.5	6.7–9.9	43–64	41–62
Moderate											
	4	100–125	50–62	287–381	150–199	266–354	100–132	11.6–13.4	10.0–11.5	65–74	63–72
	5	126–153	63–76	382–477	200–249	355–442	133–166	13.5–15.4	11.6–13.2	75–86	73–84
	6	154–179	77–89	478–572	250–299	443–531	167–199	15.5–17.3	13.3–14.9	87–96	85–94
High											
	7	180–239	90–119	573–635	300–332	532–708	200–266	17.4–19.2	15.0–16.5	97–107	95–105
	8	240–299	120–149	636–700	333–366	709–886	267–332	19.3–21.2	16.6–18.2	108–118	106–116
	9	300–359	150–179	701–763	367–399	887–1063	333–399	21.3–23.1	18.3–19.9	119–129	117–127
Very High											
	10	360 or more	180 or more	764 or more	400 or more	1064 or more	400 or more	23.2 or more	20 or more	130 or more	128 or more

*For ozone, the maximum of the 8 hourly and hourly mean was used to calculate the index value.

> **GROUP ACTIVITY**
>
> **How dangerous is radon? (30 minutes)**
>
> In groups, investigate the routes by which radon can affect human health. You may like to think about:
>
> - the routes by which radon and its products can enter the human body
> - human activities which might increase exposure
> - evidence for effects on human health
> - means of reducing exposure.

Water-borne pollution

Water-borne pollution arises from a number of human activities, especially where appropriate controls or monitoring is not in place. Examples include industrial chemical waste or accident, deliberate or accidental discharge of waste from shipping and run-off from agriculture. The last of these is of particular concern in rural parts of the UK, where, for example, nitrate run-off has affected aquatic environments.

> **CASE STUDY**
>
> **English Nitrate Vulnerable Zones (NVZs)**
>
> (Note: Water quality in the UK is devolved to the constituent nations. In this section, we will be looking at the English situation; similar actions exist for Wales, Scotland and Northern Ireland.)
>
> A UK government research report in 2014 (http://researchbriefings.parliament.uk/ResearchBriefing/Summary/POST-PN-478) highlights the issue that almost 80% of English ground and surface water is failing to meet good ecological standards as defined by EU Water Frameworks. As part of the process to address this issue, DEFRA and the Environment Agency defined a series of Nitrate Vulnerable Zones to better manage fertiliser and livestock-related nitrate levels in water.
>
> Figure 14.3 shows the scope of NVZs in England, as of 2013. The seriousness of this is clear, as this map covers most of the agricultural land in England, especially East Anglia and the East of England, where significant arable farming takes place.
>
> ▲ Figure 14.3 DEFRA and EA Nitrate Vulnerable Zones in England

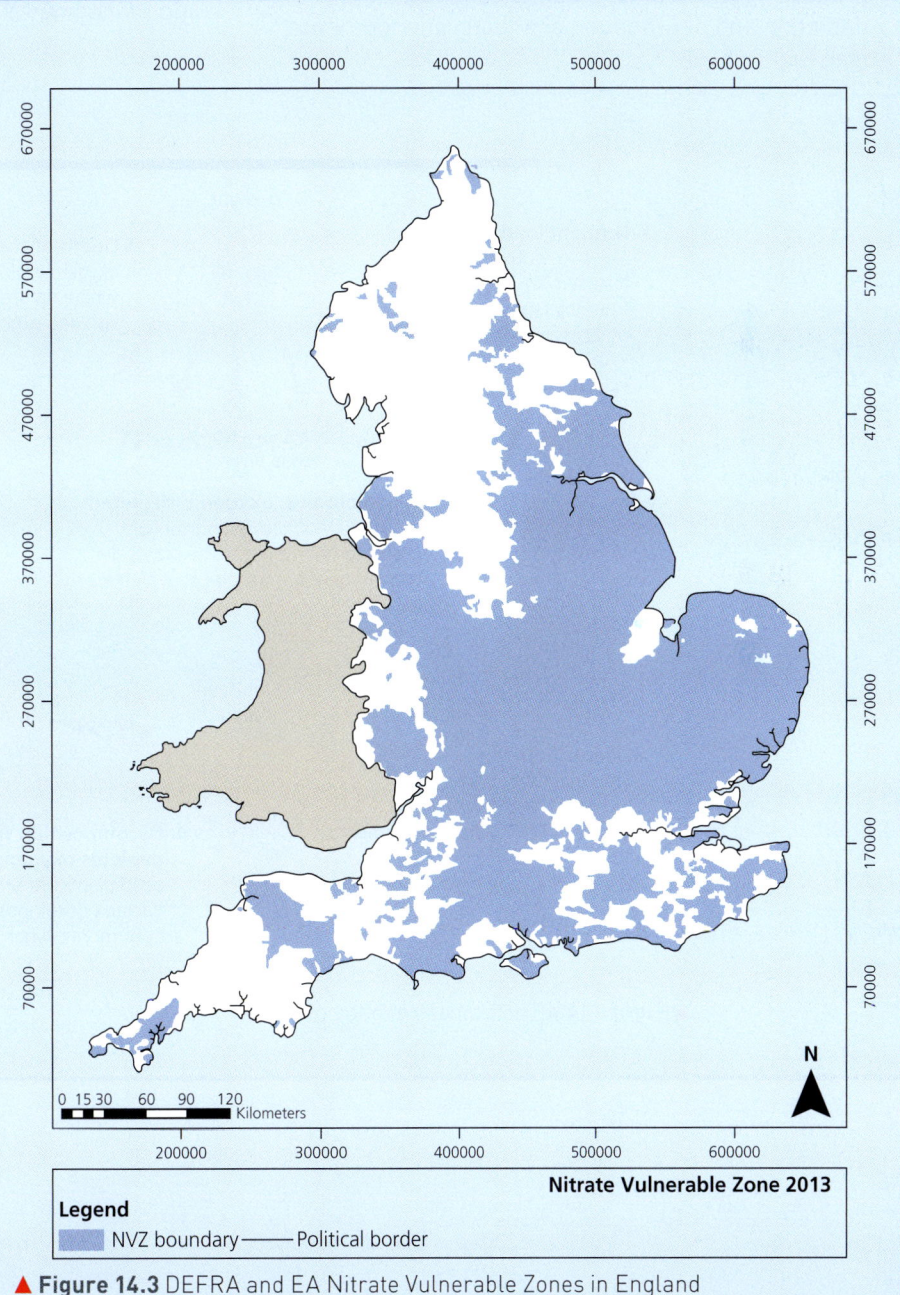

Nitrates are a particular focus since:
- around one million tonnes of nitrogen fertiliser is used annually for arable farming in the UK (see www.agindustries.org.uk/latest-documents/aic-fertiliser-statistics-report-2015-presentation/fertiliser-statistics-report-2015-presentation.pdf)
- livestock farming generates significant amounts of nitrogen-rich slurry, which requires controlled handling to prevent it entering the water supply
- significant run-off of nutrients from farming can lead to eutrophication of aquatic ecosystems, a process by which algal blooms lead to oxygen depletion and degradation of aquatic environments (see Figure 14.4).

The principal roles of the NVZs are therefore to:
- reduce the effects of run-off by limiting times when nitrate-based fertiliser can be applied to maximise uptake of nutrients by crops
- provide clear guidance on level of fertiliser application, minimising wastage
- provide clear information on storage of organic fertilisers
- provide protocols for farmers in respect of monitoring storage and application of fertilisers, and managing risks.

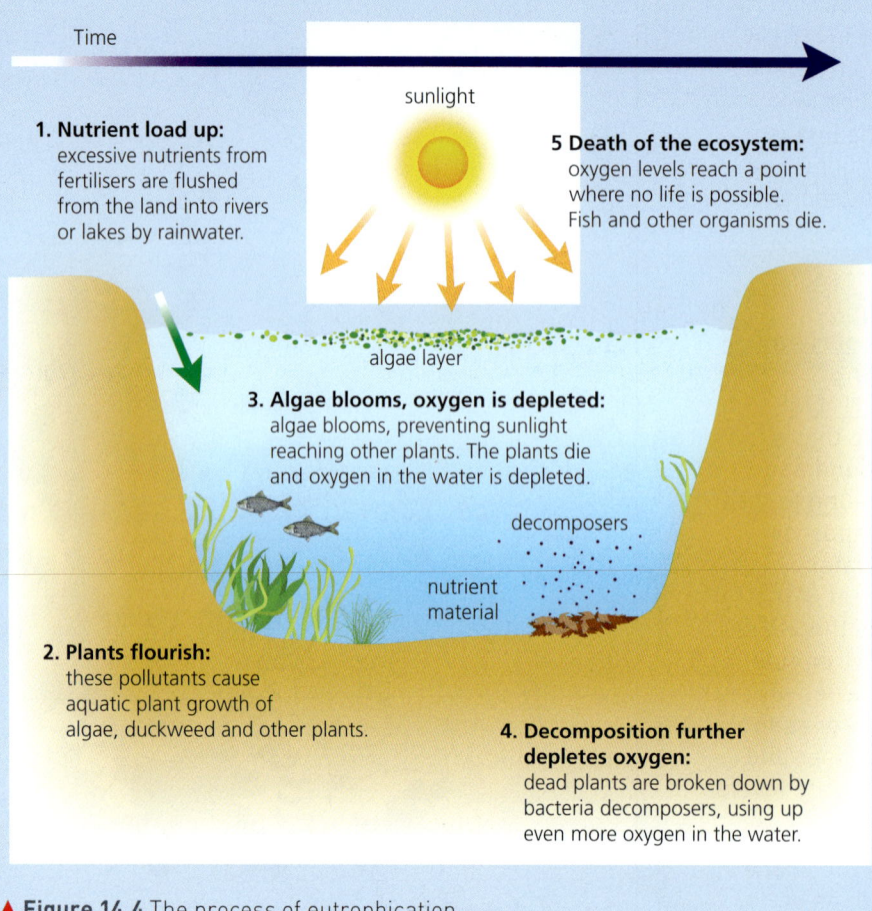

Time →

1. **Nutrient load up:** excessive nutrients from fertilisers are flushed from the land into rivers or lakes by rainwater.

sunlight

algae layer

3. **Algae blooms, oxygen is depleted:** algae blooms, preventing sunlight reaching other plants. The plants die and oxygen in the water is depleted.

decomposers

nutrient material

2. **Plants flourish:** these pollutants cause aquatic plant growth of algae, duckweed and other plants.

4. **Decomposition further depletes oxygen:** dead plants are broken down by bacteria decomposers, using up even more oxygen in the water.

5 **Death of the ecosystem:** oxygen levels reach a point where no life is possible. Fish and other organisms die.

▲ Figure 14.4 The process of eutrophication

Soil-borne pollution

A final major route for pollution is soil, with contamination taking many forms and from many sources. Factors affecting soil-borne pollution include:

- Previous use of land – for example, industrial and mining sites abandoned before rigorous environmental protection came into force can be contaminated by heavy metals.
- Current use of land – for example, land currently used for livestock farming may contain microbial or other biohazardous contaminants.
- Underlying geology – for example, as seen in the section on air-borne pollution, soil underlain by granite can contain higher than normal levels of natural radioactivity, as well as contamination from heavy metals and toxic chemicals weathered from the underlying rocks.

GROUP ACTIVITY

Soil contamination and geology (20 minutes)

The British Geological Survey (BGS) provides an interactive resource to investigate some soil contaminants (arsenic, cadmium, copper and lead) in relation to underlying geology, with data available for England. The resource is available at: http://mapapps2.bgs.ac.uk/bccs/home.html and a screenshot is shown in Figure 14.5.

Use this resource to investigate the connection between geology and soil contamination.

Consider the following:

- Are there elements which you can associate predominantly with underlying geology?
- Can you identify areas where contaminants are as a result of industry or human activity?
- Are there any results you find surprising? (Hint: Take a close look at the Lake District in the north west of England and at the South Downs in the south east of England, in relation to arsenic and cadmium.)

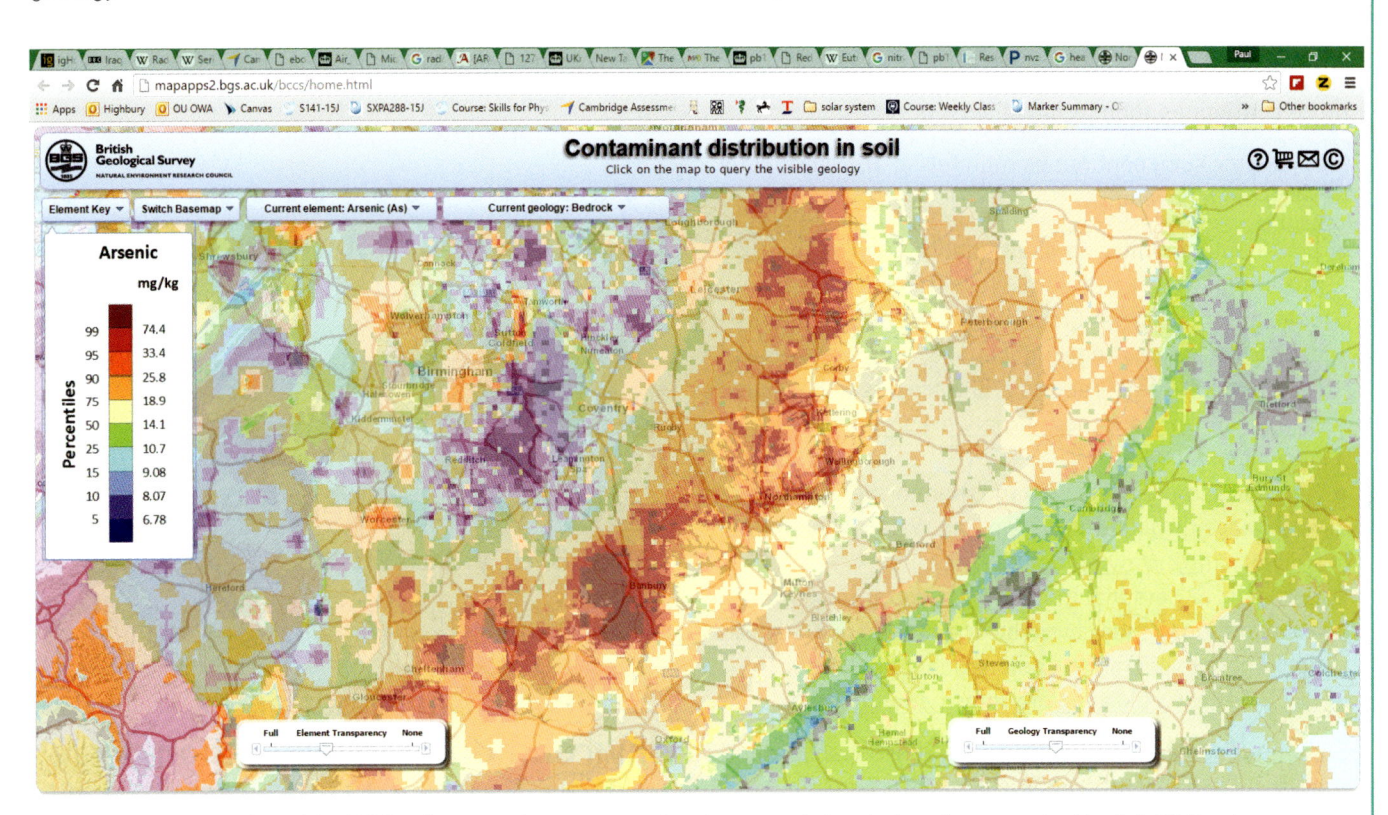

▲ Figure 14.5 A section of the BGS soil contaminants map, showing arsenic levels in soil across the English Midlands

From the above activity, you can see that natural and industrial contamination of soil can be closely connected, but there are exceptions which may be of importance when planning activities such as housing developments or agriculture.

2.2 To apply safe working practices in the field and laboratory

Most of the practices in respect of health and safety have been covered in earlier units, and do not need to be repeated here. For general aspects of health and safety in science, see Unit 2. In addition, health and safety in field work also needs to be considered, and you should review section 3.2 in Unit 13.

2.3 Data collection techniques and 2.4 Recording information

In terms of investigative field and laboratory work and keeping records, the methods and techniques for environmental management uses those discussed in earlier units, in particular Unit 13. In addition to one-off studies though, environmental management may use more continuous monitoring of sites or areas.

Methods for such continuous monitoring are varied, but include the following:

- Continuous monitoring of air quality, such as the UK's Automatic Urban and Rural Network (AURN), which uses automated instruments, collects and transmits data from a wide range of locations, allowing pollution forecasts, reports and warnings to be issued (https://uk-air.defra.gov.uk/networks/network-info?view=aurn).
- Regular sample collection and monitoring of UK waterways through the UK Upland Waters Monitoring Network (http://awmn.defra.gov.uk/).
- Domestic or personal monitoring, for example of radon levels through readily available monitoring and analysis equipment (e.g. ordering radon monitoring kits via DEFRA at www.ukradon.org/information/ukmaps).

Such methods are, inevitably, ground-based, and provide information at a local level. For a more global view, remote sensing, in particular, space-based satellites monitoring, is a vital tool. Examples of environmental monitoring satellites include:

- weather observation satellites such as the GOES (NASA) and Meteosat (European Space Agency, ESA) series of satellites
- land mapping satellites, such as the NASA Landsat series
- specialist environmental monitoring satellites such as:
  - TOPEX/Posiedon (NASA, France), which monitored ocean topography and properties
  - Cryosat 2 (ESA), monitoring the polar regions, especially ice cover
  - Sentinel 2 (ESA), providing land usage information for forestry, agriculture and natural disaster management.

The techniques used for remote sensing range from passive methods, such as visible and infra-red imaging used in weather satellites, to more active measurements such as radar used by the Cryosat 2 satellite to monitor ice thickness and elevation.

KNOW IT

1. State three types of air-borne pollution related to human activity.
2. Explain how radon pollution is related to underlying geology and soil.
3. Describe two means of monitoring environments at local and global scales.

LO2 Assessment activities

Below are suggested assessment activities that have been directly linked to the Pass and Merit criteria in LO2 to help with assignment preparation and include top tips on how to achieve best results.

Activity 1 Conduct safety assessments of field activity and laboratory activities P3

When planning your field and laboratory investigations, record safety assessments of each part in your log book.

Activity 2 Carry out an environmental investigation, to include field and laboratory work which produces both qualitative and quantitative data P4

As a class, you will define an environmental investigation. You should participate fully in the investigation your class defines, both as an individual and as part of a group.

Your investigation could include a survey of pollution or contamination levels in your local area, such as measurements of nitrates or other chemical pollutants in air and water or levels of metals such as lead or cadmium in soils. You could also consider monitoring radon levels if you are in an area where such matters are of concern.

You could design your investigation to use government, environment agency or remote sensing data in support of your own observations.

Activity 3 Analyse results from the investigation in P4 *M1*

You will analyse your results choosing suitable physical, chemical and statistical techniques, allowing your outcomes to be compared to other local, national and global studies. For example, chemical methods would enable you to monitor nitrate levels in air or water, and radiological methods to monitor radon. You could apply statistical methods such as the Student's t-test to determine the significance of results.

Depending on the nature of the investigation, you could analyse and interpret your results in the context of local or global trends identified by other investigators or monitoring networks. For example, you could use British Geological Survey data to back up a soil contamination investigation, or the Automatic Urban and Rural Network to support air quality measurements.

TOP TIPS
- ✔ Remember, pollution can be naturally occurring (e.g. radon) as well as caused by human activity (e.g. excessive nitrates in water).
- ✔ Think about how mitigation of pollution affects the environment (e.g. improving house ventilation to mitigate radon might increase the need for home heating).
- ✔ Remember to think both locally and globally in terms of pollution transport and control.

LO3 Understand how legislation, regulation and agreements impact on managing natural and built environments *P5 P6*

It may seem strange to introduce a section which would be more in keeping with a law qualification, but understanding how regulatory and legal frameworks operate at regional, national and international levels will be of benefit in deciding environmental management strategies, which we'll look at in the next Learning Outcome.

As we all know, regulations are subject to frequent review and change, so it's not the purpose of this section to tell you 'chapter and verse' what these regulations are. In any case, many can be quite complex and interlinked. Rather, we will look at general principles, sources of information and hierarchies of documentation and science which inform regulations and agreements, and how these affect decisions you might need to make.

3.1 Natural and built environment legislation, regulation and agreements

The amount of regulation present for environmental management can seem overwhelming, so rather than attempt to summarise everything, we will look at the sources of information at different levels, and areas and examples where they're applicable. To indicate the relationships, we'll start at a global level, before working down to regional and national levels of regulation.

Global environmental agreements

By their very nature, many global environmental initiatives are not legally enforceable, unless adopted by national or regional governments. That said, many such policies are useful in informing best practice. To see how this works, we will look at a case study, namely the United Nations Environment Programme, and specifically, the Environmental Management Programme.

CASE STUDY

The United Nations Environment Programme (UNEP)

UNEP is a United Nations body, with a global remit to set the environmental agenda and promote adoption and implementation of environmental policies.

Specifically, UNEP's scope covers:

- assessing global, regional and national environmental conditions and trends
- developing international and national environmental procedures
- strengthening institutions for the wise management of the environment.

In the context of this unit, a particularly important aspect is 'Environmental Governance', highlighted in a recent UNEP document (UNEP Medium Term Strategy 2014–17 www.unep.org/pdf/MTS_2014-2017_Final.pdf, published January 2015).

Looking at Environmental Governance, the above report highlights UNEP achievements and performance in the period 2010–13, which include:

- agreement within the United Nations system that the United Nations would develop system-wide approaches to environmental and social safeguards
- achieving improvements in the operations of multilateral environmental agreements relating to chemical industries
- environmental sustainability fully integrated into 30 UN Development Assistance Frameworks and 18 other national developments.
- planning processes
- support for integrated environmental assessments based on the demand from different regions for regional assessments.

Building on these, the report highlights a number of objectives for the period to 2017, encompassing:

- increasing coherence and synergy of actions on environmental issues through the United Nations system and multilateral environmental agreements
- improving the capacity of countries to develop and enforce laws and strengthen institutions to achieve internationally agreed environmental objectives and goals, and to comply with related obligations
- enabling countries to increasingly place environmental sustainability in national and regional development policies and plans.

From the above, we can see that the role of UNEP has not been to enforce regulation and policies, but rather to encourage their adoption and integration into national and regional programmes and legislation. The success of this is illustrated in further UNEP documents, in particular 'Enforcement of Environmental Law: Good Practices from Africa, Central Asia, ASEAN Countries and China' www.unep.org/delc/Portals/119/publications/enforcement-environmental-laws.pdf, published 2014).

This document reports on how UNEP initiatives and guidelines have informed and enabled countries in Africa and Asia to develop and enforce environmental procedures and legislation, in particular through administrative, civil and criminal enforcements.

PAIRS ACTIVITY

Research (30 minutes)

Chapters 2, 3 and 4 of 'Enforcement of Environmental Law: Good Practices from Africa, Central Asia, ASEAN Countries and China' cover examples of administrative, civil and criminal enforcements. For example, Chapter 2 includes the following entry on page 11:

"**Uganda:** A special environmental police unit has been formed under the Ministry of Water and Environment. This has enhanced enforcement of environmental laws by enabling swifter responses to criminal acts and by shortening the prosecution process."

An internet search for 'Uganda Environmental Police' generates a number of reports in recent years, from early setbacks:

http://ugandaradionetwork.com/story/environmental-police-struggles-to-enforce-laws (accessed January 2016)

to more recent successes:

https://langojournalists.wordpress.com/2014/12/31/uganda-police-fight-to-save-environment-in-northern-uganda/ (accessed January 2016)

Briefly research the degree to which the above case has been put into practice.

Report back to your group your findings, reflecting on initial setbacks, progress made, and the effectiveness of global agreements at national and local levels.

Regional agreements

In contrast to the global agreements, most initiatives at the local or regional level tend to be enforceable legislation. As examples, we'll look at the provisions in UK and EU law, and how these relate to each other, and to the global perspective. These provisions are extensive, complex and highly interlinked, so we cannot examine them in their entirety.

CASE STUDY

The European Union Directorate-General for the Environment

The Directorate-General for the Environment (DG Environment) is a body within the European Commission with a responsibility to define new environmental legislation and enact measures to ensure such legislation is put into practice in EU member states.

The DG Environment's responsibilities can be summarised as follows:

- To maintain and improve the quality of life through protection of natural resources, effective risk assessment and management, and the implementation of Community legislation.
- To develop resource-efficiency in production, consumption and waste-disposal measures.
- To integrate environmental matters into other areas of EU policy.
- To promote growth in the EU that takes account of current and future economic, social and environmental needs.
- To address the global challenges; notably combating climate change and the international conservation of biodiversity.
- To ensure that all policies and measures involve all stakeholders in the process and are communicated in an effective way.

As with UNEP, the DG Environment produces regular general reports, in addition to specific documentation on decisions and policies. The most recent completed Annual Activity Report (http://ec.europa.eu/atwork/synthesis/aar/doc/env_aar_2014.pdf, published 2014) describes progress in environmental policies and objectives against key indicators, with progress measured against targets aimed for by 2020. For example, Objective 3 "safeguard the Union's citizens from environment-related pressures and risks to health and wellbeing" has six key indicators, namely (quoting from the above document):

1 Exposure to air pollution: Percentage of urban population resident in areas in which daily PM10 concentration exceeds the limit value (50 µg/m³ 24 hour average) over the period of a calendar year.
2 Exposure to air pollution: Percentage of urban population resident in areas in which ozone concentrations exceed the target value (120 µg O_3/m³ as daily maximum of 8 hour mean).
3 Percentage of surface water bodies in good ecological status or with good ecological potential.
4 Nitrate concentrations in ground- and surface waters: percentage of sampling points with concentration greater than 50 mg nitrate/L.
5 Percentages of total production of environmentally harmful chemicals by toxicity class (from most to least dangerous).
6 Exposure to noise: Percentage of population in urban areas exposed to more than 55 dB (day) and 50 dB (night).

Point 5, concerning environmentally harmful chemicals, is particularly interesting, since this is subject to a number of fluctuations and modifications as uses of chemicals and processes evolve. An emerging example is the development of shale gas as an energy source, requiring close examination of regulations, and determining new regulations to address specific issues. This has led to the production of a set of environmental recommendations, covering not just point 5, but including noise, biodiversity and other environmental aspects. The full recommendation can be seen at:

http://eur-lex.europa.eu/legal-content/EN/TXT/PDF/?uri=CELEX:32014H0070 (published 2014), and is referred to in the Management Plan for 2015 (http://ec.europa.eu/atwork/synthesis/amp/doc/env_mp_en.pdf published 2015, see page 16).

These, and other documents, demonstrate the scale of integration of all environmental concerns under one organisation, ensuring protection for both natural and human environments. A particular aim is the achievement of sustainability through a 'circular economy', which reuses and recycles much of its materials with little or no disposable waste.

They also look forward to consider new industries or human activities which may affect environments, such as the recognition of the shale gas industry and the regulatory and legislative needs to protect human and natural environments. A set of recommendations for the shale gas industry was recently (2014) produced, and can be found at:

http://eur-lex.europa.eu/legal-content/EN/TXT/PDF/?uri=CELEX:32014H0070

> **GROUP ACTIVITY**
>
> **Discuss (20 minutes)**
>
> An important aspect of a regional body such as DG Environment is that it has some legislative powers in respect of member states and governments.
>
> Any current actions of DG Environment towards EU member states can be found at: http://ec.europa.eu/environment/legal/law/press_en.htm
>
> Select a recent case, and discuss as a group:
>
> 1. Which aspect of DG Environment's remit is being contravened?
> 2. How does the contravention affect the human and natural environments?
> 3. Are any proposed sanctions proportionate?
>
> An example you might like to consider is a ruling in November 2015 that concerned the failure of Romania to incorporate into national legislation EU rules on sulphur in marine fuels.
>
> (Case reference IP/15/6008, http://europa.eu/rapid/press-release_IP-15-6008_en.htm)

3.2 How legislation affects the management of natural and built environments

As we have seen in the previous section, global agreements can, and do, filter down to the national level. However, long before these agreements, legislation and regulation dating back to the 13th century CE were enacted, especially in larger cities. For example, a City of London ordinance following a particularly severe fire banned the use of thatch, and required bakers and other users of fire to whitewash their buildings as a fire retardant.

We've seen how initiatives and agreements at global and regional levels can influence national level outcomes, both through support and funding in the case of the UNEP, and legislation and sanctions as in the DG Environment and European Commission.

We'll now look at national level legislation in the context of UK law, how this affects the practice of environmental management, and how it relates to the regional and global agreements we've already looked at.

The main piece of legislation in the UK is the Environmental Protection Act of 1990 and later updates. This is a large, complex document, running to over 200 pages, and many other acts are also relevant so, as mentioned earlier, we cannot examine them in their entirety. Rather, we will look at a specific case, which we mentioned earlier in the context of European regulation, and see how it is being integrated into UK practice.

> **CASE STUDY**
>
> **Onshore oil and gas in the UK**
>
> The oil and gas industry in the UK is perhaps best known for its offshore work, most notably in the North Sea. However, the onshore industry has a long history dating back to the 19th century CE, with the largest site being the Wytch Farm development in Dorset, discovered in 1973.
>
> As of 2015, the potential of so-called 'shale gas' reserves were being examined, under close public and political scrutiny, with the need for regulation being apparent, not least due to the potential environmental impacts. Such activity therefore falls under a range of regulations and bodies which include, for example:
>
> - initial permits through the government Department for Energy and Climate Change (DECC, correct in 2015)
> - Environmental Impact Assessment and obtaining of environmental permissions from the relevant Environment Agency (EA in England, SEPA in Scotland and NRW in Wales)
> - notification of the Health and Safety Executive (HSE)
> - planning permission from local authorities where required
> - advising the British Geological Survey.
>
> It's apparent from this that the regulations governing such activities are not through one source, act of Parliament or document, but rather are distributed across multiple sources. An industry body, the United Kingdom Onshore Operators Group (UKOOG) has produced documents which bring together the relevant sources, describing the general principles in an online fact sheet at:
>
> www.ukoog.org.uk/images/ukoog/pdfs/fact%20sheets/regulation.pdf
>
> with a more detailed document providing guidelines for the exploration and evaluation phase of a site, the most recent edition of which (Issue 3, March 2015) is available at:
>
> www.ukoog.org.uk/images/ukoog/pdfs/ShaleGasWellGuidelinesIssue3.pdf
>
> At this point, we can now draw together all levels of regulation and guidance, with the shale gas industry as an example.

INDEPENDENT ACTIVITY

Evaluate (60 minutes)

We've seen that there is a hierarchy of environmental agreements, regulation and legislation working from the global level, through regions and into national laws. In the context of the shale gas industry in the UK (hydraulic fracturing), review the documents we've looked at throughout this section, and evaluate:

- how well UK regulation is complying with EU and UN guidelines and requirements
- whether the regulations and guidance are suitable
- how, if at all, national and regional cultures and requirements influence global agreements.

You don't need to address all environmental aspects of the industry, but choose one, which can include:

- air quality
- water quality
- chemical controls
- noise pollution
- soil conditions
- ecological diversity
- infrastructure and local communities.

KNOW IT

1. Give examples of one global, one regional and one national environmental agreement or policy.
2. How do your chosen agreements affect and influence each other?
3. For each of your examples, give an example of where it has been used or has been effective.

LO3 Assessment activities

Below are suggested assessment activities that have been directly linked to the Pass criteria in LO3 to help with assignment preparation and include top tips on how to achieve best results.

Activity 1 Describe how domestic or EU legislation impacts on the management of an environment *P5*

We've looked, in general, at EU and domestic legislation on environmental matters. For this assessment, you could construct a chart or flow diagram of a chosen activity, identifying and illustrating the connections between the following:

- the nature of one of the environmental impacts to be considered
- relevant local and domestic legislation, and its requirements in respect of the activity
- domestic agencies involved (e.g. Environment Agency, etc.)
- EU or other regional regulations, directives and guidelines.

Activity 2 Describe how natural or built environments are influenced by Supra-national agreements *P6*

In one of the activities, you looked at a specific example of global organisations influencing local environments. For this assessment, select another example from 'Enforcement of Environmental Law: Good Practices from Africa, Central Asia, ASEAN Countries and China', and compile a report on:

- any specific local issue being addressed (these can be both environmental issues and issues of procedure or governance)
- how information is generated, shared and used by national agencies and the general population (e.g. what methods and tools are used for collecting information)
- what actions are taken to ameliorate issues and prevent reoccurrences.

You will find much of the information needed in 'Enforcement of Environmental Law: Good Practices from Africa, Central Asia, ASEAN Countries and China', but you may go deeper by conducting internet searches for specific case studies.

TOP TIPS

✔ Regulations and legislation are constantly changing to meet new circumstances. Be sure you are looking at the most current information.
✔ Remember there is a hierarchy of regulation, from global guidelines down to local by-laws.
✔ Be aware of the consequences of contravening regulations. Doing so could compromise or delay a project, and incur financial or other penalties on individuals, organisations or government bodies.

LO4 Understand environmental management assessments *P7 M2*

GETTING STARTED

Assessing your environment (20 minutes)

In this section, we'll look at a few of the techniques that are used in assessments in environmental management. Before we start, spend a bit of time discussing as a group what you think the outcomes, in general, of such assessments need to be. In particular, think about:

- What do you need to know to assess a population's impact on an environment?
- What do you need to know to protect an environment?
- How could you connect the above requirements to help you make suitable decisions?

We now move on to looking at the techniques of environmental management, and some examples of how they are applied. It's important to realise that these are more techniques around making decisions and taking actions on the basis of scientific data, rather than the techniques of collecting the data in the first place. These are matters which we looked at in the previous unit, and in the units on scientific techniques.

Key examples of areas where assessments are desirable are in:

- ecological footprinting
- product life cycle assessments and
- Environmental Impact Assessments.

First, we will look at each of these in turn, defining them clearly and their implications for environmental management, before moving on to the techniques of assessment, which include:

- Stakeholder analysis
- Driver-Pressure-State-Impact-Response (DPSIR)

4.1 Environmental management assessments

Ecological footprints usually relate to human influences and are measures of a population's environmental impact required to support current lifestyles and cultures at a given level of technology.

The ecological footprint assesses the amount of land and sea required to support a human population, its infrastructure (housing, transport, etc.), as well as the forested area required to absorb the population's carbon dioxide emissions. Part of the ecological footprint is therefore the **carbon footprint** component of a population.

A counterpart of an ecological footprint is the biocapacity of an area. If the ecological footprint is viewed as the amount *taken* from an environment by a human population, the biocapacity represents the environment's capability to regenerate resources and production capacity.

KEY TERMS

Global Hectare (gha) – Global hectares are the unit of measurement for ecological footprint and biocapacity accounting. A global hectare is a biologically productive hectare with world average biological productivity for a given year. Global hectares are needed because different land types have different productivities. A global hectare of, for example, cropland, would occupy a smaller physical area than the much less biologically productive pasture land, as more pasture would be needed to provide the same biocapacity as one hectare of cropland. Because world productivity varies slightly from year to year, the value of a global hectare may change slightly from year to year.

Land and area types – The Earth's productive land and water areas are categorised into six types: cropland, grazing land, forest land (production), forest land (CO_2 absorption), fishing ground and built-up land.

Carbon footprint – the total amount of greenhouse gases produced to directly or indirectly support human activities, usually expressed as CO_2.

As an example, we'll look at the ecological footprint of the UK across time, and how this compares with biocapacity. Figure 14.6 shows the trend in these parameters from 1961 to 2011, with the measurement of ecological footprint and biocapacity assessed in units of **global hectares** per capita.

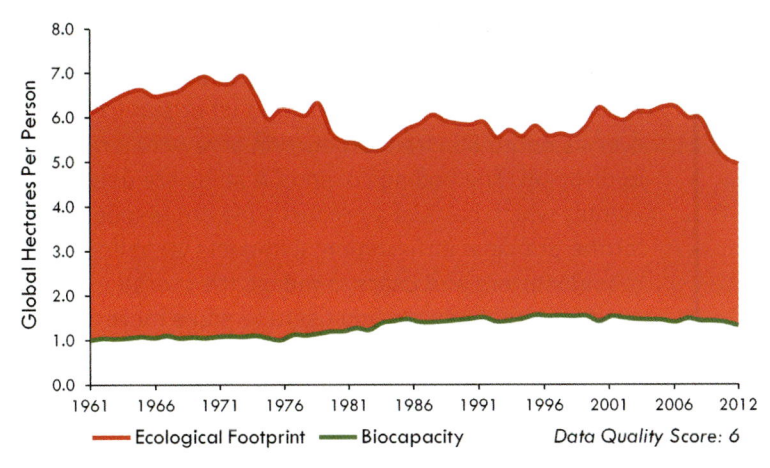

▲ Figure 14.6 Ecological footprint and biocapacity of the United Kingdom for the period 1961 to 2011

To interpret this information, we need to understand the meaning of the units global hectares per capita. As explained in the Key Terms box, a global hectare isn't simply an area of land, but is an area weighted by its level of productivity, which depends on the **land area type**. The example given in the Key Terms box, of cropland versus pasture, shows that growing plants for food is more productive than raising animals on pasture, hence, typically, a global hectare of cropland is smaller than one of pasture since a smaller real area is needed to provide the same biocapacity. The per capita element of the unit merely states that this is measured per person. Over the period of the graph then, the ecological footprint of one person is, on average, about 4.5 gha, with the corresponding biocapacity being a little over one gha for each person.

Looking first at the ecological footprint, we can see that there is a long-term drop in ecological footprint for the UK, starting in about 1971. The various dips and peaks can be correlated with economic cycles (for example, the sharp drop in the mid 1970s and a similar drop in the late 2000s reflecting economic crises), but a general, slow trend is apparent. This is reflected to a degree in the biocapacity measure, which shows a slow rise, indicating an improving capacity of the land (and surrounding sea) area to support the human population.

As of 2011, however, there remains a large gap between the ecological footprint and biocapacity, with the former being persistently the larger. This is an indicator of long-term unsustainability (whilst not allowing for importing of goods), with the human demand vastly outstripping the environment's capacity to provide.

CLASSROOM DISCUSSION

Think about (20 minutes)

We've seen how the ecological footprint per capita in the UK exceeds the biocapacity. Is this true of other countries and, indeed, globally?

Firstly, review the information for other countries available from the Global Footprint Network at: www.footprintnetwork.org/en/index.php/GFN/page/footprint_for_nations/

We already have the UK as a developed, high population density country, so consider countries which:

1 have a low population density and are developed (e.g. Sweden, Australia)
2 are undergoing rapid development (e.g. China, India).

Finally, how do your results compare to the whole world? (Select 'world' in the country list in the above link.)

By how much is the world 'overcommitted' in terms of human demand on biocapacity? What is the current trend in this overcommitment?

One benefit of ecological footprinting is that the 'answer' is a single measure of land area that can easily be understood and compared with others. But this simplicity is also one of its drawbacks, and hence ecological footprinting may be seen as an over-simplification of complex systems, processes and interactions.

The principles of ecological footprinting can also be applied at an individual or household level. A number of organisations provide online calculators to help individuals and households assess their footprints, which analyse responses to simple quizzes in terms of the main goods and services used or consumed, the waste produced and any strategies taken.

INDEPENDENT ACTIVITY

Discover (10 minutes)

The results in the previous activity reflect an average per capita ecological footprint. You might well ask yourself how you, as an individual, compare with this average, and indeed, with your colleagues. You might also be thinking how the ecological footprint can be reduced.

To help you explore these ideas, online surveys can provide estimates of your personal footprint through a series of short questions. Examples which give results in the units we've been using are:

The Bioregional Organisation calculator at:
http://calculator.bioregional.com/

(Select the short quiz – try the longer one later if you like.)

and the Anthesis Group calculator at:
http://ecologicalfootprint.com

Product life cycle assessment

Life cycle assessment (LCA) is an attempt to quantify the environmental impact of a product or service over its entire life, from production of raw materials to disposal. The findings of an LCA study are used to identify which parts of a product's life are responsible for the principal environmental impacts, enabling manufacturers and users to reduce the overall environmental impact, and users to make more informed choices of products or services.

To give an indication of the outcomes in such an assessment, we'll look at an example for a common item, a smartphone.

CASE STUDY

The Fairphone smartphone

As a relatively small Social Enterprise, the Fairphone initiative aims to provide a modular device with an open and transparent supply and production chain, with the aim of limiting environmental impacts and fairly supporting producers and manufacturers.

To this end, members of the initiative have researched and published extensively on the production phase of smartphones, on the supply and manufacture chains and on typical LCAs. An analysis published at: www.fairphone.com/2015/01/22/first-fairphones-environmental-impact/ concentrates on three key aspects which have the greatest effect in the electronics industry, namely:

- metals depletion (i.e. the extraction of raw resources via mining)
- climate change (measured as the amount of CO_2 produced) and
- human toxicity (measured as the emission of heavy metals and toxic materials).

Similarly, the four principal phases of the life of a device are considered, these being:

- production
- transportation
- use
- recycling (this can include disposal of unusable or waste material and decommissioning of facilities).

The overall analysis is summarised in Figure 14.7.

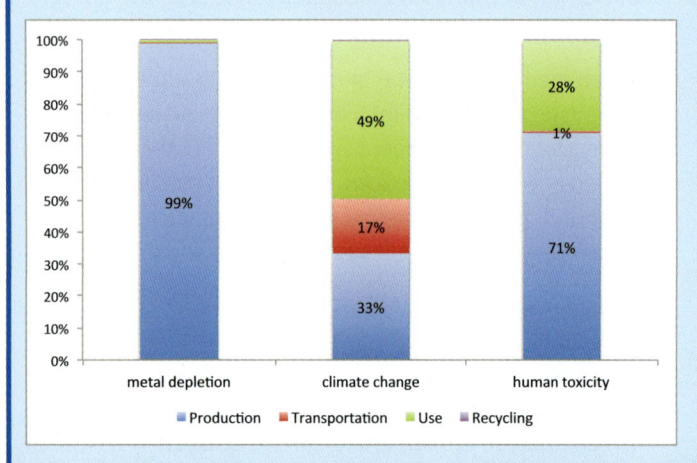

▲ **Figure 14.7** Contributions of each of the key life cycle aspects to each of the life cycle phases for the Fairphone (Figure by Merve Guvendik)

From this analysis, it is clear that metal depletion is almost exclusively in the production phase, and that nearly half the climate change contribution occurs during the use phase, essentially comprising the energy required to charge the phone during a typical use lifetime. As with metal depletion, the greatest risk of toxicity occurs in production, though with substantial amounts present in use, indicating potential exposure to replaceable batteries, storage devices, etc.

It is interesting to note that the recycling phase has very little impact, indicating a strong environmental case for a more 'circular' model of the product lifetime, whereby products are repaired and refurbished, and ultimately usable components recycled at the end of the product's practical life. Indeed, a detailed LCA for the use and recycling phases shows that use of a smartphone for two years (a typical usage time for a smartphone in Europe) leads to about 20% greater impacts in terms of metal depletion, climate change and toxicity compared to use for three years. Further, establishment of a six-year refurbishment scenario leads to increased impact reductions by up to about 30%, reducing further if a 'circular' approach is taken, involving maximum recycling and reuse of components at the end of a unit's useful lifetime.

GROUP ACTIVITY

Research (40 minutes)

The concepts described for the LCA of a smartphone can be applied to other products, activities or industries. As a group, research some of the key points in a product, activity or industry of interest to you. You might like to consider electric or hybrid vehicles, gas drilling and production or the clothing industry.

For each of the phases – production, transport, use and recycling (including disposal and decommissioning), identify any aspects of resource depletion, climate change and toxicity.

You will probably find it useful to assign smaller teams within your group, each addressing one point. You may find some of the results surprising.

Environmental Impact Assessments

Whilst ecological footprinting looks at the human impact, and life cycle analysis examines the impacts of products and activities over a period of time, an Environmental Impact Assessment (EIA) is an assessment of the impacts that a project may have on an environment, including the natural, social and economic aspects.

Internationally, practice in EIA is supported by the International Association for Impact Assessment (IAIA). This organisation defines an Environmental Impact Assessment as "the process of identifying, predicting, evaluating and mitigating the biophysical, social, and other relevant effects of development proposals prior to major decisions being taken and commitments made."

An important point of EIAs is that they do not seek a predetermined environmental outcome, but rather they require decision makers to account for environmental values in their decisions and to justify those decisions in light of detailed environmental studies and public comments on the potential environmental impacts of the proposal.

CASE STUDY

The Safir-Hadramout Road Project in Yemen

The Safir-Hadramout Road was a project supported by the World Bank in the 1990s, shortly after the reunification of the Yemen Arab Republic (North Yemen) and the People's Democratic Republic of Yemen (South Yemen). Aims were to assist the development of Yemen by connecting the east of the country with the capital, Sana'a, and to improve the transport infrastructure within the country to support the then-developing oil, agricultural and tourist industries.

The EIA used international standards pertaining at the time, and included provision of baseline (i.e. pre-project) assessment of:

- physical environment (e.g. geology, surface and groundwater, weather patterns and landscapes)
- biological environment (e.g. habitats, migration routes, livestock grazing and water sources)
- social and cultural environment (e.g. population distribution, community and tribal structure, settlements, infrastructure and services, public health and employment)
- archaeological and historical sites of significance
- legislative and regulatory considerations, taking into account local customs and traditional rights and practices
- environmental management and monitoring plans.

CLASSROOM DISCUSSION

Evaluate (30 minutes)

Review the full case study for the Safir-Hadramout Road, available from the IAIA at: www.iaia.org/pdf/case-studies/Safir.pdf

You may also find it useful to refer to the full EIA document, available from the World Bank at the following link, although this is a long and complex document: http://documents.worldbank.org/curated/en/212911468781521213/pdf/multi-page.pdf

- Does the procedure proposed for the EIA address the points highlighted above?
- Identify sections in the case study or the full EIA which support your answer.
- How did the outcomes of the EIA affect the planning for the road?
- You might like to examine the implications for the route taken, impacts on archaeological, historical and important heritage sites, and implications for local settled and nomadic populations.
- Was the EIA effective in terms of allowing the road to be constructed whilst minimising human and natural environment impacts?

The case study illustrates the positive value of the EIA as a planning tool. An important aspect of this case study is that, whilst much of the research was carried out by experts (e.g. geologists, archaeologists, etc.), local populations were involved through participation in discussions and surveys with relevant experts.

Another important observation is that, especially in the context of the developing world, a rigorous EIA is normally required by bodies supporting projects (e.g. the UN, the World Bank or the EU).

KNOW IT

1. State a benefit and a drawback of ecological footprinting as a measure of environmental impact.
2. If you were wishing to measure the environmental impact of a product or service, what techniques would you use?
3. Explain how a stakeholder analysis can benefit an Environmental Impact Assessment.

LO4 Assessment activities

Below are suggested assessment activities that have been directly linked to the Pass and Merit criteria in LO4 to help with assignment preparation and include top tips on how to achieve best results.

Activity 1 Describe a case study of the use of one environmental management assessment technique *P7*

You could prepare a short report or presentation describing an application of one of the techniques we've looked at. You could, for example, use one of the EIA examples from the International Association for Impact Assessment (IAIA) www.iaia.org/case-studies.php

Activity 2 Evaluate the use of the environmental management assessment technique in P6 in terms of its advantages, disadvantages and consequences *M2*

A good way of addressing this assessment is to include it in your answer to **P6**, for example, as an evaluation section in your short report, or as a part of your presentation.

TOP TIPS

- Remember you can apply more than one technique at a time, for example, a stakeholder analysis within an EIA.
- When selecting a technique for a particular project, think carefully about what you want to achieve.
- Many environmental management techniques are multi-disciplinary – be prepared to study reports and recommendations from experts in other fields.

LO5 Be able to carry out and report outcomes of an environmental management study
P8 P9 M3 D1

Many of the aspects of this Learning Outcome are covered in earlier outcomes as well, especially the analysis techniques. In this section, we will therefore concentrate on:

- choosing your environmental management study
- selecting your audience
- reporting or communicating your results and conclusions
- critiquing your outcomes, with a view to assessing the validity of the data and improving your results.

GETTING STARTED

Choosing your environmental management study (20 minutes)

A full environmental management study can take a long time to complete, and requires many areas of expertise. To get started, spend a little time thinking carefully about choosing the study you will carry out and report on. You may want to consider:

- choosing an aspect of a previous study to analyse – for example, you could look at part of the detailed EIA for the Safir-Hadramout Road
- selecting a product or service you're familiar with, and performing a life cycle analysis
- examining a building or infrastructure development in your local area, perhaps with a view to providing an ecological footprint.

As well as the analysis technique (ecological footprint, etc.), also think about the management technique you will apply – for example, is a stakeholder analysis suitable?

To help you further with your choice of topics, the IAIA has a set of resources at: www.iaia.org/fasttips.php and www.iaia.org/best-practice.php

Reporting environmental management studies

The findings from your environmental management study will, just like a scientific study, contain descriptive or narrative elements as well as quantitative results and outcomes. As with a scientific report, your study could be presented as

- a formal report or paper
- a presentation.

Which method you choose, and how you present the information, will depend on the target audience. Factors to consider will include:

- the use of language (formal or informal)
- the use of jargon or technical terms
- the types of figures or graphics used to display visual information.

CLASSROOM DISCUSSION

Discuss your audience (30 minutes)

Thinking about your choices of topics, which audience would you wish to direct your studies at? Examples you might like to consider are:

- planning bodies
- non-governmental organisations
- consumer groups
- local communities
- newspapers and other media
- other environmental specialists and professionals.

Think about and discuss why a particular audience is relevant, and what kind of reporting and language would be best.

The formal report

A report of this type is usually prepared as a written, formal paper or document and would normally be targeted at professional bodies.

The structure is not as constrained as that of a scientific paper, but it is still recommended that it contains the following, as shown in Table 14.4.

Preparing and giving presentations

You may need, at some point, to give a short presentation of your study. Situations where this might occur are:

- in community engagement, e.g. you might need to present the outcomes of an EIA for a local development
- a company board or committee, e.g. presenting your findings for a life cycle analysis.

There are no hard and fast rules for giving presentations, though there are plenty of guidelines available for what makes a good presentation (and, indeed, what not to do!). Look back at Unit 13, Section 4.1, for some useful presentation tips.

Table 14.4 Typical structure of a formal report

Section	Description
Title	Provide a brief, but informative, title, e.g. 'An Environmental Impact Assessment for Shale Gas Exploration'
Executive summary	This is similar in scope to a scientific paper abstract, though can go into deeper detail. The primary aim of an executive summary is to summarise a much longer document so that a reader can rapidly become acquainted with a large body of material without having to read it all.
Introduction	This section provides the context for the report. It should: • give an environmental context for the study, e.g. state why the study is important or useful • state the primary aims and objectives of the study • explain any abbreviations or special terms.
Method	Here, you set out what you did in sequence. You should: • indicate what techniques you used • explain the procedures and protocols used • describe sources of information and means of collection.
Results	This is a key section, allowing you to present your findings. You should: • provide a clear narrative of the outcomes of your study.
Discussion	Here, you explain what the outcomes mean, and actions or initiatives they suggest.
Conclusion	This need not be a long section. Its purpose is to: • briefly restate the main results and • briefly explain the significance of the findings.
References	It's important that you put your investigation into the context of the scientific field. For that reason, you should provide references to other papers and other work carried out previously. You should give this as a list at the end of the report, but you should also 'cite' the work in the text of your report, e.g. 'Smith *et al.* (2010) found that ...'

GROUP ACTIVITY

Getting the presentation right (20 minutes)

We referred to two possible audiences for a presentation – a local community and a company board or committee.

Think about what other audiences you might encounter, and, in particular, how you might tailor your presentation to:

- fit the interests and requirements of the audience (a stakeholder analysis might be relevant here)
- address any 'vested interests', e.g.
  - a community might be strongly opposed to a development, even if the recommendations of an EIA are favourable
  - a company board might be guided by short-term economics rather than long-term benefits.

Improving or adapting your investigations

It may be that your study needs to be presented to multiple audiences. So, a final element to your study should be how you would improve or adapt it. It may be that your original target was a planning committee, so it needed to be technical. You may then need to present your findings to a local community group, for which a different focus and language will be needed.

PAIRS ACTIVITY

Reflect (20 minutes)

In pairs, consider each other's studies, and reflect on how your colleague might address a different audience from that they are targeting. For example:

- Would a different medium of communication be needed?
- Are the measurement and analysis techniques still suitable?
- Are the outcomes and conclusions still relevant to the new audience?

KNOW IT

1 What is the purpose of an executive summary in an environmental report?
2 Give an example of a situation in which you would prepare a formal report of an EIA.
3 Why might different audiences require changes to your original report?

L05 Assessment activities

Below are suggested assessment activities that have been directly linked to the Pass, Merit and Distinction criteria in LO5 to help with assignment preparation and include top tips on how to achieve best results.

Activity 1 Provide a report on at least one environmental management case study for a given target audience P8

You could provide either a formal report, a presentation or use another medium to describe your chosen study and your findings and recommendations.

Activity 2 Describe how the report is made relevant to the given target audience P9

If providing a written report, the executive summary could contain a statement of how you've ensured the report is relevant to its target; if using other media, include a short document describing the presentation or other output, again, with a statement of how it meets its intended audience.

Activity 3 Justify the suitability of the report for the target audience in terms of the management technique, and the level and scope of the content M3

In **P9**, you described how you made the report suitable. For this assessment, you should justify your choices in terms of the relevance of the techniques you used for the audience (e.g. a community might be interested in a stakeholder analysis) and the scope of your study (e.g. in an EIA, the same community might want to know about how a site of historical importance will be protected).

Activity 4 Critically reflect on the report, and recommend changes so that it would present the management scenario to other audiences D1

You could use the outcomes of the final activity to help inform this assessment. In a formal report, you could include this assessment as part of the discussion section, where you could present reflections on the scope and methods, and make suitable recommendations, which could also be informed by your assessments for **P9** and **M3**.

> **TOP TIPS**
> ✔ Most of the assessments in this Learning Outcome connect with a well-written report or other communication.
> ✔ Follow guidelines for a good communication.
> ✔ Include references to show your awareness of other work.
> ✔ Make sure your methods are relevant to the outcomes and recommendations you want to achieve.

Read about it

Lists and definitions of biomes

A general discussion of biomes and their classification

https://en.wikipedia.org/wiki/Biome

A list and description of biomes according to the WWF scheme

www.worldwildlife.org/biomes

The Food and Agriculture Organization of the United Nations, providing information on specific biomes

www.fao.org/home/en/

Information on air quality monitoring

Resources for air quality monitoring in the UK from DEFRA https:

uk-air.defra.gov.uk/

Example providers and lists of air pollution monitoring equipment

www.airmonitors.co.uk/home

www.et.co.uk/products/air-quality-monitoring/

www.airqualitynews.com/suppliers/

Information on radon, radon pollution and mitigation

A general discussion of the properties of radon

https://en.wikipedia.org/wiki/Radon

Resources and information from Public Health England on radon and radon monitoring

www.ukradon.org/

Health and Safety Executive resources on radon in the workplace

www.hse.gov.uk/radiation/ionising/radon.htm

World Health Organization radon information document

www.who.int/ionizing_radiation/env/Radon_Info_sheet.pdf

Unit 18
Microbiology

ABOUT THIS UNIT

Microbiology is the study of (ology) microorganisms (microbes for short); microbes are organisms too small to be seen by the naked eye. Therefore, we study them using specialised techniques including microscopes, culturing, genetics and serology. Microbes are found everywhere, including in and out of your body, on your laptop keyboard, on your phone screen and in your food. However, they are not all harmful and are actually essential in maintaining life on earth. We use microorganisms in food production, agriculture and the production of medicines.

In this unit you will cover the structure and functions of key microorganisms, processes used for identification and classification of microbes, as well as the beneficial uses of microbes in agriculture, food and medicine.

After studying this unit you will understand the importance of microorganisms in everyday life, how to classify and identify them, and scientific uses of microbes. As well as this, you will have developed valuable practical techniques relating to microbiology, which can be used in further study and work.

LEARNING OUTCOMES

The topics, activities and suggested reading in this unit will help you to:

1. Be able to classify and identify microorganisms
2. Understand the use of microorganisms in agriculture
3. Be able to use microbiology in food production
4. Understand the action of antimicrobials on microorganisms

How will I be assessed?

You will be assessed through a variety of tasks, which may include the following:

- Carrying out practical work safely and analysing results
- Completing assignment and portfolio work
- Controlled assessment
- Completing poster work, presentations and group work

How will I be graded?

You will be graded using the following criteria:

Learning Outcome	Pass	Merit	Distinction
	The assessment criteria are the Pass requirements for this unit.	To achieve a Merit the evidence must show that, in addition to the Pass criteria, the candidate is able to:	To achieve a Distinction the evidence must show that, in addition to the Pass and Merit criteria, the candidate is able to:
1 Be able to classify and identify microorganisms	**P1** Identify the main groups of microorganisms	**M1** Use techniques to identify microorganisms	**D1** Evaluate the use of techniques used to identify microorganisms
2 Understand the use of microorganisms in agriculture	**P2** Describe the use of microorganisms in sustainable agriculture		
	P3 Describe how GM crops are produced	**M2** Describe the advantages of GM crops	**D2** Evaluate the consequences of the introduction of GM crops
3 Be able to use microbiology in food production	**P4** Describe the use of microbes in food production	**M3** Explain the optimum conditions for growth of microorganisms during a fermentation process	
	P5 Produce a microbiological food product under optimum conditions	**M4** Describe the biochemical processes involved in the production of a food from microorganisms	
4 Understand the action of antimicrobials on microorganisms	**P6** Identify the types of antimicrobial used in medicine	**M5** Describe current and projected trends in AMR	**D3** Evaluate measures to prevent future consequences of AMR
	P7 Describe the mode of action of antibiotics		
	P8 Describe the mechanism of antimicrobial resistance (AMR)		

LO1 Be able to classify and identify microorganisms *P1 M1 D1*

GETTING STARTED

Harmful microorganisms – the answer is in the word (10 minutes)

In Table 18.1 match the microorganism to the disease it causes and then write them in the correct column below. This task will introduce you to disease-causing **microorganisms** as well as the ability to understand scientific terminology. To complete these tasks use a mixture of elimination, research and prediction.

Scientific words may seem complicated; however, you will quickly learn in science that you can define a term just by breaking it down, although some are easier than others. Let's use **microbiology** as an example:

- micro = small
- bio = life
- ology = study of

In this example you can see clearly how easy it is to break down scientific terminology. Now try and apply

> **KEY TERMS**
>
> **Microorganism** – A microscopic living organism, only seen using a microscope. Includes bacteria, virus and fungi.
>
> **Microbiology** – Study of microorganisms using a variety of techniques including Gram staining, colony morphology and serological techniques.

this technique to the task on the next page and discuss your answers with a partner.

Table 18.1 Classification of bacteria

Microorganism	Disease caused
Mycobacterium tuberculosis	Vibrio Cholerae
Cholera	Tuberculosis
Herpes	Cold
Rhinovirus	Herpes simplex

> **KEY TERMS**
>
> **Morphology** – The study of the form of organisms, focusing on their shape, size, colour and characteristics.
>
> **Gram positive** – A type of bacteria which stains purple; consists of thick peptidoglycan cell wall.
>
> **Gram negative** – A type of bacteria which stains red; consists of thick peptidoglycan cell wall.
>
> **Aseptic** – Free from contamination; processes consisting of sterilisation.

1.1 Groups of microorganisms

Microorganisms can adapt and live in a range of climates and environments. But are they harmful? Some but not all: many are beneficial and we need them to survive. Evidence of the presence of microbes can be obvious, for example, in the decay of materials or food, the development of disease, or more positively fermentation of wine and beer. Microbes are divided into groups depending on their characteristics, reproduction methods, genetic make-up and **morphology**. The main groups of microbes include: viruses, archaea, bacteria, fungi, algae, protozoa and parasites.

Table 18.2 Groups of microorganisms and their functions

Microorganism	Main functions	Facts
Virus	• Takes over host cells in order to cause infection and disease.	• not considered as a living organism (non-cellular) • smaller than bacteria • cannot reproduce outside the host cell • also known as akaryotes • not always considered a microorganism
Bacteria	• Recycle nutrients and decompose. • Prevent harmful bacteria from entering the body. • Used in the biomedical and biotechnology industry. • Help digest food. • Cause infection and disease.	• unicellular prokaryote: single-celled organism without a membrane bound nucleus
Fungi	• Decomposition of dead matter in soil and recycling nutrients. • Transport minerals in plants. • Sources of antibiotics. • Production of food and drink. • Cause infection and disease.	• eukaryotic cells • multicellular • moulds and yeasts • closely related to animals • around 100 000 identified species • mostly live on land, some marine species exist
Algae	• Used as a source of agar, a growing medium. • Gelling agents in food. • Used as a fertiliser. • Energy source.	• photosynthetic organisms • unicellular and multicellular forms • examples include seaweed, charophyta (green algae)
Protoctist	• Food source for microinvertebrates. • Cause diseases such as malaria.	• unicellular • live in a wide range of habitats
Parasites	• Lives in or on another organism benefiting from stealing their nutrients. • Examples include: hookworm, tapeworm, fleas.	• can be extremely harmful • reproduce quickly • demonstrate a high level of specialisation
Archaea	• Source of enzymes. • Food processing.	• unicellular • similar to bacteria; until recently were classed as bacteria until important differences were identified • the oldest of all living organisms

1.2 Classification of bacteria

Bacteria are single-celled prokaryotes, meaning they do not contain a nucleus or membrane bound organelles. However, they do contain a control centre that appears in the form of a loop of DNA. Bacteria are found in or on multiple locations or organisms, including animals and plants, and apart from viruses they are the smallest living things on earth. Some are beneficial, whereas some can cause harm, although most people know bacteria due to their harmful effects. Bacteria demonstrate high importance through their flexibility, ability to grow rapidly and their history. Bacteria are invaluable microbes which are involved in making digestion possible in animals, fuelling the decay of dead matter and waste, and are evident in the production of many foods sold in supermarkets and shops. Like other microorganisms bacteria are classified into a variety of groups depending on their morphology. Such classification allows us to correctly identify diseases, develop medicines, eliminate harmful and non-harmful types of bacteria, and to continue our study of the evolution of bacteria. The five main groups include: spherical (cocci), spiral (spirilla), corkscrew (spirochaetes), rod (bacilli) and comma (vibrios).

Table 18.3 Characteristics and morphology of bacteria

	Characteristics	Morphology
Bacilli	Found in soil and water, produces large amounts of enzymes, making it valuable to industries.	Rod shaped, cylindrical bacteria, generally **Gram positive**, occurs in chains mainly.
Spirilla	Most live in water, as individuals as opposed to clusters or chains. Found in rodents.	Spiral shaped cells, **Gram negative**, some have flagella at both ends for movement.
Cocci	Capable of living on its own, two joined together are diplococci.	Round shape, can occur as Gram positive or Gram negative.
Spirochaetes	Capable of thriving in anaerobic conditions.	Has a double membrane, spiral shaped.
Vibrios	Associated with undercooked seafood.	Gram negative, occurring in a curved rod shape.

1.3 Techniques used to identify bacteria

The main aim of the classification of bacteria is the identification by any individual; this could be a scientist, clinician or the public. The benefits of knowing how to recognise these types of organisms is to prevent or diagnose diseases, as well as utilising them for beneficial purposes. Currently there are a number of methods scientists use to recognise bacteria, including bacterial morphology and staining methods.

PAIRS ACTIVITY

Are all bacteria dangerous? (25 minutes)

In Table 18.4 are a variety of bacteria. In pairs conduct research on the bacteria to find out the following:

1 Are they harmful or harmless?
2 What is the main function of each type?
3 Do they have a wider purpose?
4 Can they be manufactured in a laboratory?

Use the terms in the table as search criteria to help you find more information.

Table 18.4 A variety of bacteria

Lactobacillus	Streptococcus Pneumoniae	Bacillus Subtilis	Bifidobacterium

Aseptic techniques

Aseptic techniques are an essential laboratory skill in the world of microbiology, as well as other scientific divisions. The main aim of completing this technique is to prevent contamination of samples and cultures. If a sample is contaminated, it will result in an inaccurate result being produced. Contamination can arise from the individual conducting the procedure, air-borne microbes, unsterilised equipment and laboratory surfaces. Completing aseptic techniques efficiently will minimise or remove the risk of contamination.

Before and during the aseptic techniques you must ensure the following: all surfaces are wiped down, windows and doors are mostly closed, methods are conducted quickly but safely, you limit exposure of samples to the air and make sure all equipment is within immediate reach.

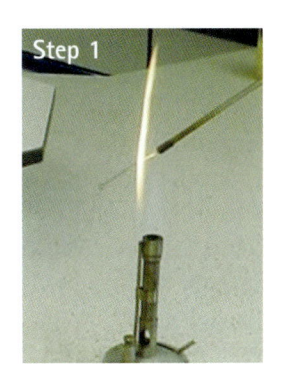

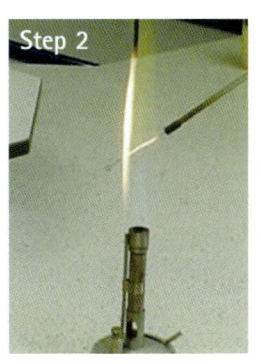

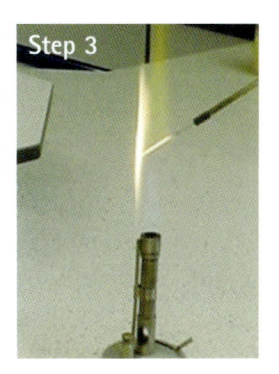

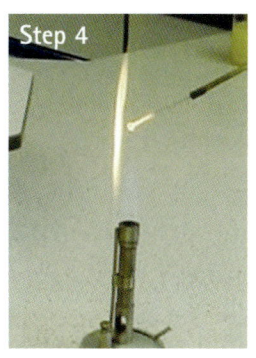

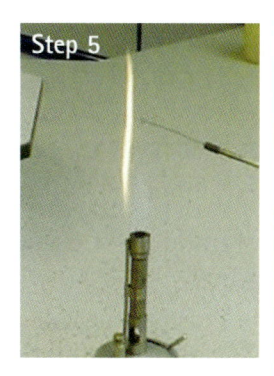

▲ Figure 18.1 Step-by-step imagery of how to sterilise a wire loop using a Bunsen burner

Using a wire loop

Wire loops are used to transfer microbial samples onto agar plates and other equipment. In order to sterilise them, they must be heated using red heat under a Bunsen burner before and after use, using a flaming procedure. The procedure is as follows:

Flaming procedure

1 Position the end of the wire into the top of the blue cone of the flame, which is the cool end of the flame.
2 Move down towards the hottest section of the flame slowly, ensuring the full extent of the wire receives efficient heating.
3 Hold there until it is red hot.
4 Allow to cool, then use immediately.
5 Do not put the loop down, and re-sterilise after use.

Flaming the neck of bottles and test tubes

The neck of test tubes and bottles must be sterilised before use to minimise or eliminate contamination.

1 Loosen the cap of the bottle or test tube if there is one, to ensure easy removal.
2 Hold the bottle/test tube towards the bottom with the left hand.
3 Remove the cap/lid with the little finger of the right hand (turn the bottle, not the cap).
4 Do not put the cap/lid down.
5 Flame the neck of the bottle/test tube by passing the neck forwards and back through the hot Bunsen burner flame.
6 Once complete, and the sample has been removed from the bottle/test tube, replace the cap using the little finger (turn the bottle, not the cap).

Colony morphology

As we mentioned earlier, morphology is the study of the appearance of organisms. Bacteria can be identified depending on their shape, colour, size, colony margins

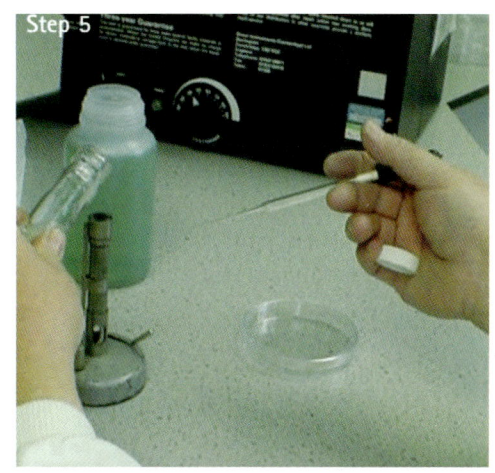

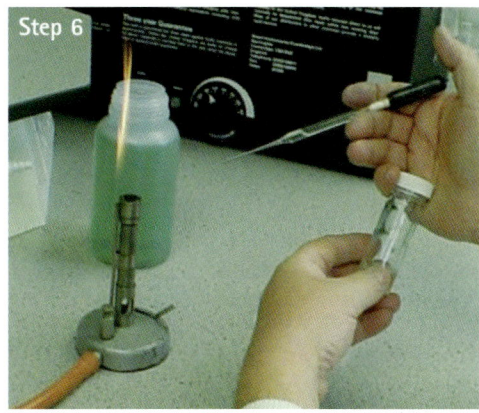

▲ Figure 18.2 Step-by-step imagery of how to sterilise a lid/cap of a test tube/bottle

and other visible characteristics. We grow bacteria on agar plates; this is known as culturing, providing them with a healthy supply of nutrients encouraging reproduction. **Colonies** are the shape, size and colour in which each type of bacteria display. Colony morphology is the method scientists used to identify bacteria; it is one of the most effective techniques used in this area of study. Scientists will undertake this technique to identify a bacterial infection, implement new diagnostic methods or create emerging treatments. The method used to grow colonies is know as plate streaking. Here is an explanation of how to complete a simple plate streaking method.

Apparatus
- Petri dish containing agar growth medium
- wire loop
- sample broth
- Bunsen burner

Method
1. Complete compulsory aseptic techniques (wire loop and bottle neck flaming).
2. Insert wire loop into broth contained within the bottle/test tube.
3. Partially lift the lid of the Petri dish.
4. Streak the wire loop backwards and forwards across the surface of the agar gently.
5. Complete aseptic techniques.
6. Complete step 4 again using the sterilised wire loop, ensuring you streak in a different section.
7. Tape the Petri dish partially and incubate.
8. What are we looking for when we observe a colony? When observing a colony you can identify the colony types:
   a. form: size, texture, colour, surface
   b. elevation
   c. margin.

Form: Circular, Irregular, Filamentous, Rhizoid

Elevation: Raised, Convex, Flat, Umbonate, Crateriform

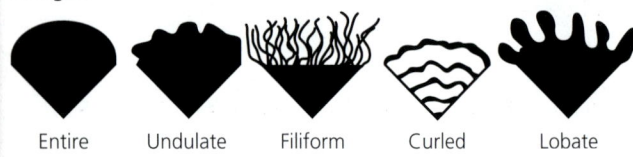

Margin: Entire, Undulate, Filiform, Curled, Lobate

▲ Figure 18.3 Image displaying colony morphology: form, elevation and margin

Gram staining

Gram staining was discovered by Hans Christian Gram in 1884, and it remains one of the most important techniques used today. It allows scientists to identify bacteria based on their morphological and varied staining properties, classifying the bacteria as either Gram positive or Gram negative. Although Gram staining is a reliable and accurate staining technique, some bacteria cannot be stained using this method. For example, Mycobacterium has a large lipid content of the peptidoglycan, which means Gram staining is not possible and alternative staining techniques must be utilised. The method itself is fairly simple: the sample slides are stained with crystal violet, iodine, de-stained with alcohol and counter stained with safranin. Gram negative will stain red and Gram positive will stain blue-purple. The result is due to the differences between the cell walls: Gram positive bacteria contain a thick multi-layered peptidoglycan (cell wall). It is a mesh-like structure, which is cross-linked with short chains of peptide. This provides the wall with a rigid structure, making it stronger than the Gram negative cell wall. This allows retention of the crystal violet stain. By constrast Gram negative consists of a thin single layer of peptidoglycan (cell wall), not retaining the crystal violet stain.

Gram staining steps
The four main steps of Gram staining are as follows:

1. Apply primary stain crystal violet to a heat fixed smear of bacteria sample.
2. Add gram's iodine.
3. Decolourisation with 95% ethyl alcohol.
4. Counterstain with safranin.

Growth media

Growth media is essential for achieving growth of bacterial colonies. It is a solid or liquid containing essential nutrients, used to support growth of microorganisms. The main types of growth media used in bacterial colony growth are:

- **Selective media**: MacConkey's for Gram negative bacteria, mannitol, salt agar. This type of media is used to grow selected types of bacteria. For example, if the aim is to grow Staphylococcus aureus, which is resistant to methicillin, methicillin is added to prevent growth of any other bacteria types. This therefore isolates *Staphylococcus aureus*.
- **Differential media**: Blood agar, Granada medium, eosin methylene blue. This type of agar distinguishes between more than one microorganism growing on one plate.

> 🔑 **KEY TERMS**
>
> **Gram staining** – A staining technique in which a violet dye is applied to distinguish between Gram positive and Gram negative bacteria.
>
> **Colony** – Two or more organisms living in close proximity to each other, usually for mutual benefit, for example to share nutrients.

Serological and genetic methods

Serological and genetic methods are highly specific and reliable techniques used in bacterial identification. Serological tests involve the reactions of microorganisms with antibodies; antibodies are formed following a reaction to an infection. This type of test can be helpful in identifying a specific strain or species of a disease, and is typically used to identify types of immune disorders or for blood typing. The main techniques used include: ELISA, agglutination, precipitation, complement fixation, western blotting and fluorescent antibodies. One example is ELISA (enzyme-linked immunosorbent assay), which includes four main steps, resulting in a coloured end product for easy identification. The higher the concentration of the antibody being tested, the stronger the colour change. Gene amplification is an important tool in the identification of bacteria. Gene sequencing and PCR (polymerase chain reaction) amplification are the most commonly used genetic methods; they are accurate and have the ability to identify a more diverse range of microorganisms. During gene sequencing, bacterial DNA is extracted and isolated; the genes are then compared to a bank of results in the hope of a match for identification. PCR is a technique used to amplify a sample of DNA; it is a cheap and easy method and allows for easy identification of microorganisms.

> ### KNOW IT
> 1 What is the meaning of 'ology'?
> 2 Name the main microorganism groups.
> 3 Why do we classify bacteria, and how is this relevant in scientific study?
> 4 How effective are the identification methods for microorganisms, and what is the importance of aseptic techniques?

LO1 Assessment activities

Below are suggested assessment activities that have been directly linked to the Pass, Merit and Distinction criteria in LO1 to help with assignment preparation and include top tips on how to achieve best results.

Activity 1 Identify the main groups of microorganisms P1

This can be completed in table form, as in Table 18.2.

Activity 2 Use techniques to identify microorganisms M1

Complete practical techniques for this criterion to identify microorganisms. The suggested technique would be Gram staining. You have the chance to identify two different types of bacteria. Colony morphology is also an efficient technique for identifying microorganisms, through observation of a range of prepared Petri dishes.

Activity 3 Evaluate the use of techniques used to identify microorganisms D1

Evaluate = to judge or determine the significance of something; in this case you will need to determine the effectiveness of the techniques used to complete M1. Do this in the form of an evaluation in a written report.

> ### TOP TIPS
> ✔ Know and learn the characteristics of all microorganisms.
> ✔ Understand and learn the need for classification of bacteria.
> ✔ Practise aseptic techniques before completing identification techniques for **M1**.
> ✔ Read examples of scientific reports before completing **D1**.
> ✔ Understand the meaning of evaluation.

LO2 Understand the use of microorganisms in agriculture P2 P3 M2 D2

During LO1 you learnt how to distinguish between various microorganisms, how to identify them and their main purposes. You now understand the presence of microbes are everywhere and not all are harmful. In LO2 you will further your understanding of the benefits of microbes and how they are utilised in agriculture for fermentation, feedstocks and more.

GETTING STARTED

The good and the bad? (25 minutes)

Agriculture is the science or practice of farming, with the aim of growing crops and rearing animals to provide food, nutrients, wool and other products. In pairs you must complete a research task. One person will be researching the advantages of agriculture and the other will be researching the disadvantages of agriculture. Discuss your findings.

Agriculture – what is it?

Now you have researched the advantages and disadvantages of agriculture, you will have an idea of the overall process. The process of agriculture involves the redirection of nature's natural food web. Traditionally, sunlight provides energy, which translates as food for plants, plants provide food for herbivores and herbivores provide food for carnivores. Decomposers or bacteria break down plants or food, which has died; this re-enters the soil and the whole process starts again. However, during agriculture this process is interrupted with the aim of producing high quality plants and food for humans. Agriculture involves microorganisms used in sustainable agriculture, energy production and **genetic engineering**. Throughout this section you will develop an understanding of each of these points, and finally be able to evaluate the development of **genetically modified** crops.

KEY TERMS

Agriculture – The science of farming, including the cultivation of soil.

Genetic engineering – The deliberate modification and manipulation of characteristics of organisms.

Genetically modified (GM) – Genetic material which has been altered within an organism.

Yield – To produce or provide.

Fertiliser – A natural or chemical product, which enhances the fertility of soil.

Pesticides – A substance used to destroy pests, usually with the aim of protecting plants.

Fermentation – The chemical breakdown of a substance.

Biofertiliser – A substance containing living microorganism, which promotes growth of soil and plants.

Bioherbicide – A substance containing microorganisms which is used to control weeds or undesirable plants.

Biopesticide – Pesticides derived from natural substances, consisting of living microorganisms, used to destroy pests.

2.1 How microorganisms are used in sustainable agriculture practice

Sustainable agriculture practice allows for the highest **yields** to be achieved, without affecting the natural resources in the surrounding areas. Throughout this process, farmers are able to limit their use of **fertilisers** and **pesticides**, saving money and protecting the environment. Table 18.5 explains the roles and facts of the four main areas of sustainable agriculture, including how microorganisms are involved.

Table 18.5 Examples of processes utilised in agriculture

	Natural fermentation	Biofertilisers	Biopesticides	Bioherbicides
Function	breaks larger molecules down into small molecules which can be absorbed	a substance containing living microorganisms with the aim of promoting growth in plants	naturally occurring substances which control pests, consisting of microorganisms	control agent targeting weeds, which compete with crops for water, nutrients and sunlight
Microorganisms	consists of specialised microorganisms: lactic acid bacteria and *lactobacillus plantarum*	consists mainly of bacteria and fungi	consists of bacteria, viruses and fungi	consists of fungi, viruses and bacteria

	Natural fermentation	Biofertilisers	Biopesticides	Bioherbicides
Facts	• occurs when microorganisms consume susceptible organic substrates as part of their own processes • examples include silage, which is special fermented food (known as fodder) for cattle, sheep and other similar animals	• all organic farming is mostly dependent on the natural microflora of the soil, which is composed of fungi and bacteria; for example, *arbuscular mycorrhiza fungi* (amf) and plant growth promoting rhizobacteria (pgpr)	• some biopesticides are incorporated directly into the plants via genetic engineering	• scientists isolate the microbes which have specific genes to target weeds and destroy them
Advantages	• can result in higher levels of nutrients • less affected by water damage • the process itself is less labour intensive	• can provide protection against drought • eco friendly and cost-effective • the microorganisms in biofertilisers restore the soil's natural nutrient cycle and build soil organic matter	• affect only the target pest • effective in small quantities • decompose quickly, therefore reducing the use of more toxic synthetic pesticides • more environmentally friendly than synthetic pesticides • increase crop yield	• cost-effective • long survival rate; not harmful to the environment • specific to their target, without harming the rest of the environment
Specific examples of microorganisms used in this process	• *Lactobacillus plantarum*	• *Rhizobium, azotobacter, Azotobacter* and Blue-green algae (BGA) • Main types: Nitrogen fixing biofertiliser (NFB) and Phosphorus solubilising biofertiliser (PBF)	• *Bacillus thuringiensis* • *Bacillus subtilis* • *Bacillus pumilus*	• *Phytophthora palmivora*

GROUP ACTIVITY

How do microorganisms work? (20 minutes)

In small groups, copy and complete Table 18.6 finding out the mode of action for each microorganism and focusing on how it is used in agricultural processes.

Use the internet and books for research, and once complete discuss your answers with the class.

Table 18.6

Lactobacillus plantarum	Plant growth promoting rhizobacteria	Bacillus pumilus	Phytophthora palmivora

1.3 Techniques used to identify bacteria

CASE STUDY

Biofertilisers – a closer look

Less developed countries (LDCs) don't have the same access to materials as developed countries. One of the most beneficial points regarding the production of biofertilisers is that anyone can make them, therefore allowing populations in LDCs to increase their crop yield naturally. Here's a closer look at a simple method for the production of biofertilisers:

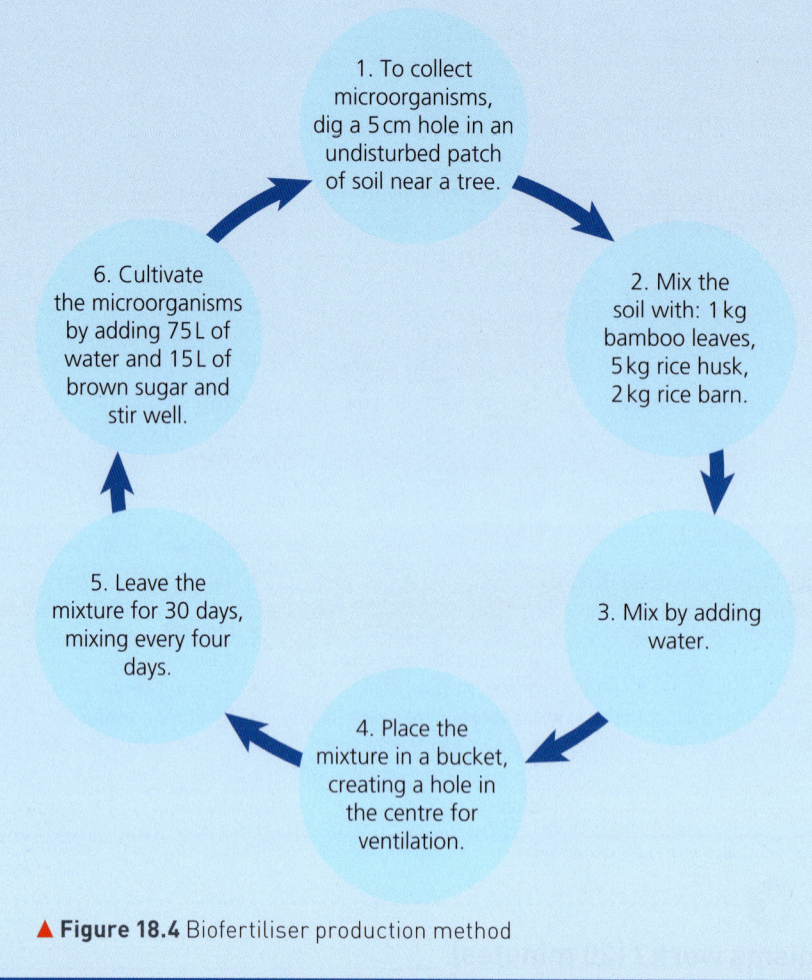

1. To collect microorganisms, dig a 5 cm hole in an undisturbed patch of soil near a tree.
2. Mix the soil with: 1 kg bamboo leaves, 5 kg rice husk, 2 kg rice barn.
3. Mix by adding water.
4. Place the mixture in a bucket, creating a hole in the centre for ventilation.
5. Leave the mixture for 30 days, mixing every four days.
6. Cultivate the microorganisms by adding 75 L of water and 15 L of brown sugar and stir well.

▲ Figure 18.4 Biofertiliser production method

? THINK ABOUT IT

1 Considering the ease of producing biofertiliser, do you think LDCs will be able to improve their crop yield, therefore improving their overall nutrition, if they are educated about the methods of producing biofertiliser?

2.2 Energy production for agriculture

Energy production is crucial for agriculture to take place. The use of farm equipment, packaging and processing, fertiliser production and other activities all require energy. Before the use of manufactured energy products, the main source of energy in agriculture was the sun for photosynthesis. The plants were used as energy sources or livestock and manure was used as fertiliser. Renewable energy is energy from a source which is not depleted when used. Currently, there are many resourceful techniques used to produce renewable energy in agriculture.

Table 18.7 Different energy sources, materials and purposes

Energy source	Materials	Purpose
Biogas	maize, grass, wheat, rye, vegetable waste, organic waste, microorganisms	• renewable energy source • production of ethanol, butanol, hydrocarbon fuels, biodiesel fuel • production of heat
Biomass		
Feedstocks		
Wind power	water	

The energy sources identified in Table 18.7 are extremely advantageous to the environment and economy:

- Staff are required to manage the production of renewable energy, providing the community with new jobs.
- Reduces pollution as natural products are used.
- Saves farmers money as they are not paying for fuel and gas, therefore increasing their net income.
- Creates new income for farmers, providing them with the opportunity to sell the energy they create to energy companies.

2.3 Genetic engineering of crops

Genetic engineering is a process by which the genetic material (DNA) of an organism is altered in a way that does not occur naturally and a set of unique genes are produced; in agriculture the aim is to introduce a new trait to a plant. This process is completed to give crops a longer shelf life and increase yield. As with many scientific processes, there are disadvantages as well as benefits. Genetic engineering is a controversial topic, which concerns many people. It can be known to result in allergic reactions for some individuals as well as containing a higher toxin level than naturally produced crops. Additionally, some people view the process of creating new life forms as ethically wrong.

Genetic engineering methods

Genetic engineering has been occurring for hundreds of years. The first method was known as artificial selection. Simple selection was and is the easiest method of modification. It involves superior plants with desired traits being selected; their seeds are sown and each time the plant grows, the seeds are saved and replanted. In modern methods it can sometimes take up to twelve years to produce a new plant via genetic engineering. There are a variety of techniques employed to do this. Overall the aim is the same; firstly the desired gene must be identified, and then cloned, and then a copy of the gene inserted into the host DNA. Figure 18.5 is an example of a genetic engineering technique using bacteria as the vector, known as microbial vectors.

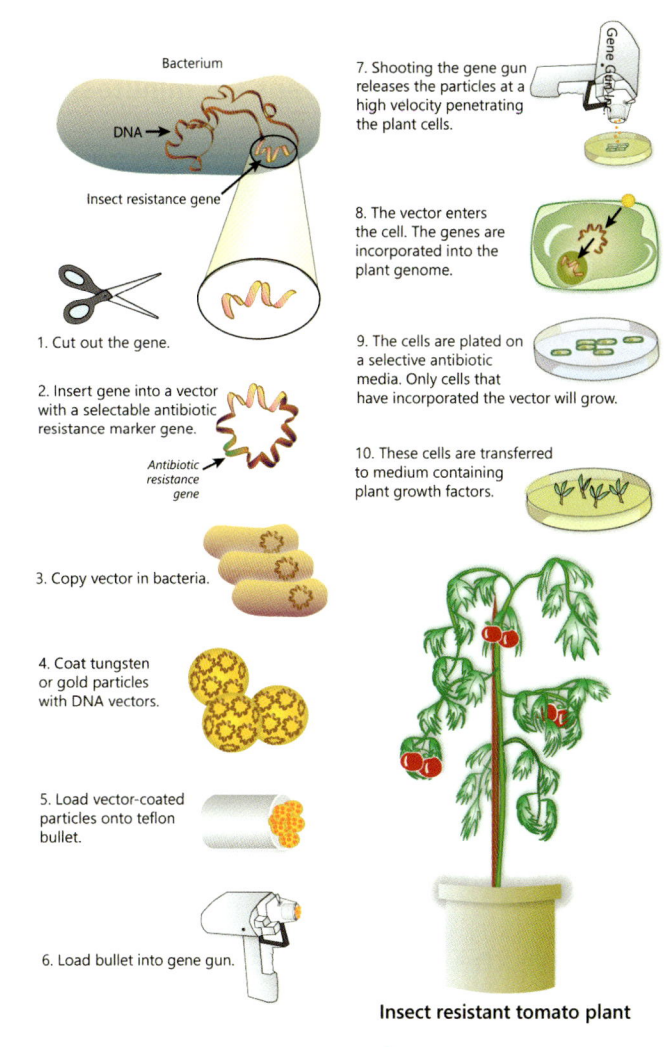

▲ Figure 18.5 Method demonstrating how to create an insect (pest) resistant tomato plant

The image shows a step-by-step method of how microorganisms are used to genetically modify a tomato plant. The microorganism used is bacteria; the gene in this bacterium is removed, inserted into another bacteria, which will be used as a **vector**, and inserted into the plant cells using a gene gun. An example of a microbial vector is Agrobacterium tumefaciens, which is a naturally occurring soil microbe. It is best known for causing disease in plant species, with the ability of transferring its own DNA into the plant cell, and remarkably the plant then reads the transferred DNA as its own. This works perfectly in genetic engineering, as scientists use the Agrobacterium by substituting its own DNA with the DNA of interest and then using it to deliver this to plant cells for modification.

> **EXTENSION ACTIVITY**
>
> **Feeding animals and humans (25 minutes)**
>
> Genetically engineered food is used in both animal and human food products. Included in Table 18.8 are examples of foods and plants used for human and animal food supply. In groups complete the table researching the uses and advantages of each product. Some answers have been completed for you.
>
> Table 18.8 Human and animal food supplies
>
> | | Human supply | Animal feed | Advantages |
> |---|---|---|---|
> | Rice | | | |
> | Maize | yes, used to produce cornstarch | yes | |
> | Potato | yes, optimised starch content | | |
> | Soybeans | improved oil profiles for healthier eating | | |
> | Wheat | | | |
> | Apples | | | |

What are the characteristics of the products we consume?

As a result of genetic engineering, there are a variety of characteristics that develop in the modified products. These characteristics display many advantages. The main aims of using modification methods are:

1. Improved crop yield.
2. Improved crop protection; virus resistance.
3. Longer shelf life; as mentioned in point number 7, potatoes and apples have a longer shelf life as a result of decreased bruising and browning.
4. Better for the environment, as they need fewer chemicals to thrive.
5. Improved flavour and nutrition; products are modified to taste good and contain specific nutrients. For example, rice known as golden rice is produced to provide greater amounts of vitamin A, although this is not yet grown commercially.
6. Cost-effective for consumer and producer; products are cheaper to make and buy as less money is spent on chemical fertilisers, etc.
7. Aesthetically pleasing; some potatoes are modified to prevent bruising, and tomatoes to be perfectly round and bright in colour. Most recently apples have been modified to prevent browning. This was achieved by silencing the gene (polyphenol oxidase) in the apple responsible for this process.
8. Resistance to insecticides and herbicides; both of these can be highly toxic and can interrupt growth of crops; therefore, plants are modified to become resistant to their toxic properties. This is achieved by the insertion of microorganisms including Agrobacterium and Bacillus thruringiensis.

> **KEY TERM**
>
> **Vector** – A carrier of disease or an organism, but does not cause disease or harm itself.

2.4 Evaluate the development of genetically-modified (GM) crops

The ethical implications of genetically modified (GM) crops are a topic of discussion relating to many controversial issues; in shops around the world you will find organic products, aimed at individuals who want to avoid consuming anything that has been genetically modified. Public support of GM products is not consistent due to ethical, safety and ecological factors. Unfortunately research is still fairly new surrounding the ethical implications in genetic engineering; however, studies are ongoing. The topics below are the most commonly researched and evidenced issues.

Disease and allergies

Through the mixing and addition of genes into GM products, there are increasing concerns relating to the development of allergies and disease following the consumption of GM products. Disease is a concern, especially in microbial vector modification, as these types of GM products are created using bacteria and virus genes. With regards to the trigger of allergies, this would be a result of allergenic genes being used in the engineering process, again as in microbial vector modification.

Environmental damage and food web risks

Agricultural processes can interrupt the natural food webs. By genetically engineering crops, ultimately food is being removed from herbivores, which will then have a direct result on carnivores. Alongside this the toxins released from herbicides and pesticides can harm animals and other organisms, and the introduction of herbicide- and pesticide-resistant crops has had an adverse effect on the environment. An example of this is GM sugar beet, which has been modified to become resistant to weeds. This is beneficial to the crop but skylarks and butterflies feed from weed flowers and seeds; and therefore producing GM sugar beet is endangering their existence.

As with anything, overuse can result in resistance. The overuse of antibiotics directly results in bacteria becoming bacterial resistant, producing superbugs. The same is thought to occur within the agriculture environment. The continued use of genetic engineering may result in the evolution of 'superweeds', which over time become resistant to GM crops as well as herbicides and pesticides.

An issue of identity preservation of crops may occur due to the cross-pollination of GM seeds onto non-GM seeds. This is a concern for farmers who certify their crops are organic.

> **KNOW IT**
> 1 Identify which microorganisms are used in agriculture practice.
> 2 What are the implications of 'superbugs' and 'superweeds'?
> 3 Identify the advantages and disadvantages of using microorganisms in agriculture.

LO2 Assessment activities

Below are suggested assessment activities that have been directly linked to the Pass, Merit and Distinction criteria in LO2 to help with assignment preparation and include top tips on how to achieve best results.

Activity 1 Describe the use of microorganisms in sustainable agriculture *P2*

Microorganisms are used in biofertilisation, bioherbicides, biopesticides and natural fermentation. For **P2** you could present this information on a professional poster (a poster completed on the computer). Include key points, the main microorganisms utilised and images.

Activity 2 Describe how GM crops are produced *P3*

An example has been included in Figure 18.5; use a similar process to describe how a GM crop of your choice is produced.

Activity 3 Describe the advantages of GM crops *M2*

This can be completed in a table: one column for advantages, another column explaining why the advantages are relevant.

Activity 4 Evaluate the consequences of the introduction of GM crops *D2*

This criterion should be completed in essay form, including the evaluation of the implications of GM crops on the environment, allergies and disease, the economy and science.

> **TOP TIPS**
> ✓ Learn the difference between natural and chemical agricultural processes.
> ✓ Know all of the examples of microorganisms utilised in the processes used for agriculture.
> ✓ Research real-life case studies relating to agriculture and the implications on the world, which will help with **D2**.

LO3 Be able to use microbiology in food production *P4 P5 M3 M4*

Microbiology is widely used in the production of food. LO2 detailed the processes of agriculture and genetic engineering, further evaluating the implications of these methods. In LO3 you will learn about the industries that utilise microorganisms in food production, and you will also be able to create your own microbiological product under optimum and safe conditions.

> **GETTING STARTED**
> **What do you know about the food you eat? (15 minutes)**
> In pairs, on a post-it note, write down what you already know about the presence of microorganisms in the food you eat. Then stick your post-it note on the whiteboard in the classroom and discuss this as a whole group.

Table 18.9 Microorganisms in food production

	Method	Microorganism used
Dairy industry	Yoghurt 1 Heat milk to around 80°C. 2 Cool to around 45°C and add the microorganisms. 3 Mix well and incubate at the same temperature for a few hours.	lactic acid bacteria: *lactobacillus, leuconostoc, lactococcus, pediococcus*
Bread industry	Bread 1 Mix flour, salt and yeast (microorganism). 2 Make a well in the centre and add water and oil; mix well. 3 Knead dough until smooth. 4 Place in a bowl and leave for an hour. 5 Place in the oven to bake.	yeast: *saccharomyces cerevisiae*
Brewing industry	Beer 1 Barley grains are soaked in water. 2 The grain is fermented for around six days and then roasted. 3 The mixture is pumped in a lauter run and sweet liquid is separated. 4 Mixture is boiled and hops are added. 5 Yeast is added and fermented. 6 The beer is matured and filtered.	yeast: *saccharomyces cerevisiae and saccharomyces uvarum*
Wine industry	Wine 1 Grapes are sorted then crushed. 2 Fermentation takes place, where yeast is added (this can take between 10 and 30 days). 3 The final mixture is filtered to remove any unwanted particles. 4 The wine is either immediately bottled or left to age.	yeast: *saccharomyces cerevisiae* bacteria: *lactobacillus*

3.1 The food industry and microorganisms

Table 18.9 identifies the basic methods for producing foods in various industries, as well as the microorganisms used in these processes.

3.2 Biochemistry of fermentation

Fermentation is possibly one of the oldest and most successful forms of food preservation. Biotechnology and genetic engineering is now being applied to food production, where yeast is one of the main driving forces of fermentation. The term fermentation was first applied to the production of wine, and it referred to the bubbling action resulting from the conversion of sugar to carbon dioxide. Later on, yeast was added, resulting in the association of microorganisms in the process. 'Fermented foods' is used as a term to describe a class of foods, which not only refer to the breakdown of carbohydrates, but also contain proteins and fats. All of these have undergone modification under the action of microorganisms.

GROUP ACTIVITY

Fermentation experiment (45 minutes)

Try out this quick and easy fermentation experiment, looking at the production of carbon dioxide in a range of conditions. You will need the following apparatus:

- two test tubes
- bored bung
- flexible delivery tube to connect test tubes
- yeast
- vegetable oil
- sugar
- limewater

Set up the apparatus folloas shown in Figure 18.6.

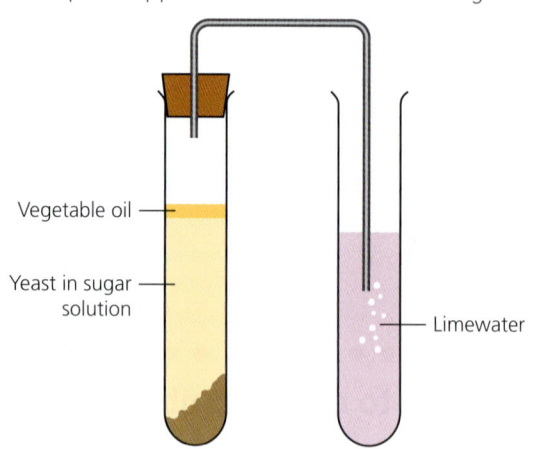

▲ Figure 18.6 Layout of the fermentation experiment

Method

1 Dissolve sugar in water which has been previously boiled.
2 Add yeast and mix; pour mixture into a test tube.
3 Add a layer of vegetable oil.

4. Connect a delivery tube from the stopper in the yeast tube to the test tube containing limewater.

The layer of vegetable oil prevents oxygen from entering the mixture, allowing carbon dioxide to escape. After about an hour bubbles will start to appear, demonstrating the fermentation process.

> **KEY TERMS**
>
> **Glycolysis** – The breakdown of glucose, producing pyruvate and energy in the form of ATP.
> **ATP** – Also known as Adenosine Triphosphate, energy source in all cells.
> **Pyruvate** – End product of glycolysis.

Growing microorganisms

In order for fermentation of food products to take place, we first must ensure we have the available microorganisms. Industrial fermenters are used to produce large amounts of microorganisms, including the examples in Table 18.9.

Industrial fermentation

Industrial fermenters are filled with nutrients and sterilised to ensure optimum growing conditions for the microorganisms. They have water jackets to keep the temperature constant so that the microorganisms grow at their optimum temperature. There is an air inlet to allow oxygen in, providing microbes with the ability to respire aerobically. The air inlets are filtered to prevent entry of other unwanted substances. Stirring paddles mix the microorganisms with the nutrients and keep the temperature constant. Finally, the nutrient inlet allows sterile nutrients in so that microorganisms can continue to grow and reproduce. The following conditions must be maintained for optimal microbe growth: oxygen levels, temperature, pH and nutrient supply.

Energy sources

Throughout the fermentation process, microorganisms require energy and nutrients: carbon, nitrogen, water, salts and micronutrients. Carbon sources are usually sugars or other carbohydrates; for example molasses, which is a by-product of sugar cane. Some fermentation may be more sensitive, and therefore use purified sugar and the carbon source. Nitrogen sources are required for the production of proteins, nucleic acids and other cellular products. Nitrogen will be provided in the form of bulk protein, pre-digested polypeptides (proteins) or as ammonia.

Glycolysis

Glycolysis is an anaerobic metabolic pathway, breaking down glucose and producing energy in the form of **ATP**, also known as anaerobic cellular respiration. Glycolysis takes place in the cytoplasm of cells, or just outside of the bacterium. In aerobic conditions, when oxygen is available, further processes follow glycolysis; these are the Krebs cycle and the electron transport chain. These steps produce many more ATP molecules than glycolysis. During glycolysis two **pyruvate** molecules are formed from carbon molecules belonging to glucose. If oxygen is present, these pyruvate molecules move into the mitochondria undergoing the Krebs cycle, producing even more ATP molecules. This doesn't occur in lactic acid and yeast fermentation; instead glycolysis occurs continuously.

Optimum fermentation conditions

In order to produce high quality food products using fermentation, conditions must be at an optimum for microorganisms to grow and function efficiently.

Table 18.10 Optimal conditions for microorganisms to grow and function

Condition	Optimal requirements
pH	Most microorganisms prefer neutral conditions, whereas others may prefer more acidic or alkaline. The pH must be set at optimum to prevent damage to the microbes occurring.
Oxygen concentration	Some microorganisms require oxygen to produce energy, via aerobic respiration.
Nutrients	All microorganisms require nutrients for metabolism, mostly carbohydrates, as the energy requirements of microorganisms are high.
Hygiene	Fermenters are sterilised to prevent contamination from occurring.
Fermentation duration	Different types of fermentation require a variety of durations. Depending on which process is being undertaken, the correct duration must be set.
Temperature	For example, most lactic acid bacteria work best at temperatures between 18 and 22°C.

KNOW IT

1. Name the most commonly used microorganisms in food production.
2. Why does the Krebs cycle and the electron transport chain not occur in lactic acid and alcohol fermentation?

essay, focus on the microorganisms utilised in these processes and why they are essential.

TOP TIPS
- ✔ Learn the methods used to produce microbiological food products.
- ✔ Watch demonstrations online of how to produce microbiological food products to familiarise yourself with the processes, which will help with **P5**.

LO3 Assessment activities

Below are suggested assessment activities that have been directly linked to the Pass and Merit criteria in LO3 to help with assignment preparation and include top tips on how to achieve best results.

Activity 1 Describe the use of microbes in food production P4

List the microorganisms used in food production, including fermentation.

Activity 2 Produce a microbiological food product under optimum conditions P5

In order to prepare yourself for this, complete the fermentation practical first. Conduct research beforehand and ensure you understand which methods will work most efficiently and safely.

Activity 3 Explain the optimum conditions for growth of microorganisms during a fermentation process M3

Complete a table including the optimum conditions. Also include why these optimum conditions are important and how they affect the microorganisms.

Activity 4 Describe the biochemical processes involved in the production of a food from microorganisms M4

This should be completed in essay format. Describe cellular respiration, focusing on glycolysis. Move into lactic acid and alcohol fermentation. Throughout the

LO4 Understand the action of antimicrobials on microorganisms
P6 P7 P8 M5 D3

KEY TERMS

Antimicrobial – A substance that destroys or inhibits microbes.

Antibiotic – A substance that destroys or inhibits bacterial cells.

Antiviral – A substance that inhibits viral cells.

Antifungal – A substance that destroys or inhibits fungi.

GETTING STARTED

One of the most important facts you must remember when studying **antimicrobials** is the difference between the main types, and which microorganism they target. For example, **antibiotics** are only used to destroy bacteria. At the end of this Learning Outcome, you will understand the difference between the antimicrobials and which microorganisms they target.

In Table 18.11 there are a list of infections and diseases. Identify whether they can be treated with antibiotics, antiretrovirals, **antivirals** or **antifungals**, as well as whether the infection or disease is bacterial, viral or fungal. The first one has been completed for you.

Table 18.11 Which antimicrobial should be used?

Infection	Type of infection	Antimicrobial
HIV	Viral	Antiretroviral
Athletes foot		
Food poisoning		
Diarrhoea		
Pneumonia		
Ringworm		

UNIT 18 MICROBIOLOGY

4.1 Types of antimicrobial

Antimicrobials are agents which kill microorganisms, or inhibit their growth. They are grouped depending on the microorganism they target:

- antibiotics: bacteria
- antifungal: fungi
- antiviral: virus.

The word antimicrobial is derived from the following Greek words: anti (against), mikros (little) and bio (life). At the start of the unit you completed a task demonstrating how you can break down scientific terminology to find out the meaning. This is a great example of that.

4.2 Major classes of antibiotics

Antibiotics are antimicrobials which fight against bacterial infections. There are many different types, which all have a varied mode of action, but they fight bacteria by either destroying them or preventing them from reproducing. Remember, antibiotics cannot be used to treat viral infections like the common cold or fungal infections like athletes foot.

PAIRS ACTIVITY

Researching antibiotics (30 minutes)

Table 18.12 consists of the major classes of antibiotics and their main uses. In pairs research the brand names, treatment uses and mode of action for each class and complete the table. The first three have been completed for you.

Table 18.12 Uses of antibiotics

Antibiotic	Use
Penicillin	• brand names: Penicillin V, Penicillin B, Amoxicillin • one of the most widely used antibiotics • treats most bacterial infections • mode of action – inhibit formation of peptidoglycan
Aminoglycoside	• brand name: Streptomycin • treats tuberculosis and other general bacterial infections • mode of action – protein synthesis inhibitor
Tetracycline	• brand names: Sumycin, Tetracyn • treats acne, rosacea, chlamydia • mode of action: inhibits protein synthesis
Glycopeptides	• cephalosporins, macrolides, quinolones, carbapenems, oxazolidinones
Monobactams	
Sulfonamides	

4.3 Mode of action of antibiotics

As mentioned previously, different classes of antibiotics have varied modes of action. It depends on a number of factors, including whether the target bacteria is Gram positive or Gram negative. As you now know, Gram positive bacteria has a thick cell wall with higher levels of peptidoglycan, making it harder to penetrate and destroy. Therefore the type of antibiotic used to treat it must have very specific methods in order to be successful.

Table 18.13 The main modes of antibiotic action

Inhibitors of cell wall synthesis	Inhibitors of protein synthesis	Inhibitors of nucleic acid synthesis
Bacterial cell walls are critical for their survival and ability to cause infection. Some antibiotics inhibit the cross linkage of peptidoglycan; by penetrating the cell wall a bacterial cell will no longer be able to survive.	Protein synthesis is necessary for the reproduction of most cells, including bacterial cells. The disruption of protein synthesis by binding to ribosomal subunits will interrupt its normal metabolic processes, therefore leading to the death or inhibition of growth.	DNA is the key to reproduction; once this has been compromised, in this case by binding and inhibiting DNA biosynthesis, bacterial cells will be unable to multiply or survive.
Examples: Penicillin, glycopeptides	Examples: Tetracyclines, aminoglycosides	Examples: Quinolones, Rifampin

4.4 Antimicrobial resistance

Antimicrobial resistance (AMR) is when a microbe becomes resistant to antimicrobials previously used to treat them. This mainly occurs in antibiotics, through evolution and overuse. Resistance is an ever-growing threat, which affects all parts of the world. Once resistant, antimicrobials are able to withstand and attack from treatments such as antibiotics. If antimicrobial resistance persists and increases, common infections will no longer be treatable, surgery will become life threatening, and our normal daily routines appear at risk.

Mechanisms of antimicrobial resistance

The main mechanisms of resistance are gene mutations or gene swapping. Gene swapping occurs when bacteria contain resistant genes; however, once this particular gene has been transferred from their forerunners, eventually that class of microbe will become resistant altogether. Genetic mutations can develop resistance or strengthen an already present resistance.

Let's take antibiotics as an example. Some bacterial cells do not carry any resistance and therefore will be destroyed immediately once faced with the antibiotic. However, bacterial cells which carry resistant genes will be able to resist the mechanisms of the antibiotic. Once they have survived, they will go on to replicate and produce stronger bacterial cells. When confronted with an antibiotic in the future, the strongest bacterial cells will survive, resulting in full antibiotic resistance. Table 18.14 shows mechanisms of antimicrobial resistance.

Antimicrobial resistance trends

Trends in AMR are consistently increasing, becoming a serious concern for national health organisations. Figure 18.7 is a graph identifying the increase of antimicrobial resistance of *Staphylococcus aureus*, which is responsible for causing skin infections, pneumonia and meningitis.

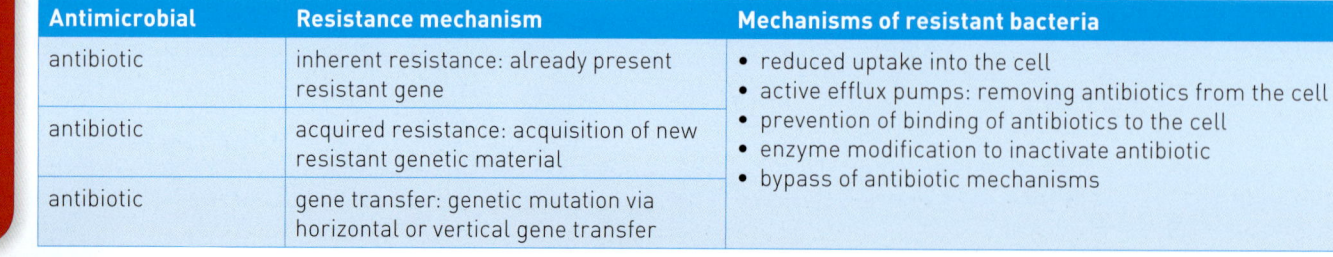

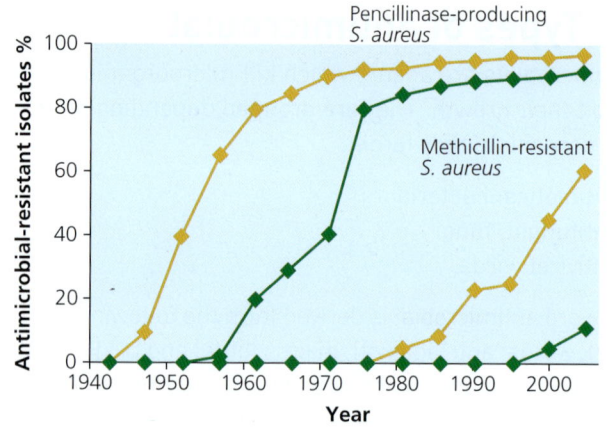

▲ **Figure 18.7** The diagram demonstrates simply how resistant bacteria can resist antibiotic effects, resulting in further reproduction and gene swapping

The green squares represent community-acquired infections and the yellow squares represent nosocomial infections (infections acquired in a hospital). *Staphylococcus aureus* is currently resistant to a type of antibiotic known as Methicillin, as well as demonstrating the ability to produce penicillinase, which inactivates natural penicillins. As you can see in the graph, there is a steady increase of both nosocomial and community-acquired infections caused by *Staphylococcus aureus*. This is due to its resistance to certain antimicrobials. MRSA (methicillin-resistant *Staphylococcus aureus*) is on the increase; from 1st April 2013 all cases of MRSA must be documented via a post infection review (PIR) that has been implemented by the UK government. This is a lengthy document, documenting exactly what the hospital patient was suffering and their journey whilst in hospital.

Projected models for AMR

The concern for AMR has become critical since 2013 and the breakout of MRSA when the UK government conducted a review on AMR. Included in this review was a future projection of what will happen if AMR is not tackled immediately. AMR threatens medical advances within microbiology. Although in developed countries there are alternative treatments, microbes will only continue to evolve and become resistant to more AMRs. Figure 18.8 shows a projected review of deaths caused by AMR in the year 2050, in comparison to other major causes of death.

Table 18.14 Antimicrobial resistance mechanisms

Antimicrobial	Resistance mechanism	Mechanisms of resistant bacteria
antibiotic	inherent resistance: already present resistant gene	- reduced uptake into the cell - active efflux pumps: removing antibiotics from the cell - prevention of binding of antibiotics to the cell - enzyme modification to inactivate antibiotic - bypass of antibiotic mechanisms
antibiotic	acquired resistance: acquisition of new resistant genetic material	
antibiotic	gene transfer: genetic mutation via horizontal or vertical gene transfer	

Deaths attributable to AMR every year compared to other major causes of death

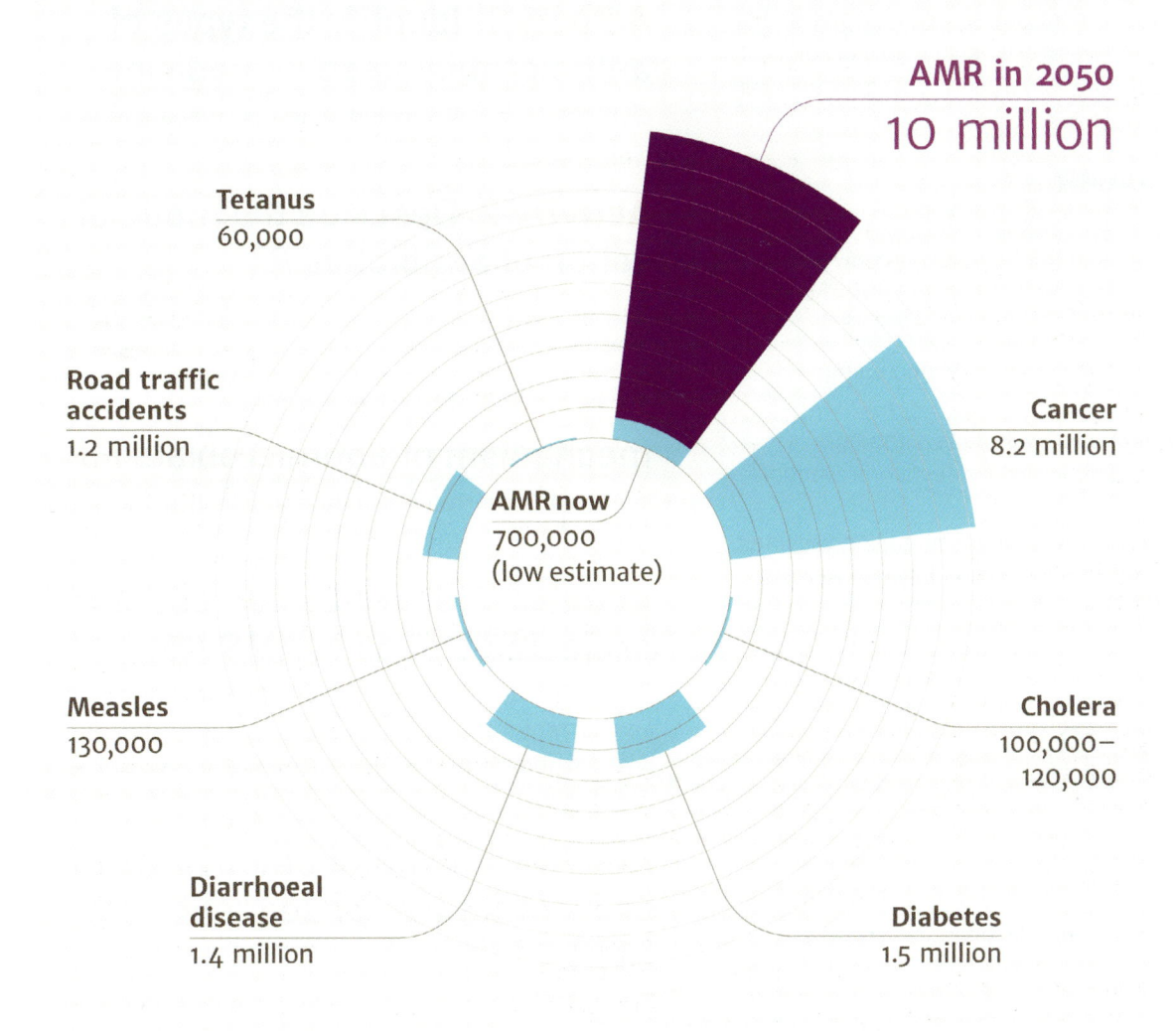

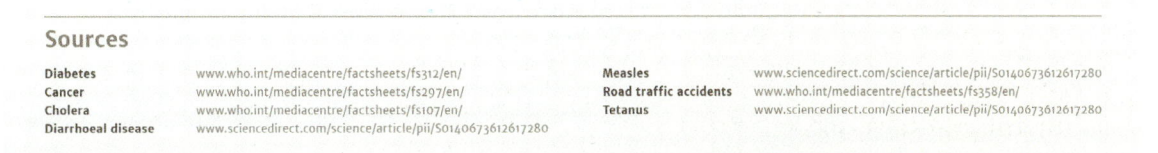

- **AMR in 2050**: 10 million
- **AMR now**: 700,000 (low estimate)
- **Tetanus**: 60,000
- **Road traffic accidents**: 1.2 million
- **Measles**: 130,000
- **Diarrhoeal disease**: 1.4 million
- **Diabetes**: 1.5 million
- **Cholera**: 100,000–120,000
- **Cancer**: 8.2 million

Sources

Diabetes	www.who.int/mediacentre/factsheets/fs312/en/
Cancer	www.who.int/mediacentre/factsheets/fs297/en/
Cholera	www.who.int/mediacentre/factsheets/fs107/en/
Diarrhoeal disease	www.sciencedirect.com/science/article/pii/S0140673612617280
Measles	www.sciencedirect.com/science/article/pii/S0140673612617280
Road traffic accidents	www.who.int/mediacentre/factsheets/fs358/en/
Tetanus	www.sciencedirect.com/science/article/pii/S0140673612617280

▲ **Figure 18.8** The chart clearly shows that if AMR isn't combated, the deaths caused by AMR will increase and become more deadly than other causes of death currently

Implications of AMR on local services

In addition to the critical effects of AMR on human lives, there are further implications of AMR on the economy, hospitals and care homes. Extra funding has been utilised for AMR research, further production of antibiotics must take place and extra treatment of patients in hospitals is already increasing. Hospitals and care homes are under pressure to take extra patients due to AMR, having a negative effect on the labour force, as more staff are required and are not always available. The UK government review stated that if AMR is not tackled, there would be a reduction of between 2% and 3.5% in gross domestic product (GDP).

Solutions to AMR

The AMR crisis can be deflected. AMR is an international issue, and the benefits will only be achieved if international action takes place. The main solutions, which will decrease AMR, include:

- Reduce misuse of antimicrobials; including in humans, animals and agriculture.
- Create a global public awareness campaign.
- Improve hygiene and improve infection prevention techniques.
- Develop alternative medications including vaccines.
- Create and develop a global fund for research and investment in new drugs.

In summary, by creating a global awareness campaign, the wider population will be educated about the risks of AMR as well as how to utilise antimicrobials efficiently in order to prevent AMR. Hygiene improvement and infection prevention will decrease the amount of resistant microbes from evolving; in addition to this the development of alternative treatments may stop this completely. As well as this, the development of new treatment methods will possibly have a beneficial impact on the economy. For example, if vaccines are implemented, there will be no need for the use of other therapeutic drugs, which will release funds from that area.

As shocking as the statistics are, AMR is a crisis that can be solved if the recommendations are completed.

KNOW IT

1. Define antimicrobials.
2. Identify the main classes of antibiotics.
3. Do you think antibiotics are more effective on Gram positive or Gram negative bacterial infections, and why?
4. Identify the mode of actions for antibiotics.
5. Identify the three most important strategies for targeting AMR and explain why you think they are the most important.

LO3 Assessment activities

Below are suggested assessment activities that have been directly linked to the Pass, Merit and Distinction criteria in LO4 to help with assignment preparation and include top tips on how to achieve best results.

Activity 1 Identify the types of antimicrobial used in medicine *P6*

List these in a table, including the microbe that the antimicrobial targets.

Activity 2 Describe the mode of action of antibiotics *P7*

You can display this in a flow chart format, and include images.

Activity 3 Describe the mechanism of antimicrobials *P8*

Include this information in the same table as P6 by adding extra columns.

Activity 4 Describe current and projected trends in AMR *M5*

Display this in essay format, describing the projected and current trends, and referring to tables and graphs. (Use the information in this unit, and the suggestions in 'Read about it'.)

Activity 5 Evaluate measures to prevent future consequences of AMR *D3*

Following on from M5, in an essay evaluate identified measures. For this section ensure you use facts that back up your work.

TOP TIPS

- ✓ Learn and understand the difference between the main types of antimicrobials.
- ✓ Understand the modes of action of specific antibiotics.
- ✓ Complete extensive research surrounding AMR in order to work towards the higher criteria.

Read about it

Journals and reviews

O'Neill, J. (May 2016) 'Antimicrobial resistance: Tackling a crisis for the health and wealth of nations'

Websites

World Health Organization

www.who.int/en/

Microbiology Online

www.microbiologyonline.org.uk

MicrobeNet

www.microbe.net

American Society for Microbiology

www.asm.org

Books

Mims, C. *et al.* (2003) *Medical Microbiology* (3rd edition), Mosby

Murray, P.R. *et al.* (2003) *Manual of Clinical Microbiology* (8th edition), American Society Microbiology

Wilkins, C.L. and Lay Jr, J.O. (eds) (2006) 'Cultural, Serological, and Genetic Methods for Identification of Bacteria' in *Identification of Microorganisms by Mass Spectrometry*, Wiley

Unit 21
Product testing techniques

ABOUT THIS UNIT

Product testing is an imperative stage during and after the production of new equipment. This unit will cover the processes involved in testing drugs, foodstuffs and cosmetics. Every consumer product must meet government regulations in order to be released; they go through rigorous testing to ensure their reliability, durability and, most importantly, safety. The testing involves a number of conditions the product may be exposed to during transportation, daily usage and storage. Specific products are also further tested for possible side-effects following inhalation and ingestion. In addition to testing the usage of products, companies can also undergo cost analysis, competitor's analysis and product improvement strategies. The full product techniques process will provide the company with product excellence over their competitors, customer satisfaction, cost-effectiveness and consistent improvement.

This unit will introduce you to the world of product testing, the regulation of consumer products and the techniques utilised in these processes. You will conduct research into regulatory bodies and undertake your own experiments into product testing.

LEARNING OUTCOMES

The topics, activities and suggested reading in this unit will help you to:

1. Understand the influence of regulatory bodies on development of consumer products
2. Understand how product testing determines the development of consumer products
3. Be able to use quantitative titration techniques on consumer products
4. Be able to use extraction and separation techniques on consumer products

How will I be assessed?

You will be assessed through a variety of tasks, which may include the following:

- Carrying out practical work safely and analysing results
- Completing assignment and portfolio work
- Controlled assessment
- Completing poster work, presentations and group work

How will I be graded?

You will be graded using the following criteria:

Learning Outcome	Pass	Merit	Distinction
	The assessment criteria are the Pass requirements for this unit.	To achieve a Merit the evidence must show that, in addition to the Pass criteria, the candidate is able to:	To achieve a Distinction the evidence must show that, in addition to the Pass and Merit criteria, the candidate is able to:
1 Understand the influence of regulatory bodies on development of consumer products	**P1** Describe the requirements of the relevant governing body on the development of consumer products	**M1** Explain how governing bodies influence quality control	
2 Understand how product testing determines the development of consumer products	**P2** Select tests to be used in product development	**M2** Explain how the effectiveness of consumer product testing is established	
	P3 Outline procedures used during formulation, production, quality control and after sale monitoring		
3 Be able to use quantitative titration techniques on consumer products	**P4** Use titrimetric techniques on consumer products	**M3** Determine the concentration of substances in consumer products using quantitative methods	**D1** Evaluate concentration of substances against those stated on product labels
4 Be able to use extraction and separation techniques on consumer products	**P5** Use solvent extraction to separate and determine the mass of the active ingredient of a consumer product		
	P6 Use TLC to investigate qualitatively the composition of a consumer product	**M4** Calculate Rf values of constituents of the consumer product to provide quantitative information of a consumer product	

LO1 Understand the influence of regulatory bodies on development of consumer products *P1 M1*

KEY TERM

Consumer – A person who purchases goods or services.

GETTING STARTED

Food Standards Agency (30 minutes)

The Food Standards Agency is one of many governing bodies responsible for monitoring **consumer** products. They were established on April 1st 2000, after taking the place of the Ministry of Agriculture, Fisheries and Food (MAFF). The MAFF suffered some controversy after its inability to reassure the consumers of the safety of the food they were purchasing. The main aim of the Food Standards Agency is to provide an honest and transparent process, in comparison to the MAFF. So far they have succeeded with their aims; their website is interactive and constantly updated with health warnings regarding food, and they also work alongside the World Health Organization conducting regular research reports revolving around improving the health and safety of the UK population.

Find and read a report written by the Food Standards Agency, 'Creating an enabling environment for population-based salt reduction strategies'. Discuss the following:

- How effective do you think the report is?
- Do you think they are working towards a healthier population by conducting these types of reports?
- Will this influence the quality of food in the UK?

> **KEY TERMS**
>
> **Pharmacopoeia** – A publication containing a list of medical drugs with their mode of action, directions for use and effects.
>
> **Quantitative** – Measuring the quantity of something.

1.1 Governing bodies

A number of governing bodies oversee the product testing for consumer products across the world. These bodies aim to keep the public safe and informed.

Consumer product requirements

Each product that is being tested must pass a number of requirements set by the company; these are different depending on the type of product.

Food labelling

There are strict policies in place regarding food labelling. If regulations are not adhered to, the individual or company can be prosecuted. The main regulations are the European Food Information to

Table 21.1 Governing bodies, their purpose and requirements

Governing body	Purpose	Requirements
UK government	The UK government has a number of organisations that work towards protecting and improving human and animal food, medicines, the environment, cosmetics and innovative research. Examples of these include: Department for Business, Innovation and Skills (BIS), Department for Environment, Food and Rural Affairs (DEFRA) and the Department of Health.	• various
Medicines and Healthcare Products Regulatory Agency (MHRA)	MHRA is an agency of the Department of Health in the UK, which regulates medicines, medical devices and blood components. Also supporting innovation in the field of medicine.	• see ABPI
Good Laboratory Practice (GLP)	Originated in the 1970s following concerns about the validity of data submitted to the FDA (see below). GLP is another agency of the Department of Health in the UK. The GLP provides regulations which must be followed by any testing facility in the UK. This includes pharmaceutical, agrochemicals, veterinary medicines, cosmetics, industrial chemicals, biocides and additives in foods.	• quality assurance • legislation and regulations adhered to • organisation and personnel • equipment maintenance • reporting laboratory data
Trading Standards	The Trading Standards Institute is an association within the UK, which safeguards the health and safety of the public by supplying information of approved consumer products and a helpline for advice, and it promotes and protects the success of the current economy.	• monitor and evaluate standards of varying organisations
Food Standards Agency	The Food Standards Agency is responsible for the monitoring of the safety of food in the UK, as well as ensuring food meets expectations of the consumer in terms of its quality. They provide regular helpful information on foods consumed by the public, as well as performing hygiene tests on restaurants, which you can find on the FSA website.	• food labelling • pesticide analysis • nutritional analysis • chemical analysis • microbiological testing
Association of the British Pharmaceutical Industry (ABPI)	The ABPI supports innovative research in biopharmaceuticals in the UK; they represent a high percentage of medicines supplied to the NHS.	• skip/periodic testing • release vs shelf life criteria • in process tests • design considerations • parametric release • **pharmacopoeia** tests • reference standard • clinical trials
Food and Drug Administration (FDA) based in the USA	Aim to protect and advance public health by approving and innovating human and veterinary drugs, biological products, medical devices, food, cosmetics and products which emit radiation.	• see Food Standards Agency and ABPI

Consumers Regulation 2011 (FIC) and the Food Information Regulations 2014 (FIR). The following information must be applied to food labels:

- the name of the food
- allergenic ingredients
- **quantitative** ingredients declaration
- list of ingredients
- nutritional information
- weight or volume
- best before or use by date
- name and address of the food business operator
- the alcoholic strength by volume for drinks containing more than 1.2% alcohol
- the mandatory information must be in an appropriate font with a minimum height of 1.2 millimetres
- storage information
- country of origin.

Laboratory food testing

Part of the product testing process for food includes ensuring it is safe from microbiological contamination, nutritional and chemical values are correct and pesticide residue is minimal.

Microbiological testing will occur at each stage of food production; ingredients/raw material, factory equipment, manufacturing process, packaging, storage and intended use. An aerobic colony count will be performed for each of these stages, to analyse the level of microbes present. As well as this, specific organisms must be tested for: *Salmonella*, *Listeria monocytogenes*, *Enterobacteriaceae* and *E. coli*. On animal carcasses, aerobic colony count is undertaken; if the microbes present are not acceptable, the meat will not be fit for sale. However, if the meat is already in store, it can remain on the market and the food business operator must review their production process.

Pesticide residue analysis is an essential part of product testing for vegetables, cereals, tea, foodstuffs, fruit, dairy, fish and water products. Farmers use pesticides to ensure produce is plentiful and aesthetically pleasing; however, if the levels of pesticides in the produce are too high, they can be harmful. Some pesticides are toxic, carcinogenic and can affect fertility. Chlormequat is a commonly used pesticide for regulating plant growth and carries no toxicity, whereas carbendazim is also a commonly used pesticide, but is known to be toxic and carcinogenic. Pesticide residue analysis must be performed almost immediately to achieve a reliable result, as pesticides are subject to degradation. Sampling techniques consist of soil sampling, water sampling, sediment sampling, vegetation sampling, tissue sampling and vertebrate sampling. The data is recorded and analysed to conclude whether the levels of pesticides are acceptable.

Nutritional and chemical testing provides information on fat, protein, carbohydrate, fibre and sodium content. In addition to this, vitamins, minerals, trace elements and pH are assessed. This type of product testing must be completed to ensure the declared levels are correct.

Drug testing

Drugs cannot be released to the public before rigorous testing has taken place; the developers must ensure the potential to cause harm has been evaluated. Drug development can take around twelve years, consisting of a number of product testing methods. The main requirements of drug development include pre-clinical research, investigational new drug (IND) resulting in clinical trials consisting of 3–4 phases, new drug application (NDA) and final approval.

PAIRS ACTIVITY

Controversy in the FDA (20 minutes)

Although the governing bodies positively work towards keeping individuals safe and informed, controversy cannot be avoided. The FDA has suffered some controversies, mainly revolving around the harmful side-effects of the drugs they have approved and deemed safe for use. Antiretroviral drugs used to treat HIV, chemotherapy used to treat cancer and antidepressants used to treat depression have all been discussed in regards to their controversial effects on patients. A number of research papers have concluded that these treatments not only cause harmful side-effects, but also death. Does this mean the FDA has failed in these cases? Do the advantages of these drugs outweigh the disadvantages? Using the internet and books, in pairs complete Table 21.2 and discuss your findings.

Table 21.2

Drug	Disease or disorder	Side-effects	Long lasting effects	Success rate
Antiretrovirals	HIV			
Chemotherapy	Cancer			
Antidepressants	Depression			

> **EXTENSION ACTIVITY**
>
> **How successful are antiretrovirals? (30 minutes)**
>
> Using your research techniques, find the most recent success rates for antiretrovirals. Use the internet to find the data. Here are some helpful websites: www.aidstruth.org and www.who.int. Evaluate the data and compare this with current HIV deaths. Take the following into consideration:
>
> - Do you think the severe side-effects caused by antiretrovirals outweigh the success of the drug?
> - Do you think the side-effects depend on the person taking the drug?
> - Is there a need for an improved version of antiretrovirals?

1.2 How government bodies influence quality control

Quality control of product testing

Quality control (QC) monitors the validity of work conducted in a laboratory or any other workspace. QC is imperative in order to reduce and correct deficiencies. It can be utilised to validate equipment, protocols and staff. In most companies there is a quality assurance team, who monitor and regulate the compliance of regulations and validity of work undertaken. Governing bodies have a strong influence on the quality of product testing; they do this by setting strict regulations and legislations. Table 21.3 demonstrates a range of regulations and legislations set by four governing bodies.

Table 21.3 Regulations and legislations set by governing bodies

Food Standards Agency	FDA	Association of the British Pharmaceutical Industry (ABPI)	Good Laboratory Practice (GLP)
European legislation	The Food And Drugs Act 1906	The International Federation of Pharmaceutical Manufacturers and Associations (IFPMA) Code of Practice	Statutory Instrument 1999
FSA policy on handling disclosures made under The Public Interest Disclosure Act (1998)	The Federal Food, Drug, And Cosmetic Act 1938	The European Federation of Pharmaceutical Industries and Associations (EFPIA) Code on the Promotion of Prescription-Only Medicines to, and Interactions with, Healthcare Professionals	Statutory Instrument 2004
The General Food Regulation	The Medical Device Amendments Act 1976	The EFPIA Code of Practice on Relationships between the Pharmaceutical Industry and Patient Organisations	The European Communities Act 1972
The Food Safety Act 1990	Food And Drug Administration Amendments Act (FDAAA) 2007	The EFPIA Code on Disclosure of Transfers of Value from Pharmaceutical Companies to Healthcare Professionals and Healthcare Organisations	REACH regulation 2006/7 (registration, evaluation, authorisation and restriction of chemicals)
The General Food Regulations 2004	Animal Drug User Fee Act 2003	The World Health Organization's Ethical Criteria for Medicinal Drug Promotion	Directive 2004/10/EC: harmonisation of laws, regulations and administrative provisions
The Food Standards Act 1990	Office of the Chief Counsel	Directive 2001/83/EC on the Community Code relating to medicinal products for human use, as amended by Directive 2004/27/EC	Directive 2004/9/EC: compliance monitoring procedures for GLP

CASE STUDY

What is really in our cosmetics?

Do you know what's in the cosmetics you use on a daily basis? Do you check the labelling regularly? Cosmetics products consist of make-up, grooming products, fragrances, skincare and haircare. Therefore, it is not just women who need to worry about the ingredients in cosmetic products. The main organisations overseeing the safety of cosmetic products are the Cosmetics Regulation and the FDA; however, they do not have the power to approve cosmetic ingredients before they are sold to the public. So who is to say they are safe? A group known as Campaign for Safe Cosmetics conducted research on the contents of fragrances. They found 14 undisclosed chemicals; among these were chemicals known as phthalates, which are used to soften plastic. Further research by the group revealed the presence of lead and other toxic metals in a large number of lipsticks in 33 well-known brands. This occurred in 2007; the FDA then failed to take action until 2009. They finally conducted their own research, and found the presence of lead to be four times higher than that found in the study undertaken in 2007. Brands involved included L'Oreal, Maybelline, MAC and Body Shop. Lead can be poisonous at even the tiniest of exposure; health concerns include decreased fertility, hormonal changes and neurotoxicity. The lead issue still continues; other toxic metals identified consist of aluminium and chromium. The FDA do not believe the amount of lead identified in the lipsticks is harmful, even though scientists state any level is harmful, and the lipsticks are still being released for sale.

1 Research the ingredients of the cosmetic product you most often use.
2 Identify any harmful effects of these ingredients.
3 How do you think product testing on cosmetics can be improved?

KNOW IT

1 What are governing bodies and why are they important?
2 What are the main consumer product requirements?
3 How does quality control ensure consumer products are safe for consumer use and high quality?

LO1 Assessment activities

Below are suggested assessment activities that have been directly linked to the Pass and Merit criteria in LO1 to help with assignment preparation and include top tips on how to achieve best results.

Activity 1 Describe the requirements of the relevant governing body on the development of consumer products *P1*

This criterion can be completed in table form or as a professional poster, identifying the main requirements and aims of each governing body.

Activity 2 Explain how governing bodies influence quality control *M2*

This should be completed as an essay, evaluating the influence the governing bodies have on quality control. Research should be completed in order to come to a clear and informed evaluation.

TOP TIPS
✔ Complete extensive research to ensure understanding is gained.
✔ Use different forms of methods to complete research.

LO2 Understand how product testing determines the development of consumer products P2 P3 M2

> 🔑 **KEY TERMS**
>
> **In-vitro** – A process performed outside of a living organism.
>
> **In-vivo** – A process performed within a living organism.
>
> **Titration** – A technique used to determine the concentration of an unknown solution.
>
> **Extraction** – The action of removing something.

GETTING STARTED

Did you know products were tested? (10 minutes)

Did you know that every product you use, whether it is cosmetics, food, medicine, toiletries, homewear, clothing, is all tested for suitability and safety? Do you think this information should be made more accessible? When products fail their testing, would you like to know? Do you think the general public should have a say on the efficiency of the testing used? Discuss in a small group your thoughts on these questions.

2.1 Types of testing

There are a variety of tests which are undertaken during product testing. Some are extremely vigorous and lengthy.

In-vitro

In-vitro testing is most commonly known as a 'test tube' experiment; it is used for a variety of purposes including product testing for cosmetics, drugs, food, toxicology and diagnostics. Before in-vitro was introduced, testing on animals was the only way to analyse the effects of these types of products; however, there is no better way of analysing the safety of products than to test them on humans, as animals are not predictive of a human reaction. In-vitro allows this to happen, performing investigations on cells, tissues and biological molecules which have been isolated outside of their normal biological environment. In-vitro testing has many benefits: it is cost-effective, reliable and more accurate than animal testing, and it prevents animal cruelty. Harvard University has created a state-of-the-art system, which mimics the structure and function of human organs, and organ systems; it is called 'organs-on-chips'. Organs-on-chips are microchips lined by human cells; they are clear in structure providing clear visibility of action, and the aim is to rapidly assess responses to new products. Harvard's invention is one to add to the in-vitro phenomenon. The in-vitro technique is revolutionary, will change the way products are tested indefinitely, and create a safer environment for everyone.

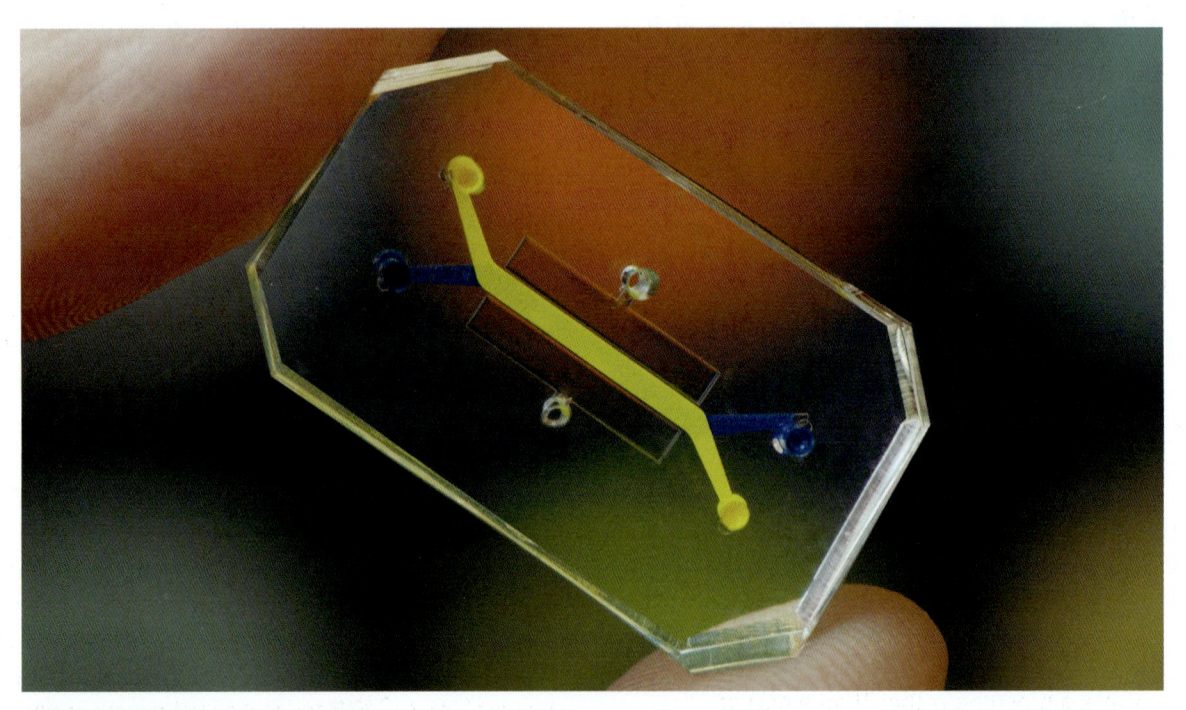

▲ **Figure 21.1** Organs-on-chips

In-vivo

In-vivo is a process which takes place within a living organism; this can be animals, plants or humans. Due to the practical and ethical concerns related to human experimentation, animal models have been utilised for the testing of new products, especially drugs. In 1933 around a dozen women were blinded and one woman died from using a permanent mascara called Lash Lure, which contained an untested chemical *p*-phenylenediamine. This disastrous event led to the increased use of animals for testing: the main organisms used are rats, mice and other rodents. For drug testing the animal is injected with disease strains and treated with new drugs to assess the success rates and side-effects. This process has been used to develop drugs such as antibiotics, antiviral drugs, chemotherapy and new surgical procedures. Examples of common product safety tests include:

- Toxicity – LD50 test (lethal dose 50%): animals ingest a high dosage of chemicals to assess the toxicity of gases and powders
- Draize test: used to measure eye irritancy
- Embryotoxicity: used to assess the toxic effects of a substance on an embryo; pregnant animals are killed prior to delivery and the fetus is studied.

Organisations including PETA (People for the Ethical Treatment of Animals) and Cruelty Free International protest against animal testing due to the harm it can cause the animals. Through this the use of animal testing for cosmetics products has been banned in the UK since 2013. Animals are still being used for drug testing, and companies must follow strict regulations to ensure this is conducted in the most humane way, although current research suggests many tests performed on animals have little validity, as animals do not respond similarly to humans. The disadvantages of animal testing are that it can be expensive, time-consuming and unpredictable, although it does have some benefits as scientists can control the level, timing and combinations of exposures.

Titration

The method of **titration** is also known as volumetric analysis; it is a common laboratory procedure used to identify the unknown concentration of a sample. The chosen sample is weighed out into a conical flask, a second chemical is used (usually an acid or alkali) and an indicator is added which will eventually indicate the end point of the experiment. It is used for a variety of reasons depending on the type of industry. For each industry the preferred method of titration is automated titration, where a machine rather than a person performs the procedure. The benefits of automated titration over manual titration are as follows:

- It is less operator dependent.
- Each sample is titrated in exactly the same way.
- The minimum sample size is smaller, increasing the accuracy of results.
- Errors in results and calculations are minimised.
- It is cost-effective.

Table 21.4 describes which types of industries use titration, the purpose and their preferred method of titration.

Table 21.4 Industrial titration methods

Industry	Purpose	Preferred method
Pharmaceutical	determine the purity of a drug	automated titration manual titration
Food	determine the quantity of a reactant in a sample, for example, salt, sugar, vitamins	automated titration manual titration
Cosmetics	assess the chemical levels	automated titration manual titration

Extraction and separation

Extraction and separation techniques are used to separate gases, liquids and solids. The most common methods include chromatography, filtration, centrifugation and electrophoresis. Although these methods are different from titration, the purpose is very similar. Chromatography is the main method of choice; a sample is dissolved first in the mobile phase, and the various compounds travel at different speeds resulting in separation. The objectives of extraction and separation are to determine purity, determine the quantity of a reactant, determine chemical levels and assess general quality of a product.

2.2 Laboratory testing during development

Throughout each stage of product development, laboratory testing is conducted to ensure the product meets the standards required. Testing continues long after its release; the safety and quality of each new product is imperative. Figure 21.2 on the next page follows the steps of laboratory testing through product development.

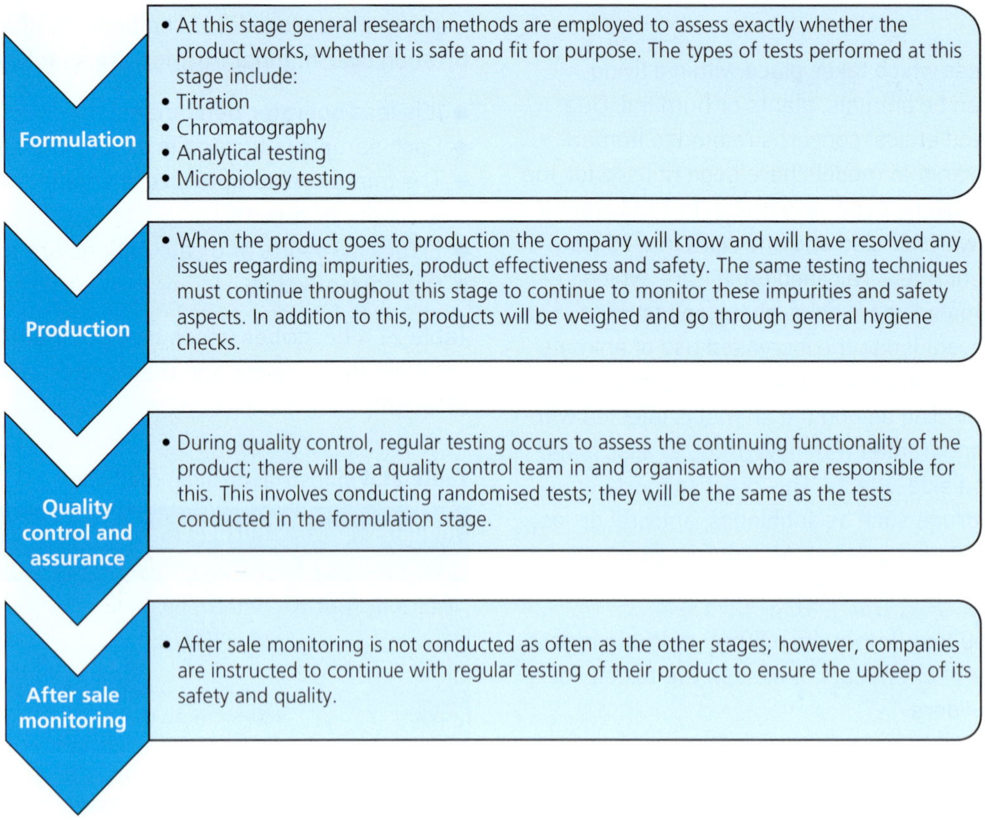

Formulation
- At this stage general research methods are employed to assess exactly whether the product works, whether it is safe and fit for purpose. The types of tests performed at this stage include:
  - Titration
  - Chromatography
  - Analytical testing
  - Microbiology testing

Production
- When the product goes to production the company will know and will have resolved any issues regarding impurities, product effectiveness and safety. The same testing techniques must continue throughout this stage to continue to monitor these impurities and safety aspects. In addition to this, products will be weighed and go through general hygiene checks.

Quality control and assurance
- During quality control, regular testing occurs to assess the continuing functionality of the product; there will be a quality control team in and organisation who are responsible for this. This involves conducting randomised tests; they will be the same as the tests conducted in the formulation stage.

After sale monitoring
- After sale monitoring is not conducted as often as the other stages; however, companies are instructed to continue with regular testing of their product to ensure the upkeep of its safety and quality.

▲ **Figure 21.2** Product development steps

CLASSROOM DISCUSSION

Dangerous testing (20 minutes)

Laboratory testing is completed throughout the product development process to ensure its safety, quality and value, with the consumer in mind. As discussed in this unit, animal testing is a controversial topic and has been banned for use in cosmetic testing, but is still used in other areas. Although there are strict regulations that must be adhered to in order to test on animals, they are still put through stressful procedures, which lead to harm and sometimes death. In-vitro testing is being developed to create an alternative to animal testing; this revolutionary technique could possibly save many animals' lives, and lead to a more effective and reliable option.

In pairs conduct research on the development of new in-vitro testing, focusing on its reliability, worldwide usage and cost-effectiveness. Use the internet to conduct your research; some useful websites are www.neavs.org and www.peta.org.

2.3 Effectiveness of product testing

Product testing is a long and intricate process made up of a number of assessments, which must adhere to strict regulations and legislations. This ensures the effectiveness and reliability of the testing, not allowing a product to be released if it doesn't successfully pass each and every test. In addition to this, follow-up testing and consumer feedback further support the effectiveness of a product. If a product is released and proves to be successful based on feedback and sales, this is proof that the product testing was effective. There are websites and organisations which provide helpful information and feedback on products; Trading Standards is one of the main organisations in the UK who not only provide details about products, but offer verbal advice for consumer complaints. If a product is deemed unsafe or unreliable, it can be pulled and further testing is issued.

KNOW IT

1. What are the main differences between in-vitro and in-vivo testing?
2. How is titration used in product testing?
3. Why does product testing have to continue after the product has been released?
4. How do we know product testing is effective?

LO2 Assessment activities

Below are suggested assessment activities that have been directly linked to the Pass and Merit criteria in LO2 to help with assignment preparation and include top tips on how to achieve best results.

Activity 1 Select tests to be used in product development P2

For this criterion create a table or poster to identify and explain the main tests used in product development: in-vitro, in-vivo, titration and extraction techniques. Include the purpose of use and which products the tests are mainly used for.

Activity 2 Outline procedures used during formulation, production, quality control and after sale monitoring P3

Create a flow chart to explain and demonstrate the procedures clearly: formulation, production, quality control and after sale monitoring.

Activity 3 Explain how the effectiveness of consumer product testing is established M2

This should be presented in essay form; conduct further research to explain the effectiveness of consumer product testing. It would be helpful to include data on success rates, advantages and disadvantages, customer complaints and any other relevant information.

TOP TIPS
- Conduct further research using journals, Google scholar, research articles and newspaper articles to broaden your understanding of the topics covered in this Learning Outcome.
- Make sure you understand and know the range of legislations and regulations; this will help you evaluate the effectiveness of product testing.
- For product testing to be effective: legislations must be adhered to, customer feedback should be positive, the product will be successful financially, the product will be sold in a range of stores.

LO3 Be able to use quantitative titration techniques on consumer products P4 M3

KEY TERM
Reagent – A substance or mixture used in chemical analysis.

GETTING STARTED

Staying safe (30 minutes)

Health and safety is critical when working in a laboratory; each experiment requires a risk assessment, COSHH assessment, and health and safety information. LO3 requires you to complete titration investigations. You must know and learn the health and safety procedures for this. In small groups create a poster identifying and explaining key health and safety actions for a titration experiment and present it to the class.

3.1 Titration techniques on consumer products

LO2 introduced you to the use of titration in industry product testing. It is used to identify the unknown concentration of a sample. The results of titration allow the successful identification of the purity, quality and chemical levels within a product. There are a variety of types of titration experiments, including the following.

Acid-base titration

This type of titration analysis determines the concentration of an acid or a base. You can analyse a number of consumer products using this method, including vinegar and disinfectants. The purpose of an acid-base titration is to assess the quality of the product, the actual acid levels and identify unidentified acids or bases. The standard method is used; however, the consumer product is used instead of an unknown one.

Precipitation titration

A **reagent** known as the titrant is slowly added to a substance until a reaction occurs. This method can be used on substances containing chloride ion, for example any saline solutions, including contact lens solutions. The chloride ion is the titrate and can be titrated with a solution of silver salt; a white precipitate is deposited on the bottom of the flask in this reaction to represent the end point.

Redox titration

A redox reaction between an **analyte** and a titrant relies on the reduction of the analyte itself, which is the unknown. The titrant is the standardised solution and the analyte is the analysed substance; the end point will be determined by a colour change. In order to calculate the concentration in this reaction, the number of moles must be calculated and then multiplied with the volume. Redox titrations are commonly used to analyse the concentration of sodium hypochlorite in bleach; however, the sodium hypochlorite must be added to potassium iodide to produce iodine in order for the titration to take place. Starch is used as an **indicator**, as it turns blue when iodine is present. The calculations will determine an equivalent value for the concentration of sodium hypochlorite.

Complexometric titration

The formation of a coloured complex is used to identify the end point in this type of titration; this method is usually used to detect calcium and magnesium levels in milk, seawater and some solids. A large molecule known as EDTA (ethylenediaminetetraacetic acid) is used; this forms a complex with magnesium and calcium. A blue dye called eriochrome black T is used as an indicator, which also forms a complex with calcium and magnesium. The end point colour should be pink.

> **KEY TERM**
>
> **Analyte** – A substance which is having its chemical structure identified or analysed.
>
> **Indicator** – A substance that gives a visible sign.

Titration procedure

Apparatus
- conical flask
- pipette
- pipette filler
- burette
- volumetric flask
- distilled water
- boss and clamp
- retort stand
- waste beaker

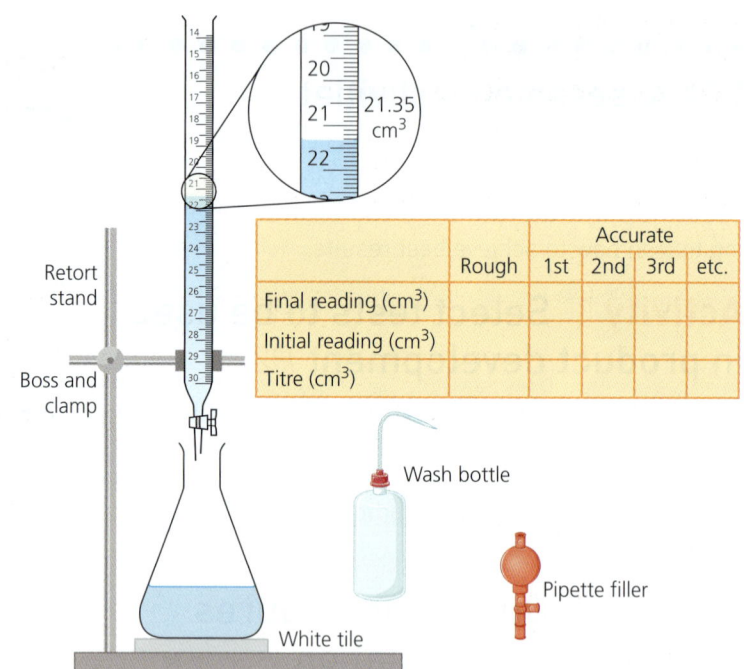

	Rough	Accurate			
		1st	2nd	3rd	etc.
Final reading (cm^3)					
Initial reading (cm^3)					
Titre (cm^3)					

▲ Figure 21.3 Titration set-up

Method
1. Set up the apparatus as in the diagram above.
2. Open the tap on the burette and rinse it by pouring distilled water down it. Make sure there are no air bubbles present and rinse again with your standard solution.
3. Close the tap and fill the burette with the prepared standard solution to the highest graduation point, then open the tap and slowly release the solution into the conical flask until the meniscus sits on the mark. Record this from the bottom of the meniscus.
4. Empty the solution into a waste beaker as you don't need it.
5. Now add the analyte you are assessing to the conical flask, and add two drops of indicator. Record the volume.
6. Now titrate, releasing the tap slowly until you see a colour change that will indicate the end point. Tip: placing a piece of white paper or white tile underneath the flask may make it easier to detect a colour change.
7. Keep swirling the flask until the colour change comes to a halt; this will be your end point.
8. The colour change should be very faint; if the colour is strong, there is too much titrant present.
9. Record the final volume of the burette, subtract the initial volume from the final volume to get the estimated volume of the titrant.
10. Repeat two more times, ensuring you record all results and errors.

PAIRS ACTIVITY

Science workbook (30 minutes)

When you perform scientific investigations, you must be prepared to record results and information regarding the success and implications of your investigation. You are to complete a range of titration experiments for LO3 to ensure you are organised and prepared to collect data. You need to prepare a scientific workbook. You can do this using a computer or by hand in an A4 workbook.

KNOW IT

1 Describe in your own words what the purpose of a titration experiment is.
2 What are the main differences between the four types of titrations?
3 What is the difference between a reagent and an analyte?
4 What are the main types of industry who benefit from using titration techniques?

LO3 Assessment activities

Below are suggested assessment activities that have been directly linked to the Pass and Merit criteria in LO3 to help with assignment preparation and include top tips on how to achieve best results.

Activity 1 Use titrimetric techniques on consumer products *P4*

Complete at least one titration technique assessing a consumer product of your choice, ensuring you complete the following:

- Health and safety
- Results table
- Prepare a workbook to take notes regarding errors, changes throughout the experiment and any other relevant information
- Compare findings to those stated on product labels.

Activity 2 Determine the concentration of substances in consumer products using quantitative methods *M3*

Complete a scientific write-up of the titration experiment completed for *P4*. Include:

- Introduction
- Aim
- Hypothesis
- Apparatus
- Method
- Results
- Conclusion
- Evaluation

TOP TIPS

✔ Follow health and safety advice.
✔ Prepare a workbook to note all errors, results and any other relevant information.
✔ Know the difference between a conclusion and an evaluation.
✔ Practise titration techniques before you conduct the real practical.

LO4 Be able to use extraction and separation techniques on consumer products *P5 P6 M4*

GETTING STARTED

Why extract? (20 minutes)

Extraction techniques are widely used in laboratories across the world; these methods are simple yet effective. In your group answer the following questions:

1 What are the main aims of extraction?
2 What are the advantages and disadvantages of extraction?
3 Where is extraction most commonly used in industry?

3.1 Titration techniques on consumer products

313

4.1 Solvent extraction techniques

Solvent extraction or liquid-liquid extraction is the removal of a substance from a solution by dissolving it into another. The two solutions must be **immiscible** meaning incapable of being mixed together, resulting in the formation of two separate layers. Usually the compound of interest will leave its original solution and enter the new one, as the compound of interest should be more soluble in this solution. Extraction is often used in industry to remove **impurities**; in order for the impurities to be removed they must be **insoluble** in the second solution to ensure they remain where they are. Take coffee as an example: coffee is a liquid containing dissolved caffeine. During extraction of caffeine an immiscible solvent is added to the coffee. The caffeine will be more **soluble** in the solvent than it is in the coffee, resulting in the movement of caffeine from the coffee to the solvent. This will create two separate layers, with the solvent now containing the caffeine. Caffeine is often extracted; it is a white solid at room temperature and is classified as an alkaloid, a nitrogen-containing compound. Caffeine is a stimulant and is present in tea, coffee, and added to fizzy drinks and energy drinks. Decaffeinating coffee and the brewing of tea are two examples of why caffeine extraction takes place.

PAIRS ACTIVITY

Health and safety (15 minutes)

As you already know, health and safety is critical when working in a laboratory. In pairs research the health and safety precautions for a solvent extraction procedure. Note these down and share with the class.

EXTENSION ACTIVITY

Accident protocols (15 minutes)

You have completed an experiment in the laboratory; you have rushed and not taken note of the health and safety actions. You spill a harmful chemical on your arm. What do you do?

Solvent extraction procedure: extraction of caffeine from tea

Apparatus
- separating funnel
- organic solvent: dichloromethane
- distilled water
- tea bags
- beaker 500 ml
- hot plate
- melting point apparatus
- conical flask

Method

1. Ensure you are wearing your PPE.
2. Add 200 cm³ of water to the beaker; place five tea bags in the water and boil on the hot plate.
3. Allow the mixture to cool for five minutes, then decant the mixture into another beaker, squeezing the tea bags to remove anything left over.
4. Leave the liquid mixture to cool, using ice to speed up the process; it must be below 20°C.
5. You will now complete the extraction three times using the dichloromethane: conduct this in a fume cupboard.
6. Pour the tea solution in the separating funnel and add 20 cm³ of dichloromethane to it; this will separate into two layers, the top being the tea and the bottom the dichloromethane as this is denser than tea.
7. Keeping your finger placed firmly on the stopper, carefully shake the separating funnel.
8. To relieve the vapour pressure, vent the separating funnel every 30 seconds or so; make sure the funnel is in an upright position whilst doing this.
9. Once the contents have been sufficiently shaken, drain the bottom layer into a conical flask.
10. Repeat steps 5–8 two more times.
11. Once complete, dry the dichloromethane solution with anhydrous sodium sulphite, add one spatula of drying agent and mix well; use vacuum filtration to filter out the drying agent.
12. Pour the dichloromethane into a conical flask, and evaporate using a heating mantle. Conduct in a fume cupboard; when all solvent is removed you will observe a green yellow-white crystalline caffeine.

4.2 Chromatographic techniques

Chromatography, meaning colour writing, is a method of separation. The separation consists of two phases: the stationary phase and the mobile phase. Although a very old technique, it is still used in industry, as this method has the ability to separate and purify a sample with great precision. Chromatography can separate a wide range of samples including any volatile, delicate or soluble sample, as long as the correct absorbent or carrier is used. Examples of its applications include cosmetic products, inks, food colourings and pharmaceuticals.

Thin layer chromatography

Thin layer chromatography (TLC) can be used to determine the number and purity of compounds in a sample; it is widely used in the pharmaceutical industries for screening unknown materials in bulk drugs and early stage drug development to provide information about impurities. The procedure involves a liquid or gas moving over a stationary paper or powder, where different compounds will move at different rates and distances. TLC consists of three main stages: spotting, development and visualisation. Spotting is the process of placing drops of the sample onto the chromatography paper, development is the movement of sample up the paper, and visualisation is the production of coloured spots at various points on the paper. There are many advantages of using TLC:

- a highly sensitive technique
- time effective: taking around 5–10 minutes to complete
- minimal clean up
- wide choice of mobile phases
- high sample loading capacity
- low cost.

TLC procedure

Here is a simple TLC procedure; use this to practise the technique. To stretch yourself go to www.rsc.org/learn-chemistry and complete the aspirin procedure.

Apparatus
- TLC plate
- test tubes
- pencil and ruler
- glass plate
- solvent; a substance that dissolves in a solution
- food dyes
- tweezers
- powdered drink mix

Method
1. Dissolve each sample in the solvent in the test tubes.
2. Draw a reference line on the chromatography paper; this should occur 2 cm from the bottom, then draw another line 1 cm from the top.
3. Draw crosses evenly spaced on the bottom line; this will be where the samples are placed.
4. Place a drop of sample onto a cross; do this for each sample. It is best to let the drop dry and add another to ensure enough sample is available.
5. Pour the solvent into a glass/beaker at 1 cm depth.
6. Roll the chromatography paper into a tube and keep in place with a safety clip; place into the glass.
7. Put the lid on the glass container and leave.
8. Once the spots have moved, take the paper out and leave to dry. Use your pencil to mark the centre of each spot.
9. Record the following: distance of X (bottom to top of pencil lines), distance of Y (bottom of pencil line to each spot).

Retention factor

Retention factor, known as the Rf value, is used to quantify the movement of samples across the chromatography paper, where you analyse the distance travelled by the substance and the distance travelled by the solvent. Each compound will have a specific Rf value; the retention factor is always the same for a particular compound if the chromatography procedure is kept constant. Use the equation below to calculate your values from the TLC experiment.

$$Rf = \frac{y}{X}$$

> **KEY TERMS**
>
> **Immiscible** – Incapable of being mixed together.
>
> **Impurity** – A compound that interferes with the purity of something.
>
> **Insoluble** – A substance incapable of being dissolved.
>
> **Soluble** – A substance capable of being dissolved.
>
> **Chromatography** – A technique of separation where components move at different rates.
>
> **Retention factor** – The relative distance that each component travels.

4.3 Micro-analytical techniques

Micro-analysis is the identification of very small amounts of samples. It is an extremely powerful tool in analysis and is used in many types of industries. The main procedures of micro-analysis include:

- High performance liquid chromatography (HPLC)
- Gas chromatography (GC)
- Inductively coupled plasma techniques (ICP)
- Mass spectrometry (MS)

High performance liquid chromatography

HPLC is an effective analysis method used in industries like pharmaceuticals, food and quality control. It is a more advanced chromatography method and can be used to separate complex mixtures, resulting in high precision and accuracy. HPLC can be used alongside mass spectrometry to analyse the metabolism of biological substances used in medical treatments. The sample in HPLC consists of very small sample sizes, therefore increasing the surface area for reaction to take place. This procedure is more complicated, consisting of a number of stages.

- The mobile phase: the solvent runs continuously through the system and pushes the solvent through the column, the solvent is contained in a reservoir, located at a higher point than the pump. A solvent filter removes any particles that could contaminate the system or the sample.
- Sample detector/injector: stores the samples and injects them when instructed.
- The column: this is the stationary phase that separates the sample. The longer the column, the higher the efficiency and resolution. The shorter the column, the faster the separation.
- Detector: receives the result of the sample separation.
- Data system: translates the signal from the detector into a chromatographic spectrum. The data system allows control of the pump, detector and sampler.

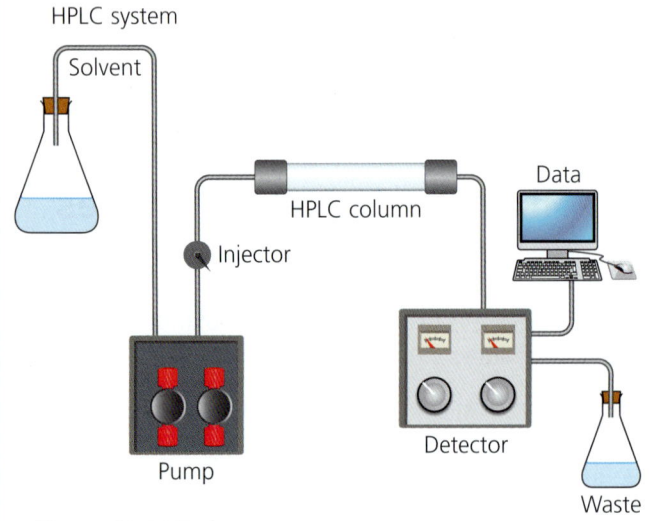

▲ **Figure 21.4** HPLC set-up

Gas chromatography

GC is a powerful separation technique for volatile organic compounds, where separation and online detection is combined. GC can be used to analyse animal fats in oils, alcohol concentrations and impurities in drugs. It is yet another highly accurate and reliable chromatography procedure, known to be more sensitive than other techniques as it not only determines what chemicals are present but also how much of the chemical is present. In GC the mobile phase is an inert gas and the stationary phase is a thin layer of an inert liquid or solid.

- The liquid is injected into the column and an inert gas is pumped into it which acts as a carrier.
- Liquidised samples are then vaporised.
- As it passes along the column it separates into different substances; as the components are removed from the column they can be quantified by the detector.

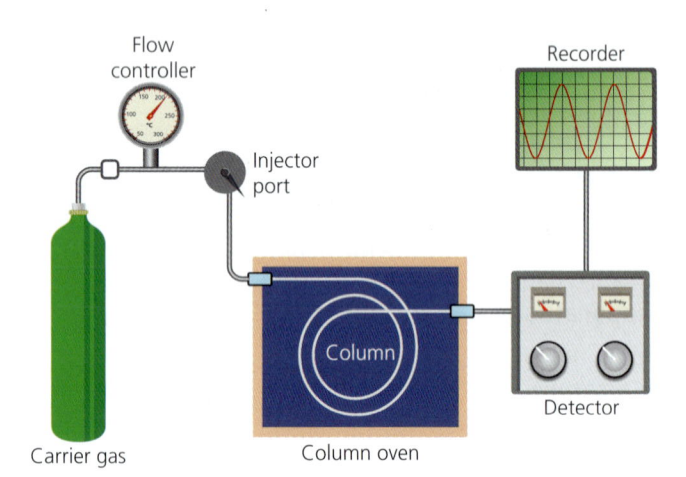

▲ **Figure 21.5** GC set-up

Inductively coupled plasma techniques (ICP)

ICP is used to identify trace concentration of elements. It has a multi-element technique and a wide range. In the process, heating the gas with an electromagnetic coil energises the plasma; directing the energy of a radio frequency generator into a suitable gas can create inductively generated plasma. Coupling is then achieved by generating a magnetic field. It is largely used in the medical field; reasons include assessing metal poisoning and biological concerns. The equipment required for this type of experiment includes:

- Sampler: converts sample into an aerosol.
- Source: high velocity inert gas produces a high energy plasma and separates the excitation region from the observation zone.
- Analyser: analyses sample.
- Detector: detects results, converting radiant energy to measurable signals.

There are various types of ICP:

- ICP-MS: inductively coupled plasma – mass spectrometry; combines a high temperature ICP with a mass spectrometer. The ICP converts the atoms in the sample into ions.
- Direct couple plasma: created by an electrical discharge between two electrodes.
- ICP-AES: atomic emission spectroscopy produces excited atoms and ions to emit electromagnetic radiation.

Mass spectrometry

MS is an analytical technique that can be used to identify the chemical properties of a gas, liquid or solid. Atoms and moles can be deflected by magnetic fields, as long as they are converted into ions first and their mass to charge ratio is measured. The sequences are as follows.

1 Ionisation

The molecule is converted into an ion so it can be manipulated easily; a high-energy beam of electrons bombards the molecules ionising them.

2 Acceleration

Ions are accelerated so they all contain the same kinetic energy.

3 Deflection

The ions are deflected by a magnetic field, the lighter they are and the higher number of positive charges results in a higher level of deflection.

4 Detection

The separated beams of ions are electronically detected whilst passing through the machine; the information is stored in the computer and analysed. The mass and abundance of separated ion fragments are measured and the results displayed.

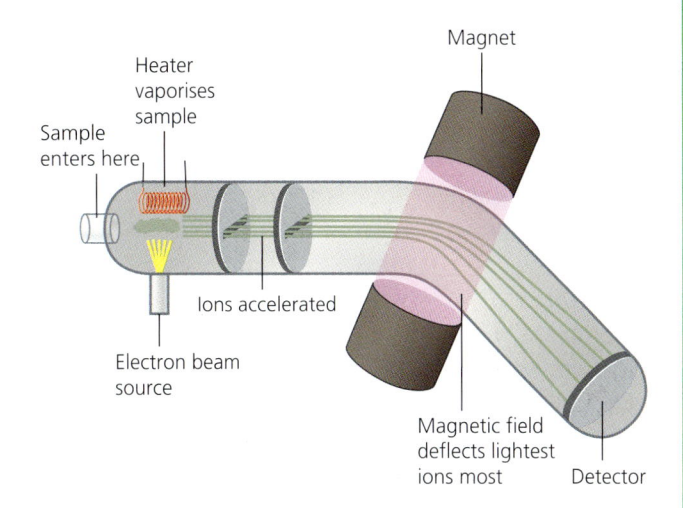

▲ Figure 21.6 Mass spectrometry set-up

KNOW IT

1 During solvent extraction, why must the two solutions be immiscible?
2 What are the main purposes of chromatography?
3 How can thin layer chromatography be used to analyse drugs?
4 What are micro-analytical techniques and why are they effective?

LO4 Assessment activities

Below are suggested assessment activities that have been directly linked to the Pass and Merit criteria in LO4 to help with assignment preparation and include top tips on how to achieve best results.

Activity 1 Use solvent extraction to separate and determine the mass of the active ingredient of a consumer product *P5*

Complete the extraction of tea experiment. Write up the experiment ensuring you include:

- Health and safety
- Aim
- Hypothesis
- Apparatus
- Method
- Results
- Conclusion

Activity 2 Use TLC to investigate qualitatively the composition of a consumer product *P6*

Conduct TLC on a consumer product of your choice. Complete a scientific write-up using the same layout as for *P4*.

Activity 3 Calculate Rf values of constituents of the consumer product to provide quantitative information of a consumer product *M4*

Calculate the Rf values of your TLC experiment; include these values in your scientific write-up.

TOP TIPS
- ✔ Follow health and safety when undertaking practical techniques.
- ✔ Practise practical techniques.
- ✔ Practise Rf calculations.

Read about it

Websites

World Health Organization

www.who.int/en/

Gov.uk www.gov.uk/

Books

Bentham, J. and Curtis, G. (2008) *AS Chemistry Unit 1: Foundation Chemistry*, HarperCollins UK

Harvey, D. (1999) *Modern Analytical Chemistry*, McGraw-Hill International

Higson, S. (2004) *Analytical Chemistry*, Oxford University Press

Glossary

Abdomen The torso below the diaphragm.
Absorption Soluble molecules are moved into the blood from the lumen of the intestine into the bloodstream.
Accurate The result is close to the true or reference value for the measurement. The accuracy of any measurement will depend on the quality of the apparatus and the skill of the person using it.
Active pharmaceutical ingredient (API) The actual active drug substance contained in the medicine.
Actual yield The amount of product actually extracted following a reaction.
ADH A pituitary hormone which stimulates the kidney to reduce the volume of water in the urine.
Adsorbent Often used to describe the stationary phase in chromatography because substances become adsorbed to it during separation.
Adsorption When a substance (e.g. a gas, liquid or solute) binds to or attaches to another, usually solid.
Adult (somatic) stem cells Cells in the adult that divide and differentiate to maintain and/or repair the tissues. The term somatic stem cell is often used instead, because adult stem cells are also found in children! In this context, Ðdult simply distinguishes this type of stem cell from embryonic stem cells.
Agent A general term that includes biological agents as well as a chemical substance (which may be a natural product such as a protein).
Agriculture The science of farming, including the cultivation of soil.
Alleles Different versions of a gene. For example, the gene for seed shape in peas has alleles for smooth and wrinkled peas. Many genes can have more than two alleles.
Allergen A type of antigen that causes an immune response called an allergic reaction.
Alternative (or experimental) hypothesis Contrary to the null hypothesis, this says that there is an effect or there is a difference.
Alveoli Where gaseous exchange happens.
Amino acid A molecule with both an amino group and a carboxyl group. There are 20 naturally occurring amino acids found in proteins and all have the amino and carboxyl groups attached to the same carbon, the β-carbon (hence β-amino acids). The same carbon also has a hydrogen and another substituent the side-chain, or R-group which is different in each different amino acid. These are building blocks of proteins.
Analogues A series of compounds with similar chemical structures.
Analyte A substance which is having its chemical structure identified or analysed or the solution of unknown concentration in a titration.
Anaphylactic shock A serious allergic reaction which can result in death.
Anion A negatively charged ion, e.g. Cl or CO32. Anions are usually formed by non-metals, or compounds of non-metals.

Anneal/annealing The process of forming a double stranded DNA molecule (dsDNA) from complementary single stranded DNA (ssDNA). The ssDNA is usually produced by heating dsDNA to separate the strands and annealing occurs when the solution is cooled. The process can be used to attach complementary primers to a DNA fragment being prepared for PCR.
Anorexia nervosa An eating disorder, where an individual keeps their body weight as low as possible.
Anti-codon The complementary sequence to the codon, following the base-pairing rules.
Antibiotic A substance that destroys or inhibits bacterial cells.
Antibody Proteins made by plasma cells. They have variable regions, which give them an immense range of shapes. Each antibody recognises and binds with a specific shape of antigen.
Antifungal A substance that destroys or inhibits fungi.
Antigen A molecule which stimulates an immune response and causes the immune system to produce antibodies against it.
Antimicrobial A substance that destroys or inhibits microbes.
Antiviral A substance that inhibits viral cells.
Artery Vessel which takes blood away from the heart.
Aseptic Free from contamination, processes consisting of sterilisation.
Assimilation A molecule becomes an integral part of the body's processes.
Atherosclerosis A condition where arteries become clogged up by fatty substances known as plaque or atheroma.
Atom This is the basic structure of all matter and consists of a nucleus surrounded by electrons.
Atomic number Also called the proton number, this is the number of protons in the nucleus of an atom. It is also the number of electrons in the atom.
ATP Also known as Adenosine Triphosphate, energy source in all cells.
Atrium Small chamber on each side at the top of the heart.
Autonomic Means 'self-governing'. The conscious part of ourselves has little control over this section of our nervous system.
Bacteria Single living cells.
Bacteriophage A virus that infects bacteria, often shortened to 'phage'.
Base pairs This is the standard unit of size (length really) for DNA. You might come across bp or even kbp (thousands of base pairs). For single stranded DNA the corresponding unit of length is b (base) or kb (1000 bases).
Bessel's Correction This is a correction to the calculation for standard deviation, introduced by the mathematician Freidrich Bessel, to account for potential bias when taking a sample from a population.
Biochemical Relating to the chemical composition of a biological substance.

Biofertiliser A substance containing living microorganism, which promotes growth of soil and plants.

Bioherbicide A substance containing microorganisms which is used to control weeds or undesirable plants.

Biological agent This term includes bacteria, viruses, fungi and other organisms, as well as toxins that they may produce.

Biological/biologics An enzyme, protein, peptide or antibody (see section 6.2 in Unit 4) used in this context as a drug.

Biome In contrast to a habitat, which may be a very small region or environmental niche, a biome is a larger region, typically on a continental or oceanic scale, which has a relatively uniform climate and types of plants and animals. There are a number of ways of defining biomes, probably the most comprehensive being that of the World Wildlife Fund (WWF), which recognises fourteen terrestrial biomes, as well as a number of freshwater and marine biomes. For the purposes of this course, we will use a smaller range of definitions, as indicated in Figure 14.1.

Biopesticide Pesticides derived from natural substances, consisting of living microorganisms, used to destroy pests.

Blinding This refers to the fact that the subjects don't know if they are receiving the placebo or the drug being tested. A double-blind trial is one where neither the subjects nor the medical staff know and so avoids bias.

BMR Basal Metabolic Rates; minimum rate of energy expenditure per unit of time by a body at rest.

Boiling The process by which a liquid converts into its gas state when the vapour pressure of the liquid is equal to the ambient pressure. For example, at the pressure of Earth's atmosphere at sea level, water boils at 100°C, meaning that at this temperature, the vapour pressure of water is 1 atmosphere.

British Nutrition Foundation A registered charity that provides impartial evidence-based information on food and nutrition.

Bronchoconstriction The smooth muscles in the walls of the bronchioles contract, narrowing the lumen.

Bronchodilation The smooth muscles in the walls of the bronchioles relax, widening the lumen.

Bulimia A mental health and eating disorder. Individuals binge eat and often use laxatives or forced vomiting to get rid of excess food.

Calorimeter An object for measuring the heat of chemical reactions.

Campylobacter A bacterial foodborne disease.

Capillaries Very narrow vessels which supply blood to tissues and which connect arteries to veins.

Capture antibody An antibody bound to a well in a microtitre plate used to capture an antigen added to the well.

Carcinogen An agent that is directly involved in causing cancer.

Cardiovascular disease (CVD) Disease involving the heart or blood vessels.

Carrier A heterozygous individual in which the recessive allele causes a disease in the homozygote. The carrier does not show symptoms, but can pass the disease on to their children.

Cation A positively charged ion, e.g. Na^+ or $NH4^+$ (the ammonium ion). Cations are usually formed by metals, but not always, as you can see from the ammonium ion.

Cell membrane This separates the interior of individual cells from their environment.

Cell signalling pathways The role of hormones, growth factors and other regulatory substances in communication between cells. The term is also used to apply to the various biochemical events inside a cell that follow binding of a signalling molecule to the cell surface.

Cell signalling The mechanisms of communication between single-cell organisms or between cells in a multicellular organism.

Charge The electric charge is a fundamental property of many subatomic particles, and is a measure of how strongly a particle feels the electromagnetic force. By convention, electrons are regarded as having a negative charge. The unit of charge is the coulomb (C); the charge of a single electron is 1.6 x 10-19 C. Charge is often given the symbol Q.

Chemical receptors A molecule in a cell's surface membrane, which will bind with a messenger molecule, such as a hormone. This is not the same as the 'receptor' in the nervous system.

Cholesterol A lipid molecule synthesised by animal cells.

Chromatids These are the product of DNA replication, so that after replication a single chromosome consists of two identical DNA strands that are called (sister) chromatids held together by the centromere until the time comes for them to separate. Once separated, they are no longer referred to as chromatids they are now chromosomes.

Chromatography A technique of separation where components move at different rates.

Climate change Changes in the Earth's climate have occurred throughout its history, from high temperatures and little or no ice cover, as at the end of the Cretaceous period 65 million years ago, to extremely cold periods such as the most recent glaciation which ended about 10 000 years ago. The causes of climate change are varied and can be slow, such as those caused by changes in the Earth's orbit, or they can be rapid, due to, for example, massive volcano events or meteor impacts. In recent times, evidence has emerged which links human activities to possible climate change, in large part caused by CO_2 emissions from transport and industry.

Clinical phase This involves testing of the drug in humans; there are three phases of clinical testing before the drug is approved for use as a medicine.

Co-factor A non-protein 'helper' for an enzyme that is involved in the enzyme mechanism in some way. Co-factors can be inorganic (usually metal ions) or organic molecules (or metallo-organic if they contain metal atoms or ions).

Codominant Neither allele is dominant or recessive and both alleles contribute to the phenotype.

Codon A sequence of three bases in DNA or mRNA that codes for a single amino acid. There are four bases, so there are 4 x 4 x 4 = 64 possible combinations, meaning 64 different codons more than enough to code for all 20 naturally occurring amino acids.

Colony Two or more organisms living in close proximity to each other, usually for mutual benefit, for example to share nutrients.

Complement A system of plasma proteins that can be activated directly by pathogens.

Complementary The shape of one molecule fits exactly around the shape of a second molecule.

Compound A substance that contains more than one type of atom. Those atoms can be held together in two main ways by ionic or covalent bonds.

Concentration The number of molecules of a substance within a set volume.

Confounding variables These are similar to extraneous variables, but potentially more serious. Whereas the extraneous variable is something external to the investigation, a confounding variable changes along with the variables we are trying to control and measure.

Consumer A person who purchases goods or services.

Control A scientific control is a method for minimising the effect of variables in an experiment. In an environmental context, it could be as simple as using a well-known sample to calibrate or verify the correct working of a procedure, or it might be more complex, such as splitting an area under investigation into sections where interventions take place (the experiment, for example, cultivating a field) and those where no interventions are carried out (the controls, for example, leaving a field fallow).

Coronary arteries Arteries which supply the heart muscle with blood.

Current Era (CE) and Before Current Era (BCE) These are the preferred ways to express dating for historical purposes. They are equivalent to the older terms Anno Domini (AD) and Before Christ (BC). As examples, the Great Pyramid of Giza was built in the decades around 2570 BCE, the Roman Empire annexed Britain in 43 CE and this book was written in 2016 CE.

Current This is a measure of the rate at which electric charge is flowing past a particular point in a circuit. The unit of current is the ampere (A), often shortened to amp, and can be expressed as C s1 (coulombs per second). A current of 1 A therefore means that in 1 s, 1 C of charge flows past a given point in a conductor. Current is often given the symbol I.

Cytokines A broad category of small proteins involved in cell signalling. They are released by some cells and act on other cells, often as part of the response to infection.

Cytology The study of the structure, function and chemistry of cells.

Decontamination Cleaning an object or substance so as to remove microorganisms or other hazardous materials.

Dementia A brain disorder that causes a decrease in the ability to think.

Department of Health This is the part of the government responsible for the policy on health.

Dependent variable A variable (often denoted by y) whose value depends on that of another variable. In an experiment, we usually measure the independent variable.

Detection antibody An antibody that is specific for a target antigen and will bind to it once captured on a plate.

Diabetes A metabolic disease in which the body has a high level of glucose (blood sugar).

Differentiate Cells become specialised for a particular job, changing in features and abilities.

Digestion Large, insoluble molecules are broken down into much smaller, soluble molecules.

Diploid A cell that has two sets of chromosomes, two of each homologous pair.

Discontinuous data A set of data is said to be discontinuous (or discrete) if the values are unconnected, i.e. distinct and separate, e.g. shoe size or number of organisms (you can't have fractions of an organism).

Disinfection The destruction, inhibition or removal of microbes that may cause disease or other problems.

Displayed formula Shows the arrangement of atoms in a molecule. Each atom is shown by its chemical symbol and covalent bonds are shown by single, double or triple straight lines. You may also see simplified displayed formulae where not every single bond is shown, e.g. using CH_3 to represent a methyl group.

Diuretic Any substance that promotes the production of urine.

Diverticular disease A digestive disease affecting the large intestine.

Dominant An allele that is expressed even if only one copy is present is known as dominant. In other words, the allele is expressed in heterozygous or homozygous individuals.

Drug A chemical or biological compound used to treat disease. Most drugs are available only with a doctor's prescription and are strictly regulated. Don't confuse pharmaceutical drugs with recreational drugs. which

Drug screening This is a model system that aims to identify compounds (hits) with the necessary activity against the drug target.

Drug target This is usually a protein (enzyme or receptor) or gene implicated in the disease that is the target for a new drug.

***E. coli* O157** Bacterium that lives in the gut of animals; the toxins it produces are harmful to humans.

Effector A muscle, gland or organ which can make a response when it is stimulated by an electrical impulse from a neurone.

Efficacy A measure of how well a drug works at its required dose.

Elastic Can stretch and then return to its original size and shape.

Electrolytes Substances that produce electrical conducting solutions when dissolved in fluid (for example water or blood).

Electron A negatively charged particle with a mass about 1/1836th that of a proton.

Eluate The mobile phase, containing dissolved substances, as it emerges from a column.

Eluent The solvent (mobile phase) used to wash substances out of a column.

Elution To wash out. In column chromatography this means washing out a substance that has become adsorbed to the column (stationary phase).

Embryonic stem cells (ESCs) Cells in the embryo that can give rise to all the different types of cells in the adult.

Empirical formula The simplest whole number ratio of the different atoms in the compound. The empirical formula of any compound can be obtained from the percentage by mass of each element in the compound. Different compounds can have the same empirical formula. For example, methanal (the simplest aldehyde), ethanoic acid and glucose all have the empirical formula CH_2O.

Emulsification When the surface tension of a mass of lipid in watery surroundings is broken down, splitting it into smaller droplets which mix more easily with the water.

End point The point in a titration where the indicator changes colour.

Endoparasite A parasite that lives insides its host and obtains its nutrition directly from the host.

Enteric administration A route for drug delivery that involves absorption through the intestines.

Environmental management In the context of this course, we will define environmental management as the process of managing human-environment relationships. This doesn't mean hands-on looking after an environment, but relates to ideas and development of regulation, policy and practice.

Environments This is a very broad term, but for the purposes of this unit, we can understand it as all natural and human habitats on Earth (air, oceans and land environments) as well as the sub-surface environment of underlying geology.

Enzyme A protein which acts as a catalyst. Each one makes a particular chemical reaction happen at the temperatures found in the body, and at a much faster rate. Enzymes are not used up, so an enzyme molecule catalyses its reaction many millions of times.

Equivalence point The point of neutralisation where the number of moles of acid and base are equal. This should ideally be the point at which the indicator changes colour.

Essential proteins Proteins that the body cannot produce itself.

Excipient A component of a formulation other than the active ingredient. Excipients include bulking agents (tablets) or diluents (liquid medicines) as well as other substances that assist in manufacture or delivery of the drug.

Excretion Removal of a waste product made by the body itself, e.g. carbon dioxide.

Extraction The action of removing something.

Extraneous variables These are variables which we are not directly interested in, but still affect the outcome.

Fast protein liquid chromatography (FPLC) A type of HPLC (high performance liquid chromatography see section 2.2 in Unit 2) that uses separation methods and materials more suitable for use with proteins than small molecules.

Fermentation The chemical breakdown of a substance.

Fertiliser A natural or chemical product, which enhances the fertility of soil.

Fibrillation When the contractions of the heart are unsynchronised, irregular and rapid.

Fissure vents, where eruptions occur along a linear crack, rather than at a single point. Examples include Laki and Holuhraun, both in Iceland.

Folic acid A B vitamin, synthetically produced; can be found in some foods and supplements.

Food Standards Agency Responsible for protecting public health in relation to food.

Formulation The process by which different chemical substances, including the active drug (API) are combined to produce a medicine. Formulation is a verb, i.e. 'to formulate' a medicine, but you will often see it used as a noun, as in 'sustained release formulation'.

Functional group A group of atoms that give a molecule its particular properties. Compounds with the same functional group have similar properties and often form a homologous series. Examples of functional groups are carbonyl, alcohol, acid, ester and halogen (e.g. chlorine, bromine).

Gametes Specialised sex cells such as egg (female gamete) and sperm (male gamete).

Gas chromatography (GC) A separation technique using an inert carrier gas as the mobile phase and a thin layer of liquid or polymer on an inert solid support as the stationary phase.

Gaseous exchange Passing oxygen and carbon dioxide in opposite directions. Oxygen enters the body, while carbon dioxide leaves it.

Gel electrophoresis Use of an electric field for separation of compounds based on differences in their charge and size.

Gene A section of DNA that codes for a protein, although some genes code for RNA or regulate other genes.

Gene technology This is the study and manipulation of DNA to provide practical applications.

Genetic engineering The deliberate modification and manipulation of characteristics of organisms.

Genetically modified (GM) Genetic material which has been altered within an organism.

Genome The full set of genes contained in an organism. A better definition is that it includes all the genetic material (DNA or RNA) in an organism, which means genes and non-coding sequences of DNA or RNA.

Genomics The study of the genome, its sequence, structure and function.

Genotype The genetic makeup of an individual, specifically the alleles it contains. We usually describe the genotype as the alleles for a particular characteristic present in an individual.

Global Hectare (gha) Global hectares are the unit of measurement for ecological footprint and biocapacity accounting. A global hectare is a biologically productive hectare with world average biological productivity for a given year. Global hectares are needed because different land types have different productivities. A global hectare of, for example, cropland, would occupy a smaller physical area than the much less biologically productive pasture land, as more pasture would be needed to provide the same biocapacity as one hectare of cropland. Because world productivity varies slightly from year to year, the value of a global hectare may change slightly from year to year.

Glucose A type of sugar.

Glycogen An insoluble polysaccharide carbohydrate, which is chemically similar to starch, made from monomers of glucose. Animals and fungi use glycogen as an energy store.

Glycolysis The breakdown of glucose, producing pyruvate and energy in the form of ATP.

Good manufacturing practice (GMP) The principles of GMP include following clearly defined standard procedures that are validated to ensure consistency and compliance with specifications. Detailed record keeping of all steps in manufacture, including documentation of quality of raw materials and intermediates, is essential. The intention of GMP is to reduce the risk to consumers and patients, particularly in ways that might not be detected through analysis of the final product.

Gram negative A type of bacteria which stains red, consists of thick peptidoglycan cell wall.

Gram positive A type of bacteria which stains purple; consists of thick peptidoglycan cell wall.

Gram staining A staining technique in which a violet dye is applied to distinguish between Gram positive and Gram negative bacteria.

Group This refers to the columns in the periodic table. Elements in the same group have similar chemical properties.

Habitat This is an environmental area inhabited by humans or other organisms. Habitats aren't necessarily natural, or geographical. For example, the habitat of a human might be a city, town or village; that of a flea, the skin of a cat or a dog.

Haploid A cell that has only one set of chromosomes, just one of each homologous pair.

Heterozygous When an organism has two different alleles for a particular gene.

High performance liquid chromatography (HPLC) A type of column chromatography that uses very small particles and high pressures to achieve better separation. It can be used for analysis or, on a larger scale, for purification.

Histamine Apart from being part of the body's immune response, it also regulates physiological function in the gut.

Histology The study of the anatomy of cells and tissues using microscopy.

Hit compound A substance with promising characteristics or activity worthy of further development, sometimes just called 'hit'.

Homeostasis Keeping the conditions in the environment around each cell constant, and at the levels that the cell needs to work effectively.

Homologous chromosomes A pair of chromosomes, one inherited from each parent. Homologous chromosomes carry the same genes, but may carry different alleles of those genes.

Homologous series A series of compounds with the same general formula, but differing in one respect usually each one has an extra CH_2 than the one before it in the series, i.e. the carbon chain is longer. All of the groups of compounds we will study in this section form their own homologous series.

Homozygous When an organism has two identical alleles for a particular gene.

Hormones Chemical messengers made by glands and carried in the bloodstream.

Human activity These are actions by humans which affect environments. These can be immediate, as in land clearance for agriculture or building, or as a consequence of activity (for example, mining activities might affect water and air quality). Human activities need not be negative, but can be positive as well, for example where action is taken to conserve or repair environments.

Hygrophobic Intolerant of moisture.

Hygroscopic Hygroscopic substances absorb water molecules from the surroundings. In extreme cases this leads to them dissolving in the water they absorb.

Hypothalamus A region at the base of the brain.

Hypothesis A proposed explanation for a phenomenon or observation. In science a hypothesis is no use unless we can test it, i.e. design an experiment.

Immiscible Incapable of being mixed together.

Immune Resistant to a particular infection.

Immune response The reaction of the body to something which is not recognised as part of the body itself.

Immunogenic Produces an immune response.

Immunological response A process that protects the body against disease and substances which can cause it harm.

Impurity A compound that interferes with the purity of something. In normal speech, the terms stress and strain are often used interchangeably. However, in science and engineering, they have well-defined meanings.

In-vitro A process performed outside of a living organism. reading or measurement

In-vivo A process performed within a living organism.

Inclusions Non-living components of cells such as glycogen granules, lipid droplets, crystals and pigments.

Independent assortment This is also known as random assortment or independent segregation and refers to the fact that each daughter nucleus produced in meiosis I receives at random either a maternal or paternal member of each homologous pair, i.e. the daughter nuclei don't contain all maternal or all paternal chromosomes. This is an important source of variation that will be discussed in the section 'The importance of meiosis'.

Independent variable A variable (often denoted by *x*) whose value does not depend on that of another variable. In an experiment, the independent variable is usually what we change.

Independent variable This is typically a variable which the investigator has control over. For example, in an investigation to look at carbon dioxide emissions from an industrial site, the independent variable might be time of day.

Indicator A substance that gives a visible sign.

Inflammation In inflammation, capillaries become more permeable, more fluid and white blood cells than usual escape, and the tissues become swollen.

Inoculate Introduction of a microorganism into a culture medium (e.g. in a flask) or onto an agar plate.

Inoculum The substance, usually a microorganism, used to inoculate a culture.

Insoluble A substance incapable of being dissolved.

Ionisation energy This is the energy needed to remove an electron from an atom in the gas phase. It is a measure of how tightly the electron is held by the nucleus the strength of attraction between the positive nucleus and the negative electron.

Isotopes These are atoms of the same element (i.e. they have the same number of protons) but with different numbers of neutrons.

Joint The structure where bones meet.

Kinase An enzyme that phosphorylates (adds a phosphate group to) another protein, usually causing that protein to become activated. Many of the components of signalling pathways are kinases you can usually recognise them because their abbreviated name contains at least one letter K.

Lactose intolerance When the body cannot easily digest lactose.

Land and area types The Earth's productive land and water areas are categorised into six types: cropland, grazing land, forest land (production), forest land (CO_2 absorption), fishing ground and built-up land.

Lead optimisation (Lead is pronounced the same as a dog lead, not like lead the metal.) The process of modifying a hit compound, through several rounds of screening different analogues, to produce one or two compounds with high activity.

Lever A rigid bar with a pivot, used to transmit a force.

Ligand A general term for a substance, such as a hormone, growth factor or other regulator, that binds to a receptor.

Lipid bilayer A double layer of phospholipids with the hydrophobic tails arranged towards the middle and the hydrophilic head groups on the outside; it forms the basis of all biological membranes.

Lipid Fats and oils, mostly composed of three fatty acid molecules bonded with one glycerol molecule.

Lipoprotein A biochemical of proteins and lipids.

Listeria A type of bacteria from contaminated food.

Locus A specific position on a chromosome, corresponding to a particular gene or section of DNA. The plural is loci.

Lungs A pair of air-filled bags in the thorax.

Lymphocyte A family of white blood cells with large nuclei found in the lymph system. Types of lymphocyte include T cells, B cells and natural killer cells.

Lysis The disintegration of the cell caused by rupture or the cell membrane (or cell wall in plant cells).

Lysogenic cycle This involves integration of the viral nucleic acid into the genome of the host cell. The viral nuclei acid is replicated along with the host genome. The term was originally used in the context of bacteriophage infection of bacterial cells the integrated viral nucleic acid is known as a prophage. The term is now also used to describe the life cycle of retroviruses.

Lytic cycle The main method of viral reproduction inside cells. It results in lysis of the cell and release of progeny viruses that go on to infect other cells.

Macromolecule A very large molecule, containing thousands or more atoms. They are often, but not always, polymers and the term is used particularly to describe large biological molecules.

Malnutrition A condition from the result of not eating enough and/or not getting the required nutrients.

Maltose Disaccharide composed of two glucose molecules.

Mass spectrometry (MS) A technique that creates positive ions from a sample and then separates them according to their mass-to-charge ratio.

Me-too drug A drug that is very similar to one or more existing drugs that treat the same disease.

Median The median is similar to the mean, except it splits the data set in half, so that there are the same number of quantities above the median as below. In our example above, if we arrange our numbers in order, we have 2, 2, 3, 6, 7. The number 3 now sits in the middle of the set, splitting it in half, so 3 is the median.

Medicinal chemistry The process where organic chemists make analogues hoping for greater activity.

Meiosis Nuclear division that produces four genetically different daughter nuclei.

Melting The process by which a solid converts to a liquid. Normally, this means that the inter-atomic or inter-molecular forces either reduce or are no longer sufficient, such that individual atoms or molecules are able to move freely. The forces are still sufficient though to prevent the liquid boiling into the vapour phase.

Membrane All membranes consist of a lipid bilayer together with proteins and other components. They are selectively permeable and can control movement of substances across the membrane as well as being the sites of many important processes in the cell.

Metabolised The breaking down of substances and their reorganisation into another form.

Metabolism The chemical reactions going on inside the body.

Microbiology Study of microorganisms using a variety of techniques including Gram staining, colony morphology and serological techniques.

Microorganism A microscopic living organism, only seen using a microscope. Includes bacteria, virus and fungi.

Microtitre plate A plastic plate with 96 wells arranged in an 8 12 layout used for ELISA and other types of assay.

Miscarriage A natural death of an embryo or fetus.

Mitosis The division of a nucleus to form two genetically identical daughter nuclei.

Mode The mode is simply the most common quantity in a set of data. In our example it is 2.

Molecular biology The study of DNA and its application. This includes topics such as DNA sequencing as well as gene technology and molecular genetics the understanding of genetics at the level of DNA, RNA and proteins.

Molecular formula Lists the number of each type of atom in a compound. If you know the empirical formula of a compound and its mass (obtained by mass spectrometry) you can calculate the molecular formula. So, the molecular formula for methanal is CH_2O (the same as the empirical formula), for ethanoic acid it is C_2H_4O and for glucose it is $C_6H_{12}O_6$.

Monomer The individual molecule or molecules that form a polymer.

Morphology The study of the form of organisms, focusing on their shape, size, colour and characteristics.

Mutagen An agent that changes the genetic material (DNA in humans); as many mutations cause cancer, a mutagen may also be a carcinogen.

Myelin This is a fatty substance which forms a sheath around the axon of some nerves.

Natural processes In contrast to human activity, natural processes arise as a result of non-human activity and events. This can include weather events, earthquakes, volcanoes and water transport and action (e.g. erosion, flooding) both at the surface and through underlying rocks.

Natural product A drug substance extracted from or produced by an animal, plant or microorganism.

Negative feedback To control an internal condition. In negative feedback, a condition changes from its optimum value.

Neurone Very long threadlike cells that carry information from place to place quickly around the body in the form of electrochemical impulses.

Neurotransmitters Chemicals that transmit signals between nerve cells.

Neutron A nucleon with no net electric charge and a relative mass of approximately 1, or more accurately 1.0087 so the mass is slightly more than that of a proton.

NHS Choices An NHS website with information and advice on health.

Nitrogen A chemical element found in all proteins.

Non-essential proteins Proteins that the body can produce itself.

Normal or Gaussian distribution This is a common distribution of quantities about a mean, and has the useful property that it is symmetric about the mean, so that the mean, median and mode are identical. It is also commonly known as a bell curve.

$$\frac{2 \times 10^{-3} \text{ m}}{2\text{m}}$$

(note : because strain is a simple ratio, it has no units)

Norovirus A virus causing nausea, vomiting and diarrhoea, passed on through contaminated food and water.

Nucleic acid A group of complex polymers found in living cells.

Nucleon A general term describing both protons and neutrons.

Nucleoprotein A substance made up of a nucleic acid and a simple protein.

Nucleus This contains protons and neutrons surrounded by electrons.

Null hypothesis In statistics, this is the hypothesis that says there is no effect or no difference between two measurements.

Osteoporosis A condition where bones become brittle, normally as a result of lack of vitamin D, lack of calcium or hormonal issues.

Parenteral administration A route for drug delivery that involves injection, inhalation or absorption through the skin, i.e. a route other than the intestines.

Pathogen / Pathogenic A pathogen is a disease-causing organism, usually a microorganism.

Pathogen An infectious agent such as a bacterium, virus, fungus or parasite that causes disease in its host.

Peptide A compound containing two or more amino acids joined together by peptide bonds.

Percentage error The ratio of the difference between the theoretical and actual yields, and the theoretical yield, expressed as a percentage.

Percentage yield The ratio of actual yield to percentage yield, expressed as a percentage.

Period This refers to the 'rows' in the periodic table. Each period has one more shell of electrons than the previous period.

Pesticides A substance used to destroy pests, usually with the aim of protecting plants.

pH A logarithmic scale that measures acidity (concentration of hydrogen ions) of a solution. Pure water has a pH of 7.0 at room temperature. A strong acid has a pH of 12 whereas a strong base will have a pH of 1314. Reducing the pH by one whole unit means the hydrogen ion concentration increases ten times.

Pharmacogenetics The study of how individual genes affect a person's response to drugs.

Pharmacogenomics The study of how a person's entire genetic makeup (genome) affects their response to drugs.

Pharmacokinetics The study of how a drug is absorbed by the body, how it reaches the target tissues, how it is metabolised (broken down) and excreted.

Pharmacopoeia A publication containing a list of medical drugs with their mode of action, directions for use and effects.

Phenotype The characteristics expressed in an organism. In the case of the gene for eye colour, the phenotypes might be blue eyes or brown eyes. However, in many cases the phenotype depends not just on the genes (alleles) that are expressed, but also on environmental factors.

Phosphatase An enzyme that removes a phosphate group from another protein, usually reducing its activity. Many signal transduction pathways involve both kinases and phosphatases working in opposition.

Phospholipid A molecule containing a glycerol molecule covalently bound to two fatty acid molecules and a phosphate. It has a hydrophilic head group (because of the phosphate) and a hydrophobic tail (because of the fatty acids).

Phosphoprotein A protein modified by the attachment of at least one phosphate group.

Phosphoric acid This is a mineral acid.

Pituitary A pea-sized gland on a stalk of tissue below the hypothalamus.

Placebo A medically ineffectual treatment used as a control in clinical trials. A placebo should represent the drug being tested as closely as possible, with the exception of the active ingredient.

Planning In the context of development and construction, planning is the legal process which ensures the need for a development and the suitability of the site and construction methods. It also ensures that impacts on local people, facilities and the human and natural environment are minimised.

Pluripotent stem cells Originate from totipotent cells and can differentiate into nearly all types of cell.

Polar A polar (hydrophilic) substance will dissolve in or mix with water.

Polymer A large molecule formed from one or more smaller molecules or monomers.

Polymorphism The existence of two or more alleles of a gene, i.e. variants of a particular DNA sequence. Generally refers to the existence of two or more forms of a drug target (usually a protein).

Population In statistics, a population is the collection of all objects or measurements. For example, all the people in the UK represent a population.

Potency Compounds with high potency (or activity) can be given in lower doses, which helps to reduce side-effects.

Pre-clinical phase This starts with identifying a hit compound and finishes when a potential drug is ready to be tested in humans.

Precise Data points are close together, although there could be a random error.

Preservatives These are added to food to lengthen the shelf life.

Pressure Force exerted over an area of surface. For example 10 N/m² is 10 Newtons of force exerted over every 1 m² of surface. An alternative unit, mmHg, is often used in medicine.

Primary antibody An antibody raised against a specific target antigen; primary antibodies can be polyclonal or monoclonal. Monoclonal antibodies are more specific and show much less cross-reactivity.

Primary data The results that we obtain from an experiment or measurement that we carry out ourselves.

Primary recovery In the first phase of production, natural processes within the well are sufficient to drive oil and gas to the surface. This can simply be pressure in the well, but can also include flows from other parts of the reservoir and the release of dissolved gases (by analogy, think about how a fizzy drink behaves when the bottle or can is rapidly opened).

Primary standard A reference material of high purity, such as sodium carbonate or potassium hydrogen phthalate, is used to prepare a standard solution of known concentration.

Process development The application of chemistry to the scale up of new synthetic processes from the laboratory, through pilot plant to full scale commercial manufacture. It brings together disciplines such as synthetic organic chemistry, process technology and chemical engineering.

Produced water Water is invariably trapped within the pore spaces of rocks when they are formed, along with the hydrocarbon reservoir. Produced water is water extracted at the same time as, and along with, the hydrocarbon resource. Produced water is normally saline with a high temperature by nature of its long residence time in the rocks and its depth. It is also heavily contaminated with both free and dissolved hydrocarbons, and may contain chemicals used in the extraction process, heavy metals, and naturally occurring radioactive material.

Prosthetic group A co-factor that is tightly bound to the enzyme, sometimes even by covalent bonds.

Protein Long chains of amino acids chemically bonded together. Proteins have many different shapes, depending on the sequence of their amino acids. A polypeptide with a recognisable three dimensional structure.

Protein synthesis How individual cells construct proteins.

Proton A nucleon with a single positive charge and a relative mass of approximately 1, or more accurately 1.0073 so the mass is slightly less than that of a neutron.

Pulmonary Involving the lungs.

Pumps A protein complex embedded in a membrane that uses energy from ATP to move specific substances across the membrane.

Pyrogenic Causes fever.

Pyruvate End product of glycolysis.

Quantitative Measuring the quantity of something.

Reagent A substance or mixture used in chemical analysis.

Receptor A cell which converts a change in the environment into an electrical impulse in a neurone. For example, a touch receptor in the skin.

Recessive An allele that is only expressed when two identical copies are present, i.e. only in a heterozygous individual.

Recombine/recombination In the context of meiosis, this means the exchange of genetic material (i.e. alleles) between maternal and paternal chromosomes by formation of chiasmata between homologous chromosomes.

Reference value A value accepted as being very close to the true value, for example a standard weight could have been measured on a balance with very little error and so is very close to the true value.

Regenerative medicine The replacement, regeneration or re-engineering of human cells, tissues or organs in order to treat disease.

Relative atomic mass The mass of an atom relative to 1/12th the mass of an atom of carbon-12. It is more convenient to use this relative unit rather than trying to determine the mass of individual atoms or particles in grams; it also becomes more useful when calculating amounts of substance (you will cover this in Unit 2). The same principle is used to define relative molecular or formula mass.

Reliability The reproducibility of a measurement when repeated at random in the same subject or specimen. It is often confused with validity. Reliability can be assessed as the proportion of all variation in a clinical trial that is not due to errors in measurement; it can be estimated from replicate measurements.

Repeat unit A way of showing the pattern of monomers in the polymer.

Resting Energy Expenditures (REE) The energy that your resting body uses to maintain bodily functions.

Restriction endonuclease Sometimes called restriction enzymes, these are enzymes that cut (cleave) DNA at a specific sequence of bases, usually 48 base pairs long (the restriction sequence or site for that particularly enzyme). Some restriction endonucleases produce a 'staggered' cut, so that the resulting fragments have 'sticky' ends.

Retention factor The relative distance that each component travels.

Retrovirus A virus with single stranded RNA that infects a cell and uses its own reverse transcriptase to make a DNA copy that becomes integrated into the host cell genome.

Reverse transcriptase A viral enzyme that converts single stranded viral RNA into double stranded cDNA.

Rounding This is the process of simplifying a number while keeping its value close to the original value. For example, 63 rounded to the nearest ten is 60.

Salmonella A group of bacteria that can cause food poisoning.

Sample A sample is simply a subset of a population selected for a particular investigation. How this sample is chosen can impact the outcomes of an investigation. For example, a **complete sample** is a subset of all the items or events in a population which have a particular characteristic (e.g. 'All black cats selected from the population of cats'). Conversely, a **unbiased sample** draws items or events from a population, irrespective of their particular properties.

Scientific Advisory Committee on Nutrition (SACN) This is a committee that advises the government on nutrition on public health issues.

Second messenger The intra-cellular signalling molecule produced when a ligand (the first messenger) binds to its receptor. There are a number of second messengers such as cAMP (cyclic adenosine monophosphate), cGMP (cyclic guanosine monophosphate), inositol phosphate, or even inorganic ions such as Ca^{2+}.

Secondary antibody An antibody raised against the antibody type of the species used to make the primary antibody. If we raise a primary antibody in rabbits against the human cell-surface protein Fas (rabbit anti-human Fas) and it is an IgG, then the secondary antibody could be raised in sheep against rabbit IgG (sheep anti-rabbit IgG).

Secondary antibody An antibody that binds to the detection antibody.

Secondary data The results obtained by others in independent but comparable experiments or measurements, for example published in textbooks or research papers.

Secondary recovery As the processes driving oil and gas to the surface become less effective over time (for example, the reservoir pressure falls as oil and gas are extracted) production moves into the secondary phase. Additional methods are now needed to drive oil and gas to the surface.

Secondary standard A solution used as a standard in titration that has been calibrated against a primary standard.

Seismic activity/Earthquake This is a rapid release of energy, typically deep in the Earth's surface, resulting in surface movements from imperceptible to extremely violent. They are most common in areas where sections of the Earth's crust are moving relative to one another, especially around the Pacific rim ('The Ring of Fire'), Southern Europe and the Middle East, and the Himalayas. An earthquake results when stresses built up by sections of crust being unable to move smoothly are suddenly released; the movement may be vertical, horizontal or a combination.

Shield volcanoes, formed by eruptions of lava which flows long distances before solidifying. This gives these volcanoes their characteristic gentle slopes and low profiles. Examples include the volcanoes of the Hawaiian Islands.

Short tandem repeats Also known as microsatellites, these are sequences that consist of between five and fifty repeats of, usually, four nucleotides.

SI Commonly known as the metric system, SI (Système International d'Unités) is an international set of standards of measurement and units, adopted by most countries, at least for science and engineering purposes.

Signal transduction The events inside the cell following binding of a signalling molecule to the cell surface.

Significant figures We can be more formal in rounding by specifying a number of significant figures, or the number of figures which carry real meaning. For example, you might calculate the distance of a journey to be 1343.348 metres, but it can only be measured to the nearest 10 metres. This means that it's only sensible to quote the calculation to no better than this, i.e. only the first three figures in the calculation are significant. Our distance is therefore 1340 metres to three significant figures.

Single nucleotide polymorphism (SNP) A change in a single base in a DNA sequence, e.g. C replaces G. This means that the two variants are alleles.

Sinoatrial node Specialised muscle cells in the wall of the right atrium.

Site of Special Scientific Interest (SSSI) A Site of Special Scientific Interest (Area of Special Scientific Interest in England, Scotland and Wales, or ASSI in Northern Ireland) is a legal designation protecting an area from development or change of use. The SSSI designation may be connected with any aspect of the area for example, the area may contain a habitat that supports unique or rare species, have important monuments, contain interesting landscapes and geology, or be historically important.

Skeletal formulae These show just the bonds as lines and leave out the carbons.

Smooth muscle A type of muscle over which we have no conscious control.

Soluble A substance capable of being dissolved.

Specific When a molecule will bind to only one shape of a second molecule.

Spore A small, highly resistant form of bacterial cell produced under adverse conditions. Bacterial spores can resist extreme dryness and many are not killed by freezing, high temperatures or chemical disinfectants. When conditions improve, they can begin to divide again.

Stakeholder In the context of environmental management, a stakeholder is any group of people, organised or unorganised, who share a common interest or stake in a particular matter; they can be at any level or position in society, from global, national and regional bodies (e.g. UN, EU, national governments, commercial companies) down to the level of local communities, households or individuals.

Standard deviation In statistics, this is a measure of the spread of a set of data about the mean value, and is simply the square root of the variance.

Standard Form or Scientific Notation This is a means of writing quantities enabling (1) a clear statement of significant figures and (2) both large and small quantities to be easily represented and understood. A number in scientific notation is written as:

Standard solution The solution of known concentration in a titration.

Starch An insoluble polysaccharide carbohydrate. Its monomers are glucose molecules. Found in foods made from plants, such as bread and rice.

Stem cells Non-specialised cells that can give rise to one or more types of specialised (differentiated) cell.

Sterilisation A process that eliminates or kills all forms of life; including viruses and spores.

Still birth Defined as fetal death after 20 weeks of gestation.

Strain The ratio of the extension of an object divided by its total length. In the example above, if the wire were 2 m in length, and extended by 2 mm (2×10^{-3} m) when 1 N is applied, the strain would be: $\frac{2 \times 10^{-3} m}{2m}$ (note – : because strain is a simple ratio, it has no units)

Stratovolcanoes, which form hills or mountains of alternating layers of lava and other materials such as ash. These are perhaps the most dramatic-appearing volcano type, and are responsible for many events damaging to human environments. Examples include Vesuvius in Italy and Fuji in Japan.

Stress The force applied per unit cross-sectional area of material. For example, if we have a wire with a cross-sectional area of 1mm² (1×10^{-6} m²) and the wire is stretched by a force of 1N, the stress is: $\frac{1N}{1 \times 10^{-6} m^2} = 1 \times 10^6\ Nm^{-2}$

Striated muscle Muscles which are under conscious control. Viewed under a microscope, they appear striated, meaning stripey.

Stroke volume The volume of blood pushed out by the left ventricle in one heartbeat.

Structural formula Shows the structure of the molecule. There are a number of conventions that you need to follow. The structural formula of methanal is HCHO, ethanoic acid is CH3COOH and glucose is complicated! One form has the structural formula CH$_2$OHCHOHCHOHCHOHCHOHCHO, which is why displayed formulae are often more useful!

Student's t-test This is a statistical test developed by the mathematician and chemist William Gosset (who published his work under the pen-name 'Student') in 1908. It's commonly used to test whether the means of two sets of values are equal, and place a probability on the outcome occurring by chance.

Sublimation A process not often seen in everyday circumstances. Sublimation occurs when a material goes straight from the solid phase into the vapour, without melting into a liquid. Examples of where this occurs are frozen carbon dioxide (dry ice), naphthalene and water under reduced pressure (e.g. water ice on the planet Mars).

System In an environmental management context, we can define a system as an interconnected and interdependent set of components with coherent organisation. For example, we can consider a natural environment as a system, comprising all living and non-living components, connected by their interactions, e.g. predator-prey relationships, seasonal cycles, etc. Likewise, a building project is also a system, whose components include planning and regulation, building materials, workforces and support industries.

Target organ The organ which will make a response when it encounters a hormone.

Teratogen An agent that can disturb the development of an embryo or fetus. This may cause a birth defect or even spontaneous abortion (miscarriage).

Thoracic cavity The chamber of the body protected by the thoracic wall.

Thorax The part of the torso above the diaphragm.

Thyroid The thyroid gland controls the rate of use of energy and protein synthesis.

Tissue fluid The internal environment that the cell lives in.

Titration A technique used to determine the concentration of an unknown solution.

Titre The volume of **standard solution** needed to neutralise the analyte (i.e. to reach the end point of the titration).

Totipotent stem cells Differentiate into embryonic and extraembryonic (placenta) cell types. Such cells can construct a complete, viable organism. The zygote is, by definition, totipotent as are the cells produced by the first few divisions of the zygote.

Toxicity The degree to which a substance can poison or harm an organism.

Toxin A toxic (poisonous) substance produced by a living organism.

Transition metal A metal in the d block. Strictly speaking, either the metal or at least one of its ions must have an incomplete d orbital. Without going into the reasons for this, it helps to know that transition metals can have multiple oxidation states. They can also accept dative covalent bonds (co-ordinate bonds), where another atom donates a pair of electrons to form the bond. These two features make transition metals useful as catalysts in chemistry and the same features make them useful as cofactors in biochemistry enzymes are biological catalysts, after all.

True value An ideal or perfect value of a measurement (mass, volume, temperature, etc.) that can never be known exactly.

Tsunami A tsunami (from the Japanese 'harbour wave') is a surface water wave caused by the displacement of large volumes of water, often due to rapid movements of the sea bed in an earthquake, though other events such as meteor impacts and landslides can have the same effect. Tsunamis are sometimes called tidal waves, but it is important to be clear that they have nothing to do with the normal tides caused by the sun and moon.

Uncertainty Is the amount of error that is inherent in any measurement. It will depend on the sensitivity of the equipment you are using.

Unmet medical need A medical condition for which there is, as yet, no available treatment.

Urea A chemical compound produced during metabolism.

Vaccination A person's immune system is presented with the antigen without the risk of being harmed by the pathogen.

Validity The extent to which the variable being measured actually does measure the underlying trait being investigated. It is the ability to provide correct answers to the questions being asked.

Vapour pressure All liquids are evaporating all the time that is, atoms or molecules in the liquid are escaping into the gas phase. The vapour pressure of a liquid is the pressure exerted by these atoms or molecules when in equilibrium with the liquid. In other words, the rate at which atoms or molecules are escaping from the liquid is equal to the rate at which they are returning.

Variables In science, a variable is any quantity which we can measure or control. There are several types of variable we need to think about such as the ones below.

Vector A carrier of disease or an organism, but does not cause disease or harm itself.

Vein Vessel which takes blood towards the heart.

Ventricle Large lower chamber at each side of the heart.

Vesicle A structure inside the cell consisting of liquid enclosed by a membrane. Different types of vesicle are formed in different ways or have different functions, such as secretory vesicles and phagocytic vesicles.

Viruses These are strands of genetic material surrounded by protein coats.

Volcano There are many types of volcano, but all of them result from deep fractures in the upper layers of the Earth, allowing material such as lava and volcanic ash to reach the surface, along with often toxic gases such as sulfur dioxide and hydrogen fluoride.

Voltage Fundamentally, the voltage is a measure of the change of energy of a charged particle when it moves through an electric field, and, scientifically, is often referred to as potential difference. The unit of voltage is the volt (V), which can be expressed as $J\,C^{-1}$ (joules per coulomb). This means that if a material with a charge of 1 C moves through a potential difference of 1 V, it will gain 1 J of energy. Voltage is often given the symbol V.

Yield To produce or provide.

Zygote The diploid cell produced when haploid gametes (sperm and egg) fuse during fertilisation.

Index

3-methylbutanoic acid 16
abdomen 112
ABPI *see* Association of the British Pharmaceutical Industry
absorption 100
accuracy 84–5
accurate 38
acid (or base) concentration using titration
    alternative techniques 55
    appropriate indicators 53
    calculating 51–2
    calculation of mass required to make solution 53–4
    measuring equipment 52
    performing titration 54–5
    techniques 52–5
activation energy 10
active pharmaceutical ingredient (API) 215
active transport 190
actual yield 72
adaptation
    of chromatographic techniques 88–9
    as quantitative techniques 91
adaptive immune system 121–2
Adenosine Triphosphate (ATP) 295
ADH 118
adrenal gland 118
adsorbent 46, 47
adsorption 46
adult stem cells 202
aerobic exercise 177
AES *see* atomic emissions spectroscopy
agent 148
agriculture 287, 288
    energy production 290
    genetic engineering of crops 291
    implications of genetic engineering 292–3
    use of microorganisms in sustainable practice 288–90
air-borne pollution 261–2
airflow cabinets, controlled 61
alcohols 15
aldehydes 15
alkanes, alkenes, alkynes 15
alleles 127
allergens 182
allergies 182, 292
alloys 6, 7–8
    composition/uses of common alloys 7
alternative hypotheses 245
alveoli 112
amino acid 19, 169, 170

anaerobic exercise 177
analogue 209, 210
anaphylactic shock 182, 184
anion 9
    chemical tests 59
anneal/annealing 138
anomaly/anomalous result 84
anorexia nervosa 181, 183
anti-codon 20
Anti-terrorism, Crime and Security Act (2001) 165
antibiotic 296, 297
    aminoglycoside 297
    glycopeptides 297
    mode of action 297
    monobactams 297
    penicillin 297
    sulfonamides 297
    tetracycline 297
    uses 297
antibody 122
antifungal 296
antigen 120, 150
antimicrobial 296, 297
antimicrobial resistance 298
    implications on local services 299
    mechanisms 298
    projected models 298-9
    solutions 299
    trends 298
antiviral 296
API *see* active pharmaceutical ingredient
arithmetic 66
artery 107
arthritis 105
aseptic 60, 282, 284
    purpose of working in aseptic/clean room 61
    streak a plate procedure 62
    tissue culture procedure 62
aseptic techniques 284
    colony morphology 285–6
    flaming neck of bottles/test tubes 284
    Gram staining 286
    growth media 286
    serological/genetic methods 287
    wire loop 285
assimilation 102
Association of the British Pharmaceutical Industry (ABPI) 304
asthma 114
atherosclerosis 184
atmosphere 260
atom 2
atomic emissions spectroscopy (AES) 60

atomic numbers 3
atomic radius 2
ATP *see* Adenosine Triphosphate
atrium 107
autoimmune diseases 122–3
autonomic 116
autotitration 55

bacteria 120, 171, 283
    bacilli 284
    classification 284
    cocci 284
    identifying 284
    spirilla 284
    spirochaetes 284
    vibrios 284
bacterial cell 151, 154
bacterial identification
    colony morphology 91
    growth/behaviour on differential, selective, enriched media 92
    staining techniques 91–2
bacterial population growth 153
bacteriophage 149, 154
bar chart 77–8
Basal Metabolic Rates (BMR) 175, 177
base pairs 136
Before Current Era (BCE) 230
Bessel's Correction 75
binary fission 153
binomial nomenclature 83
bioavailability 222
biochemical 18–19, 171
biochemical screening 218
biofertiliser 288, 290
biohazard symbol 156
biohazards 155–6
bioherbicide 289
bioinorganic chemistry 23
biological agent 147, 148
biological samples
    alternative techniques offering improved separation, sensitivity, quantification 60
    alternative techniques to study microscopic features 57
    calculating magnification/scale 56–7
    identify cations/anions 58–60
    recording observation 56–7
    use of graticule 56–7
biological/biologics 209, 210
biome 256
biopesticide 288
blinding 224
blood 106
blood vessels 107–8

Index

BMR *see* Basal Metabolic Rates
BODMAS (Brackets, Order, Division, Multiplication, Addition, Subtraction) 71
boiling 28
boiling a kettle
    latent heat 29
    specific heat capacity 29
boiling point 29
British Nutrition Foundation 182
bronchoconstriction 117
bronchodilation 116
built environment 259
    industrial 259
    transport networks 259
    urban 259
bulimia 181, 183

calibration 38, 39–40
    producing standards to enable quantitative analysis 49–50
calories 177–8
calorimeter 178
cancer 123, 200
capillaries 107
capture antibody 219
carbohydrates 19, 170–1
    types 20
carbon 14
    alkanes, alkenes, alkynes 15
    atom 15
    chain 18
carbon compounds
    empirical/structural formulae 16
    isomer formation 17–18
    large complex molecule formation 18–19
carbon footprint 272
carboxylic acids 15
carcinogen 147, 148
cardiovascular disease (CVD) 184
cardiovascular system 106
    blood 106
    blood vessels 107–8
    common disorders 108–9
    heart 107
    monitoring 108–10
carrier 129, 130
case studies
    biofertilisers 290
    cosmetic ingredients 307
    English Nitrate Vulnerable Zones (NVZs) 263
    European Union Directorate-General for the Environment 269
    Fairphone smartphone 274–5
    London sewer system 232
    onshore oil and gas in UK 270
    P2X7 knockout mice 212
    Safir-Hadramout Road Project (Yemen) 275
    United Nations Environment Programme (UNEP) 268
catalysts 10
cations 90
    chemical tests 58–9
    flame tests 58
cell
    components/role 11–12
    counting techniques 196
    cycle 198
    division (checkpoints/clinical techniques) 199–200
    DNA/RNA 13
    membrane 171
    movement into/out of 190–2
    organisation/structure 10–13
    signalling 190–2
    signalling pathways 191, 192
    types 11
    wall 12
cell differentiation 201
    gene expression/repression 202
    undifferentiated cells (stem cells) 202
cell wall variation 151
chain-termination 139
charge 31
chemical
    digestion 101–2
    receptors 117
    signalling 192
chemical reaction 6, 8
    addition 9
    displacement 9
    factors affecting rate of 9–10
    oxidation/reduction (or redox) 8
    polymerisation 9
    radical reactions 9
    substitution 9
chemistry 6
chloride 89
chloroplasts 12
cholesterol 171, 184
chromatids 126
chromatogram 49–50
chromatography 46, 314, 315
    adaptation 88–9
    gas (GC) 48, 316
    high performance liquid chromatography (HPLC) 48, 316
    paper 48
    positive identification of components of mixture when linked to mass spectrometer 50
    stationary/mobile phases 47–8
    techniques 314–15
    thin layer (TLC) 47, 315
chromosomes 135
classification system 82–3
climate change 242
clinical phase 224, 225
clinical trials 225
    licensing approval 226
    statistical interpretation of results 226
    validity/reliability 225–6
CLP (classification, labelling and packaging of substances/mixtures) Regulation 40
co-factor 23
codominant 128
codon 20
colloids 8
colony 285, 286
colorimetric methods 218
column chromatography 88–9
comfort eating 184
complement 121
complementary 122
compound 5
compulsive eating 181–2
concentration 112
conclusion 86–7
confocal microscopy 195–6
confounding variable 245, 246
consumer products 303
contamination 61
continuous data 77, 78
control 245, 246
conventional current 30–1
coronary arteries 107
coronary heart disease 110–11
    angina 111
    heart attack 110–11
COSHH Regulations 40, 157
Coulter counter 196
covalent bonding 5
current 31
Current Era (CE) 230
CVD *see* cardiovascular disease
cystic fibrosis 115
cytokines 152
cytokinesis 198
cytology 194
cytoplasm 12

dametes 126
data
    collection/analysis 43, 247–8, 266–7
    conclusion given and justified 86–7
    identification of conflicting evidence 87
    level of uncertainty 84–5
    primary/secondary 82, 84, 85, 86–7
    quality 83–6
    set 74
    statistics/interpreting 248–9
data recording 92
    3D representations 93
    environmental management 266–7
    graphs 93

modelling 93
notebooks/logbooks 92
  photographs, sketches, video, audio 93
  tables 92–3
data reporting 93–4
  chosen audience 94
  evaluating 94–5
  findings in appropriate format 44–5
  nature of data/scientific findings reported 95
  publication/information source 95
  quality of 95
  scientific media 94
decontamination 159
defibrillators 110
dehydration 120, 184
dementia 179, 180
densitometry 89
Department of Health 169, 170
dependent variable 77–8, 245, 246
detection antibody 219
DG Environment *see* European Union Directorate-General for the Environment
diabetes 119, 181, 183, 184
dichotomous 81, 82
dideoxy method 139
dietary fibre 170
dietary reference values (DRV) 187
differentiate 104
diffusion 190
digestion 100, 171
  eyes and nose 172
  large intestine 172
  mouth 172
  small intestine 172
  stomach 172
digestive system 100–1
  chemical 101–2
  common disorders 103
  lower gastrointestinal tract 101
  mechanical 102
  peristalsis 102
  upper gastrointestinal tract 101
dihybrid inheritance
  expected/observed data in a cross 132–3
  gene linkage/epistasis 133–4
  genetic maps 134–5
  predicting genotype/phenotype ratios 131
  test cross 131
diploid 125, 126
disaccharides 21
discontinuous data 77, 78
disease development 204
disease-causing organisms 149
  bacterial growth through binary fission 153

binary fission 153
capsules to block phagocytosis 152
endotoxins in cell wall 152
link between toxin production/symptoms 152
lytic/lysogenic cycles as related to diseases 150–1
plasmids 152
presence of antigens 150
presence of pili for adhesion 152
structure of bacterial cell and how bacteria reproduce 151
structure of virus as nucleic acid, capsid, viral enzymes 149–50
disinfection 61, 62
displayed formula 16
diuretic 175
diverticular disease 179, 180
DNA (deoxyribonucleic acid) 13, 19, 136–7
  next-generation sequencing 139–40
  principles of genetic profiling 140–2
  Sanger sequencing 139
  use of electrophoresis 48–9
DNA-sequencing project 143
  ethical, legal, social implications 144
  implications of genomics 144
  increasing understanding 144
dominant 128
DPSIR *see* driver-pressure-state-impact-response
drug target 210
drugs 208
  analytical techniques in assessing purity 217–19
  biological 210
  categories 209–10
  clinical trials 225
  computational techniques 213
  consumer target 210
  design/optimisation 212
  development planning 225–6
  drug discovery 212
  effectiveness 210–11
  environmental pressures 211
  extraction 215
  legislative requirements 211
  licensing approval 226
  modelling techniques 211–13
  new/emerging technologies/materials 211
  packaging, labelling, storage 223
  pharmacogenomics approach 213–14
  pre-clinical development 212
  product formulation/dosage form 222
  production 215–17
  purification 216–17

quantitative structure-activity relationships 211–13
research stages 210–11
route of administering 222
screening 211, 212
synthesis 216
synthetic (small molecule) 209
target validation 212
testing 204, 303
DRV *see* dietary reference values

EAR *see* Estimated Average Requirements
earthquake 238, 240
eating disorders 181
Eatwell Guide 175
E.coli 0157 179, 180
EDTA 90
effector 116
efficacy 211
EIA *see* Environmental Impact Assessment
elastic 107
electric signalling 192
electrical circuits 33
electrical resistance 31
electrolytes 173, 174
electromagnetic radiation 10
electrophoresis 48–9
elements 3–4
  groups 4
  organisation within periodic table 4
  periods 4
  react together to form compounds 5
ELISA *see* enzyme-linked immunosorbent assay
ELS *see* ethical, legal, social implications
eluate 46, 47
eluent 46, 47
elution 46, 47, 89
embryonic stem cells (ESCs) 201, 202
emphysema 115
empirical formula 16
emulsification 102
ENCODE Project 143
end point 53
endocrine system 117–18
endocytosis 190
endomembrane systems 190–1
endoparasites 155–6
endoplasmic reticulum (ER) 12
endotoxins 152
enteric administration 222
environment 231
  aquatic 260
  built 259
  chemical 260
  climate 260
  geographical location 259
  geology/soil 260

global agreements 267
human impact 230–5
management assessments 272–3
national legislation 270
natural 258
natural processes 236
plant/animal life 260
pollution 261–6
population growth 230–1
regional agreements 268
seasonality 260
small/local scale 259
volcanic/seismic activity 236–8
water use 229–33
environment management 258
data collection/recording 266–7
formal report 277
improving/adapting investigations 278
legislation, regulation, agreements 267–8
preparing/giving presentations 22–8
Environmental Impact Assessment (EIA) 275
environmental survey 239
chemical tests/indicators for common pollutants 248
data collection/analysis 248–9
improving investigations 252
objectives 244–5
planning 241
regional/global conservation 242–3
reporting science 251–2
safe working practices in field/laboratory 245–6
enzyme 101
enzyme-linked immunosorbent assay (ELISA) 218–19
enzymes 10, 169, 170
epistasis 134
equivalence point 51, 52
error bars (range bars) 79
error, sources of 84–5
ESCs see embryonic stem cells
essential proteins 169, 170
esters 15
Estimated Average Requirements (EAR) 177
ethical, legal, social implications (ELS) 144
eukaryotic cells 190–1
European Union Directorate-General for the Environment (DG Environment) 269
excipient 222, 223
excretion 118
expiration 112, 223
extraction 215, 306
extraneous variable 245, 246
extrapolation 80

facilitated diffusion 190
FACS see Fluorescence Activated Cell Sorting
fast protein liquid chromatography (FPLC) 217
fat 171
FDA see Food and Drug Administration
fermentation 294–5
energy sources 295
glycolysis 295
growing microorganisms 295
industrial 295
optimum conditions 295
fertiliser 288
fibrillation 110
flaming procedure 25
fluids 174–5
Fluorescence Activated Cell Sorting (FACS) 196
folic acid 179
food
allergies/intolerances 182, 291
biochemistry of fermentation 294–5
bread industry 294
brewing industry 294
dairy industry 293
implications of genetic engineering 292–3
labelling 185–6, 304–5
microorganisms in food production 294
supplies 292
wine industry 294
Food and Drug Administration (FDA) 303
Food Standards Agency (FSA) 175, 183, 302, 304
formulation 222
FPLC see fast protein liquid chromatography
FSA see Food Standards Agency
functional group 9

gametes 125
gas chromatography (GC) 48, 316
retention times 49
gaseous exchange 111, 112
Gaussian (normal) distribution 76
GC see gas chromatography
gel electrophoresis 48
gene 128
epistasis 134
expression/repression 202
linkage 133–4
technology 137
genetic
counselling 130
method (bacterial identification) 287
profiling 142
genetic engineering 288

crops 291–2
disease/allergies 292
environmental damage/food web risks 292–3
implications 291–2
genetic maps 134
limitations 135
genetically modified (GM) 288
Genetically Modified Organisms (Contained Use) Regulations (2000) 165
genetically modified organisms (GMOs) 165
genome 136, 150
whole genome sequencing 136–7
genomics 137
implications 144
geometric progression 73–4
gha see global hectare
GHS see Globally Harmonized System
GIS (geographical information system) 93
global hectare (gha) 272
Globally Harmonized System (GHS) 40
GLP see good laboratory practice
glucose 170
glycogen 102
glycolysis 295
GM see genetically modified
GMOs see genetically modified organisms
GMP see good manufacturing practice
Golgi apparatus 12, 191
good laboratory practice (GLP) 38, 304
Good Manufacturing Practice (GMP) 215, 216
governing bodies 304
influence on quality control 306
gradient of a line 81
gram positive/negative 151, 283, 284
gram stain 91–2, 286
graphs
accuracy/precision 78–9
calculating gradient of a line 81
choosing 77–8
determining intercepts 81
interpolation/extrapolation 80
interpreting data 80–1
linear/non-linear 78
greenhouse effect 230
group 3
growth media 286
differential 286
selective 286

habitat 258
haemocytometer 196
haploid 125
hazardous agents 148
biological agents 148
chemical 149

compounds listed as carcinogens, mutagens, teratogens 148
physical 149
HCI *see* high content imaging
health & safety legislation 40
Health and Safety at Work Act (HSWA, 1974) 41
Health and Safety Executive (HSE) 41
heart 107
failure 119
monitors 109
rate during exercise/rest 109
heterozygous 128
high content imaging (HCI) 196
high performance liquid chromatography (HPLC) 48, 314
retention times 49
Hindhead Tunnel 241–2
histamine 182, 183
histogram 78
histology 194
hit compound 211
homeostasis
autonomic nervous system 116–17
common disorders 119–20
concept 116
endocrine system 117–18
liver 118–19
structure/function of kidneys 118
homotogous chromosomes 126
homotogous series 15, 16
homozygous 129, 130
Hooke's Law 26–7
hormones 117
HPLC *see* high performance liquid chromatography
HSE *see* Health and Safety Executive
human activity 230
Human Genome Project 143
hydracids 23
hygrophobic 222
hygroscopic 222
hyperclycaemia 119
hypertension 110
hypoglycaemia 119
hypothalamus 117, 118
hypothesis 86, 245, 246

IBS *see* irritable bowel syndrome
ICC *see* immunocytochemistry
ICP *see* inductively coupled plasma (ICP)
identification 81–2
IHC *see* immunohistochemistry
Illumina method 139
immiscible 314, 315
immune 120
immune response 122
immune system 120–1
adaptive 121–2
cellular barriers 121
complementary systems 121
disorders 122–3
natural killer cells 121
surface barriers 121
immunisation 156–7
immunocytochemistry (ICC) 194
immunodeficiency 123
immunogenic 152
immunohistochemistry (IHC) 194–5
immunological response 182, 183
immunological screening 218
impurities 314, 315
in-vitro 308
in-vivo 309
inclusions 194
independent assortment 126
independent variable 77, 245, 246
indicator 51, 52, 310
induced pluripotent stem cells (iPSCs) 203–4
inductively coupled plasma (ICP) techniques 314–15
inductively coupled plasma-atomic emission spectroscopy (ICP-AES) 60
infection/s 103, 148, 149
infectious agents 155–6
inflammation 103
inflammatory diseases 123
infra-red (IR) spectroscopy 217
inheritance *see* dihybrid inheritance; monohybrid inheritance
inoculate 62
inoculum 62
inorganic chemistry 22
inorganic compounds 22, 23
metals/metal ions 22
oxidation state 22
insoluble 314, 315
inspiration 112
instrument error 85
insulators 31
interactions 6
metals/alloys 6–8
suspensions/colloids 8
intercepts 81
interpolation 80
interval 85
iodine solution 89–90
ion chromatography 60
ion torrent/ion proton sequencing 140
ionic bonding 5
ionisation energy 4
ions 90
iPSCs *see* induced pluripotent stem cells
Iron(II) by spectrophotometry (1, 10-phenanthroline) 91
Iron(III) by colorimetry (thiocyanate) 91
irritable bowel syndrome (IBS) 103
isomer
geometric 17–18
optical 18
structural 17, 18
isotopes 2

joint 104
components of synovial joint 105
limitations of movement 104–5
muscles around 105

ketones 15
kidneys 118
kinase 191, 192
Kirchhoff's Current Law 32–3
kite diagram 78

labelling 224
laboratories
air management/fume control 161
apparatus 284
changing facilities 161
computer hardware/software 161
design specifications 165
ergonomics/aesthetics 161
food testing 305
furniture 160–1
initial planning 159–61
laboratory equipment 161
legislation 164–5
personal protective equipment (PPE) 162
planning 159–61
product development/testing 309–10
resources 161–2
security, access, containment 159–60
services/utilities 160
storage space 160
testing equipment 161–2
utilisation of space/technician workspace position 159
waste disposal 161
laboratory hazards 40, 41, 155
biohazard symbol 156
COSHH regulations 156–7
identifying 245
infectious agents 155–6
instrumentation/electrical equipment 157
procedures 162–4
RIDDOR regulations 157
risk assessment 156
standard operating procedures 157
laboratory procedures 162
accident procedures 163
disinfection 163–4
disposal of waste 164
fire precautions 163
fume cupboards/fume control 163
hazard recognition 162
incident procedures 163
risk assessments 162
safe working practices 245–6

safety awareness 162
storage/security 163
working under supervision 162
lactose intolerance 181, 183
land area type 272, 273
Law of Gravity (Newton's) 2
LCA *see* life cycle assessment
lead optimisation 211, 212
legislation
    codes of practice 164
    control of toxic/flammable substances 40–1
    COSHH regulations 156–7
    environmental management 267–9
    food 304
    health and safety at work 164
    microbiological hazards 165
    pharmaceutical industry 211
    set by governing bodies 306
lever 104
life cycle assessment (LCA) 274
ligand 191, 192
light intensity 10
light microscopy 194
    extending the limits 195–6
    immunohistochemistry 194–5
    limits 194
    microscopical/differential staining techniques 194
    staining of organelles, inclusions, secretions 194
line graph 77
linkage 133–4
lipid 19, 102
lipid bilayer 11
lipoprotein 170
listeria 179, 180
liver 118–19
locus 141
low resolution and high-resolution proton nuclear magnetic resonance spectroscopy (1H-NMR) 217–18
lungs 111
lymphocyte 121
lysis 149, 150
lysogenic cycle 150
lysosome 12
lytic cycle 149, 150

macromolecule 19, 190
macronutrients 168–73
malnutrition 179, 180
maltose 101
Management of Health and Safety at Work Regulations 40
mass spectrometry (MS) 50, 317
materials, electrical properties 30
    atomic scale model of conduction 30–1
    energy/power in circuits 33
    Kirchhoff's Current Law/combining resistors 32–3
    voltage, current, resistance, Ohm's Law 31–2
materials, mechanical properties 24
    atomic models 25–6
    definitions 25
    quantifying 26–8
materials, physico-chemical properties 28–9
mathematical techniques 71
    calculating variance and standard deviation 74–7
    calculation of surface area and volume 72
    geometric progression (serial dilutions) 73–4
    percentage yields and errors 72
me-too drug 211
mean 66
measurement error 85
    instrument error 85
    random error 85
    range and interval 85
    systematic error 85
measuring equipment 38
mechanical digestion 102
median 66
medicinal chemistry 209
Medicines and Healthcare Products Regulatory Agency (MHRA) 304
meiosis 125, 126–7
melting 28
melting point 29
membrane 11
metabolised 170
metabolism 118
metal ions, biological functions 23, 24
metals 6–8
metric system 68–71
MHRA *see* Medicines and Healthcare Products Regulatory Agency
micro-analysis 315
    gas chromatography 316
    high performance liquid chromatography 317
    inductively coupled plasma techniques 316–17
    mass spectrometry 317
microbiology 282
    agriculture practice 287–94
    antibiotics 297
    antimicrobials 297, 298–9
    aseptic techniques 284–8
    classification of bacteria 284
    classifying/identifying microorganisms 283
    food industry 294–5
micronutrients 168–73
microorganism 282
    in agriculture 289–91
    algae 283
    archaea 283
    bacteria 283
    classifying/identifying 283
    in fermentation 295
    fungi 283
    parasites 283
    protoctist 283
    virus 283
microscopy
    alternative staining procedures 88
    analytical techniques according to need 88
    preparation of permanent slides 88
microtitre plate 219
minerals 23, 173–4
    major 174
    trace 174
miscarriage 179, 180
mitochondria 12
mitosis 125, 126, 197–201
    cell cycle 198–9
    importance 199
    mechanisms that arrest cell division 199–200
    stages 199
mode 66
model cells 204
molecular biology 139, 140
molecular formula 16
monohybrid inheritance 128
    codominant alleles 129
    genetic counselling 130
    incomplete inheritance (or incomplete/partial dominance) 129
    normal trait 128–9
    sex-linked inheritance 130
    single gene disorder 129
monomer 16, 17
monosaccharides 21
morphology 282
MS *see* mass spectrometry
musculoskeletal system 104
    bone and muscle 104
    common disorders 105
    joints and limitations of movement 104–5
    muscle action around joint 105
    synovial joint 105
mutagens (physical and chemical) 147, 148
myelin 171

natural processes 230
natural product 215
negative feedback 116
neurone 116
neurotransmitters 169, 170
neutron 2
next-generation DNA sequencing (NGA/high-throughput sequencing) 139–40

NHS 100,000 Genomes Project 143
NHS Choices 185
Nitrate Vulnerable Zones (NVZs) 263–4
nitrates 23
nitrogen 170
NMR *see* nuclear magnetic resonance
non-essential proteins 169
non-polar molecules 190
normal (Gaussian) distribution 76
norovirus 185
nuclear magnetic resonance (NMR) 217–18
nuclear notation 2
nucleic acid 169, 170
nucleon 2
nucleoprotein 169, 170
nucleus 2
    attractive/repulsive forces 2–3
    cell 12
null hypothesis 132, 245
nutrition
    activity levels 181
    baby's needs 180
    calculating intake/output 176–7
    calory requirements 177–8
    dietary reference values 187
    factors affecting eating habits 181–3
    fluid intake 174–5
    food intake imbalance 183–4
    importance/principles of balanced diet 175
    labelling of food 185–6
    lifestyle/education 182
    macronutrients/micronutrients 168–73
    measuring energy content of food 178
    older adults 180
    prisons, hospitals, care homes, schools 180
    school age children/teenagers 180
    sources of nutrients 168
    variations in dietary needs 179–81
    variations in individual requirements 176
NVZs *see* Nitrate Vulnerable Zones

obesity 181–2, 184
oil immersion microscopy 195
organic food 186
osteoporosis 105, 179, 180
outliers 79
oxides 23
oximeter 113
pacemakers 109–10
pancreas 117
parathyroid 117–18
parenteral administration 222
Parkinson's disease 205
pathogen/pathogenic 60, 61, 120, 148

PCR *see* polymerase Chain Reaction
peptide 19
percentage error 39, 72
percentage yield 72
period 3
periodic table 3–4
peristalsis 102
peroxide 23
personal protective equipment (PPE) 162
pesticides 288
pH 51, 52
pH meter 55
phagocytosis 152
pharmaceutical formulation 222
pharmaceutics 222
pharmacogenetics 213
pharmacogenomics 213
    genomic profiling of different diseases/identification of argets 213
    identification of sub-groups of patients that will benefit from targeted therapy 213–14
pharmacopoeia 304
phenotype 128, 151
phosphatases 191, 192
phosphates 23
phospholipid 11
phospholipid bilayer 12, 190
phosphoproteins 170
phosphoric acid 170
physical state 10
pie chart 78
pineal gland 117
pituitary 118
placebo 225
planning 241
plasma membrane 12
pluripotent 205
polar 46, 47
polar molecules 190
pollutants
    air-borne 261–3
    chemical tests/indicators 248
    natural/human generated 261–3
    soil-borne 265–6
    water-borne 263–4
polymer 16, 17
polymerase Chain Reaction (PCR) 137–8
    of short tandem repeats 140–2
polymorphism 213
polysaccharides 21
population 41, 74, 75
potassium dichromate 89
potency 222
PPE *see* personal protective equipment
pre-clinical phase 211, 212
precise 38
precision 84–5
presentations

    environment management 277
    scientific 44–5, 251–2
preservatives 186
pressure 10, 107
primary
    antibody 195, 219
    data 82, 84, 85, 86–7
    standard 54
primary recovery 234, 235
produced water 234, 235
product testing
    chromatography techniques 312–13
    consumer requirements 302
    drugs 303
    effectiveness 308
    extraction/separation 307, 312
    food labelling 302–3
    governing bodies 302
    in-vitro 308
    in-vivo 309
    laboratory 305, 309–10
    micro-analytical techniques 315–7
    quality control 306
    regulations/legislation 306
    titration 309, 311–12
    types 308–9
properties of materials 25
    brittleness 25
    density 25
    elastic modulus 25
    elasticity 25
    flexibility 25
    hardness 25
    plastic deformation 25
    relative density 25
    stiffness 25
    strength (compression/tension) 25
prosthetic group 24
protein 19, 102, 169
    synthesis 19, 170
proton 2
Public Health (Control of Disease) Act (1984) 165
pulmonary 108
pulse rates 108
pumps 190
Punnett square 128
purification 216–17
    chromatographic methods 216–17
    precipitation/crystallisation 216
    solvent extraction 216
pyrogenic 152
pyruvate 295
QSAR *see* quantitative structure-activity relationship
quantitative 305, 306
quantitative structure-activity relationship (QSAR) 212–13

radon 261–2

random error 85
range 85
range bars (error bars) 79
reactions *see* chemical reactions
reagent 306–7
    analyte 312
receptor 116
recessive 128
recombine/recombination 126
redox (oxidation/reduction) 8
REE *see* resting energy expenditure
reference value 38
regenerative cells 203
relative atomic mass 3
reliability 38, 225
repeat unit 16, 17
repeatability 38, 85–6
reproducibility 38, 85–6
resistors 31
respiratory system 111–12
    common disorders 112–13
    gaseous exchange 112
    inspiration/expiration 112
respiratory system monitoring
    comparison of different populations 113–14
    lung volumes/capacities 113
    oxygen saturation 113
    peak flow 113
resting energy expenditure (REE) 177, 178
restriction endonuclease 137
retention factor 315
retrovirus 149, 150
    implications for human disease 151
reverse transcriptase 149, 150
ribosomes 12
Richter Scale 238
RIDDOR regulations 157
    accident 157
    reportable injuries 157
    work-related 157
risk assessment 41, 156
    field study 247
RNA 13
    messenger, ribosomal, transfer 19
Roche 454 sequencing 140
rounding 67–8

SACN *see* Scientific Advisory Committee on Nutrition
Safir-Hadramout Road Project (Yemen) 275
salmonella 179, 180
sample 41, 75, 245
    complete 41
    integrity 43, 44
    labelling, storing, transporting 43–4
    random 42
    recording procedures/information in data collection 43
    representative 42
    unbiased 41
    whole 41
Sanger sequencing process (chain-termination/dideoxy method) 139, 140
scatter graph 77
scientific
    notation 70
    presentation 44–5, 251–2
    report 44, 45, 251
Scientific Advisory Committee on Nutrition (SACN) 177, 178
Second Generation Multiplex Plus SGMPlus® 141–2
second messenger 191, 192
secondary
    antibody 195, 219
    data 82, 84, 85, 86–7
    standard 54
secondary recovery 234, 235
seismic activity 236
semiconductors 31
separation methods 46
    alternative qualitative/quantitative techniques 48–50
    techniques 46–8
serial dilution 73–4
serological method (bacterial identification) 286
shelf life 222–3
short tandem repeats (STRs) 140–2
SI system 69–70
SI units 68
signal transduction 191, 192
significant figures 67–8
    decimal places 68
single nucleotide polymorphism (SNP) 213
sinoatrial node 107
Site of Special Scientific Interest (SSSI) 241
skeletal formulae 16
smoking 113–14, 115
smooth muscle 102
SNP *see* single nucleotide polymorphism
sodium thiosulfate 90
soil-borne pollution 265–6
soluble 314, 315
solvent extraction 314
solvents 10
SOPs *see* standard operating procedures
specific 122
spirometer 113
spore 153
standard deviation 74
    of data set 75–6
    significance of 76
standard form 70
standard operating procedures (SOPs) 157
standard solution 51, 52, 54
starch 101
stem cell research 204
    disease development 204
    drug testing 204
    model cells 204
stem cell therapies 204–5
    clinical studies 205
    moral/ethical issues 205
    potential 203
    scientific issues 205
stem cells 201, 202
    adult 203
    embryonic (ESCs) 203
    induced pluripotent 203–4
sterilisation 61, 62, 153
still birth 179, 180
stoichiometry 90
storage 223
storage of materials 46
strain 27
stress 27
striated muscle 104
STRs *see* short tandem repeats
structural formula 16
student's t-test 249
sublimation 28
sulfates 23
surface area 10
suspensions 8
synthesis 216
systematic error 38, 85

t-test 248
target organ 117
temperature 10
teratogen 148
theoretical yield 72
thin layer chromatography (TLC) 47, 88–9, 315
    procedure 315
thoracic cavity 111
thorax 111
thymus 117
thyroid 118, 178
tissue 13
    bone 14
    connective 14
    epithelial 13
    fluid 116
    muscle 14
    nerve 14
    ovary/testes (gonads) 14
titration 308
titration techniques 89, 311
    acid-base 311
    complexometric 90, 312
    precipitation 89, 311–13
    procedure 312
    redox 89–90, 312

titre 54
TLC *see* thin layer chromatography
totipotent stem cells 201, 202
toxicity 152
toxin 148
Trading Standards 304
transition metal 24
true value 39
tsunami 238, 239

uncertainty 38
UNEP *see* United Nations Environment Programme
United Nations Environment Programme (UNEP) 268
unmet medical need 210
urban pollution 261
urea 170

vaccination 122
validity 224, 225
vapour pressure 28
variables 84, 246
variance 74
varicose veins 111
VEI *see* Volcanic Explosivity Index
vein 108
ventricle 107
vesicle 190
viruses 120
    infect bacteria 154
    structure as nucleic acid, capsid, viral enzymes 149–50
vitamins 172–3
    lifestyle/medication effects 184
Volcanic Explosivity Index (VEI) 237
volcano 236
    fissure vents 236
    shield volcanoes 236
    stratovolcanoes 236
voltage 31

water stress 229–33
water-borne pollution 263–4

X^2 test 132

yield 288
Young's modulus 27–8

zygote 127